T0234241

Birkhäuser

Oberwolfach Seminars

Volume 49

The workshops organized by the *Mathematisches Forschungsinstitut Oberwolfach* are intended to introduce students and young mathematicians to current fields of research. By means of these well-organized seminars, also scientists from other fields will be introduced to new mathematical ideas. The publication of these workshops in the series *Oberwolfach Seminars* (formerly *DMV seminar*) makes the material available to an even larger audience.

Willy Dörfler • Marlis Hochbruck • Jonas Köhler •
Andreas Rieder • Roland Schnaubelt •
Christian Wieners

Wave Phenomena

Mathematical Analysis and Numerical Approximation

Birkhäuser

Willy Dörfler
Institute for Applied and Numerical Mathematics
Karlsruhe Institute of Technology
Karlsruhe, Germany

Marlis Hochbruck
Institute for Applied and Numerical Mathematics
Karlsruhe Institute of Technology
Karlsruhe, Germany

Jonas Köhler
Institute for Applied and Numerical Mathematics
Karlsruhe Institute of Technology
Karlsruhe, Germany

Andreas Rieder
Institute for Applied and Numerical Mathematics
Karlsruhe Institute of Technology
Karlsruhe, Germany

Roland Schnaubelt
Institute for Analysis
Karlsruhe Institute of Technology
Karlsruhe, Germany

Christian Wieners
Institute for Applied and Numerical Mathematics
Karlsruhe Institute of Technology
Karlsruhe, Germany

ISSN 1661-237X ISSN 2296-5041 (electronic)
Oberwolfach Seminars
ISBN 978-3-031-05792-2 ISBN 978-3-031-05793-9 (eBook)
https://doi.org/10.1007/978-3-031-05793-9

This book is published under the imprint Birkhäuser, www.birkhauser-science.com by the registered company Springer Nature Switzerland AG
The registered company address is: Gewerbestrasse 11, 6330 Cham, Switzerland

Preface

The research on wave-type problems is a fascinating and emerging field in mathematical research with many challenging applications in sciences and engineering. Profound investigations on waves require a strong interaction of several mathematical disciplines including functional analysis, partial differential equations, mathematical modeling, mathematical physics, numerical analysis, and scientific computing.

The goal of these lecture notes is to present a comprehensive introduction to the research on wave phenomena. Starting with basic models for acoustic, elastic, and electromagnetic waves, we will consider the existence of solutions for linear and some nonlinear material laws, efficient discretizations and solution methods in space and time, and the application to inverse parameter identification problems.

Karlsruhe, Germany
January 2022

Willy Dörfler
Marlis Hochbruck
Jonas Köhler
Andreas Rieder
Roland Schnaubelt
Christian Wieners

Acknowledgements

Funded by the Deutsche Forschungsgemeinschaft (DFG, German Research Foundation)—
Project-ID 258734477—SFB 1173.

Contents

About the Authors

Willy Dörfler Institute for Applied and Numerical Mathematics, Karlsruhe Institute of Technology, Karlsruhe, Germany

Marlis Hochbruck Institute for Applied and Numerical Mathematics, Karlsruhe Institute of Technology, Karlsruhe, Germany

Jonas Köhler Institute for Applied and Numerical Mathematics, Karlsruhe Institute of Technology, Karlsruhe, Germany

Andreas Rieder Institute for Applied and Numerical Mathematics, Karlsruhe Institute of Technology, Karlsruhe, Germany

Roland Schnaubelt Institute for Analysis, Karlsruhe Institute of Technology, Karlsruhe, Germany

Christian Wieners Institute for Applied and Numerical Mathematics, Karlsruhe Institute of Technology, Karlsruhe, Germany

Part I

Space-Time Approximations for Linear Acoustic, Elastic, and Electro-Magnetic Wave Equations

Willy Dörfler and Christian Wieners

We introduce and analyze variational approximation schemes in space and time for linear wave equations. The discretization is based on a discontinuous Galerkin approximation in space and a Petrov–Galerkin method in time. This is applied to models for acoustic and elastic waves in solids and for electro-magnetic waves. For the corresponding linear hyperbolic first-order systems we show the existence of strong and weak solutions. Then, based on the exact solution of Riemann problems the upwind flux is constructed. Discrete inf-sup stability is established for the nonconforming space-time discretization, which builds the basis for a priori and a posteriori error estimates.

Modeling of Acoustic, Elastic, and Electro-Magnetic Waves

1

Mathematical modeling of physical processes yields a system of partial differential equations that describes the behavior of a system physically correct and allows for analytical and numerical predictions of the system behavior. Here we start by shortly summarizing modeling principles which are illustrated for simple linear models in one space dimension. Then this is specified for different types of wave equations.

1.1 Modeling in Continuum Mechanics

Describing a model in continuum mechanics is a complex process combining physical principles, parameters and data. For a mathematical framework, we introduce the following terminology:

- **Geometric configuration**
  We select a domain in space $\Omega \subset \mathbb{R}^d$ ($d \in \{1, 2, 3\}$) and a time interval $I \subset \mathbb{R}$, and for the specification of boundary conditions we select boundary parts $\Gamma_k \subset \partial\Omega$, $k = 1, \ldots, m$, where m is the number of components of the variables which describe the current state of the physical system.
- **Constituents**
  Which physical quantities determine the model?
  Which quantities directly depend on these primary quantities?
  For the mathematical formulation it is required to select a set of primary variables.
- **Parameters**
  Which material data are required for the model?
  Which properties do these parameters have in order to be physically meaningful?

W. Dörfler et al., *Wave Phenomena*, Oberwolfach Seminars 49,
https://doi.org/10.1007/978-3-031-05793-9_1

- **Balance relations**
  This collects relations between the physical quantities (and external sources) which are derived from basic energetic or kinematic principles. These relations are independent of specific materials and applications.
- **Material laws**
  This collects relations between the physical quantities that have to be determined by measurements and depend on the specific material and application.
- **External forces, boundary and initial data**
  The system behavior is controlled by the initial state at $t = 0$, by external forces in the interior of the space-time domain $I \times \Omega$, and by conditions on the boundary $I \times \partial\Omega$.

1.2 The Wave Equation in 1d

This formalism is now specified for the most simple wave model in 1d with constant coefficients. Therefore, we assume that all quantities are sufficiently smooth, so that all derivatives and integrals are well-defined.

Configuration We consider an interval $\Omega = (0, X) \subset \mathbb{R}$ in space and a time interval $I = (0, T) \subset \mathbb{R}$ for given $X, T > 0$.

Constituents Here, we consider the simplified situation that material points in $\mathbb{R}^2$ move up and down vertically. The state of this physical system is then determined by the vertical *displacement*

$$u \colon [0, T] \times \overline{\Omega} \longrightarrow \mathbb{R}$$

describing the position of the material point $\big(x, u(t, x)\big) \in \mathbb{R}^2$ at time t, and the *tension*

$$\sigma \colon [0, T] \times \overline{\Omega} \longrightarrow \mathbb{R}$$

describing the forces between the points $x \in \Omega$. In this simplified 1d setting with vertical displacements the tension corresponds to the *shear stress* in higher dimensions.

Depending on the primal variable u, we define the *velocity* $v = \partial_t u$, the *acceleration* $a = \partial_t v = \partial_t^2 u$, the *strain* $\varepsilon = \partial_x u$, and the *strain rate* $\partial_t \varepsilon = \partial_x v = \partial_x \partial_t u$.

Material Parameters This simple model only depends on the *mass density* $\rho > 0$ and the *stiffness* $\kappa > 0$; together, this defines $c = \sqrt{\kappa/\rho}$. We will see that c is the *wave speed* which characterizes this model.

Balance of Momentum Depending on the velocity v and the mass density ρ we define the *momentum* ρv. Newton's law states that the temporal change of the momentum in time

equals the sum of all driving forces. Here, without any external forces, this balance relation reads as follows:
for all $0 < x_1 < x_2 < X$ and $0 < t_1 < t_2 < T$ we have

$$\int_{x_1}^{x_2} \rho(x)\big(v(t_2, x) - v(t_1, x)\big)\, \mathrm{d}x = \int_{t_1}^{t_2} \big(\sigma(t, x_2) - \sigma(t, x_1)\big)\, \mathrm{d}t\,.$$

For smooth functions this yields

$$\int_{x_1}^{x_2} \int_{t_1}^{t_2} \rho(x)\partial_t v(t, x)\, \mathrm{d}t\, \mathrm{d}x = \int_{t_1}^{t_2} \int_{x_1}^{x_2} \partial_x \sigma(t, x)\, \mathrm{d}x\, \mathrm{d}t\,,$$

and since this holds for all $0 < x_1 < x_2 < X$ and $0 < t_1 < t_2 < T$, this holds point-wise, i.e.,

$$\rho(x)\partial_t v(t, x) = \partial_x \sigma(t, x)\,, \qquad (t, x) \in (0, T) \times (0, X)\,. \tag{1.1}$$

Material Law One observes that the tension $\sigma(t, x)$ only depends on the strain $\varepsilon(t, x) = \partial_x u(t, x)$. This is formulated as a material law: a material is by definition *elastic*, if a function Σ exists such that $\sigma = \Sigma(\partial_x u)$, and it is *linear elastic*, if $\sigma = \kappa\varepsilon$ with stiffness $\kappa > 0$. In a homogeneous material, the stiffness κ is independent of $x \in (0, X)$.

Boundary and Initial Data The actual physical state at time t of the system depends on its state at the beginning $t = 0$ and on constraints at the boundary. Here, assume that at $t = 0$ the system is given by the initial displacement $u(0, x) = u_0(x)$ and velocity $v(0, x) = v_0(x)$ for $x \in \Omega$, and we use homogeneous boundary conditions $u(t, 0) = u(t, X) = 0$ for $t \in [0, T]$ corresponding to a string that is fixed at the endpoints.

Inserting $v = \partial_t u$ and $\varepsilon = \partial_x u$ in (1.1) we obtain the second-order formulation of the wave equation

$$2\partial_t^2 u(t, x) - c^2 \partial_x^2 u(t, x) = 0 \qquad \text{for } (t, x) \in (0, T) \times (0, X)\,, \tag{1.2a}$$

$$u(0, x) = u_0(x) \qquad \text{for } x \in (0, X) \text{ at } t = 0\,, \tag{1.2b}$$

$$\partial_t u(0, x) = v_0(x) \qquad \text{for } x \in (0, X) \text{ at } t = 0\,, \tag{1.2c}$$

$$u(t, x) = 0 \qquad \text{for } x \in \{0, X\} \text{ and } t \in (0, T)\,. \tag{1.2d}$$

Note that the same equation can be derived for a 1d wave with horizontal displacement, corresponding to an actual position of the material point $x + u(x) \in \mathbb{R}$.

The Solution of the Linear Wave Equation in 1d in Homogeneous Media The equation (1.2) with constant wave speed $c > 0$ can be solved explicitly. For given initial values (1.2b) and (1.2c) the solution is given within the cone

$$C = \Big\{(t, x) \in (0, T) \times (0, X) \colon 0 < x - ct < x + ct < X\Big\}$$

by the d'Alembert formula

$$u(t, x) = \frac{1}{2}\left(u_0(x - ct) + u_0(x + ct) + \frac{1}{c}\int_{x-ct}^{x+ct} v_0(\xi)\,\mathrm{d}\xi\right), \qquad (t, x) \in C\,.$$

Now we consider the solution in the bounded interval $\Omega = (0, X)$ of length $X = \pi$ with homogeneous Dirichlet boundary conditions (1.2d). The solution can be expanded into eigenmodes of the operator $-\partial_x^2 u$ in $\mathrm{H}_0^1(\Omega) \cap \mathrm{H}^2(\Omega)$, so that we obtain

$$u(t, x) = \sum_{n=1}^{\infty} \Big(\alpha_n \cos(cnt) + \beta_n \sin(cnt)\Big) \sin(nx)\,,$$

where the coefficients are determined by the initial values (1.2b) and (1.2c). For the special example with initial values $u_0(x) = 1$, $v_0(x) = 0$ for $x \in (0, \pi)$, and wave speed $c = 1$, we obtain the explicit Fourier representation

$$\begin{aligned} u(t, x) &= \frac{4}{\pi}\sum_{n=0}^{\infty} \frac{1}{2n+1} \cos\big((2n+1)t\big) \sin\big((2n+1)x\big) \\ &= \frac{1}{2}\Big(u_0(x+t) + u_0(x-t)\Big)\,, \end{aligned} \tag{1.3}$$

where the initial function u_0 is extended to the periodic function

$$u_0(x) = \begin{cases} 1 & x \in (0, \pi) + 2\pi\mathbb{Z}\,, \\ 0 & x \in \pi\mathbb{Z}\,, \\ -1 & x \in (-\pi, 0) + 2\pi\mathbb{Z}\,, \end{cases}$$

cf. Fig. 1.1. We observe that this solution solves the wave equation only in a weak sense since it is discontinuous along linear characteristics $x \pm ct = \text{const}$.

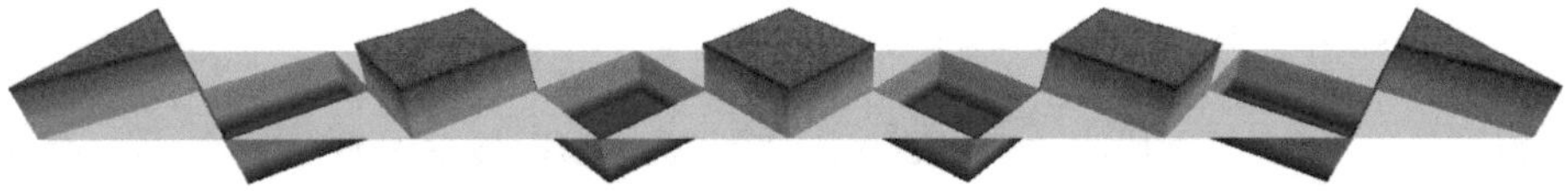

Fig. 1.1 Weak solution $u \in L_2\big((0,8) \times (0,\pi)\big)$ with initial values for $u(0,\cdot) = 1$, $\partial_t u(0,\cdot) = 0$, and homogeneous Dirichlet boundary values $u(\cdot,0) = u(\cdot,\pi) = 0$

1.3 Harmonic, Anharmonic and Viscous Waves

Special solutions of the linear wave equation (1.2) can be derived by the ansatz

$$u(t,x) = \exp(-i\omega t)a(x)$$

with a fixed frequency $\omega \in \mathbb{R}$. This yields in case of constant wave speed $c = \sqrt{\kappa/\rho}$

$$\partial_t^2 u(t,x) - c^2\partial_x^2 u(t,x) = -\Big(\omega^2 a(x) + c^2\partial_x^2 a(x)\Big)\exp(-i\omega t)\,.$$

The equation $\omega^2 a(x) + c^2\partial_x^2 a(x) = 0$ is solved by $a(x) = a_0 \exp(ikx)$ with $k = \omega/c$ and $a_0 \in \mathbb{R}$, cf. Table 1.1.

Interaction with Material: Anharmonic Waves The harmonic wave with constant amplitude is an idealistic model. This contradicts to observations: a wave traveling through material interacts with the particles in some sense, so that the amplitude is decreasing in time. A simple ansatz are waves of the form

$$u(t,x) = a(t)\exp\big(i(kx - \omega t)\big)\,, \qquad a(t) = a_0\exp(-\tau t) \tag{1.4}$$

depending on wave number k, angular frequency ω, and relaxation time $\tau > 0$. Then, we observe for (1.4) in case of constant ρ and κ

$$(\rho\partial_t^2 - \kappa\partial_x^2)u(x,t) = \big(\rho(\tau + i\omega)^2 + \kappa k^2\big)u(x,t)\,, \qquad \partial_t u(x,t) = -(\tau + i\omega)u(x,t)$$

which yields with the angular frequency

$$\omega = \sqrt{k^2\kappa/\rho + \tau^2} \in \mathbb{R} \tag{1.5}$$

Table 1.1 Characteristic quantities for harmonic waves $u(t,x) = a_0\exp\big(i(kx - \omega t)\big)$

Wave number	k	Angular frequency	ω	Frequency	$\nu = \omega/2\pi$
Wave speed	$c = \omega/k$	Wave length	$\lambda = c/\nu$	Amplitude	a_0

a solution of the wave equation with attenuation

$$\rho\partial_t^2 u(t,x) - \kappa\partial_x^2 u(t,x) + 2\tau\rho\,\partial_t u(t,x) = 0\,. \tag{1.6}$$

In general, one observes that the wave speed depends on the frequency of the wave, i.e., the wave is *dispersive*. For the case of constant parameters this is characterized by the *dispersion relation* $\omega = \omega(k)$. In this example, we find the dispersion relation (1.5) for the wave equation with attenuation (1.6). For the general description of real media this approach is too simple and applies only for the wave propagation within a limited frequency range, in particular since the relaxation time also depends on the frequency. For viscous waves suitable material laws are constructed where the parameters can be determined from measurements of the dispersion relation at sample frequencies which are relevant for the application. This is now demonstrated for a specific example.

A Model for Viscous Waves One approach to characterize waves with dispersion is to use a linear superposition of the constitutive law for a harmonic wave with several relations for anharmonic waves. In this ansatz the material law for the stress is based on a decomposition $\sigma = \sigma_0 + \sigma_1 + \cdots + \sigma_r$ with Hooke's law for σ_0, i.e.,

$$\sigma_0 = \kappa_0\varepsilon\,, \tag{1.7a}$$

and several Maxwell bodies for $\sigma_1, \ldots, \sigma_r$ described by the relations

$$\partial_t\sigma_j + \tau_j^{-1}\sigma_j = \kappa_j\partial_t\varepsilon\,, \qquad j = 1, \ldots, r\,. \tag{1.7b}$$

This model depends on the stiffness of the components $\kappa_0, \ldots, \kappa_r$ and relaxation times $\tau_1, \ldots, \tau_r$. Solving the linear ODE (1.7b) with initial value $\sigma_j(0) = 0$ and inserting $\partial_t\varepsilon = \partial_x v$ yields

$$\sigma_j(t) = \int_0^t \kappa_j \exp\Big(-\frac{1}{\tau_j}(t-s)\Big)\partial_x v(s)\,\mathrm{d}s\,,$$

and together with (1.7a) we obtain the *retarded material law*

$$\sigma(t) = \kappa_0\partial_x u(t) + \int_0^t \sum_{j=1}^r \kappa_j \exp\Big(-\frac{1}{\tau_j}(t-s)\Big)\partial_x v(s)\,\mathrm{d}s\,.$$

This can be summarized to

$$\begin{aligned}\partial_t\sigma(t) &= \kappa_0\partial_x v(t) + \sum_{j=1}^r \kappa_j\partial_x v(t) - \int_0^t \sum_{j=1}^r \frac{\kappa_j}{\tau_j}\exp\Big(-\frac{1}{\tau_j}(t-s)\Big)\partial_x v(s)\,\mathrm{d}s\\ &= \kappa\partial_x v(t) + \int_0^t \dot\kappa(t-s)\partial_x v(s)\,\mathrm{d}s\end{aligned}$$

with the total stiffness $\kappa = \kappa_0 + \kappa_1 + \cdots + \kappa_r$ and the *retardation kernel*

$$\dot\kappa(s) = -\sum_{j=1}^r \frac{\kappa_j}{\tau_j}\exp\Big(-\frac{s}{\tau_j}\Big).$$

Together with the balance relation $\rho\partial_t v = \partial_x\sigma$ this is a model for viscous waves.

1.4 Elastic Waves

In the next step we derive equations for waves in solids. We consider heterogeneous media where the material parameters depend on the position, and we assume that the wave energy is sufficiently small, so that the material law can be approximated by a linear relation.

Configuration We consider an elastic body in the spatial domain $\Omega \subset \mathbb{R}^3$ and we fix a time interval $I = (0, T)$. The boundary $\partial\Omega = \Gamma_V \cup \Gamma_S$ is decomposed into parts corresponding to dynamic boundary conditions for the velocity and static boundary conditions for the stress.

Constituents The current state of the body is described by the *deformation* or by the *displacement*

$$\boldsymbol{\varphi} = \mathrm{id} + \mathbf{u}\colon [0, T] \times \overline{\Omega} \longrightarrow \mathbb{R}^3, \qquad \mathbf{u}\colon [0, T] \times \overline{\Omega} \longrightarrow \mathbb{R}^3,$$

i.e., $\boldsymbol{\varphi}(t, \mathbf{x}) = \mathbf{x} + \mathbf{u}(t, \mathbf{x})$ is the actual position of the point $\mathbf{x} \in \Omega$ at time t. Depending on the displacement, we define the *velocity* $\mathbf{v} = \partial_t\mathbf{u}$, the *strain* $\boldsymbol{\varepsilon}(\mathbf{u}) = \mathrm{sym}(\mathrm{D}\mathbf{u})$, the *acceleration* $\mathbf{a} = \partial_t\mathbf{v} = \partial_t^2\mathbf{u}$, and the *strain rate* $\boldsymbol{\varepsilon}(\mathbf{v}) = \mathrm{sym}(\mathrm{D}\mathbf{v}) = \partial_t\boldsymbol{\varepsilon}(\mathbf{u})$.

The internal forces in the material are described by the *stress tensor*

$$\boldsymbol{\sigma}\colon [0, T] \times \overline{\Omega} \longrightarrow \mathbb{R}^{3\times3}_{\mathrm{sym}}.$$

Material Parameters Measurements are required to determine the distribution of the mass density

$$\rho\colon \overline{\Omega} \longrightarrow (0,\infty)$$

and to determine the material stiffness in all directions which are collected in Hooke's tensor

$$\mathbf{C}\colon \overline{\Omega} \longrightarrow \mathcal{L}(\mathbb{R}^{3\times 3}_{\mathrm{sym}}, \mathbb{R}^{3\times 3}_{\mathrm{sym}})\,.$$

Balance of Momentum Newton's law postulates equality between the temporal change of the momentum $\rho\mathbf{v}$ in any time interval $(t_1, t_2) \in (0, T)$ within any subvolume $K \subset \Omega$ and the driving forces on the boundary ∂K described by the stress in direction of the outer normal vector $\mathbf{n}$ on ∂K. This results in the balance relation (without external loads)

$$\int_K \rho(\mathbf{x})\big(\mathbf{v}(t_2,\mathbf{x}) - \mathbf{v}(t_1,\mathbf{x})\big)\,\mathrm{d}\mathbf{x} = \int_{t_1}^{t_2}\int_{\partial K} \boldsymbol{\sigma}(t,\mathbf{x})\mathbf{n}(\mathbf{x})\,\mathrm{d}\mathbf{a}\,\mathrm{d}t\,. \tag{1.8}$$

For smooth functions we obtain by the Gauß theorem

$$\int_K\int_{t_1}^{t_2} \rho(\mathbf{x})\partial_t\mathbf{v}(t_2,\mathbf{x})\,\mathrm{d}t\,\mathrm{d}\mathbf{x} = \int_{t_1}^{t_2}\int_K \operatorname{div}\boldsymbol{\sigma}(t,\mathbf{x})\,\mathrm{d}\mathbf{x}\,\mathrm{d}t\,,$$

and since this holds for all time intervals and subvolumes, we get the pointwise relation

$$\rho\partial_t\mathbf{v} = \operatorname{div}\boldsymbol{\sigma} \qquad \text{in } (0,T)\times\Omega\,. \tag{1.9}$$

Remark 1.1 In the balance relation (1.8) only the normal stress $\boldsymbol{\sigma}(t,\mathbf{x})\mathbf{n}$ for all directions $\mathbf{n} \in S^2$ on the boundary of a subvolume $K \subset \Omega$ is included. This describes the force between material points left and right from $x \in \partial K$ with respect to the direction $\mathbf{n}$. The existence of such a vector for all directions and all points is postulated by the Cauchy axiom, and by the Cauchy theorem a tensor representing this force exists; moreover, the symmetry of this tensor is a consequence of the balance of angular momentum.

Material Law Since the forces between the material points $\mathbf{x}_1$ and $\mathbf{x}_2$ only depend on the difference of the actual positions $\mathbf{u}(t,\mathbf{x}_2) - \mathbf{u}(t,\mathbf{x}_1)$, the stress $\boldsymbol{\sigma}(t,\mathbf{x})$ only depends on the deformation gradient $\mathrm{D}\boldsymbol{\varphi}$.

By definition, a material is *elastic*, if a function Σ exists such that $\boldsymbol{\sigma} = \Sigma(\mathrm{D}\boldsymbol{\varphi})$. Then, $\partial_t\boldsymbol{\sigma} = \mathrm{D}\Sigma(\mathrm{D}\boldsymbol{\varphi})[\mathrm{D}\mathbf{v}]$. In the limit of small strains the material response can be approximated by a linear model, i.e., we assume $\mathrm{D}\boldsymbol{\varphi} \approx \mathbf{I}$, and we use the linear relation $\partial_t\boldsymbol{\sigma} = \mathrm{D}\Sigma(\mathbf{I})[\mathrm{D}\mathbf{v}]$. In addition, we assume that the stress response is *objective*, i.e., it is independent of the observer's position; then it can be shown that it only depends on the symmetric strain $\boldsymbol{\varepsilon}(\mathbf{u}) = \mathrm{sym}(\mathrm{D}\mathbf{u})$. Together, we obtain Hooke's law

$$\partial_t \boldsymbol{\sigma} = \mathbf{C}\boldsymbol{\varepsilon}(\mathbf{v})\,. \tag{1.10}$$

Boundary and Initial Data We start with $\mathbf{u}(0) = \mathbf{u}_0$ and $\mathbf{v}(0) = \mathbf{v}_0$ in Ω at $t = 0$, and for $t \in (0, T)$ we use the boundary conditions for the displacement $\mathbf{u}(t) = \mathbf{u}_\mathrm{V}(t)$ or the velocity $\mathbf{v}(t) = \mathbf{v}_\mathrm{V}(t)$ on the dynamic boundary Γ_V, and for the stress $\boldsymbol{\sigma}(t)\mathbf{n} = \mathbf{g}_\mathrm{S}$ on the static boundary Γ_S.

Including external body forces $\mathbf{f}$, we obtain the second-order formulation of the linear wave equation

$$\rho\partial_t^2\mathbf{u} - \operatorname{div}\mathbf{C}\boldsymbol{\varepsilon}(\mathbf{u}) = \mathbf{f} \qquad \text{in } (0,T)\times\Omega\,, \tag{1.11a}$$

$$\mathbf{u}(0) = \mathbf{u}_0 \qquad \text{in } \Omega \text{ at } t = 0\,, \tag{1.11b}$$

$$\partial_t\mathbf{u}(0) = \mathbf{v}_0 \qquad \text{in } \Omega \text{ at } t = 0\,, \tag{1.11c}$$

$$\mathbf{u}(t) = \mathbf{u}_\mathrm{V}(t) \qquad \text{on } \Gamma_\mathrm{V} \text{ for } t \in (0,T)\,, \tag{1.11d}$$

$$\mathbf{C}\boldsymbol{\varepsilon}(\mathbf{u})\mathbf{n} = \mathbf{g}_\mathrm{S}(t) \qquad \text{on } \Gamma_\mathrm{S} \text{ for } t \in (0,T)\,, \tag{1.11e}$$

and, equivalently, the first-order formulation

$$\rho\partial_t\mathbf{v} - \operatorname{div}\boldsymbol{\sigma} = \mathbf{f} \qquad \text{in } (0,T)\times\Omega\,, \tag{1.12a}$$

$$\partial_t\boldsymbol{\sigma} - \mathbf{C}\boldsymbol{\varepsilon}(\mathbf{u}) = \mathbf{0} \qquad \text{in } (0,T)\times\Omega\,, \tag{1.12b}$$

$$\mathbf{v}(0) = \mathbf{v}_0 \qquad \text{in } \Omega \text{ at } t = 0\,, \tag{1.12c}$$

$$\boldsymbol{\sigma}(0) = \mathbf{C}\boldsymbol{\varepsilon}(\mathbf{u}_0) \qquad \text{in } \Omega \text{ at } t = 0\,, \tag{1.12d}$$

$$\mathbf{v}(t) = \partial_t\mathbf{u}_\mathrm{V}(t) \qquad \text{on } \Gamma_\mathrm{V} \text{ for } t \in (0,T)\,, \tag{1.12e}$$

$$\boldsymbol{\sigma}(t)\mathbf{n} = \mathbf{g}_\mathrm{S}(t) \qquad \text{on } \Gamma_\mathrm{S} \text{ for } t \in (0,T)\,. \tag{1.12f}$$

1.5 Visco-Elastic Waves

The balance of momentum (1.9) together with Hooke's law $\partial_t\boldsymbol{\sigma} = \mathbf{C}\boldsymbol{\varepsilon}(\mathbf{v})$ describes linear elastic waves. We observe

$$\boldsymbol{\sigma}(t) = \boldsymbol{\sigma}(0) + \int_0^t \partial_t\boldsymbol{\sigma}(s)\,\mathrm{d}s = \boldsymbol{\sigma}(0) + \int_0^t \mathbf{C}\boldsymbol{\varepsilon}(\mathbf{v}(s))\,\mathrm{d}s\,.$$

General linear visco-elastic waves are described by a *retarded material law*

$$\boldsymbol{\sigma}(t) = \boldsymbol{\sigma}(0) + \int_0^t \mathbf{C}(t-s)\boldsymbol{\varepsilon}(\mathbf{v}(s))\,\mathrm{d}s$$

implying

$$\partial_t\boldsymbol{\sigma}(t) = \mathbf{C}(0)\boldsymbol{\varepsilon}(\mathbf{v}(t)) + \int_0^t \dot{\mathbf{C}}(t-s)\boldsymbol{\varepsilon}\big(\mathbf{v}(s)\big)\,\mathrm{d}s$$

with a time-dependent extension $\dot{\mathbf{C}}$ of the elasticity tensor $\mathbf{C}$.

In analogy to the 1d model (1.7), one defines *Generalized Standard Linear Solids* with the *relaxation tensor*

$$\dot{\mathbf{C}}(s) = -\sum_{j=1}^{r}\frac{1}{\tau_j}\exp\Big(-\frac{s}{\tau_j}\Big)\mathbf{C}_j\,, \qquad \mathbf{C}(0) = \mathbf{C}_0 + \mathbf{C}_1 + \cdots + \mathbf{C}_r\,.$$

Introducing the corresponding stress decomposition $\boldsymbol{\sigma} = \boldsymbol{\sigma}_0 + \cdots + \boldsymbol{\sigma}_r$ with

$$\boldsymbol{\sigma}_j(t) = \int_0^t \exp\Big(\frac{s-t}{\tau_j}\Big)\mathbf{C}_j\boldsymbol{\varepsilon}(\mathbf{v}(s))\,\mathrm{d}s\,, \qquad j = 1,\ldots,r,$$

results in the first-order system for visco-elastic waves

$$\rho\,\partial_t\mathbf{v} - \nabla\cdot\big(\boldsymbol{\sigma}_0 + \cdots + \boldsymbol{\sigma}_r\big) = \mathbf{f}\,, \tag{1.13a}$$

$$\partial_t\boldsymbol{\sigma}_0 - \mathbf{C}_0\boldsymbol{\varepsilon}(\mathbf{v}) = \mathbf{0}\,, \tag{1.13b}$$

$$\partial_t\boldsymbol{\sigma}_j - \mathbf{C}_j\boldsymbol{\varepsilon}(\mathbf{v}) + \tau_j^{-1}\boldsymbol{\sigma}_j = \mathbf{0}\,, \qquad j = 1,\ldots,r\,. \tag{1.13c}$$

This is complemented by initial and boundary conditions for the velocity $\mathbf{v}$ and the total stress $\boldsymbol{\sigma}$, which are the observable quantities. The stress components $\boldsymbol{\sigma}_1,\ldots,\boldsymbol{\sigma}_r$ are inner variables describing the retarded material law; they can be replaced, e.g., by memory variables encoding the material history.

1.6 Acoustic Waves in Solids

In isotropic media, Hooke's tensor only depends on two parameters, e.g., the *Lamé parameters* μ, λ

$$\begin{aligned}\mathbf{C}\boldsymbol{\varepsilon} &= 2\mu\boldsymbol{\varepsilon} + \lambda\,\mathrm{tr}(\boldsymbol{\varepsilon})\mathbf{I}\\ &= 2\mu\,\mathrm{dev}(\boldsymbol{\varepsilon}) + \kappa\,\mathrm{tr}(\boldsymbol{\varepsilon})\mathbf{I}\,, \qquad \mathrm{dev}(\boldsymbol{\varepsilon}) = \boldsymbol{\varepsilon} - \frac{1}{3}\mathrm{tr}(\boldsymbol{\varepsilon})\mathbf{I}\,.\end{aligned}$$

For the wave dynamics, one uses a decomposition into components corresponding to shear waves depending on the *shear modulus* μ, and compressional waves depending on the *compression modulus* $\kappa = \frac{2}{3}\mu + \lambda$. Then, the linear second order elastic wave equation (1.11a) in isotropic and homogeneous media takes the form

$$\rho \partial_t^2 \mathbf{u} + \mu \nabla \times \nabla \times \mathbf{u} - 3\kappa \nabla(\nabla \cdot \mathbf{u}) = \mathbf{f}.$$

A vanishing shear modulus $\mu \to 0$ leads to the *linear acoustic wave equation* for the *hydrostatic pressure* $p = \frac{1}{3}\operatorname{tr}(\boldsymbol{\sigma})$ and the velocity, described by the first-order system

$$\begin{aligned}
\rho\, \partial_t \mathbf{v} - \nabla p &= \mathbf{f} && \text{in } (0, T) \times \Omega\,, && (1.14a)\\
\partial_t p - \kappa \nabla \cdot \mathbf{v} &= 0 && \text{in } (0, T) \times \Omega\,, && (1.14b)\\
\mathbf{v}(0) &= \mathbf{v}_0 && \text{in } \Omega \text{ at } t = 0\,, && (1.14c)\\
p(0) &= p_0 && \text{in } \Omega \text{ at } t = 0\,, && (1.14d)\\
\mathbf{n} \cdot \mathbf{v}(t) &= g_{\mathrm{V}}(t) && \text{on } \Gamma_{\mathrm{V}} \text{ for } t \in (0, T)\,, && (1.14e)\\
p(t) &= p_{\mathrm{S}}(t) && \text{on } \Gamma_{\mathrm{S}} \text{ for } t \in (0, T)\,, && (1.14f)
\end{aligned}$$

where we set $p_{\mathrm{S}} = \mathbf{n} \cdot \mathbf{g}_{\mathrm{S}}$ for the static boundary condition and $g_{\mathrm{V}} = \mathbf{n} \cdot \mathbf{v}_{\mathrm{V}}$ for the dynamic boundary condition. For acoustics, this corresponds to Dirichlet and Neumann boundary conditions, for elasticity this is reversed.

In homogeneous media and for $\mathbf{f} = \mathbf{0}$, (1.14a) and (1.14b) combine to the linear second-order acoustic wave equation

$$\partial_t^2 p - c^2 \Delta p = 0\,, \qquad c = \sqrt{\kappa/\rho}\,.$$

Remark 1.2 Simply neglecting the shear component is only an approximation and not fully realistic for waves in solids, in particular since by reflections compressional waves split in compressional and shear components. Nevertheless, in applications the acoustic wave equation is used also in solids since the system is much smaller so that computations are much faster.

Remark 1.3 One obtains the same acoustic wave equations describing compression waves in a fluid or a gas. Note that, historically, the sign conventions for pressure and stress are different in fluid and solid mechanics.

Visco-Acoustic Waves Generalized Standard Linear Solids can be reduced to acoustics. The corresponding retarded material law for the hydrostatic pressure takes the form

$$\partial_t p(t) = \kappa \nabla \cdot \mathbf{v}(t) + \int_0^t \dot{\kappa}(t-s) \nabla \cdot \mathbf{v}(s)\, \mathrm{d}s\,, \qquad \dot{\kappa}(s) = -\sum_{j=1}^{r} \frac{\kappa_j}{\tau_j} \exp\Big(-\frac{s}{\tau_j}\Big)\,.$$

Defining $\kappa = \kappa_0 + \kappa_1 + \cdots + \kappa_r$ and $p = p_0 + p_1 + \cdots + p_r$ with

$$p_j(t) = \int_0^t \exp\Big(\frac{s-t}{\tau_j}\Big)\kappa_j \nabla\cdot \mathbf{v}(s)\,\mathrm{d}s\,, \qquad j = 1,\dots,r$$

results in the first-order system for linear visco-acoustic waves

$$\begin{aligned} \rho\,\partial_t \mathbf{v} - \nabla(p_0 + \cdots + p_r) &= \mathbf{f}\,,\\ \partial_t p_0 - \kappa_0 \nabla\cdot\mathbf{v} &= 0\,,\\ \partial_t p_j - \kappa_j\nabla\cdot\mathbf{v} + \tau_j^{-1} p_j &= 0\,, \qquad j = 1,\dots,r\,. \end{aligned}$$

This is complemented by initial and boundary conditions (1.14c)–(1.14f).

1.7 Electro-Magnetic Waves

Electric fields induce magnetic fields and vice versa. This is formulated by Maxwell's equations describing the propagation of electro-magnetic waves.

Configuration We consider a spatial domain $\Omega \subset \mathbb{R}^3$, a time interval $I = (0,T)$, and a boundary decomposition $\partial\Omega = \Gamma_{\mathrm{E}} \cup \Gamma_{\mathrm{I}}$ corresponding to perfect conducting or transmission boundaries.

Constituents Electro-magnetic waves are determined by the *electric field* and the *magnetic field intensity*

$$\mathbf{E}\colon \overline{I\times\Omega} \to \mathbb{R}^3\,, \qquad \mathbf{H}\colon \overline{I\times\Omega} \to \mathbb{R}^3\,,$$

and by the *electric flux density* and *magnetic induction*

$$\mathbf{D}\colon \overline{I\times\Omega} \to \mathbb{R}^3\,, \qquad \mathbf{B}\colon \overline{I\times\Omega} \to \mathbb{R}^3\,.$$

Further quantities are the *electric current density* and the *electric charge density*

$$\mathbf{J}\colon I\times\Omega \to \mathbb{R}^3\,, \qquad \rho\colon I\times\Omega \to \mathbb{R}\,.$$

Balance Relations Faraday's law states that the temporal change of the magnetic induction through a two-dimensional subset $A \subset \Omega$ induces an electric field along the boundary ∂A, so that for all $0 < t_1 < t_2 < T$

$$\int_A \big(\mathbf{B}(t_2) - \mathbf{B}(t_1)\big)\cdot \mathbf{da} = -\int_{t_1}^{t_2}\int_{\partial A} \mathbf{E}\cdot \mathrm{d}\boldsymbol{\ell}\,\mathrm{d}t\,.$$

Ampere's law states that the temporal change of the electric flux density together with the electric current density through a two-dimensional manifold $A \subset \Omega$ induces a magnetic field intensity along the boundary ∂A, i.e.,

$$\int_A \big(\mathbf{D}(t_2) - \mathbf{D}(t_1)\big) \cdot \, \mathbf{da} + \int_{t_1}^{t_2} \int_A \mathbf{J} \cdot \, \mathbf{da} \; \mathrm{d}t = \int_{t_1}^{t_2} \int_{\partial A} \mathbf{H} \cdot \mathrm{d}\boldsymbol{\ell} \; \mathrm{d}t \, .$$

Here, we use $\mathbf{u} \cdot \, \mathbf{da} = \mathbf{u} \cdot \mathbf{n} \; \mathrm{da}$ and $\mathbf{u} \cdot \mathrm{d}\boldsymbol{\ell} = \mathbf{u} \cdot \boldsymbol{\tau} \, \mathrm{d}\ell$, the normal vector field $\mathbf{n} \colon A \to \mathbb{R}^3$ and the tangential vector field $\boldsymbol{\tau} \colon \partial A \to \mathbb{R}^3$ (where the orientation of ∂A is given by $\mathbf{n}$).

The Gauß laws state for all subvolumes $K \subset \Omega$ the conservation of the magnetic induction

$$\int_{\partial K} \mathbf{B} \cdot \, \mathbf{da} = 0$$

and the equilibrium of electric charge density in the volume with electric flux density across the boundary ∂K

$$\int_{\partial K} \mathbf{D} \cdot \, \mathbf{da} = \int_K \rho \; \mathrm{d}\mathbf{x} \, .$$

Together, by the integral theorems of Stokes and Gauß we obtain

$$\int_A \int_{t_1}^{t_2} \partial_t \mathbf{B} \cdot \, \mathbf{da} \; \mathrm{d}t = - \int_{t_1}^{t_2} \int_A \nabla \times \mathbf{E} \cdot \, \mathbf{da} \; \mathrm{d}t \, ,$$

$$\int_K \nabla \cdot \mathbf{B} \; \mathrm{d}\mathbf{x} = 0 \, ,$$

$$\int_A \int_{t_1}^{t_2} \partial_t \mathbf{D} \cdot \, \mathbf{da} \; \mathrm{d}t + \int_{t_1}^{t_2} \int_A \mathbf{J} \cdot \, \mathbf{da} \; \mathrm{d}t = \int_{t_1}^{t_2} \int_A \nabla \times \mathbf{H} \cdot \, \mathbf{da} \; \mathrm{d}t,$$

$$\int_K \nabla \cdot \mathbf{D} \; \mathrm{d}\mathbf{x} = \int_K \rho \; \mathrm{d}\mathbf{x} \, ,$$

and since this holds for all $(t_1, t_2) \subset I$ and all $A, K \subset \Omega$, it results in the Maxwell system

$$\partial_t \mathbf{B} + \nabla \times \mathbf{E} = \mathbf{0} \, , \quad \partial_t \mathbf{D} - \nabla \times \mathbf{H} = -\mathbf{J} \, , \quad \nabla \cdot \mathbf{B} = 0 \, , \quad \nabla \cdot \mathbf{D} = \rho \, . \tag{1.15}$$

Note that a combination of the second and fourth equation implies the conservation of charge $\partial_t \rho + \nabla \cdot \mathbf{J} = 0$.

Material Laws in Vacuum Without the interaction with matter, electric field and the electric flux density, $\mathbf{D} = \varepsilon_0 \mathbf{E}$, and magnetic induction and magnetic field intensity, $\mathbf{B} = \mu_0 \mathbf{H}$, are proportional by multiplication with the constant *permittivity* ε_0 and *permeability* μ_0, respectively, which together results in the linear second-order Maxwell equation for $\mathbf{E}$

$$\partial_t^2 \mathbf{E} - c^2 \nabla \times \nabla \times \mathbf{E} = \mathbf{0}$$

with speed of light $c = 1/\sqrt{\varepsilon_0 \mu_0}$. A corresponding equation holds for $\mathbf{H}$. In vacuum, in the absence of electric currents and electric charges, we find $\mathbf{J} = \mathbf{0}$ and $\rho = 0$.

Effective Material Laws for Electro-Magnetic Waves in Matter The interaction of electro-magnetic waves with the atoms in matter are described by the *polarization* $\mathbf{P}$ and the *magnetization* $\mathbf{M}$ depending on the electric field $\mathbf{E}$ and the magnetic induction $\mathbf{B}$. For the electric flux density holds

$$\mathbf{D} = \varepsilon_0 \mathbf{E} + \mathbf{P}(\mathbf{E}, \mathbf{B}) ,$$

and the magnetic field intensity is given by

$$\mu_0 \mathbf{H} = \mathbf{B} - \mathbf{M}(\mathbf{E}, \mathbf{B}) .$$

The electric current density depends on the conductivity σ (*Ohm's law*) and the external current $\mathbf{J}_0$, so that

$$\mathbf{J} = \sigma \mathbf{E} + \mathbf{J}_0 .$$

In case of linear materials with instantaneous response, the polarization is proportional to the electric field

$$\mathbf{P} = \varepsilon_0 \chi \mathbf{E}$$

with the *susceptibility* χ, that yields $\mathbf{D} = \varepsilon_r \mathbf{E}$ with relative permittivity $\varepsilon_r = \varepsilon_0(1 + \chi)$.

Linear materials with retarded response are given by

$$\mathbf{P}(t) = \varepsilon_0 \int_{-\infty}^{t} \chi(t - s) \mathbf{E}(s) \, \mathrm{d}s . \tag{1.16}$$

A special case is the Debye model with $\chi(t) = \exp\Big(-\frac{t}{\tau}\Big)\frac{\varepsilon_s - \varepsilon_\infty}{\tau}$, so that the polarization is determined by

$$\tau \partial_t \mathbf{P} + \mathbf{P} = \varepsilon_0 (\varepsilon_s - \varepsilon_\infty) \mathbf{E} .$$

This model is dispersive with a dispersion relation similar to the model for viscous elastic waves.

The relation (1.16) extends to nonlinear materials by, e.g.,

$$\mathbf{P}(t) = \varepsilon_0 \int_{-\infty}^{t} \chi_1(t-s)\mathbf{E}(s)\, \mathrm{d}s$$
$$+ \int_{-\infty}^{t}\int_{-\infty}^{t}\int_{-\infty}^{t} \chi_3(t-s_1, t-s_2, t-s_3)\big(\mathbf{E}(s_1), \mathbf{E}(s_2), \mathbf{E}(s_3)\big)\, \mathrm{d}s_1 \mathrm{d}s_2 \mathrm{d}s_3 \,.$$

For materials of Kerr-type this response is instantaneous, i.e.,

$$\mathbf{P} = \chi_1 \mathbf{E} + \chi_3 |\mathbf{E}|^2 \mathbf{E}\,.$$

In more complex material models, the Maxwell system (1.15) is coupled to evolution equations for polarization or magnetization. E.g., in the Maxwell–Lorentz system the evolution of the polarization is determined by

$$\partial_t^2 \mathbf{P} = \frac{1}{\varepsilon_0^2}(\mathbf{E} - \mathbf{P}) + |\mathbf{P}|^2 \mathbf{P}\,.$$

In the Landau–Lifshitz–Gilbert (LLG) equation the magnetization $\mathbf{M}$ is given by

$$\partial_t \mathbf{M} - \alpha \mathbf{M} \times \partial_t \mathbf{M} = -\mathbf{M} \times \mathbf{H}_{\text{eff}}\,, \qquad |\mathbf{M}| = 1\,,$$

where $\alpha > 0$ is a damping factor, and the effective field $\mathbf{H}_{\text{eff}}$ is a combination of the external magnetic field and the demagnetizing field, which is a magnetic field due to the magnetization.

Boundary Conditions The Maxwell system is complemented by conditions on $\partial\Omega$. On a perfectly conducting boundary Γ_{E}, we have

$$\mathbf{E} \times \mathbf{n} = \mathbf{0} \quad \text{and} \quad \mathbf{B} \cdot \mathbf{n} = 0\,,$$

and on the impedance (or Silver–Müller) boundary Γ_{I}, we prescribe

$$\mathbf{H} \times \mathbf{n} + \zeta\, (\mathbf{E} \times \mathbf{n}) \times \mathbf{n} = \mathbf{0}$$

depending on the given impedance ζ.

Together, we obtain for general nonlinear instantaneous material laws $\mathbf{D}(\mathbf{E},\mathbf{H})$ and $\mathbf{B}(\mathbf{E},\mathbf{H})$ the first-order system

$$2\partial_t \mathbf{D}(\mathbf{E},\mathbf{H}) - \nabla\times\mathbf{H} + \sigma\mathbf{E} = -\mathbf{J}_0, \quad \text{in } (0,T)\times\Omega\,, \tag{1.17a}$$

$$\partial_t \mathbf{B}(\mathbf{E},\mathbf{H}) + \nabla\times\mathbf{E} = \mathbf{0} \quad \text{in } (0,T)\times\Omega\,, \tag{1.17b}$$

$$\mathbf{E}(0) = \mathbf{E}_0 \quad \text{in } \Omega \text{ at } t=0\,, \tag{1.17c}$$

$$\mathbf{H}(0) = \mathbf{H}_0 \quad \text{in } \Omega \text{ at } t=0\,, \tag{1.17d}$$

$$\mathbf{E}\times\mathbf{n} = \mathbf{0} \quad \text{on } \Gamma_{\mathrm{E}} \text{ for } t\in(0,T)\,, \tag{1.17e}$$

$$\mathbf{H}\times\mathbf{n} + \zeta\,(\mathbf{E}\times\mathbf{n})\times\mathbf{n} = \mathbf{g} \quad \text{on } \Gamma_{\mathrm{I}} \text{ for } t\in(0,T)\,. \tag{1.17f}$$

In nonlinear optics, for the special case of an instantaneous nonmagnetic material law $\mathbf{D}(\mathbf{E}) = \varepsilon_0\mathbf{E}+\mathbf{P}(\mathbf{E})$ and $\mathbf{M}\equiv\mathbf{0}$, the Maxwell system reduces to the second-order equation

$$\partial_t^2\mathbf{D}(\mathbf{E}) + \mu_0^{-1}\nabla\times\nabla\times\mathbf{E} + \sigma\mathbf{E} = -\partial_t\mathbf{J}_0$$

complemented by initial and boundary conditions.

Bibliographic Comments

The mathematical foundations of modeling elastic solids (including a detailed discussion and a proof of the Cauchy theorem) is given in [27], and more physical background is given in [37]. For generalized standard linear solids we refer to [70]. An overview on modeling of electro-magnetic waves is given in [100], the mathematical aspects of photonics are considered in [47]. The example (1.3) is taken from [119, Example 3.4]. Dispersion relations and the analogy in the modeling of elastic and electro-magnetic waves are collected in [23, Chap. 2 and Chap. 8].

2 Space-Time Solutions for Linear Hyperbolic Systems

The linear wave equation can be analyzed in the framework of symmetric Friedrichs systems as a special case of linear hyperbolic conservation laws. Here, we introduce a general framework for the existence and uniqueness of strong and weak solutions in space and time which applies to general linear wave equations.

We consider operators in space and time of the form $L = M\partial_t + A$ describing a linear hyperbolic system, where A is a first-order operator in space. All results transfer to operators of the form $L = M\partial_t + A + D$ with an additional positive semi-definite operator D; this applies to visco-acoustic and visco-elastic models, to mixed boundary conditions of Robin type and impedance boundary conditions.

In the following, we use standard notations: for open domains $G \subset \mathbb{R}^d$ in space or $G \subset \mathbb{R}^{1+d}$ in space-time and functions $v, w\colon G \to \mathbb{R}$ we define the inner product $(v, w)_G = \int_G vw \,\mathrm{d}\mathbf{x}$, the norm $\|v\|_G = \sqrt{(v, v)_G}$ and the Hilbert space $\mathrm{L}_2(G)$ of measurable functions $v\colon G \to \mathbb{R}$ with $\|v\|_G < \infty$.

2.1 Linear Hyperbolic First-Order Systems

Let $\Omega \subset \mathbb{R}^d$ be a domain in space with Lipschitz boundary, $I = (0, T)$ a time interval, and we denote the space-time cylinder by $Q = (0, T) \times \Omega$. Boundary conditions will be imposed on $\Gamma_k \subset \partial\Omega$ for $k = 1, \ldots, m$, depending on the model, so that the corresponding equations are well-posed.

We consider a linear operator in space and time of the form $L = M\partial_t + A$ with a uniformly positive definite operator M defined by $M\mathbf{y}(\mathbf{x}) = \underline{M}(\mathbf{x})\mathbf{y}(\mathbf{x})$ with a matrix valued function $\underline{M} \in \mathrm{L}_\infty(\Omega; \mathbb{R}^{m\times m}_{\mathrm{sym}})$, and a differential operator $A\mathbf{y} = \sum_{j=1}^d \underline{A}_j \partial_j \mathbf{y}$ with matrices $\underline{A}_j \in \mathbb{R}^{m\times m}_{\mathrm{sym}}$. Moreover, we define the matrix $\underline{A}_{\mathbf{n}} = \sum_{j=1}^d n_j \underline{A}_j \in \mathbb{R}^{m\times m}_{\mathrm{sym}}$ for $\mathbf{n} \in \mathbb{R}^d$ and the corresponding boundary operator $(A_{\mathbf{n}}\mathbf{y})(\mathbf{x}) = \underline{A}_{\mathbf{n}}\mathbf{y}(\mathbf{x})$.

W. Dörfler et al., *Wave Phenomena*, Oberwolfach Seminars 49,
https://doi.org/10.1007/978-3-031-05793-9_2

In the first step, we consider the properties of the operators A and L for smooth functions. Then the operators are extended to Hilbert spaces and, by specifying boundary conditions, we define maximal domains for the operators.

Example 2.1 This applies to the linear acoustic wave equation (1.14) with $m = d + 1$ and

$$\mathbf{y} = \begin{pmatrix} \mathbf{v} \\ p \end{pmatrix}, \quad M\mathbf{y} = \begin{pmatrix} \rho\mathbf{v} \\ \kappa^{-1}p \end{pmatrix}, \quad A\mathbf{y} = \begin{pmatrix} -\nabla p \\ -\nabla\cdot\mathbf{v} \end{pmatrix}, \quad \underline{A}_{\mathbf{n}}\mathbf{y} = \begin{pmatrix} -p\mathbf{n} \\ -\mathbf{n}\cdot\mathbf{v} \end{pmatrix}. \tag{2.1}$$

For linear elastic waves with $\mathbf{v} = \partial_t\mathbf{u}$ and $\boldsymbol{\sigma} = \mathbf{C}\boldsymbol{\varepsilon}(\mathbf{u})$ we have

$$\mathbf{y} = \begin{pmatrix} \mathbf{v} \\ \boldsymbol{\sigma} \end{pmatrix}, \; M\mathbf{y} = \begin{pmatrix} \rho\mathbf{v} \\ \mathbf{C}^{-1}\boldsymbol{\sigma} \end{pmatrix}, \; A\mathbf{y} = \begin{pmatrix} -\operatorname{div}\boldsymbol{\sigma} \\ -\boldsymbol{\varepsilon}(\mathbf{v}) \end{pmatrix}, \; \underline{A}_{\mathbf{n}}\mathbf{y} = \begin{pmatrix} -\boldsymbol{\sigma}\mathbf{n} \\ -\frac{1}{2}(\mathbf{n}\mathbf{v}^\top + \mathbf{v}\mathbf{n}^\top) \end{pmatrix}, \tag{2.2}$$

and $\frac{1}{2}M\mathbf{y}\cdot\mathbf{y} = \frac{1}{2}\big(\rho|\mathbf{v}|^2 + \boldsymbol{\sigma}\cdot\mathbf{C}^{-1}\boldsymbol{\sigma}\big) = \frac{1}{2}\big(\rho|\partial_t\mathbf{u}|^2 + \boldsymbol{\varepsilon}(\mathbf{u})\cdot\mathbf{C}\boldsymbol{\varepsilon}(\mathbf{u})\big)$ is the kinetic and potential energy.

For linear electro-magnetic waves we have

$$\mathbf{y} = \begin{pmatrix} \mathbf{E} \\ \mathbf{H} \end{pmatrix}, \quad M\mathbf{y} = \begin{pmatrix} \varepsilon_0\mathbf{E} \\ \mu_0\mathbf{H} \end{pmatrix}, \quad A\mathbf{y} = \begin{pmatrix} -\nabla\times\mathbf{H} \\ \nabla\times\mathbf{E} \end{pmatrix}, \quad \underline{A}_{\mathbf{n}}\mathbf{y} = \begin{pmatrix} -\mathbf{n}\times\mathbf{H} \\ \mathbf{n}\times\mathbf{E} \end{pmatrix}, \tag{2.3}$$

and $\frac{1}{2}M\mathbf{y}\cdot\mathbf{y} = \frac{1}{2}\big(\varepsilon_0|\mathbf{E}|^2 + \mu_0|\mathbf{H}|^2\big)$ is the electro-magnetic energy.

Linear Conservation Laws Defining

$$\underline{A} = (\underline{A}_1, \ldots, \underline{A}_d) = (A_{j,kl})_{j=1,\ldots,d,\; k,l=1,\ldots,m} \in \mathbb{R}^{d\times m\times m}$$

we observe for the operator $A\mathbf{y} = \operatorname{div}(\underline{A}\mathbf{y})$ and the matrix $\underline{A}_{\mathbf{n}} = \mathbf{n}\cdot\underline{A}$, so that the system $L\mathbf{y} = \mathbf{f}$ takes the form of a *linear conservation law*

$$\underline{M}\partial_t\mathbf{y} + \operatorname{div}(\underline{A}\mathbf{y}) = \mathbf{f}.$$

Integration by parts and using the symmetry of $\underline{A}_j$ yields for differentiable functions with compact support in Ω

$$\begin{aligned}
(A\mathbf{y},\mathbf{z})_\Omega &= \sum_{j=1}^d \int_\Omega \underline{A}_j\partial_j\mathbf{y}\cdot\mathbf{z}\,\mathrm{d}\mathbf{x} = \sum_{j=1}^d\sum_{k,l=1}^m \int_\Omega A_{j,kl}(\partial_j y_l)z_k\,\mathrm{d}\mathbf{x}\\
&= -\sum_{j=1}^d\sum_{k,l=1}^m \int_\Omega A_{j,kl}y_l\,\partial_j z_k\,\mathrm{d}\mathbf{x} = -\sum_{j=1}^d\sum_{k,l=1}^m \int_\Omega y_l A_{j,lk}\,\partial_j z_k\,\mathrm{d}\mathbf{x}\\
&= -\sum_{j=1}^d \int_\Omega \mathbf{y}\cdot\underline{A}_j\partial_j\mathbf{z}\,\mathrm{d}\mathbf{x} = -(\mathbf{y},A\mathbf{z})_\Omega\,, \qquad \mathbf{y},\mathbf{z}\in \mathrm{C}^1_\mathrm{c}(\Omega;\mathbb{R}^m)\,,
\end{aligned}$$

so that $A^* = -A$ on $\mathrm{C}^1_\mathrm{c}(\Omega;\mathbb{R}^m)$. On the boundary $\partial\Omega$ with outer unit normal $\mathbf{n}$, integration by parts yields for $\mathbf{y},\mathbf{z}\in\mathrm{C}^1(\Omega;\mathbb{R}^m)\cap\mathrm{C}^0(\overline{\Omega};\mathbb{R}^m)$

$$\begin{aligned}
(A\mathbf{y},\mathbf{z})_\Omega + (\mathbf{y},A\mathbf{z})_\Omega &= \sum_{j=1}^d\sum_{k,l=1}^m \int_\Omega \Big(A_{j,kl}(\partial_j y_l)z_k + y_l A_{j,lk}\,\partial_j z_k\Big)\,\mathrm{d}\mathbf{x}\\
&= \sum_{j=1}^d\sum_{k,l=1}^m \int_\Omega \partial_j\big(A_{j,kl}y_l z_k\big)\,\mathrm{d}\mathbf{x} = \sum_{j=1}^d\sum_{k,l=1}^m \int_{\partial\Omega} n_j\,A_{j,kl}y_l z_k\,\mathrm{d}\mathbf{a}\\
&= \int_{\partial\Omega} \underline{A}_\mathbf{n}\mathbf{y}\cdot\mathbf{z}\,\mathrm{d}\mathbf{a} = (\underline{A}_\mathbf{n}\mathbf{y},\mathbf{z})_{\partial\Omega}\,.
\end{aligned}$$

Together, we obtain in space and time for $L = M\partial_t + A$ and its adjoint $L^* = -L$

$$\begin{aligned}
&(L\mathbf{v},\mathbf{w})_Q - (\mathbf{v},L^*\mathbf{w})_Q\\
&\qquad = \big(M\mathbf{v}(T),\mathbf{w}(T)\big)_\Omega - \big(M\mathbf{v}(0),\mathbf{w}(0)\big)_\Omega + (\underline{A}_\mathbf{n}\mathbf{v},\mathbf{w})_{(0,T)\times\partial\Omega}
\end{aligned}\tag{2.4}$$

for $\mathbf{v},\mathbf{w}\in\mathrm{C}^1(Q;\mathbb{R}^m)\cap\mathrm{C}^0(\overline{Q};\mathbb{R}^m)$.

Example 2.2 For linear acoustic waves (2.1) we have

$$\begin{aligned}
\big(L(\mathbf{v},p),(\mathbf{w},q)\big)_Q + \big((\mathbf{v},p),L(\mathbf{w},q)\big)_Q &= \big(\rho\mathbf{v}(T),\mathbf{w}(T)\big)_\Omega + \big(\kappa^{-1}p(T),q(T)\big)_\Omega\\
&\quad - \big(\rho\mathbf{v}(0),\mathbf{w}(0)\big)_\Omega - \big(\kappa^{-1}p(0),q(0)\big)_\Omega\\
&\quad - (p,\mathbf{n}\cdot\mathbf{w})_{(0,T)\times\partial\Omega} - (\mathbf{n}\cdot\mathbf{v},q)_{(0,T)\times\partial\Omega}\,.
\end{aligned}$$

For linear elastic waves (2.2) we have

$$\begin{aligned}\big(L(\mathbf{v},\boldsymbol{\sigma}),(\mathbf{w},\boldsymbol{\tau})\big)_Q+\big((\mathbf{v},\boldsymbol{\sigma}),L(\mathbf{w},\boldsymbol{\tau})\big)_Q &= \big(\rho\mathbf{v}(T),\mathbf{w}(T)\big)_\Omega+\big(\mathbf{C}^{-1}\boldsymbol{\sigma}(T),\boldsymbol{\tau}(T)\big)_\Omega\\ &\quad -\big(\rho\mathbf{v}(0),\mathbf{w}(0)\big)_\Omega-\big(\mathbf{C}^{-1}\boldsymbol{\sigma}(0),\boldsymbol{\tau}(0)\big)_\Omega\\ &\quad -(\boldsymbol{\sigma}\mathbf{n},\mathbf{w})_{(0,T)\times\partial\Omega}-(\mathbf{v},\boldsymbol{\tau}\mathbf{n})_{(0,T)\times\partial\Omega}\,.\end{aligned}$$

For linear electro-magnetic waves (2.3) we have

$$\begin{aligned}\big(L(\mathbf{E},\mathbf{H}),(\mathbf{e},\mathbf{h})\big)_Q+\big((\mathbf{E},\mathbf{H}),L(\mathbf{e},\mathbf{h})\big)_Q &= \big(\varepsilon_0\mathbf{E}(T),\mathbf{e}(T)\big)_\Omega+\big(\mu_0\mathbf{H}(T),\mathbf{h}(T)\big)_\Omega\\ &\quad -\big(\varepsilon_0\mathbf{E}(0),\mathbf{e}(0)\big)_\Omega-\big(\mu_0\mathbf{H}(0),\mathbf{h}(0)\big)_\Omega\\ &\quad -(\mathbf{E}\times\mathbf{n},\mathbf{h})_{(0,T)\times\partial\Omega}+(\mathbf{H}\times\mathbf{n},\mathbf{e})_{(0,T)\times\partial\Omega}\,.\end{aligned}$$

Here we use the following calculus: for vectors $\mathbf{a},\mathbf{b},\mathbf{c}\in\mathbb{R}^3$ we have $\mathbf{a}\cdot(\mathbf{b}\times\mathbf{c}) = (\mathbf{a}\times\mathbf{b})\cdot\mathbf{c} = (\mathbf{c}\times\mathbf{a})\cdot\mathbf{b}$, and for vector fields $\mathbf{u},\mathbf{v}\colon\Omega\to\mathbb{R}^3$ we have $\nabla\cdot(\mathbf{u}\times\mathbf{v}) = \mathbf{v}\cdot(\nabla\times\mathbf{u})-\mathbf{u}\cdot(\nabla\times\mathbf{v})$. Thus, the Gauß theorem gives

$$\begin{aligned}\int_\Omega \mathbf{v}\cdot(\nabla\times\mathbf{u})\,\mathrm{d}\mathbf{x}-\int_\Omega \mathbf{u}\cdot(\nabla\times\mathbf{v})\,\mathrm{d}\mathbf{x} &= \int_\Omega \nabla\cdot(\mathbf{u}\times\mathbf{v})\,\mathrm{d}\mathbf{x}\\ &= \int_{\partial\Omega}(\mathbf{u}\times\mathbf{v})\cdot\mathbf{n}\,\mathrm{d}a=\int_{\partial\Omega}\mathbf{u}\cdot(\mathbf{v}\times\mathbf{n})\,\mathrm{d}a\,.\end{aligned}$$

The formulation in our examples of wave equations as Friedrichs systems yields symmetric matrices of the form $\underline{A}_j=\begin{pmatrix}0 & \underline{\tilde A}_j\\ \underline{\tilde A}_j^\top & 0\end{pmatrix}$ with $\underline{\tilde A}_j\in\mathbb{R}^{m_1\times m_2}$ and $m=m_1+m_2$. In order to obtain a well-posed problem with a unique solution, boundary conditions are required. Here we select $\Gamma_1=\ldots=\Gamma_{m_1}\subset\partial\Omega$ and the complement $\Gamma_k=\partial\Omega\setminus\overline{\Gamma}_1$ for $k=m_1+1,\ldots,m$, as it is specified in the next section for acoustics in Example 2.3.

2.2 Solution Spaces

We define the Hilbert spaces

$$\begin{aligned}\mathrm{H}(A,\Omega) &= \big\{\mathbf{y}\in\mathrm{L}_2(\Omega;\mathbb{R}^m)\colon \mathbf{z}\in\mathrm{L}_2(\Omega;\mathbb{R}^m)\text{ exists with}\\ &\qquad (\mathbf{z},\mathbf{w})_\Omega=(\mathbf{y},A^*\mathbf{w})_\Omega\text{ for all }\mathbf{w}\in\mathrm{C}^1_\mathrm{c}(\Omega;\mathbb{R}^m)\big\}\,,\\ \mathrm{H}(L,Q) &= \big\{\mathbf{v}\in\mathrm{L}_2(Q;\mathbb{R}^m)\colon \mathbf{z}\in\mathrm{L}_2(Q;\mathbb{R}^m)\text{ exists with}\\ &\qquad (\mathbf{z},\mathbf{w})_Q=(\mathbf{v},L^*\mathbf{w})_Q\text{ for all }\mathbf{w}\in\mathrm{C}^1_\mathrm{c}(Q;\mathbb{R}^m)\big\}\,,\end{aligned}$$

so that for $\mathbf{y} \in \mathrm{H}(A, \Omega)$ and $\mathbf{v} \in \mathrm{H}(L, Q)$ the weak derivatives $A\mathbf{y} \in \mathrm{L}_2(\Omega; \mathbb{R}^m)$ and $L\mathbf{v} \in \mathrm{L}_2(Q; \mathbb{R}^m)$ exist; the corresponding norms are

$$\|\mathbf{y}\|_{\mathrm{H}(A,\Omega)} = \sqrt{\|\mathbf{y}\|_\Omega^2 + \|A\mathbf{y}\|_\Omega^2}\,, \qquad \|\mathbf{v}\|_{\mathrm{H}(L,Q)} = \sqrt{\|\mathbf{v}\|_Q^2 + \|L\mathbf{v}\|_Q^2}\,.$$

Depending on homogeneous boundary conditions on $\Gamma_k \subset \partial\Omega$, $k = 1, \dots, m$, we define

$$\mathcal{Z} = \big\{\mathbf{w} \in \mathrm{C}^1(\Omega; \mathbb{R}^m) \cap \mathrm{C}^0(\overline{\Omega}; \mathbb{R}^m) \colon (\underline{A}_{\mathbf{n}}\mathbf{w})_k = 0 \text{ on } \Gamma_k\,, k = 1, \dots, m\big\}, \tag{2.5a}$$

$$\mathcal{V} = \big\{\mathbf{w} \in \mathrm{C}^1(Q; \mathbb{R}^m) \cap \mathrm{C}^0(\overline{Q}; \mathbb{R}^m) \colon \mathbf{w}(0) = \mathbf{0}\,, \tag{2.5b}$$
$$(\underline{A}_{\mathbf{n}}\mathbf{w})_k = 0 \text{ on } (0, T) \times \Gamma_k\,,\ k = 1, \dots, m\big\}\,,$$

$$\mathcal{V}^* = \big\{\mathbf{z} \in \mathrm{C}^1(Q; \mathbb{R}^m) \cap \mathrm{C}^0(\overline{Q}; \mathbb{R}^m) \colon \mathbf{z}(T) = \mathbf{0}\,, \tag{2.5c}$$
$$(\underline{A}_{\mathbf{n}}\mathbf{z})_k = 0 \text{ on } (0, T) \times \Gamma_k^*\,,\ k = 1, \dots, m\big\}\,,$$

where the sets $\Gamma_k \subset \partial\Omega$ are chosen such that

$$(A\mathbf{z}, \mathbf{z})_\Omega = 0\,, \qquad \mathbf{z} \in \mathcal{Z}\,, \tag{2.6}$$

and such that for the sets $\Gamma_k^* \subset \partial\Omega$ in the definition of the test space holds

$$(\underline{A}_{\mathbf{n}}\mathbf{w}, \mathbf{z})_{(0,T)\times\partial\Omega} = \sum_{k=1}^m \big((\underline{A}_{\mathbf{n}}\mathbf{w})_k, z_k\big)_{(0,T)\times\Gamma_k}\,, \qquad \mathbf{w} \in \mathrm{C}^1(Q; \mathbb{R}^m)\,,\ \mathbf{z} \in \mathcal{V}^*\,. \tag{2.7}$$

This is obtained by taking $\Gamma_k^* \subset \partial\Omega$ minimal such that for homogeneous boundary conditions in $\mathcal{V}$ and $\mathcal{V}^*$

$$(\underline{A}_{\mathbf{n}}\mathbf{w}, \mathbf{z})_{(0,T)\times\partial\Omega} = 0\,, \qquad \mathbf{w} \in \mathcal{V}\,,\ \mathbf{z} \in \mathcal{V}^*\,. \tag{2.8}$$

The choice of Γ_k and Γ_k^* is essential in order to obtain a well-posed problem; this will be explained for our examples in Sect. 2.7. Since we have $A^* = -A$, this implies $(A\mathbf{z}, \mathbf{z})_\Omega = \frac{1}{2}(\underline{A}_{\mathbf{n}}\mathbf{z}, \mathbf{z})_{\partial\Omega}$, and we observe $\Gamma_k^* = \Gamma_k$. Note that this is specific for our applications to wave problems but does not apply to general linear hyperbolic systems.

Let $Z \subset \mathrm{H}(A, Q)$ be the closure of $\mathcal{Z}$ with respect to the norm $\|\cdot\|_{\mathrm{H}(A,Q)}$, let $V \subset \mathrm{H}(L, Q)$ be the closure of $\mathcal{V}$ with respect to the norm $\|\cdot\|_{\mathrm{H}(L,Q)}$, and let $V^* \subset \mathrm{H}(L^*, Q)$ be the closure of $\mathcal{V}^*$ with respect to the norm $\|\cdot\|_{\mathrm{H}(L^*,Q)}$. Then, we obtain from (2.4) and (2.7)

$$(L\mathbf{v}, \mathbf{w})_Q - (\mathbf{v}, L^*\mathbf{w})_Q = 0\,, \qquad \mathbf{v} \in V\,,\ \mathbf{w} \in V^*\,. \tag{2.9}$$

Example 2.3 For linear acoustic waves (2.1) we have $\mathrm{H}(A, \Omega) = \mathrm{H}(\mathrm{div}, \Omega) \times \mathrm{H}^1(\Omega)$, and for $d = 2$ the boundary parts $\Gamma_1 = \Gamma_2 = \Gamma_S$ and $\Gamma_3 = \Gamma_V$ with $\partial\Omega = \Gamma_S \cup \Gamma_V$ in Example 2.2 yields that (2.9) holds with $\Gamma_k = \Gamma_k^*$, and we obtain

$$
\begin{aligned}
Z &= \big\{(\mathbf{v}, p) \in \mathrm{H}(\mathrm{div}, \Omega) \times \mathrm{H}^1(\Omega) \colon \mathbf{v} \cdot \mathbf{n} = 0 \text{ on } \Gamma_V\,,\ p = 0 \text{ on } \Gamma_S\big\}\,,\\
V &\supset \big\{(\mathbf{v}, p) \in \mathrm{H}^1(0, T; \mathrm{L}_2(\Omega; \mathbb{R}^m)) \cap \mathrm{L}_2(0, T; \mathrm{H}(\mathrm{div}, \Omega) \times \mathrm{H}^1(\Omega))\colon\\
&\qquad \mathbf{v}(0) = \mathbf{0}\,,\ p(0) = 0\,, \mathbf{v} \cdot \mathbf{n} = 0 \text{ on } (0, T) \times \Gamma_V\,,\ p = 0 \text{ on } (0, T) \times \Gamma_S\big\}\,,\\
V^* &\supset \big\{(\mathbf{w}, q) \in \mathrm{H}^1(0, T; \mathrm{L}_2(\Omega; \mathbb{R}^m)) \cap \mathrm{L}_2(0, T; \mathrm{H}(\mathrm{div}, \Omega) \times \mathrm{H}^1(\Omega))\colon\\
&\qquad \mathbf{w}(T) = \mathbf{0}\,,\ q(T) = 0\,, \mathbf{w} \cdot \mathbf{n} = 0 \text{ on } (0, T) \times \Gamma_V\,,\ q = 0 \text{ on } (0, T) \times \Gamma_S\big\}\,.
\end{aligned}
$$

In $Y = \mathrm{L}_2(\Omega; \mathbb{R}^m)$ and $W = \mathrm{L}_2(Q; \mathbb{R}^m)$ we use the energy norms

$$
\|\mathbf{y}\|_Y = \sqrt{(M\mathbf{y}, \mathbf{y})_\Omega}\,, \quad \mathbf{y} \in Y\,, \qquad \|\mathbf{w}\|_W = \sqrt{(M\mathbf{w}, \mathbf{w})_Q}\,, \quad \mathbf{w} \in W\,,
$$

and for the L_2 adjoints

$$
\|\mathbf{y}\|_{Y^*} = \sup_{\mathbf{z} \in Y \setminus \{\mathbf{0}\}} \frac{(\mathbf{y}, \mathbf{z})_\Omega}{\|\mathbf{z}\|_Y} = \sqrt{(M^{-1}\mathbf{y}, \mathbf{y})_\Omega}\,, \qquad \|\mathbf{w}\|_{W^*} = \sqrt{(M^{-1}\mathbf{w}, \mathbf{w})_Q}\,.
$$

In V and V^* we use the weighted norms

$$
\|\mathbf{v}\|_V = \sqrt{\|\mathbf{v}\|_W^2 + \|L\mathbf{v}\|_{W^*}^2}\,, \qquad \|\mathbf{z}\|_{V^*} = \sqrt{\|\mathbf{z}\|_W^2 + \|L^*\mathbf{z}\|_{W^*}^2}\,, \qquad \mathbf{v} \in V\,,\ \mathbf{z} \in V^*\,.
$$

Remark 2.4 For the extension to visco-acoustic and visco-elastic models the same solution spaces can be used. For mixed boundary conditions of Robin type or impedance boundary conditions a modification is required to include additional conditions on the boundary, see Remark 2.19. This relies on the fact that traces are well-defined for smooth test functions in $\mathcal{V}^*$, but in general not in V, where traces on mixed boundaries are only defined in distributional sense.

2.3 Solution Concepts

We consider different solution spaces of the equation $L\mathbf{u} = \mathbf{f}$ with initial and boundary conditions.

Definition 2.5 Depending on regularity of the data, we define:

(a) $\mathbf{u} \in \mathrm{C}^1(Q; \mathbb{R}^m) \cap \mathrm{C}^0(\overline{Q}; \mathbb{R}^m)$ is a *classical solution*, if

$$\begin{aligned} L\mathbf{u} &= \mathbf{f} && \text{in } Q = (0,T) \times \Omega\,, \\ \mathbf{u}(0) &= \mathbf{u}_0 && \text{in } \Omega \text{ at } t = 0\,, \\ (\underline{A}_{\mathbf{n}}\mathbf{u})_k &= g_k && \text{on } (0,T) \times \Gamma_k\,,\ k = 1,\dots,m\,, \end{aligned}$$

for $\mathbf{f} \in \mathrm{C}^0(Q; \mathbb{R}^m)$, $\mathbf{u}_0 \in \mathrm{C}^0(\Omega; \mathbb{R}^m)$, $g_k \in \mathrm{C}^0((0,T) \times \Gamma_k)$.

(b) $\mathbf{u} \in \mathrm{H}(L, Q)$ is a *strong solution*, if

$$\begin{aligned} L\mathbf{u} &= \mathbf{f} && \text{in } Q = (0,T) \times \Omega\,, \\ \mathbf{u}(0) &= \mathbf{u}_0 && \text{in } \Omega \text{ at } t = 0\,, \\ (\underline{A}_{\mathbf{n}}\mathbf{u})_k &= g_k && \text{on } (0,T) \times \Gamma_k\,,\ k = 1,\dots,m\,, \end{aligned}$$

for $\mathbf{f} \in \mathrm{L}_2(Q; \mathbb{R}^m)$, $\mathbf{u}_0 \in \mathrm{L}_2(\Omega; \mathbb{R}^m)$, $g_k \in \mathrm{L}_2((0,T) \times \Gamma_k)$.

(c) $\mathbf{u} \in \mathrm{L}_2(Q; \mathbb{R}^m)$ is a *weak solution*, if

$$\big(\mathbf{u}, L^*\mathbf{z}\big)_Q = \langle \ell, \mathbf{z}\rangle\,, \qquad \mathbf{z} \in \mathcal{V}^*,$$

with the linear functional ℓ defined by

$$\langle \ell, \mathbf{z}\rangle = (\mathbf{f}, \mathbf{z})_Q + \big(M\mathbf{u}_0, \mathbf{z}(0)\big)_\Omega - (\mathbf{g}, \mathbf{z})_{(0,T)\times\partial\Omega}$$

for data $\mathbf{f} \in \mathrm{L}_2(Q; \mathbb{R}^m)$, $\mathbf{u}_0 \in \mathrm{L}_2(\Omega; \mathbb{R}^m)$, and $g_k \in \mathrm{L}_2((0,T) \times \Gamma_k)$.
We set $\mathbf{g} = (g_k)_{k=1,\dots,m} \in \mathrm{L}_2((0,T) \times \partial\Omega; \mathbb{R}^m)$ with $g_k = 0$ on $\partial\Omega \setminus \Gamma_k$.

Remark 2.6 For the variational definition of weak solutions we use smooth test functions $\mathcal{V}^*$ so that the space-time traces on $\{0\} \times \Omega \subset \partial Q$ and $(0,T) \times \partial\Omega \subset \partial Q$ are well defined; with additional assumptions in Theorems 2.8 and 4.10 this extends to test functions in V^*.

Example 2.7 A weak solution $(v, \sigma) \in \mathrm{L}_2((0,T)\times(0,X); \mathbb{R}^2)$ of the linear wave equation (1.2) in 1d with wave speed $c = \sqrt{\kappa/\rho}$ and homogeneous Dirichlet boundary conditions satisfies

$$\begin{aligned} \big(v, -\rho\partial_t w + \partial_x\tau\big)_{(0,T)\times(0,X)} &+ \big(\sigma, -\kappa^{-1}\partial_t\tau + \partial_x w\big)_{(0,T)\times(0,X)} \\ &= \big(v_0, w(0)\big)_{(0,X)} + \big(\sigma_0, \tau(0)\big)_{(0,X)} \end{aligned}$$

for all test functions $w, \tau \in \mathrm{C}^1([0,T] \times [0,X])$ with $w(T,x) = \tau(T,x) = 0$ for $x \in (0,X)$ and $w(t,0) = w(t,X) = 0$ for $t \in (0,T)$. This allows for discontinuities of the solution along the characteristics

$$\left\{\begin{pmatrix} t \\ x_0 \pm ct \end{pmatrix} \in (0,T) \times \mathbb{R} : x_0 \pm ct \in \Omega\right\} = \left\{\begin{pmatrix} t \\ x \end{pmatrix} \in (0,T) \times \Omega : \begin{pmatrix} t \\ x - x_0 \end{pmatrix} \cdot \begin{pmatrix} \pm c \\ 1 \end{pmatrix} = 0\right\}$$

for some $x_0 \in \mathbb{R}$. Here we illustrate this for a simple example: consider a piecewise constant function

$$\begin{pmatrix} v(t,x) \\ \sigma(t,x) \end{pmatrix} = \begin{cases} \begin{pmatrix} v_{\mathrm{L}} \\ \sigma_{\mathrm{L}} \end{pmatrix} & \text{for } x < x_0 + ct\,, \qquad [v] = v_{\mathrm{R}} - v_{\mathrm{L}}\,, \\ \begin{pmatrix} v_{\mathrm{R}} \\ \sigma_{\mathrm{R}} \end{pmatrix} & \text{for } x > x_0 + ct\,, \qquad [\sigma] = \sigma_{\mathrm{R}} - \sigma_{\mathrm{L}}\,. \end{cases}$$

Then, we have for all $(w, \tau) \in \mathrm{C_c}([0,T] \times [0,X], \mathbb{R}^2)$

$$\begin{aligned}
&\int_0^T \int_0^X \begin{pmatrix} v \\ \sigma \end{pmatrix} \cdot \begin{pmatrix} -\rho \partial_t w + \partial_x \tau \\ -\kappa^{-1} \partial_t \tau + \partial_x w \end{pmatrix} \mathrm{d}x\ \mathrm{d}t \\
&= \int_{x<x_0+ct} \begin{pmatrix} \partial_t \\ \partial_x \end{pmatrix} \cdot \begin{pmatrix} -\rho v_{\mathrm{L}} w - \kappa^{-1} \sigma_{\mathrm{L}} \tau \\ v_{\mathrm{L}} \tau + \sigma_{\mathrm{L}} w \end{pmatrix} \mathrm{d}x\ \mathrm{d}t \\
&\quad + \int_{x>x_0+ct} \begin{pmatrix} \partial_t \\ \partial_x \end{pmatrix} \cdot \begin{pmatrix} -\rho v_{\mathrm{R}} w - \kappa^{-1} \sigma_{\mathrm{R}} \tau \\ v_{\mathrm{R}} \tau + \sigma_{\mathrm{R}} w \end{pmatrix} \mathrm{d}x\ \mathrm{d}t \\
&= \int_{x=x_0+ct} \frac{1}{\sqrt{1+c^2}} \begin{pmatrix} -c \\ 1 \end{pmatrix} \cdot \begin{pmatrix} -\rho v_{\mathrm{L}} w - \kappa^{-1} \sigma_{\mathrm{L}} \tau \\ v_{\mathrm{L}} \tau + \sigma_{\mathrm{L}} w \end{pmatrix} \mathrm{d}a \\
&\quad + \int_{x=x_0+ct} \frac{1}{\sqrt{1+c^2}} \begin{pmatrix} c \\ -1 \end{pmatrix} \cdot \begin{pmatrix} -\rho v_{\mathrm{R}} w - \kappa^{-1} \sigma_{\mathrm{R}} \tau \\ v_{\mathrm{R}} \tau + \sigma_{\mathrm{R}} w \end{pmatrix} \mathrm{d}a \\
&= -\frac{1}{\sqrt{1+c^2}} \int_{x=x_0+ct} \begin{pmatrix} c \\ 1 \end{pmatrix} \cdot \begin{pmatrix} -\rho [v] w - \kappa^{-1} [\sigma] \tau \\ [v] \tau + [\sigma] w \end{pmatrix} \mathrm{d}a \\
&= \frac{1}{\sqrt{1+c^2}} \int_{x=x_0+ct} \left(c \begin{pmatrix} \rho & 0 \\ 0 & \kappa^{-1} \end{pmatrix} \begin{pmatrix} [v] \\ [\sigma] \end{pmatrix} - \begin{pmatrix} 0 & 1 \\ 1 & 0 \end{pmatrix} \begin{pmatrix} [v] \\ [\sigma] \end{pmatrix} \right) \cdot \begin{pmatrix} w \\ \tau \end{pmatrix} \mathrm{d}a \\
&= \frac{1}{\sqrt{1+c^2}} \int_{x=x_0+ct} \left(c \underline{M} \begin{pmatrix} [v] \\ [\sigma] \end{pmatrix} + \underline{A} \begin{pmatrix} [v] \\ [\sigma] \end{pmatrix} \right) \cdot \begin{pmatrix} w \\ \tau \end{pmatrix} \mathrm{d}a\,.
\end{aligned}$$

We observe, that (v, σ) is a weak solution if the jump $([v], [\sigma])^\top$ is an eigenvector of

$$\underline{A}\begin{pmatrix}[v]\\ [\sigma]\end{pmatrix} = -c\underline{M}\begin{pmatrix}[v]\\ [\sigma]\end{pmatrix}.$$

This is equivalent to the jump conditions $[\sigma] - c\rho[v] = 0$ and $[v] - c\kappa^{-1}[\sigma] = 0$.

Based on the jump conditions we construct a weak solution

$$(v, \sigma) \in \mathrm{L}_2\big((0, T) \times (0, X), \mathbb{R}^2\big)$$

with $X = cT$ that is discontinuous along the characteristics $(t, j\Delta x \pm ct)$ on a special mesh in space and time depending of the wave speed c with $\Delta x = c\Delta t$ and $\Delta t = T/N$, $N \in \mathbb{N}$, cf. Fig. 2.1. Starting with $v(0, x) = v^0_{j-\frac12}$ and $\sigma(0, x) = \sigma^0_{j-\frac12}$ for $(j-1)\Delta x < x < j\Delta x$, we obtain from the jump condition recursively for $n = 1, 2, \dots, N$

$$\begin{aligned}
v_j^{n-\frac12} &= \frac12\Big(v^{n-1}_{j+\frac12} + v^{n-1}_{j-\frac12} + \sigma^{n-1}_{j+\frac12} - \sigma^{n-1}_{j-\frac12}\Big), & & v^n_{-\frac12} = -v^n_{\frac12},\\
\sigma_j^{n-\frac12} &= \frac12\Big(v^{n-\frac12}_{j+\frac12} - v^{n-\frac12}_{j-\frac12} + \sigma^{n-\frac12}_{j+\frac12} + \sigma^{n-\frac12}_{j-\frac12}\Big), & j = 0, \dots, N, \quad & v^n_{N+\frac12} = -v^n_{N-\frac12},\\
v^n_{j-\frac12} &= \frac12\Big(v_j^{n-\frac12} + v^{n+\frac12}_{j-1} + \sigma_j^{n-\frac12} - \sigma^{n-\frac12}_{j-1}\Big), & & \sigma^n_{-\frac12} = \sigma^n_{\frac12},\\
\sigma^n_{j-\frac12} &= \frac12\Big(v_j^{n-\frac12} - v^{n-\frac12}_{j-1} + \sigma_j^{n-\frac12} + \sigma^{n-\frac12}_{j-1}\Big), & j = 1, \dots, N, \quad & \sigma^n_{N+\frac12} = \sigma^n_{N-\frac12},
\end{aligned}$$

with suitable extensions for homogeneous Dirichlet boundary conditions for v, see Fig. 2.1 for an example.

2.4 Existence and Uniqueness of Space-Time Solutions

Now we construct strong and weak solutions by a least squares approach. Therefore, we define the quadratic functionals

$$\begin{aligned}
J(\mathbf{v}) &= \frac12\|L\mathbf{v} - \mathbf{f}\|^2_{W^*}, \qquad \mathbf{v} \in \mathrm{H}(L, Q),\\
J^*(\mathbf{z}) &= \frac12\|L^*\mathbf{z}\|^2_{W^*} - \langle \ell, \mathbf{z}\rangle, \quad \mathbf{z} \in \mathcal{V}^*.
\end{aligned}$$

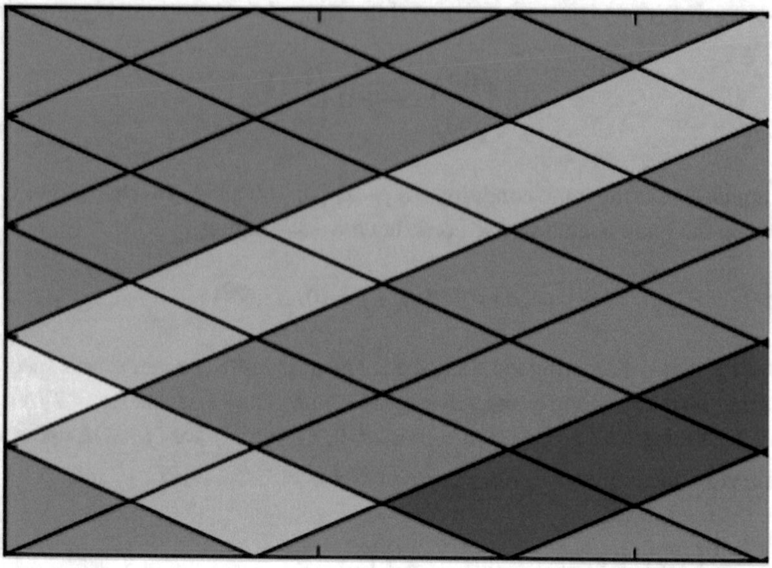

Fig. 2.1 Illustration of a piecewise constant weak solution in 1d of the wave equation in space and time with jumps along the characteristics. The solution is computed by the explicit time stepping scheme in Example 2.7

Theorem 2.8 *Depending on the regularity of the data, we obtain:*

(a) *Assume that* $C_L > 0$ *exists with*

$$\|\mathbf{v}\|_W \leq C_L \|L\mathbf{v}\|_{W^*}\,, \qquad \mathbf{v} \in V\,. \tag{2.10}$$

Then, a unique minimizer $\mathbf{u} \in V$ *of* $J(\cdot)$ *exists, and if* $L(V) = W$*, the minimizer* $\mathbf{u} \in V$ *is the unique strong solution of*

$$(L\mathbf{u}, \mathbf{w})_Q = (\mathbf{f}, \mathbf{w})_Q\,, \qquad \mathbf{w} \in W \tag{2.11}$$

with homogeneous initial and boundary data.

(b) *Assume that* $C_{L^*} > 0$ *and* $C_\ell > 0$ *exists with*

$$\|\mathbf{z}\|_W \leq C_{L^*} \|L^*\mathbf{z}\|_{W^*}\,, \qquad |\langle \ell, \mathbf{z}\rangle| \leq C_\ell \|\mathbf{z}\|_{V^*}\,, \qquad \mathbf{z} \in \mathcal{V}^*\,. \tag{2.12}$$

Then, $J^*(\cdot)$ *extends to* V^**, a unique minimizer* $\mathbf{z}^* \in V^*$ *of* $J^*(\cdot)$ *exists, and if* $L^*(\mathcal{V}^*) \subset W$ *is dense,* $\mathbf{u} = L^*\mathbf{z}^* \in \mathrm{L}_2(Q; \mathbb{R}^m)$ *is the unique weak solution of*

$$(\mathbf{u}, L^*\mathbf{z})_Q = \langle \ell, \mathbf{z}\rangle\,, \qquad \mathbf{z} \in V^*\,. \tag{2.13}$$

Proof ad (a) The functional $J(\cdot) > 0$ is bounded from below, and any minimizing sequence $\{\mathbf{u}_n\}_{n\in\mathbb{N}} \subset V$ with

$$\lim_{n\to\infty} J(\mathbf{u}_n) = \inf_{\mathbf{v}\in V} J(\mathbf{v}) := J_{\inf}$$

satisfies

$$\begin{aligned}\frac{1}{4}\|L\mathbf{u}_n - L\mathbf{u}_k\|_{W^*}^2 &= \frac{1}{2}\|L\mathbf{u}_n - \mathbf{f}\|_{W^*}^2 + \frac{1}{2}\|L\mathbf{u}_k - \mathbf{f}\|_{W^*}^2 - \Big\|L\frac{1}{2}(\mathbf{u}_n + \mathbf{u}_k) - \mathbf{f}\Big\|_{W^*}^2 \\ &= J(\mathbf{u}_n) + J(\mathbf{u}_k) - 2J\Big(\frac{1}{2}(\mathbf{u}_n + \mathbf{u}_k)\Big) \\ &\le J(\mathbf{u}_n) + J(\mathbf{u}_k) - 2J_{\inf} \longrightarrow 0 \quad \text{for } n, k \longrightarrow \infty .\end{aligned}$$

Condition (2.10) implies the norm equivalence

$$\|L\mathbf{v}\|_{W^*} \le \|\mathbf{v}\|_V = \sqrt{\|\mathbf{v}\|_W^2 + \|L\mathbf{v}\|_{W^*}^2} \le \sqrt{1 + C_L^2}\,\|L\mathbf{v}\|_{W^*}, \qquad \mathbf{v} \in V, \tag{2.14}$$

so that the minimizing sequence is a Cauchy sequence converging to $\mathbf{u} \in V$. Since $J(\cdot)$ is strictly convex, the minimizer is unique. Moreover, since $J(\cdot)$ is differentiable, $\mathbf{u}$ is a critical point, i.e.,

$$0 = \partial J(\mathbf{u})[\mathbf{v}] = (L\mathbf{u} - \mathbf{f}, L\mathbf{v})_{W^*} = (L\mathbf{u} - \mathbf{f}, M^{-1}L\mathbf{v})_Q, \qquad \mathbf{v} \in V .$$

If L is surjective, this implies (2.11) by inserting $\mathbf{w} = M^{-1}L\mathbf{v} \in M^{-1}L(V) = W$.

ad (b) By assumption (2.12), $J^*(\cdot)$ and ℓ are continuous in $\mathcal{V}^*$ with respect to the norm in V^*, so they extend to V^*, and we observe that $J^*(\cdot)$ is bounded from below by

$$J^*(\mathbf{z}) = \frac{1}{2}\|L^*\mathbf{z}\|_{W^*}^2 - \langle \ell, \mathbf{z}\rangle \ge \frac{1}{2(1 + C_{L^*}^2)}\|\mathbf{z}\|_{V^*}^2 - C_\ell\,\|\mathbf{z}\|_{V^*} \ge -\frac{1}{2}C_\ell^2\big(1 + C_{L^*}^2\big).$$

By the same arguments as above a unique minimizer $\mathbf{z}^* \in V^*$ exists characterized by

$$0 = \partial J^*(\mathbf{z}^*)[\mathbf{z}] = (L^*\mathbf{z}^*, L^*\mathbf{z})_{W^*} - \langle \ell, \mathbf{z}\rangle, \qquad \mathbf{z} \in V^* .$$

Inserting $\mathbf{u} = L^*\mathbf{z}^*$ implies (2.13). Now assume that $\tilde{\mathbf{u}}$ also solves (2.13); then, $(\mathbf{u} - \tilde{\mathbf{u}}, L^*\mathbf{z})_Q = 0$ for all $\mathbf{z} \in \mathcal{V}^*$. Since $L^*(\mathcal{V}^*)$ is dense in W, this implies $\mathbf{u} = \tilde{\mathbf{u}}$, so that the weak solution is unique. □

Remark 2.9 Strong solutions with inhomogeneous initial and boundary data exist, if the initial function $\mathbf{u}_0$ in Ω can be extended to a function $\mathbf{u}_0 \in \mathrm{H}(L, Q)$ satisfying the boundary conditions.

2.5 Mapping Properties of the Space-Time Operator

Lemma 2.10 $\|\mathbf{v}\|_W \le C_L \|L\mathbf{v}\|_{W^*}$ *for* $\mathbf{v} \in V$ *holds with* $C_L = 2T$.

Proof For $\mathbf{v} \in \mathcal{V}$ we have $\mathbf{v}(0) = \mathbf{0}$, and using (2.6) we obtain

$$\begin{aligned}
\|\mathbf{v}\|_W^2 &= \int_0^T \big(M\mathbf{v}(t), \mathbf{v}(t)\big)_\Omega \,\mathrm{d}t = \int_0^T \Big(\big(M\mathbf{v}(t), \mathbf{v}(t)\big)_\Omega - \big(M\mathbf{v}(0), \mathbf{v}(0)\big)_\Omega\Big) \,\mathrm{d}t \\
&= \int_0^T \int_0^t \partial_s \big(M\mathbf{v}(s), \mathbf{v}(s)\big)_\Omega \,\mathrm{d}s\,\mathrm{d}t = 2\int_0^T \int_0^t \big(M\partial_s\mathbf{v}(s), \mathbf{v}(s)\big)_\Omega \,\mathrm{d}s\,\mathrm{d}t \\
&= 2\int_0^T \int_0^t \Big(\big(M\partial_s\mathbf{v}(s), \mathbf{v}(s)\big)_\Omega + \big(A\mathbf{v}(s), \mathbf{v}(s)\big)_\Omega\Big) \,\mathrm{d}s\,\mathrm{d}t \\
&= 2\int_0^T \int_0^t \big(L\mathbf{v}(s), \mathbf{v}(s)\big)_\Omega \,\mathrm{d}s\,\mathrm{d}t = 2\int_0^T (T-t)\big(L\mathbf{v}(t), \mathbf{v}(t)\big)_\Omega \,\mathrm{d}t \\
&\le 2T\, \|L\mathbf{v}\|_{W^*} \|\mathbf{v}\|_W \,.
\end{aligned}$$

Since $\mathcal{V}$ is dense in V, this extends to V. □

As a consequence of Lemma 2.10, the operator $L\colon V \to \mathrm{L}_2(Q;\mathbb{R}^m)$ is injective and continuous, i.e., $L \in \mathcal{L}(V, W)$.

Corollary 2.11 $L(V) \subset \mathrm{L}_2(Q;\mathbb{R}^m)$ *is closed.*

Proof For any sequence $(\mathbf{w}_n)_{n\in\mathbb{N}} \subset V$ with $\lim\limits_{n\to\infty} L\mathbf{w}_n = \mathbf{f} \in W$ we have

$$\|\mathbf{w}_n - \mathbf{w}_k\|_W + \|L\mathbf{w}_n - L\mathbf{w}_k\|_{W^*} \le (C_L + 1)\, \|L\mathbf{w}_n - L\mathbf{w}_k\|_{W^*} \longrightarrow 0\,, \quad n, k \to \infty\,,$$

so that $(\mathbf{w}_n)_n$ is a Cauchy sequence in V; since $V \subset \mathrm{H}(L, Q)$ is closed, the limit $\mathbf{w} = \lim \mathbf{w}_n \in V$ with $L\mathbf{w} = \mathbf{f}$ exists. □

Let the domain $\mathcal{D}(A) = Z \subset \mathrm{H}(A, \Omega)$ of the operator A be the closure of $\mathcal{Z}$ defined in (2.5a). Then, (2.6) gives $\big((M + \tau A)\mathbf{z}, \mathbf{z}\big)_\Omega = \big(M\mathbf{z}, \mathbf{z}\big)_\Omega > 0$ for all $\mathbf{z} \ne \mathbf{0}$ and $\tau \in \mathbb{R}$, i.e., $M + \tau A$ is injective on Z. Moreover, we require that $M + \tau A$ is surjective on Z, which is achieved in our applications in Sect. 2.7 by a suitable balanced selection of $\Gamma_k \subset \partial\Omega$.

Lemma 2.12 *Assume that* $M + \tau A\colon Z \to \mathrm{L}_2(\Omega;\mathbb{R}^m)$ *is surjective for all* $\tau > 0$.
Then, $L(V) \subset \mathrm{L}_2(Q;\mathbb{R}^m)$ *is dense.*

Proof For $\mathbf{f} \in \mathrm{L}_2(Q;\mathbb{R}^m)$, $N \in \mathbb{N}$ and $t_{N,n} = n\frac{T}{N}$ let $\mathbf{f}_N \in \mathrm{L}_2(Q;\mathbb{R}^m)$ be piecewise constant in time with $\mathbf{f}_{N,n} = \mathbf{f}_N|_{(t_{N,n-1},t_{N,n})}$ so that $\lim_{N\to\infty} \|\mathbf{f}_N - \mathbf{f}\|_Q = 0$. Since the operator $M + \frac{T}{N}A \colon Z \longrightarrow \mathrm{L}_2(\Omega;\mathbb{R}^m)$ is surjective, starting with $\mathbf{u}_{N,0} = \mathbf{0}$ we find $\mathbf{u}_{N,n} \in Z$ with

$$\Big(M + \frac{T}{N}A\Big)\mathbf{u}_{N,n} = \mathbf{u}_{N,n-1} + \frac{T}{N}\mathbf{f}_{N,n}\,, \qquad n = 1,\dots,N\,.$$

Let $\mathbf{u}_N \in \mathrm{H}^1(0,T;Z) \subset V$ be the piecewise linear interpolation: for $n = 1,\dots,N$ set

$$\mathbf{u}_N(t) = \frac{t_{N,n} - t}{t_{N,n} - t_{N,n-1}}\mathbf{u}_{N,n-1} + \frac{t - t_{N,n-1}}{t_{N,n} - t_{N,n-1}}\mathbf{u}_{N,n}\,, \qquad t \in (t_{N,n-1}, t_{N,n})\,.$$

Then, we observe by construction $L\mathbf{u}_N = \mathbf{f}_N$ and thus $\lim_{N\to\infty} \|L\mathbf{u}_N - \mathbf{f}\|_Q = 0$. □

Remark 2.13 Together with Corollary 2.11 we observe $L(V) = \mathrm{L}_2(Q;\mathbb{R}^m)$, i.e., the operator $L\colon V \longrightarrow \mathrm{L}_2(Q;\mathbb{R}^m)$ is surjective.

A corresponding result can be achieved for $L^*(V^*)$ as the same arguments as in Lemma 2.10, 2.12, and Corollary 2.11 hold for L^* and V^*. We obtain

$$\|\mathbf{z}\|_W \le C_L\|L^*\mathbf{z}\|_{W^*}\,, \qquad \mathbf{z} \in V^*$$

i.e., $C_L = C_{L^*}$. By the assumption of Lemma 2.12, $L^*(V^*) \subset \mathrm{L}_2(Q;\mathbb{R}^m)$ is dense which implies $L^*(V^*) = \mathrm{L}_2(Q;\mathbb{R}^m)$.

Remark 2.14 Since $L(\mathcal{V})$ and $L^*(\mathcal{V}^*)$ are dense in W, we have

$$V = \big\{\mathbf{v} \in \mathrm{H}(L,Q) : (L\mathbf{v},\mathbf{z})_Q = (\mathbf{v}, L^*\mathbf{z})_Q \text{ for } \mathbf{z} \in \mathcal{V}^*\big\}\,,$$
$$V^* = \big\{\mathbf{z} \in \mathrm{H}(L^*,Q) : (L^*\mathbf{z},\mathbf{v})_Q = (\mathbf{z}, L\mathbf{v})_Q \text{ for } \mathbf{v} \in \mathcal{V}\big\}\,,$$

i.e., V^* is the Hilbert adjoint space of V, and V is the Hilbert adjoint space of V^*.

Lemma 2.15 *For $\mathbf{z} \in V^*$ holds*

$$\|\mathbf{z}(0)\|_Y^2 \le \|\mathbf{z}\|_{V^*}^2\,.$$

Proof We obtain, using $\mathbf{z}(T) = \mathbf{0}$,

$$\begin{aligned}\|\mathbf{z}(0)\|_Y^2 &= \|\mathbf{z}(0)\|_Y^2 - \|\mathbf{z}(T)\|_Y^2 = -\int_0^T \partial_t\|\mathbf{z}(t)\|_Y^2\,\mathrm{d}t = -2(M\partial_t\mathbf{z},\mathbf{z})_Q\\ &= -2(M\partial_t\mathbf{z},\mathbf{z})_Q - 2(A\mathbf{z},\mathbf{z})_Q = 2(L^*\mathbf{z},\mathbf{z})_Q \le \|\mathbf{z}\|_{V^*}^2\,.\end{aligned}$$

□

2.6 Inf-Sup Stability

From the previous section we directly obtain the following results.

Theorem 2.16 *The bilinear form $b\colon V \times W \to \mathbb{R}$, $b(\mathbf{v}, \mathbf{w}) = (L\mathbf{v}, \mathbf{w})_Q$, is inf-sup stable satisfying*

$$\inf_{\mathbf{v}\in V\setminus\{\mathbf{0}\}} \sup_{\mathbf{w}\in W\setminus\{\mathbf{0}\}} \frac{b(\mathbf{v}, \mathbf{w})}{\|\mathbf{v}\|_V \|\mathbf{w}\|_W} \geq \beta := \frac{1}{\sqrt{1 + C_L^2}}\,.$$

Thus, for all $\mathbf{f} \in \mathrm{L}_2(Q, \mathbb{R}^m)$ a unique Petrov–Galerkin solution $\mathbf{u} \in V$ of

$$b(\mathbf{u}, \mathbf{w}) = \big(\mathbf{f}, \mathbf{w}\big)_Q\,, \qquad \mathbf{w} \in W,$$

exists, and the solution is bounded by $\|\mathbf{u}\|_V \leq \beta^{-1}\|\mathbf{f}\|_{W^*}$.

Proof For $\mathbf{v} \in V \setminus \{\mathbf{0}\}$ we test with $\mathbf{w} = M^{-1}L\mathbf{v}$, so that with (2.14)

$$\sup_{\mathbf{w}\in W\setminus\{\mathbf{0}\}} \frac{b(\mathbf{v}, \mathbf{w})}{\|\mathbf{w}\|_W} \geq \frac{b(\mathbf{v}, M^{-1}L\mathbf{v})}{\|M^{-1}L\mathbf{v}\|_W} = \|M^{-1}L\mathbf{v}\|_W \geq \frac{1}{\sqrt{1 + C_L^2}}\|\mathbf{v}\|_V\,.$$

The existence and the a priori bound are now an easy consequence. □

Corollary 2.17 *Due to our previous results on the adjoint operator L^* we find correspondingly that for all $\mathbf{d} \in \mathrm{L}_2(Q, \mathbb{R}^m)$ the dual problem $L^*\mathbf{z} = \mathbf{d}$ admits a unique solution $\mathbf{z} \in V^*$ which is bounded by $\|\mathbf{z}\|_{V^*} \leq \beta^{-1}\|\mathbf{d}\|_{W^*}$.*

Corollary 2.18 *Additional regularity for the right-hand side $\mathbf{f} \in \mathrm{H}^1(0, T; \mathrm{L}_2(\Omega; \mathbb{R}^m))$ implies for the solution the regularity $\mathbf{u} \in \mathrm{H}^1(0, T; \mathrm{L}_2(\Omega; \mathbb{R}^m))$ and the estimate $\|\partial_t\mathbf{u}\|_W \leq C_L\|\partial_t\mathbf{f}\|_{W^*}$.*

Proof This simply follows from $L\mathbf{u} = \mathbf{f}$, which formally gives for the derivative in time $L\partial_t\mathbf{u} = \partial_t\mathbf{f}$. If $\partial_t\mathbf{f} \in W$, a solution $\mathbf{v} \in V$ solving $L\mathbf{v} = \partial_t\mathbf{f}$ exists, and since the solution is unique, $\mathbf{v} = \partial_t\mathbf{u}$. □

2.7 Applications to Acoustics and Visco-Elasticity

Acoustic Waves In the setting of Example 2.3 we have $A(\mathbf{v}, p) = -(\nabla p, \nabla\cdot\mathbf{v})$ and

$$\big(A(\mathbf{v}, p), (\mathbf{w}, q)\big)_\Omega + \big((\mathbf{v}, p), A(\mathbf{w}, q)\big)_\Omega = -(p, \mathbf{n}\cdot\mathbf{w})_{\partial\Omega} - (\mathbf{n}\cdot\mathbf{v}, q)_{\partial\Omega}\,.$$

We now show that the assumption in Lemma 2.12 is satisfied. For all $(\mathbf{f}, g) \in \mathrm{L}_2(\Omega; \mathbb{R}^{d+1})$ and $\tau > 0$ we define in the first step $p \in \mathrm{H}^1(\Omega)$ with $p = 0$ on Γ_S by solving the elliptic equation

$$\tau\big(\rho^{-1}\nabla p, \nabla\phi\big)_\Omega + \big(\kappa^{-1}p, \phi\big)_\Omega = \big(g, \phi\big)_\Omega - \big(\rho^{-1}\mathbf{f}, \nabla\phi\big)_\Omega \tag{2.15}$$

for $\phi \in \mathrm{H}^1(\Omega)$ with $\phi = 0$ on Γ_S. Then, we define $\mathbf{v} = \rho^{-1}(\tau\nabla p + \mathbf{f}) \in \mathrm{L}_2(\Omega; \mathbb{R}^d)$, and inserting (2.15), we observe

$$\big(\mathbf{v}, \nabla\phi\big)_\Omega = \big(g, \phi\big)_\Omega - \big(\kappa^{-1}p, \phi\big)_\Omega, \qquad \phi \in \mathrm{C}^1_\mathrm{c}(\Omega),$$

i.e., $\nabla\cdot\mathbf{v} = -g + \kappa^{-1}p \in \mathrm{L}_2(\Omega)$, and thus

$$0 = \big(\mathbf{v}, \nabla\phi\big)_\Omega + \big(\nabla\cdot\mathbf{v}, \phi\big)_\Omega = \langle\mathbf{n}\cdot\mathbf{v}, \phi\rangle_{\partial\Omega}, \qquad \phi \in \mathrm{C}^1(\overline{\Omega}),\ \phi = 0 \text{ on } \Gamma_\mathrm{S},$$

so that $\mathbf{n}\cdot\mathbf{v} = 0$ on $\partial\Omega \setminus \Gamma_\mathrm{S} = \Gamma_\mathrm{V}$. Together, $(\mathbf{v}, p) \in Z$ and

$$(M + \tau A)(\mathbf{v}, p) = (\mathbf{f}, g).$$

Moreover, the solution is unique, so that $M + \tau A$ is injective and surjective.

Visco-Elastic Waves For the system (1.13) we set $\mathbf{y} = \big(\mathbf{v}, \boldsymbol{\sigma}_0, \dots, \boldsymbol{\sigma}_r\big)^\top$ and

$$\underline{M} = \begin{pmatrix} \rho & 0 & \cdots & 0 \\ 0 & \mathbf{C}_0^{-1} & & \\ \vdots & & \ddots & \\ 0 & & & \mathbf{C}_r^{-1} \end{pmatrix}, \quad A = -\begin{pmatrix} 0 & \operatorname{div} & \cdots & \operatorname{div} \\ \boldsymbol{\varepsilon} & 0 & & \\ \vdots & & \ddots & \\ \boldsymbol{\varepsilon} & 0 & & 0 \end{pmatrix}, \quad \underline{D} = \underline{M}\begin{pmatrix} 0 & 0 & 0 & \cdots & 0 \\ 0 & 0 & 0 & \cdots & 0 \\ 0 & 0 & \tau_1^{-1} & & \\ \vdots & \vdots & & \ddots & \\ 0 & 0 & & & \tau_r^{-1} \end{pmatrix}$$

with $m = 2 + 3(1 + r)$ components for $d = 2$ and $m = 3 + 6(1 + r)$ for $d = 3$, and where $\underline{D} \in \mathrm{L}_\infty(\Omega; \mathbb{R}^{m\times m}_\mathrm{sym})$ is a positive semi-definite matrix function. This defines the operator $D\mathbf{y}(x) = \underline{D}(x)\mathbf{y}(x)$, and we have $(D\mathbf{y}, \mathbf{y})_\Omega \geq 0$ for all $\mathbf{y} \in \mathrm{L}_2(\Omega; \mathbb{R}^m)$.

The space-time setting is extended to the operator $L = M\partial_t + A + D$, and formally the adjoint operator is $L^* = -M\partial_t - A + D$. The assumption in Lemma 2.12 can be verified analogously to the acoustic case.

Remark 2.19 The extension to mixed boundary conditions on $\Gamma_\mathrm{R} \subset \partial\Omega$ requires L_2 regularity of the traces on the boundary part Γ_R. Then, extending the norm $\|\cdot\|_V$ by a corresponding boundary term again defines V as closure of $\mathcal{V}$ with respect to this stronger norm, and the space-time operator L is extended by a dissipative boundary operator D.

Bibliographic Comments

Least squares for linear first-order systems for finite elements are considered in [21, 22], where also the LL^* technique is established which is used to prove Theorem 2.8 (b). Here this is applied to the space-time setting, see [48, 49, 65, 66]. The extension to mixed boundary conditions is considered in [50].

The inf-sup constant β in Theorem 2.16 is not optimal for the continuous problem; for an improved estimate see [65, Lem. 1]. Here, it relies on the estimate for C_L in Lemma 2.10 which is generalized in Theorem 4.1 for the approximation. The suitable choice of boundary conditions for general Friedrichs systems is discussed in [42, Chap. 7.2].

3 Discontinuous Galerkin Methods for Linear Hyperbolic Systems

We develop a space-time method with a discontinuous Galerkin discretization in space for linear wave problems. For the ansatz space we use piecewise polynomials in every cell, where the traces on the cell interfaces can be different from the two sides. Therefore, we need to extend the first-order operator A to discontinuous finite element spaces. Here, we introduce the discrete operator A_h with upwind flux, where the evaluation of the upwind flux is based on solving Riemann problems, i.e., by construction of piecewise constant solutions in space and time. We start with simple examples for interface and transmission problems, and then consider the general case for waves in heterogeneous media.

3.1 Traveling Wave Solutions in Homogeneous Media

We consider linear hyperbolic first-order systems $L = M\partial_t + A$ introduced in Sect. 2.1, and we start with the case of homogeneous material parameters, so that the operator M is represented by a symmetric positive definite matrix $\underline{M} \in \mathbb{R}^{m\times m}_{\mathrm{sym}}$ which is constant in $\overline{\Omega}$.

Let $(\lambda, \mathbf{w}) \in \mathbb{R} \times \mathbb{R}^m$ be an eigenpair of $\underline{A}_{\mathbf{n}}\mathbf{w} = \lambda\underline{M}\mathbf{w}$, and let $a \in \mathrm{C}^1(\mathbb{R})$ be an amplitude function describing the shape of the traveling wave. Then, we observe for $\mathbf{y}(t, \mathbf{x}) = a(\mathbf{n}\cdot\mathbf{x} - \lambda t)\,\mathbf{w}$

$$\begin{aligned}
\partial_t \mathbf{y}(t, \mathbf{x}) &= -\lambda a'(\mathbf{n}\cdot\mathbf{x} - \lambda t)\mathbf{w}\,,\\
\partial_{x_j} \mathbf{y}(t, \mathbf{x}) &= n_j a'(\mathbf{n}\cdot\mathbf{x} - \lambda t)\mathbf{w}\,,\\
L\mathbf{y}(t, \mathbf{x}) &= \underline{M}\partial_t \mathbf{y}(t, \mathbf{x}) + A\mathbf{y}(t, \mathbf{x})
\end{aligned}$$

W. Dörfler et al., *Wave Phenomena*, Oberwolfach Seminars 49,
https://doi.org/10.1007/978-3-031-05793-9_3

$$= a'(\mathbf{n} \cdot \mathbf{x} - \lambda t)\left(-\lambda \underline{M} + \sum_{j=1}^{d} n_j \underline{A}_j\right)\mathbf{w}$$

$$= a'(\mathbf{n} \cdot \mathbf{x} - \lambda t)\left(\underline{A}_{\mathbf{n}} - \lambda \underline{M}\right)\mathbf{w} = \mathbf{0},$$

so that $\mathbf{y}$ solves $L\mathbf{y} = \mathbf{0}$ for all $t \in \mathbb{R}$ in $\Omega = \mathbb{R}^d$.

Example 3.1 For acoustic waves with wave speed $c = \sqrt{\kappa/\rho}$ we have

$$\mathbf{y} = \begin{pmatrix} \mathbf{v} \\ p \end{pmatrix}, \quad \underline{M}\mathbf{y} = \begin{pmatrix} \rho \mathbf{v} \\ \kappa^{-1} p \end{pmatrix}, \quad \underline{A}_{\mathbf{n}}\mathbf{y} = -\begin{pmatrix} p\mathbf{n} \\ \mathbf{v} \cdot \mathbf{n} \end{pmatrix}, \quad \lambda \in \{0, \pm c\}, \quad \mathbf{w} = \begin{pmatrix} \mp c\mathbf{n} \\ \kappa \end{pmatrix}.$$

For elastic waves with wave speeds $c_{\mathrm{p}} = \sqrt{(2\mu + \lambda)/\rho}$ for compressional waves and $c_{\mathrm{s}} = \sqrt{\mu/\rho}$ for shear waves, we have

$$\mathbf{y} = \begin{pmatrix} \mathbf{v} \\ \boldsymbol{\sigma} \end{pmatrix}, \quad \underline{M}\mathbf{y} = \begin{pmatrix} \rho \mathbf{v} \\ \mathbf{C}^{-1}\boldsymbol{\sigma} \end{pmatrix}, \quad \underline{A}_{\mathbf{n}}\mathbf{y} = -\begin{pmatrix} \boldsymbol{\sigma}\mathbf{n} \\ \frac{1}{2}(\mathbf{n}\mathbf{v}^{\mathsf{T}} + \mathbf{v}\mathbf{n}^{\mathsf{T}}) \end{pmatrix},$$

$$\lambda \in \{0, \pm c_{\mathrm{p}}, \pm c_{\mathrm{s}}\}, \quad \mathbf{w}_{\mathrm{p}} = \begin{pmatrix} \mp c_{\mathrm{p}}\mathbf{n} \\ 2\mu\mathbf{n}\mathbf{n}^{\mathsf{T}} + \lambda\mathbf{I} \end{pmatrix}, \quad \mathbf{w}_{\mathrm{s}} = \begin{pmatrix} \mp c_{\mathrm{s}}\boldsymbol{\tau} \\ \mu(\mathbf{n}\boldsymbol{\tau}^{\mathsf{T}} + \boldsymbol{\tau}\mathbf{n}^{\mathsf{T}}) \end{pmatrix},$$

where $\boldsymbol{\tau} \in \mathbb{R}^d$ is a tangential unit vector, i.e., $\boldsymbol{\tau} \cdot \mathbf{n} = 0$ and $|\boldsymbol{\tau}| = 1$.

For linear electro-magnetic waves with wave speed $c = 1/\sqrt{\varepsilon\mu}$ we have

$$\mathbf{y} = \begin{pmatrix} \mathbf{E} \\ \mathbf{H} \end{pmatrix}, \quad \underline{M}\mathbf{y} = \begin{pmatrix} \varepsilon\mathbf{E} \\ \mu\mathbf{H} \end{pmatrix}, \quad \underline{A}_{\mathbf{n}}\mathbf{y} = -\begin{pmatrix} \mathbf{n} \times \mathbf{H} \\ -\mathbf{n} \times \mathbf{E} \end{pmatrix},$$

$$\lambda \in \{0, \pm c\}, \quad \mathbf{w}_1 = \begin{pmatrix} \sqrt{\varepsilon}\mathbf{n} \times \boldsymbol{\tau} \\ \pm\sqrt{\mu}\boldsymbol{\tau} \end{pmatrix}, \quad \mathbf{w}_2 = \begin{pmatrix} \pm\sqrt{\varepsilon}\boldsymbol{\tau} \\ \sqrt{\mu}\mathbf{n} \times \boldsymbol{\tau} \end{pmatrix}.$$

3.2   Reflection of Traveling Acoustic Waves at Boundaries

In the next step we consider solutions of the acoustic wave equation in the half space

$$\begin{pmatrix} \rho\partial_t \mathbf{v} - \nabla p \\ \kappa^{-1}\partial_t p - \nabla \cdot \mathbf{v} \end{pmatrix} = \begin{pmatrix} \mathbf{0} \\ 0 \end{pmatrix} \qquad \text{in } \Omega_{\mathrm{R}} = \{\mathbf{x} \in \mathbb{R}^d : \mathbf{n} \cdot \mathbf{x} > 0\}$$

with initial value

$$\begin{pmatrix} \mathbf{v}(0,\mathbf{x}) \\ p(0,\mathbf{x}) \end{pmatrix} = a(\mathbf{n}\cdot\mathbf{x}) \begin{pmatrix} c\mathbf{n} \\ \kappa \end{pmatrix}$$

depending on $a \in \mathrm{C}^1(\mathbb{R})$ with $a(\mathbf{n}\cdot\mathbf{x}) = 0$ for $\mathbf{n}\cdot\mathbf{x} < ct_0$ and $t_0 > 0$, i.e., $\operatorname{supp} a \subset [ct_0, \infty]$.

The wave starts traveling from right to left, and at time $t = t_0$ it reaches the boundary. In case of a homogeneous Neumann boundary condition $\mathbf{v}\cdot\mathbf{n} = 0$ it is reflected, i.e.,

$$\begin{pmatrix} \mathbf{v}(t,\mathbf{x}) \\ p(t,\mathbf{x}) \end{pmatrix} = \begin{cases} a(ct+\mathbf{n}\cdot\mathbf{x}) \begin{pmatrix} c\mathbf{n} \\ \kappa \end{pmatrix} & 0 < c(t_0 - t) < \mathbf{n}\cdot\mathbf{x}\,, \\ a(ct+\mathbf{n}\cdot\mathbf{x}) \begin{pmatrix} c\mathbf{n} \\ \kappa \end{pmatrix} + a(ct-\mathbf{n}\cdot\mathbf{x}) \begin{pmatrix} -c\mathbf{n} \\ \kappa \end{pmatrix} & 0 < \mathbf{n}\cdot\mathbf{x} < c(t-t_0)\,. \end{cases}$$

This is illustrated in Fig. 3.1. Otherwise, with homogeneous Dirichlet boundary conditions $p = 0$ the reflection also changes sign, i.e.,

$$\begin{pmatrix} \mathbf{v}(t,\mathbf{x}) \\ p(t,\mathbf{x}) \end{pmatrix} = \begin{cases} a(ct+\mathbf{n}\cdot\mathbf{x}) \begin{pmatrix} c\mathbf{n} \\ \kappa \end{pmatrix} & 0 < c(t_0 - t) < \mathbf{n}\cdot\mathbf{x}\,, \\ a(ct+\mathbf{n}\cdot\mathbf{x}) \begin{pmatrix} c\mathbf{n} \\ \kappa \end{pmatrix} - a(ct-\mathbf{n}\cdot\mathbf{x}) \begin{pmatrix} -c\mathbf{n} \\ \kappa \end{pmatrix} & 0 < \mathbf{n}\cdot\mathbf{x} < c(t-t_0)\,. \end{cases}$$

For smooth amplitude functions this is a classical solution.

3.3 Transmission and Reflection of Traveling Waves at Interfaces

Now we consider solutions of the acoustic wave equation in $\mathbb{R}^d$ with an interface

$$\begin{pmatrix} \rho\partial_t\mathbf{v} - \nabla p \\ \kappa^{-1}\partial_t p - \nabla\cdot\mathbf{v} \end{pmatrix} = \begin{pmatrix} \mathbf{0} \\ 0 \end{pmatrix} \quad \text{in } \Omega_\mathrm{L} \cup \Omega_\mathrm{R}\,, \qquad \begin{cases} \Omega_\mathrm{L} = \{\mathbf{x}\in\mathbb{R}^d : \mathbf{n}\cdot\mathbf{x} < 0\}, \\ \Omega_\mathrm{R} = \{\mathbf{x}\in\mathbb{R}^d : \mathbf{n}\cdot\mathbf{x} > 0\} \end{cases}$$

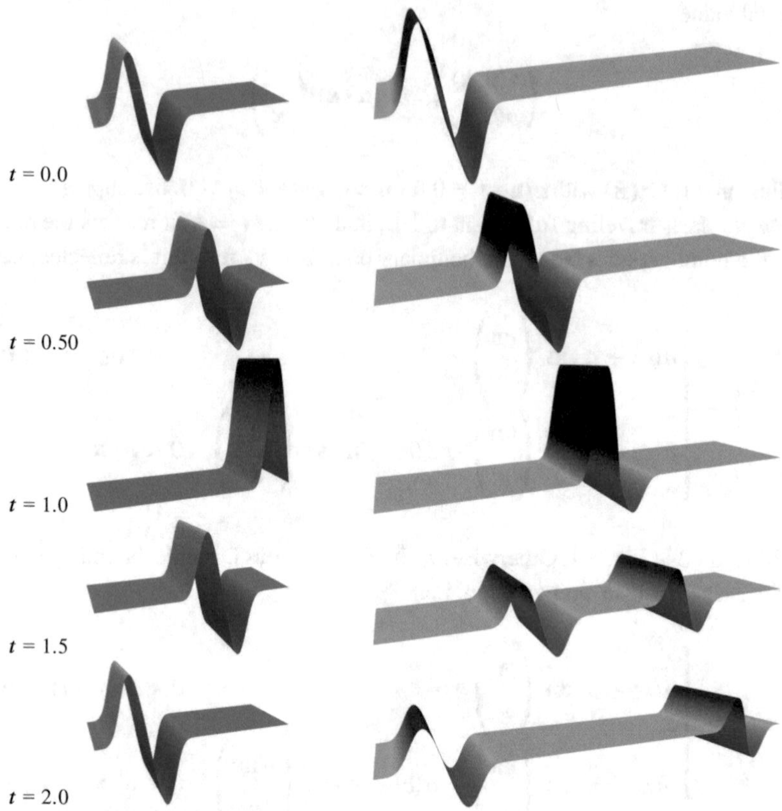

Fig. 3.1 The evolution of the pressure distribution with reflection at a fixed boundary (left, cf. Sect. 3.2), and reflection and transmission at an interface (right, cf. Sect. 3.3) of traveling waves

with constant coefficients (ρ_L, κ_L) in Ω_L and (ρ_R, κ_R) in Ω_R defining $\underline{M}_L$ and $\underline{M}_R$, starting in Ω_R with

$$\begin{pmatrix} \mathbf{v}(0,\mathbf{x}) \\ p(0,\mathbf{x}) \end{pmatrix} = a(\mathbf{n}\cdot\mathbf{x}/c_R)\begin{pmatrix} \mathbf{n} \\ Z_R \end{pmatrix}, \qquad a(\mathbf{n}\cdot\mathbf{x}) = 0 \text{ for } \mathbf{n}\cdot\mathbf{x} < c_R t_0\,,\ t_0 > 0\,,$$

where $Z_L = \sqrt{\kappa_L \rho_L}$, $Z_R = \sqrt{\kappa_R \rho_R}$ are the left and right impedances, and where $c_L = \sqrt{\kappa_L/\rho_L}$, $c_R = \sqrt{\kappa_R/\rho_R}$ are the left and right wave speeds. Note that we use a different scaling of the eigenvectors for the transmission problem.

We state continuity at the interface to determine a classical solution and obtain

$$\begin{pmatrix} \mathbf{v}(t,\mathbf{x}) \\ p(t,\mathbf{x}) \end{pmatrix} = \begin{cases} a(t+\mathbf{n}\cdot\mathbf{x}/c_{\mathrm{R}}) \begin{pmatrix} \mathbf{n} \\ Z_{\mathrm{R}} \end{pmatrix} & 0 < c_{\mathrm{R}}(t_0 - t) < \mathbf{n}\cdot\mathbf{x}, \\ a(t+\mathbf{n}\cdot\mathbf{x}/c_{\mathrm{R}}) \begin{pmatrix} \mathbf{n} \\ Z_{\mathrm{R}} \end{pmatrix} \\ \qquad + \beta_{\mathrm{R}}\, a(t-\mathbf{n}\cdot\mathbf{x}/c_{\mathrm{R}}) \begin{pmatrix} -\mathbf{n} \\ Z_{\mathrm{R}} \end{pmatrix} & 0 < \mathbf{n}\cdot\mathbf{x} < c_{\mathrm{R}}(t - t_0), \\ \beta_{\mathrm{L}}\, a(t+\mathbf{n}\cdot\mathbf{x}/c_{\mathrm{L}}) \begin{pmatrix} \mathbf{n} \\ Z_{\mathrm{L}} \end{pmatrix} & c_{\mathrm{L}}(t_0 - t) < \mathbf{n}\cdot\mathbf{x} < 0 \end{cases}$$

with transmission and reflection coefficients

$$\beta_{\mathrm{L}} = \frac{2Z_{\mathrm{R}}}{Z_{\mathrm{R}} + Z_{\mathrm{L}}}, \qquad \beta_{\mathrm{R}} = \frac{Z_{\mathrm{L}} - Z_{\mathrm{R}}}{Z_{\mathrm{R}} + Z_{\mathrm{L}}}$$

derived from the interface condition $\underline{A_{\mathbf{n}}}[\mathbf{y}] = \mathbf{0}$, see Fig. 3.1. By the interface condition we obtain $L(\mathbf{v}, p) \in \mathrm{L}_{2,\mathrm{loc}}(\mathbb{R} \times \mathbb{R}^d; \mathbb{R}^{d+1})$, so that $(\mathbf{v}, p)$ is a strong solution.

We observe that no wave is reflected if the impedance $Z_{\mathrm{L}} = Z_{\mathrm{R}}$ is continuous. This property can be used to design absorbing boundary layers.

3.4 The Riemann Problem for Acoustic Waves

Now we consider weak solutions in $\mathrm{L}_{2,\mathrm{loc}}(\mathbb{R}^d; \mathbb{R}^{d+1})$ of the acoustic wave equation

$$\begin{pmatrix} \rho\partial_t \mathbf{v} - \nabla p \\ \kappa^{-1}\partial_t p - \nabla\cdot\mathbf{v} \end{pmatrix} = \begin{pmatrix} \mathbf{0} \\ 0 \end{pmatrix} \quad \text{in } \Omega_{\mathrm{L}} \cup \Omega_{\mathrm{R}}, \quad \begin{cases} \Omega_{\mathrm{L}} = \{\mathbf{x} \in \mathbb{R}^d : \mathbf{n}\cdot\mathbf{x} < 0\} \\ \Omega_{\mathrm{R}} = \{\mathbf{x} \in \mathbb{R}^d : \mathbf{n}\cdot\mathbf{x} > 0\} \end{cases},$$

with constant coefficients $(\rho_{\mathrm{L}}, \kappa_{\mathrm{L}})$ in Ω_{L} and $(\rho_{\mathrm{R}}, \kappa_{\mathrm{R}})$ in Ω_{R}, and with piecewise constant initial values

$$\begin{pmatrix} \mathbf{v}(0,\mathbf{x}) \\ p(0,\mathbf{x}) \end{pmatrix} = \begin{pmatrix} \mathbf{v}_{\mathrm{L}} \\ p_{\mathrm{L}} \end{pmatrix}, \quad \mathbf{x} \in \Omega_{\mathrm{L}}, \qquad \begin{pmatrix} \mathbf{v}(0,\mathbf{x}) \\ p(0,\mathbf{x}) \end{pmatrix} = \begin{pmatrix} \mathbf{v}_{\mathrm{R}} \\ p_{\mathrm{R}} \end{pmatrix}, \quad \mathbf{x} \in \Omega_{\mathrm{R}},$$

called *Riemann problem*. The weak solution is of the form

$$\begin{pmatrix} \mathbf{v}(t,\mathbf{x}) \\ p(t,\mathbf{x}) \end{pmatrix} = \begin{cases} \begin{pmatrix} \mathbf{v}_\mathrm{L} \\ p_\mathrm{L} \end{pmatrix} & \mathbf{x}\cdot\mathbf{n} < -c_\mathrm{L} t \\ \begin{pmatrix} \mathbf{v}_\mathrm{L} \\ p_\mathrm{L} \end{pmatrix} + \beta_\mathrm{L} \begin{pmatrix} \mathbf{n} \\ Z_\mathrm{L} \end{pmatrix} & -c_\mathrm{L} t < \mathbf{x}\cdot\mathbf{n} < 0 \\ \begin{pmatrix} \mathbf{v}_\mathrm{R} \\ p_\mathrm{R} \end{pmatrix} + \beta_\mathrm{R} \begin{pmatrix} \mathbf{n} \\ -Z_\mathrm{R} \end{pmatrix} & 0 < \mathbf{x}\cdot\mathbf{n} < c_\mathrm{R} t \\ \begin{pmatrix} \mathbf{v}_\mathrm{R} \\ p_\mathrm{R} \end{pmatrix} & c_\mathrm{R} t < \mathbf{x}\cdot\mathbf{n} \end{cases}$$

depending on $\beta_\mathrm{L}, \beta_\mathrm{R} \in \mathbb{R}$ determined by the flux condition

$$\underline{A}_\mathbf{n}\left(\begin{pmatrix} \mathbf{v}_\mathrm{L} \\ p_\mathrm{L} \end{pmatrix} + \beta_\mathrm{L}\begin{pmatrix} \mathbf{n} \\ Z_\mathrm{L} \end{pmatrix}\right) = \underline{A}_\mathbf{n}\left(\begin{pmatrix} \mathbf{v}_\mathrm{R} \\ p_\mathrm{R} \end{pmatrix} + \beta_\mathrm{R}\begin{pmatrix} \mathbf{n} \\ -Z_\mathrm{R} \end{pmatrix}\right), \tag{3.1}$$

which yields $\beta_\mathrm{L} = \dfrac{[p] + Z_\mathrm{R}\mathbf{n}\cdot[\mathbf{v}]}{Z_\mathrm{L} + Z_\mathrm{R}}$, $\beta_\mathrm{R} = \dfrac{[p] - Z_\mathrm{L}\mathbf{n}\cdot[\mathbf{v}]}{Z_\mathrm{L} + Z_\mathrm{R}}$ depending on $[p] = p_\mathrm{R} - p_\mathrm{L}$, $[\mathbf{v}] = \mathbf{v}_\mathrm{R} - \mathbf{v}_\mathrm{L}$.

For discontinuous initial values the solution is discontinuous along the characteristic linear manifolds $\mathbf{x}\cdot\mathbf{n} + c_\mathrm{L} t = 0$ and $\mathbf{x}\cdot\mathbf{n} - c_\mathrm{R} t = 0$ in the space-time domain, so that we only obtain a weak solution.

3.5 The Riemann Problem for Linear Conservation Laws

We now construct a weak solution of the Riemann problem for general linear conservation laws, i.e., a piecewise constant weak solution of $L\mathbf{y} = \mathbf{0}$ in $\mathrm{L}_{2,\mathrm{loc}}(\mathbb{R}^d;\mathbb{R}^m)$ with discontinuous initial values

$$\mathbf{y}_0(\mathbf{x}) = \begin{cases} \mathbf{y}_\mathrm{L} & \text{in } \Omega_\mathrm{L} = \{\mathbf{x}\in\mathbb{R}^d : \mathbf{n}\cdot\mathbf{x} < 0\}, \\ \mathbf{y}_\mathrm{R} & \text{in } \Omega_\mathrm{R} = \{\mathbf{x}\in\mathbb{R}^d : \mathbf{n}\cdot\mathbf{x} > 0\}, \end{cases} \qquad \mathbf{y}_\mathrm{L}, \mathbf{y}_\mathrm{R} \in \mathbb{R}^m, \quad \underline{M}_\mathrm{L}, \underline{M}_\mathrm{R} \in \mathbb{R}^{m\times m}_{\mathrm{sym}}.$$

Let $\{(\lambda_j^\mathrm{L}, \mathbf{w}_j^\mathrm{L})\}_{j=1,\dots,m}$ and $\{(\lambda_j^\mathrm{R}, \mathbf{w}_j^\mathrm{R})\}_{j=1,\dots,m}$ be eigensystems, i.e.,

$$\underline{A}_\mathbf{n}\mathbf{w}_j^\mathrm{L} = \lambda_j^\mathrm{L}\underline{M}_\mathrm{L}\mathbf{w}_j^\mathrm{L}, \ \underline{A}_\mathbf{n}\mathbf{w}_j^\mathrm{R} = \lambda_j^\mathrm{R}\underline{M}_\mathrm{R}\mathbf{w}_j^\mathrm{R}, \quad \mathbf{w}_k^\mathrm{L}\cdot\underline{M}_\mathrm{L}\mathbf{w}_j^\mathrm{L} = \mathbf{w}_k^\mathrm{R}\cdot\underline{M}_\mathrm{R}\mathbf{w}_j^\mathrm{R} = 0 \text{ for } j \neq k.$$

A solution is constructed by a superposition of traveling waves

$$\mathbf{y}(t,\mathbf{x}) = \begin{cases} \mathbf{y}_\mathrm{L} + \sum\limits_{j:\,\mathbf{x}\cdot\mathbf{n}>\lambda_j^\mathrm{L} t} \beta_j^\mathrm{L}\mathbf{w}_j^\mathrm{L} & \mathbf{x}\in\Omega_\mathrm{L},\\ \mathbf{y}_\mathrm{R} + \sum\limits_{j:\,\mathbf{x}\cdot\mathbf{n}<\lambda_j^\mathrm{R} t} \beta_j^\mathrm{R}\mathbf{w}_j^\mathrm{R} & \mathbf{x}\in\Omega_\mathrm{R}, \end{cases}$$

and by solving the equation for β_j^L, β_j^R (only depending on $[\mathbf{y}_0] = \mathbf{y}_\mathrm{R} - \mathbf{y}_\mathrm{L}$)

$$\underline{A}_\mathbf{n}\Big(\mathbf{y}_\mathrm{L} + \sum_{j:\,\lambda_j^\mathrm{L}<0} \beta_j^\mathrm{L}\mathbf{w}_j^\mathrm{L}\Big) = \underline{A}_\mathbf{n}\Big(\mathbf{y}_\mathrm{R} + \sum_{j:\,\lambda_j^\mathrm{R}>0} \beta_j^\mathrm{R}\mathbf{w}_j^\mathrm{R}\Big) \quad \text{on} \quad \partial\Omega_\mathrm{L}\cap\partial\Omega_\mathrm{R}\,. \tag{3.2}$$

Then, the flux $\underline{A}_\mathbf{n}\mathbf{y}$ is continuous for $t > 0$, and the piecewise constant function $\mathbf{y}$ is the unique weak solution of $L\mathbf{y} = \mathbf{0}$ with initial value $\mathbf{y}(0) = \mathbf{y}_0$.

In summary, the solution of the Riemann problem defines the *upwind flux*

$$A_\mathbf{n}^\mathrm{upw}\mathbf{y}_0 = \underline{A}_\mathbf{n}\Big(\mathbf{y}_\mathrm{L} + \sum_{j:\,\lambda_j^\mathrm{L}<0} \beta_j^\mathrm{L}\mathbf{w}_j^\mathrm{L}\Big)\,. \tag{3.3}$$

On the boundary, depending on the boundary conditions, a system corresponding to (3.2) is solved defining an operator $\underline{A}_\mathbf{n}^\mathrm{bnd}$ with

$$A_\mathbf{n}^\mathrm{upw}\mathbf{y}_0 = \underline{A}_\mathbf{n}\mathbf{y}_\mathrm{L} + \underline{A}_\mathbf{n}^\mathrm{bnd}\mathbf{g} \tag{3.4}$$

depending on the boundary data $\mathbf{g}$. This is specified in the following sections for our examples.

3.6 The DG Discretization with Full Upwind

Let $\mathcal{K}_h$ be a set of open convex cells $K \in \mathcal{K}_h$ with $K \subset \Omega \subset \mathbb{R}^d$ such that $\Omega_h = \bigcup_{K\in\mathcal{K}_h} K$ is a decomposition of Ω with skeleton $\partial\Omega_h = \overline{\Omega}\setminus\Omega_h = \bigcup_{K\in\mathcal{K}_h} \partial K$.

Let $\mathcal{F}_K$ be the set of faces $F \subset \partial K$, such that $\overline{F} = \partial K \cap \partial\Omega$ for boundary faces, and such that for inner faces $F \subset \partial K \cap \Omega$ the neighboring cell $K_F \in \mathcal{K}_h$ exists with $\overline{F} = \partial K \cap \partial K_F$. Let $\mathcal{F}_h = \bigcup_K \mathcal{F}_K$ be the set of all faces. For the boundary conditions on $\Gamma_k \subset \partial\Omega$ we assume compatibility of the decomposition so that $\overline{\Gamma}_k = \bigcup_{F\in\mathcal{F}_h\cap\Gamma_k} \overline{F}$.

Let $Y_h \subset \mathbb{P}(\Omega_h;\mathbb{R}^m) = \prod_{K\in\mathcal{K}_h} \mathbb{P}(K;\mathbb{R}^m)$ be a discontinuous piecewise polynomial finite element space, where $\mathbb{P}(K)$ denotes the space of polynomials of any degree in K.

For $\mathbf{y}_h \in Y_h$, let $\mathbf{y}_{h,K} \in \mathbb{P}(\overline{K};\mathbb{R}^m)$ be the continuous extension of $\mathbf{y}_h|_K$ to $\overline{K}$. On inner faces $F \in \mathcal{F}_h \cap \Omega$, we define by $[\mathbf{y}_h]_{K,F} = \mathbf{y}_{h,K_F} - \mathbf{y}_{h,K}$ the *jump across* F.

Lemma 3.2 *We have for* $\mathbf{y}_h \in Y_h$*:*

$$\mathbf{y}_h \in \mathrm{H}(A, \mathbb{R}^m) \iff \underline{A}_{\mathbf{n}_K}[\mathbf{y}_h]_{K,F} = \mathbf{0} \quad \textit{for all } F \in \mathcal{F}_K \cap \Omega\,, \ K \in \mathcal{K}_h\,.$$

Proof We define $\mathbf{f}_{h,K} = A\mathbf{y}_{h,K}$ in K, and since $|\Omega \setminus \Omega_h|_d = 0$, this defines a function $\mathbf{f}_h \in \mathrm{L}_2(\Omega; \mathbb{R}^m)$ by $\mathbf{f}_h|_K = \mathbf{f}_{h,K}$. Now we observe for test functions $\mathbf{z} \in \mathrm{C}^1_\mathrm{c}(\Omega, \mathbb{R}^m)$

$$\begin{aligned}(\mathbf{f}_h, \mathbf{z})_\Omega + (\mathbf{y}_h, A\mathbf{z})_\Omega &= \sum_{K\in\mathcal{K}_h} \Big((\mathbf{f}_{h,K} - A\mathbf{y}_{h,K}, \mathbf{z})_K + (\underline{A}_{\mathbf{n}_K}\mathbf{y}_{h,K}, \mathbf{z})_{\partial K} \Big) \\ &= -\frac{1}{2} \sum_{K\in\mathcal{K}_h} \sum_{F\in\mathcal{F}_K\cap\Omega} (\underline{A}_{\mathbf{n}_K}[\mathbf{y}_h]_{K,F}, \mathbf{z})_F\end{aligned}$$

using $\underline{A}_{\mathbf{n}_{K_F}} = -\underline{A}_{\mathbf{n}_K}$. Thus, $\mathbf{y}_h \in \mathrm{H}(A, \mathbb{R}^m)$ and $A\mathbf{y}_h = \mathbf{f}_h \in \mathrm{L}_2(\Omega; \mathbb{R}^m)$ if and only if $\underline{A}_{\mathbf{n}_K}[\mathbf{y}_h]_{K,F}$ vanishes on all inner faces. □

For $\mathbf{y}_h, \mathbf{z}_h \in Y_h$ we observe

$$\begin{aligned}(A\mathbf{y}_h, \mathbf{z}_h)_\Omega &= \sum_{K\in\mathcal{K}_h} \big(\operatorname{div} \underline{A}\mathbf{y}_{h,K}, \mathbf{z}_{h,K}\big)_K \\ &= \sum_{K\in\mathcal{K}_h} \Big((\underline{A}_{\mathbf{n}_K}\mathbf{y}_{h,K}, \mathbf{z}_{h,K})_{\partial K} - (\mathbf{y}_{h,K}, A\mathbf{z}_{h,K})_K \Big)\,.\end{aligned}$$

Inserting the upwind flux (3.3) defines the DG approximation A_h, where $\underline{A}_{\mathbf{n}_K}$ is replaced by $\underline{A}^{\mathrm{upw}}_{\mathbf{n}_K}\mathbf{y}_h$, i.e.,

$$\begin{aligned}(A_h\mathbf{y}_h, \mathbf{z}_h)_\Omega &= \sum_{K\in\mathcal{K}_h} \Big((\underline{A}^{\mathrm{upw}}_{\mathbf{n}_K}\mathbf{y}_h, \mathbf{z}_{h,K})_{\partial K} - (\mathbf{y}_{h,K}, A\mathbf{z}_{h,K})_K \Big) \qquad (3.5) \\ &= \sum_{K\in\mathcal{K}_h} \Big((A\mathbf{y}_{h,K}, \mathbf{z}_{h,K})_K + \sum_{F\in\mathcal{F}_K} (\underline{A}^{\mathrm{upw}}_{\mathbf{n}_K}\mathbf{y}_h - \underline{A}_{\mathbf{n}_K}\mathbf{y}_{h,K}, \mathbf{z}_{h,K})_F \Big)\,.\end{aligned}$$

For inhomogeneous boundary conditions, using (3.4), the corresponding right-hand side is defined by

$$\langle \ell_h, \mathbf{z}_h \rangle = (\mathbf{f}, \mathbf{z}_h)_\Omega - \sum_{F\in\mathcal{F}_h\cap\partial\Omega} (\underline{A}^{\mathrm{bnd}}_{\mathbf{n}}\mathbf{g}, \mathbf{z}_h)_F\,. \qquad (3.6)$$

As we see in our examples, the boundary term is consistent with

$$(\underline{A}^{\mathrm{bnd}}_{\mathbf{n}}\mathbf{g}, \mathbf{z})_{\partial\Omega} = \sum_{k=1}^m (g_k, z_k)_{\Gamma_k} \qquad (3.7)$$

for all test functions $\mathbf{z} \in \mathcal{D}(A)$ with homogeneous boundary conditions $z_k = 0$ on $\partial\Omega \setminus \Gamma_k$, $k = 1, \dots, m$.

Consistent Extension of the Discrete DG Operator

For sufficiently smooth functions $\mathbf{y} \in \mathrm{H}^1(\Omega_h; \mathbb{R}^m)$ traces on the skeleton $\partial\Omega_h$ exist in L_2, so that the discrete operator A_h extends to $A_h \in \mathcal{L}(\mathrm{H}^1(\Omega_h; \mathbb{R}^m), Y_h)$ by

$$\big(A_h\mathbf{y}, \mathbf{z}_h\big)_\Omega = \sum_{K\in\mathcal{K}_h} \Big(\big(A\mathbf{y}, \mathbf{z}_{h,K}\big)_K + \sum_{F\in\mathcal{F}_K} \big(\underline{A}_{\mathbf{n}_K}^{\mathrm{upw}}\mathbf{y} - \underline{A}_{\mathbf{n}_K}\mathbf{y}, \mathbf{z}_{h,K}\big)_F\Big) \tag{3.8}$$

for $\mathbf{y} \in \mathrm{H}^1(\Omega_h; \mathbb{R}^m)$ and $\mathbf{z}_h \in Y_h$. Since in the conforming case by construction

$$\underline{A}_{\mathbf{n}_K}^{\mathrm{upw}}\mathbf{y} = \underline{A}_{\mathbf{n}_K}\mathbf{y}_K\,, \qquad K \in \mathcal{K}_h\,,\ \mathbf{y} \in \mathrm{H}^1(\Omega; \mathbb{R}^m)\,,$$

we obtain consistency for sufficiently smooth functions, i.e.,

$$\big(A_h\mathbf{y}, \mathbf{z}_h\big)_\Omega = \big(A\mathbf{y}, \mathbf{z}_h\big)_\Omega\,, \qquad \mathbf{y} \in \mathrm{H}^1_0(\Omega; \mathbb{R}^m)\,,\ \mathbf{z}_h \in Y_h\,. \tag{3.9}$$

In case of homogeneous boundary conditions, this extends to $\mathbf{y} \in Y_h \cap \mathcal{D}(A)$; this will be proved for acoustics in Lemma 3.3.

3.7 The Full Upwind Discretization for the Wave Equation

We evaluate (3.3) using the eigensystems in Example 3.1. Therefore, we assume that the material parameters are constant in every cell K and the possible material interfaces are aligned with the mesh.

For acoustics, we obtain on inner faces $F \in \mathcal{F}_h \cap \Omega$ from (3.1)

$$\begin{aligned}
&\underline{A}_{\mathbf{n}_K}\left(\begin{pmatrix}\mathbf{v}_{h,K}\\ p_{h,K}\end{pmatrix} + \beta_K\begin{pmatrix}\mathbf{n}_K\\ Z_K\end{pmatrix}\right) = \underline{A}_{\mathbf{n}_K}\left(\begin{pmatrix}\mathbf{v}_{h,K_F}\\ p_{h,K_F}\end{pmatrix} + \beta_{K_F}\begin{pmatrix}\mathbf{n}_K\\ -Z_{K_F}\end{pmatrix}\right)\\
\Longrightarrow\quad & 0 = \begin{pmatrix}\mathbf{n}_K\\ Z_{K_F}\end{pmatrix}\cdot\underline{A}_{\mathbf{n}_K}\left(\begin{pmatrix}[\mathbf{v}_h]_{K,F}\\ [p_h]_{K,F}\end{pmatrix} - \beta_K\begin{pmatrix}\mathbf{n}_K\\ Z_K\end{pmatrix}\right)\\
\Longrightarrow\quad & \underline{A}_{\mathbf{n}_K}^{\mathrm{upw}}\begin{pmatrix}\mathbf{v}_h\\ p_h\end{pmatrix} = \underline{A}_{\mathbf{n}_K}\begin{pmatrix}\mathbf{v}_{h,K}\\ p_{h,K}\end{pmatrix} - \frac{[p_h]_{K,F} + Z_{K_F}\mathbf{n}_K\cdot[\mathbf{v}_h]_{K,F}}{Z_K + Z_{K_F}}\begin{pmatrix}Z_K\mathbf{n}_K\\ 1\end{pmatrix}
\end{aligned} \tag{3.10}$$

by solving the equation for β_K. This extends to the boundary by defining the jump terms depending on the boundary conditions. On boundary faces $F \in \mathcal{F}_h \cap \partial\Omega$, we obtain from

$$\underline{A}_{\mathbf{n}_K}\left(\begin{pmatrix}\mathbf{v}_K\\ p_K\end{pmatrix} + \beta_K\begin{pmatrix}\mathbf{n}_K\\ Z_K\end{pmatrix}\right) = \begin{pmatrix}p_K\mathbf{n}_K\\ \mathbf{n}_K\cdot\mathbf{v}_K\end{pmatrix} - \beta_K\begin{pmatrix}Z_K\mathbf{n}_K\\ 1\end{pmatrix} \tag{3.11}$$

in case of Dirichlet boundary conditions $\beta_K = \frac{1}{Z_K}$, which corresponds to the numerical fluxes $[p_h]_{K,F} = -2p_h$ and $\mathbf{n}_K \cdot [\mathbf{v}_h]_{K,F} = 0$. This applies to the static boundary Γ_S for the pressure (1.14f).

In case of Neumann boundary conditions we obtain $\beta_K = 1$ corresponding to $\mathbf{n}_K \cdot [\mathbf{v}_h]_{K,F} = -2\mathbf{n}_K \cdot \mathbf{v}_h$ and $[p_h]_{K,F} = 0$, which applies to the dynamic boundary Γ_V for the velocity (1.14e). In both cases we extend the impedance on boundary faces F by $Z_{K_F} = Z_K$.

The DG Operator for Acoustics and Visco-Acoustics The operator $A_h \in \mathcal{L}(Y_h, Y_h)$, $A_h = \sum_{K\in\mathcal{K}_h} A_{h,K}$ for acoustics (with $r = 0$) and visco-acoustics ($r \geq 1$) with full upwind (3.10) on inner faces and (3.11) on the boundary is explicitly given by

$$\begin{aligned}
\big(A_{h,K}\mathbf{y}_h, \mathbf{z}_h\big)_K = &- \big(\nabla\cdot\mathbf{v}_{h,K}, q_{h,K}\big)_K - \big(\nabla p_{h,K}, \mathbf{w}_{h,K}\big)_K \\
&- \sum_{F\in\mathcal{F}_K} \frac{1}{Z_K + Z_{K_F}}\big([p_h]_{K,F} + Z_{K_F}\mathbf{n}_K\cdot[\mathbf{v}_h]_{K,F}, q_{h,K} + Z_K\mathbf{n}_K\cdot\mathbf{w}_{h,K}\big)_F \\
= &- \big(\nabla\cdot\mathbf{v}_{h,K}, q_{h,K}\big)_K - \big(\nabla p_{h,K}, \mathbf{w}_{h,K}\big)_K \\
&- \sum_{F\in\mathcal{F}_K\cap\Omega} \frac{1}{Z_K + Z_{K_F}}\big([p_h]_{K,F} + Z_{K_F}\mathbf{n}_K\cdot[\mathbf{v}_h]_{K,F}, q_{h,K} + Z_K\mathbf{n}_K\cdot\mathbf{w}_{h,K}\big)_F \\
&+ \sum_{F\in\mathcal{F}_K\cap\Gamma_S} \frac{1}{Z_K}\big(p_{h,K}, q_{h,K} + Z_K\mathbf{n}_K\cdot\mathbf{w}_{h,K}\big)_F \\
&+ \sum_{F\in\mathcal{F}_K\cap\Gamma_V} \big(\mathbf{n}_K\cdot\mathbf{v}_{h,K}, q_{h,K} + Z_K\mathbf{n}_K\cdot\mathbf{w}_{h,K}\big)_F
\end{aligned} \tag{3.12}$$

for $\mathbf{y}_h = (\mathbf{v}_h, p_{0,h}, \ldots, p_{r,h}), \mathbf{z}_h = (\mathbf{w}_h, q_{0,h}, \ldots, q_{r,h}) \in Y_h$ with $p_h = p_{0,h} + \cdots + p_{r,h}$, $q_h = q_{0,h} + \cdots + q_{r,h}$.

For inhomogeneous boundary conditions we obtain the right-hand side (3.6) by $\ell_h = \sum_{K\in\mathcal{K}_h} \ell_{h,K}$ with

$$\langle \ell_{h,K}, \mathbf{z}_{h,K}\rangle = \big(\mathbf{f}_h, \mathbf{z}_{h,K}\big)_K + \sum_{F\in\mathcal{F}_K\cap\Gamma_S} \big(p_S, Z_K^{-1} q_{h,K} + \mathbf{n}_K\cdot\mathbf{w}_{h,K}\big)_F$$
$$+ \sum_{F\in\mathcal{F}_K\cap\Gamma_V} \big(g_V, q_{h,K} + Z_K\mathbf{n}_K\cdot\mathbf{w}_{h,K}\big)_F\,. \tag{3.13}$$

Lemma 3.3 *The DG discretization* (3.12) *is*

(a) consistent, i.e.,

$$\big(A_h\mathbf{y}, \mathbf{z}_h\big)_\Omega = \big(A\mathbf{y}, \mathbf{z}_h\big)_\Omega\,, \qquad \mathbf{y}\in Y_h\cap\mathcal{D}(A)\,,\ \mathbf{z}_h\in Y_h\,,$$
$$\big(A_h\mathbf{y}_h, \mathbf{z}\big)_\Omega = -\big(\mathbf{y}_h, A\mathbf{z}\big)_\Omega\,, \qquad \mathbf{y}_h\in Y_h\,,\ \mathbf{z}\in Y_h\cap\mathcal{D}(A)\,;$$

(b) monotone / dissipative satisfying

$$\big(A_h\mathbf{y}_h, \mathbf{y}_h\big)_\Omega = \frac{1}{2}\sum_{K\in\mathcal{K}_h}\sum_{F\in\mathcal{F}_K} \frac{1}{Z_K + Z_{K_F}}\Big(\big\|[p_h]_{K,F}\big\|_F^2 + Z_K Z_{K_F}\big\|\mathbf{n}_K\cdot[\mathbf{v}_h]_{K,F}\big\|_F^2\Big)$$
$$\geq 0\,, \qquad \mathbf{y}_h\in Y_h\,.$$

Proof It is sufficient to consider $r = 0$. For $\mathbf{y} = (\mathbf{v}, p)\in Y_h\cap\mathcal{D}(A)$ we obtain $\mathbf{n}_K\cdot[\mathbf{v}]_{K,F} = [p]_{K,F} = 0$ for $F\in\mathcal{F}_K\cap\Omega$, $p_h = 0$ on $F\in\mathcal{F}_K\cap\Gamma_V$, and $\mathbf{n}_K\cdot\mathbf{v} = 0$ on $F\in\mathcal{F}_K\cap\Gamma_S$, so that consistency is obtained by

$$\big(A_h\mathbf{y}, \mathbf{z}_h\big)_\Omega = \sum_{K\in\mathcal{K}_h}\big(A_{h,K}\mathbf{y}_K, \mathbf{z}_{h,K}\big)_K = \sum_{K\in\mathcal{K}_h}\big(A\mathbf{y}_K, \mathbf{z}_{h,K}\big)_K = \big(A\mathbf{y}, \mathbf{z}_h\big)_\Omega$$

for $\mathbf{z}_h\in Y_h$ since all flux terms on the faces are vanishing.

Integration by parts for $\mathbf{y}_h = (\mathbf{v}_h, p_h)\in Y_h$ and $\mathbf{z} = (\mathbf{w}, q)\in Y_h\cap\mathcal{D}(A)$ yields dual consistency by

$$\big(A_h\mathbf{y}_h, \mathbf{z}\big)_\Omega = \sum_{K\in\mathcal{K}_h}\Big(- \big(\nabla\cdot\mathbf{v}_{h,K}, q_K\big)_K - \big(\nabla p_{h,K}, \mathbf{w}_K\big)_K$$
$$- \sum_{F\in\mathcal{F}_K\cap\Omega}\frac{1}{Z_K + Z_{K_F}}\big([p_h]_{K,F} + Z_{K_F}\mathbf{n}_K\cdot[\mathbf{v}_h]_{K,F}, q_K + Z_K\mathbf{n}_K\cdot\mathbf{w}_K\big)_F$$

$$
\begin{aligned}
&+ \sum_{F\in\mathcal{F}_K\cap\Gamma_S} \big(p_{h,K}, \mathbf{n}_K\cdot\mathbf{w}_K\big)_F + \sum_{F\in\mathcal{F}_K\cap\Gamma_V} \big(\mathbf{n}_K\cdot\mathbf{v}_{h,K}, q_K\big)_F\Big) \\
&= \sum_{K\in\mathcal{K}_h} \Big(\big(\mathbf{v}_{h,K}, \nabla q_K\big)_K + \big(p_{h,K}, \nabla\cdot\mathbf{w}_K\big)_K \\
&\quad - \sum_{F\in\mathcal{F}_K\cap\Omega} \Big(\frac{1}{Z_K+Z_{K_F}}\big([p_h]_{K,F} + Z_{K_F}\mathbf{n}_K\cdot[\mathbf{v}_h]_{K,F}, q_K + Z_K\mathbf{n}_K\cdot\mathbf{w}_K\big)_F \\
&\qquad\qquad + \big(p_{h,K}, \mathbf{n}_K\cdot\mathbf{w}_K\big)_F + \big(\mathbf{n}_K\cdot\mathbf{v}_{h,K}, q_K\big)_F\Big)\Big) \\
&= \sum_{K\in\mathcal{K}_h} \Big(-\big(\mathbf{y}_{h,K}, A\mathbf{z}_K\big)_K \\
&\quad + \sum_{F\in\mathcal{F}_K\cap\Omega} \frac{1}{Z_K+Z_{K_F}}\Big(\big([p_h]_{K,F}, q_K\big)_F + Z_K Z_{K_F}\big(\mathbf{n}_K\cdot[\mathbf{v}_h]_{K,F}, \mathbf{n}_K\cdot\mathbf{w}_K\big)_F\Big)\Big) \\
&= -\big(\mathbf{y}_h, A\mathbf{z}\big)_\Omega .
\end{aligned}
$$

For $\mathbf{y}_h = (\mathbf{v}_h, p_h) \in Y_h$ we obtain the identity

$$
\begin{aligned}
&\big(A_h\mathbf{y}_h, \mathbf{y}_h\big)_\Omega = \sum_{K\in\mathcal{K}_h} \big(A_{h,K}\mathbf{y}_h, \mathbf{y}_h\big)_K \\
&= \sum_{K\in\mathcal{K}_h} \Big(-\big(\nabla\cdot\mathbf{v}_{h,K}, p_{h,K}\big)_K - \big(\nabla p_{h,K}, \mathbf{v}_{h,K}\big)_K \\
&\qquad - \sum_{F\in\mathcal{F}_K\cap\Omega} \frac{1}{Z_K+Z_{K_F}}\big([p_h]_{K,F} + Z_{K_F}\mathbf{n}_K\cdot[\mathbf{v}_h]_{K,F}, p_{h,K} + Z_K\mathbf{n}_K\cdot\mathbf{v}_{h,K}\big)_F \\
&\qquad + \sum_{F\in\mathcal{F}_K\cap\Gamma_S} \frac{1}{Z_K}\big(p_{h,K}, p_{h,K} + Z_K\mathbf{n}_K\cdot\mathbf{v}_{h,K}\big)_F \\
&\qquad + \sum_{F\in\mathcal{F}_K\cap\Gamma_V} \big(\mathbf{n}_K\cdot\mathbf{v}_{h,K}, p_{h,K} + Z_K\mathbf{n}_K\cdot\mathbf{v}_{h,K}\big)_F\Big) \\
&= \frac{1}{2}\sum_{K\in\mathcal{K}_h}\sum_{F\in\mathcal{F}_K} \frac{1}{Z_K+Z_{K_F}}\Big(\big\|[p_h]_{K,F}\big\|_F^2 + Z_K Z_{K_F}\big\|\mathbf{n}_K\cdot[\mathbf{v}_h]_{K,F}\big\|_F^2\Big)
\end{aligned}
$$

since we obtain, using $\big(\nabla\cdot\mathbf{v}_{h,K}, p_{h,K}\big)_K + \big(\nabla p_{h,K}, \mathbf{v}_{h,K}\big)_K = \big(\mathbf{n}_K\cdot\mathbf{v}_{h,K}, p_{h,K}\big)_{\partial K}$, for the remaining terms

$$\sum_{K\in\mathcal{K}_h}\Bigg(-\big(\mathbf{n}_K\cdot\mathbf{v}_{h,K}, p_{h,K}\big)_{\partial K}$$

$$-\sum_{F\in\mathcal{F}_K\cap\Omega}\frac{1}{Z_K+Z_{K_F}}\Big(Z_K\big([p_h]_{K,F}, \mathbf{n}_K\cdot\mathbf{v}_{h,K}\big)_F + Z_{K_F}\big(\mathbf{n}_K\cdot[\mathbf{v}_h]_{K,F}, p_{h,K}\big)_F\Big)$$

$$+\sum_{F\in\mathcal{F}_K\cap\Gamma_{\mathrm{V}}}\big(p_{h,K}, \mathbf{n}_K\cdot\mathbf{v}_{h,K}\big)_F + \sum_{F\in\mathcal{F}_K\cap\Gamma_{\mathrm{S}}}\big(\mathbf{n}_K\cdot\mathbf{v}_{h,K}, p_{h,K}\big)_F\Bigg) = 0\,.$$

□

The DG Operator for Visco-Elasticity The operator $A_h = \sum_{K\in\mathcal{K}_h} A_{h,K} \in \mathcal{L}(Y_h, Y_h)$ with full upwind is defined by

$$\big(A_{h,K}\mathbf{y}_h, \mathbf{z}_h\big)_K = -\big(\nabla\cdot\boldsymbol{\sigma}_{h.K}, \boldsymbol{\psi}_{h,K}\big)_K - \big(\boldsymbol{\varepsilon}(\mathbf{v}_{h,K}), \boldsymbol{\eta}_{h,K}\big)_K$$

$$-\sum_{F\in\mathcal{F}_K}\frac{1}{Z^{\mathrm{P}}_K+Z^{\mathrm{P}}_{K_F}}\Big(\mathbf{n}_K\cdot\big([\boldsymbol{\sigma}_h]_{K,F}\mathbf{n}_K + Z^{\mathrm{P}}_{K_F}[\mathbf{v}_h]_{K,F}\big), \mathbf{n}_K\cdot\big(\boldsymbol{\eta}_{h,K}\mathbf{n}_K + Z^{\mathrm{P}}_K\mathbf{w}_{h,K}\big)\Big)_F$$

$$-\sum_{F\in\mathcal{F}_K}\frac{1}{Z^{\mathrm{S}}_K+Z^{\mathrm{S}}_{K_F}}\Big(\mathbf{n}_K\times\big([\boldsymbol{\sigma}_h]_{K,F}\mathbf{n}_K + Z^{\mathrm{S}}_{K_F}[\mathbf{v}_h]_{K,F}\big), \mathbf{n}_K\times\big(\boldsymbol{\eta}_{h,K}\mathbf{n}_K + Z^{\mathrm{S}}_K\mathbf{w}_{h,K}\big)\Big)_F$$

for $\mathbf{y}_h = (\mathbf{v}_h, \boldsymbol{\sigma}_{0,h},\dots,\boldsymbol{\sigma}_{r,h})$, $\mathbf{z}_h = (\mathbf{w}_h, \boldsymbol{\eta}_{0,h},\dots,\boldsymbol{\eta}_{r,h}) \in Y_h$, $\boldsymbol{\sigma}_h = \sum\boldsymbol{\sigma}_{j,h}$, and $\boldsymbol{\eta}_h = \sum\boldsymbol{\eta}_{j,h}$. The coefficients $Z^{\mathrm{P}}_K = \sqrt{(2\mu+\lambda)\rho}\,|_K$ and $Z^{\mathrm{S}}_K = \sqrt{\mu\rho}\,|_K$ are the impedance of compressional waves and shear waves, respectively.

On boundary faces $F\in\mathcal{F}_h\cap\Gamma_{\mathrm{V}}$, we set $[\mathbf{v}_h]_{K,F} = -2\mathbf{v}_h$ and $[\boldsymbol{\sigma}_h]_{K,F}\mathbf{n}_K = \mathbf{0}$, and on $F\in\mathcal{F}_h\cap\Gamma_{\mathrm{S}}$ we set $[\mathbf{v}_h]_{K,F} = \mathbf{0}$ and $[\boldsymbol{\sigma}_h]_{K,F}\mathbf{n}_K = -2\boldsymbol{\sigma}_h\mathbf{n}_K$. We have

$$\big(A_h\mathbf{y}_h, \mathbf{y}_h\big)_\Omega = \frac{1}{2}\sum_{K\in\mathcal{K}_h}\sum_{F\in\mathcal{F}_K}\Bigg(\frac{\big\|\mathbf{n}_K\cdot\big([\boldsymbol{\sigma}_h]_{K,F}\mathbf{n}_K\big)\big\|_F^2 + Z^{\mathrm{P}}_K Z^{\mathrm{P}}_{K_F}\big\|\mathbf{n}_K\cdot[\mathbf{v}_h]_{K,F}\big\|_F^2}{Z^{\mathrm{P}}_K+Z^{\mathrm{P}}_{K_F}}$$

$$+\frac{\big\|\mathbf{n}_K\times\big([\boldsymbol{\sigma}_h]_{K,F}\mathbf{n}_K\big)\big\|_F^2 + Z^{\mathrm{S}}_K Z^{\mathrm{S}}_{K_F}\big\|\mathbf{n}_K\times[\mathbf{v}_h]_{K,F}\big\|_F^2}{Z^{\mathrm{S}}_K+Z^{\mathrm{S}}_{K_F}}\Bigg) \ge 0\,.$$

The DG Operator for Linear Electro-Magnetic Waves
For $(\mathbf{H}_h, \mathbf{E}_h), (\boldsymbol{\varphi}_h, \boldsymbol{\psi}_h) \in Y_h$ we have

$$\begin{aligned}\big(A_h(\mathbf{H}_h, \mathbf{E}_h), (\boldsymbol{\varphi}_{h,K}, \boldsymbol{\psi}_{h,K})\big)_{0,K} &= (\operatorname{curl} \mathbf{E}_{h,K}, \boldsymbol{\varphi}_{h,K})_{0,K} - (\operatorname{curl} \mathbf{H}_{h,K}, \boldsymbol{\psi}_{h,K})_{0,K}\\ &- \sum_{F\in\mathcal{F}_K} \frac{1}{Z_K + Z_{K_F}}\Big(\big(Z_{K_F}[\mathbf{E}_h]_{K,F} + \mathbf{n}_K \times [\mathbf{H}_h]_{K,F}, \mathbf{n}_K \times \boldsymbol{\varphi}_{h,K}\big)_F\\ &\qquad + \big(Z_{K_F}\mathbf{n}_K \times [\mathbf{E}_h]_{K,F} - [\mathbf{H}_h]_{K,F}, Z_K \mathbf{n}_K \times \boldsymbol{\psi}_{h,K}\big)_F\Big)\\ &+ \sum_{F\in\mathcal{F}_K\cap\Gamma_{\mathrm{I}}} \big(\zeta\, \mathbf{n}_K \times \mathbf{E}_{K,h}, \mathbf{n}_K \times \boldsymbol{\varphi}_{h,K}\big)_F\end{aligned}$$

with coefficient $Z_K = \sqrt{\varepsilon_K/\mu_K}$ and impedance ζ.

On boundary faces $F \in \mathcal{F}_h \cap \Gamma_{\mathrm{E}}$, perfect conducting boundary conditions are modeled by the (only virtual) definition of $\mathbf{n}_K \times \mathbf{E}_h = -\mathbf{n}_K \times \mathbf{E}_h$ and $\mathbf{n}_K \times \mathbf{H}_h = \mathbf{n}_K \times \mathbf{H}_h$, i.e., $\mathbf{n}_K \times [\mathbf{E}]_{K,F} = -2\mathbf{n}_K \times \mathbf{E}_h$ and $\mathbf{n}_K \times [\mathbf{H}_h]_{K,F} = \mathbf{0}$. On impedance boundary faces $F \in \mathcal{F}_h \cap \Gamma_{\mathrm{I}}$, we set $\mathbf{n}_K \times [\mathbf{E}]_{K,F} = \mathbf{0}$ and $\mathbf{n}_K \times [\mathbf{H}_h]_{K,F} = -2\mathbf{n}_K \times \mathbf{H}_h$.

With the same arguments as for the acoustic case we obtain

$$\begin{aligned}&\big(A_h(\mathbf{H}_h, \mathbf{E}_h), (\mathbf{H}_h, \mathbf{E}_h)\big)_{0,\Omega}\\ &\quad= \frac{1}{2}\sum_{K\in\mathcal{K}_h}\sum_{F\in\mathcal{F}_K} \frac{Z_K Z_{K_F}\big\|\mathbf{n}_K \times [\mathbf{E}_h]_{K,F}\big\|_F^2 + \big\|\mathbf{n}_K \times [\mathbf{H}_h]_{K,F}\big\|_F^2}{Z_K + Z_{K_F}}\\ &\qquad + \sum_{F\in\mathcal{F}_h\cap\Gamma_{\mathrm{I}}} \zeta\, \big\|\mathbf{n}_K \times \mathbf{E}_h\big\|_F^2 \ge 0\,.\end{aligned}$$

Bibliographic Comments
An introduction to discontinuous Galerkin methods for hyperbolic conservation laws is given, e.g., in [85, 86]. The numerical flux for wave equations is evaluated in [92] and extended to viscous waves in [169]. For the explicit evaluation of the numerical flux for inhomogeneous boundary conditions we refer to [49].

4 A Petrov–Galerkin Space-Time Approximation for Linear Hyperbolic Systems

We introduce and analyze a variational discretization in space and time by extending the discontinuous Galerkin method in space to a Petrov–Galerkin method in time and space. We verify discrete inf-sup stability, and this yields well-posedness, stability and convergence for strong solutions. By duality, this extends to convergence for weak solutions. Finally, we address a posteriori error bounds. For a given error functional, the corresponding dual solution is computed and an error indicator is defined by weighted residuals. A reliable error estimator of weak solutions is obtained computing local conforming reconstructions.

4.1 Decomposition of the Space-Time Cylinder

For the discretization, we use tensor product space-time cells combining the mesh in space (see Sect. 3.6) with a decomposition in time. For $0 = t_0 < t_1 < \cdots < t_N = T$, we define

$$I_h = (t_0, t_1) \cup \cdots \cup (t_{N-1}, t_N) \subset I = (0, T)\,.$$

Together with the decomposition in space $\Omega_h = \bigcup_{K \in \mathcal{K}_h} K$ into open cells $K \subset \Omega \subset \mathbb{R}^d$ we obtain a decomposition $Q_h = I_h \times \Omega_h = \bigcup_{R \in \mathcal{R}_h} R$ of the space-time cylinder $Q = I \times \Omega \subset \mathbb{R}^{1+d}$, so that $\overline{Q} = Q_h \cup \partial Q_h$, where ∂Q_h is the space-time skeleton.

For every space-time cell $R = (t_{n-1}, t_n) \times K$ we select polynomial degrees $p_R = p_{n,K} \geq 1$ in time and $q_R = q_{n,K} \geq 0$ in space. This defines the discontinuous test space

W. Dörfler et al., *Wave Phenomena*, Oberwolfach Seminars 49,
https://doi.org/10.1007/978-3-031-05793-9_4

in the space-time cylinder

$$W_h = \prod_{R \in \mathcal{R}_h} \mathbb{P}_{p_R - 1} \otimes \mathbb{P}_{q_R}(K; \mathbb{R}^m) \subset \mathbb{P}(I_h \times \Omega_h; \mathbb{R}^m) = \prod_{R \in \mathcal{R}_h} \mathbb{P}(R; \mathbb{R}^m) \subset \mathrm{L}_2(Q; \mathbb{R}^m),$$

where $\mathbb{P}_p$ are the polynomials up to order p, $\mathbb{P}_q(K)$ are the polynomials up to order q in K, and $\mathbb{P}(R)$ are polynomials of any degree in R.

Defining the discontinuous spaces

$$Y_{n,h} = \prod_{K \in \mathcal{K}_h} \mathbb{P}_{q_{n,K}}(K; \mathbb{R}^m) \subset \mathbb{P}(\Omega_h; \mathbb{R}^m) \subset \mathrm{L}_2(\Omega; \mathbb{R}^m), \quad Y_h = Y_{1,h} + \cdots + Y_{N,h},$$

we observe $W_h \subset \mathrm{L}_2(0,T;Y_h)$, and in every time slice $\mathbf{w}_h(t) \in Y_{n,h} \subset Y_h$ for all $t \in (t_{n-1}, t_n)$ and $\mathbf{w}_h \in W_h$.

4.2 The Petrov–Galerkin Setting

Let $L_h = M_h\partial_t + D_h + A_h \colon \mathrm{H}^1(0,T;Y_h) \longrightarrow \mathrm{L}_2(0,T;Y_h)$ be the linear mapping approximating the differential operator $L = M\partial_t + D + A$ with the following properties:

(a) $M_h \in \mathcal{L}(Y_h, Y_h)$ is uniformly positive definite, i.e., $c_M > 0$ exists with

$$(M_h\mathbf{y}_h, \mathbf{y}_h)_\Omega \geq c_M \|\mathbf{y}_h\|_W^2, \qquad \mathbf{y}_h \in Y_h; \tag{4.1a}$$

(b) $D_h \in \mathcal{L}(Y_h, Y_h)$ is monotone, i.e.,

$$(D_h\mathbf{y}_h, \mathbf{y}_h)_\Omega \geq 0, \qquad \mathbf{y}_h \in Y_h; \tag{4.1b}$$

(c) $A_h \in \mathcal{L}(Y_h, Y_h)$ is monotone and consistent, i.e.,

$$(A_h\mathbf{y}_h, \mathbf{y}_h)_\Omega \geq 0, \qquad \mathbf{y}_h \in Y_h, \tag{4.1c}$$

$$(A_h\mathbf{z}, \mathbf{y}_h)_\Omega = (A\mathbf{z}, \mathbf{y}_h)_\Omega,$$

$$(A_h\mathbf{y}_h, \mathbf{z})_\Omega = -(\mathbf{y}_h, A\mathbf{z})_\Omega, \qquad \mathbf{z} \in Y_h \cap \mathcal{D}(A). \tag{4.1d}$$

The operators M_h, D_h, A_h do not depend on the time variable $t \in (0,T)$, i.e., they are defined in $Y_h \subset \mathbb{P}(\Omega_h; \mathbb{R}^m)$.

In the next step we construct a suitable ansatz space $V_h \subset \mathbb{P}(Q_h; \mathbb{R}^m)$. In every time slice (t_{n-1}, t_n) let

$$\Pi_{n,h} \colon \mathrm{L}_2(\Omega; \mathbb{R}^m) \longrightarrow Y_{n,h}$$

be the weighted L_2-projection defined by

$$\big(M_h \Pi_{n,h} \mathbf{y}, \mathbf{z}_h\big)_\Omega = \big(M_h \mathbf{y}, \mathbf{z}_h\big)_\Omega\,, \qquad \mathbf{y} \in \mathrm{L}_2(\Omega; \mathbb{R}^m)\,,\ \mathbf{z}_h \in Y_{n,h}$$

corresponding to the norm

$$\|\mathbf{y}_h\|_{Y_h} = \sqrt{\big(M_h \mathbf{y}_h, \mathbf{y}_h\big)_\Omega}\,, \qquad \mathbf{y}_h \in Y_h\,.$$

For $\mathbf{v}_h \in \mathbb{P}(I_h \times \Omega_h; \mathbb{R}^m)$ let $\mathbf{v}_{n,h} \in \mathbb{P}([t_{n-1}, t_n] \times \Omega_h; \mathbb{R}^m)$ be the extension of $\mathbf{v}_h|_{(t_{n-1},t_n)\times\Omega_h}$ to $[t_{n-1}, t_n]$. Then, we define

$$V_h = \Big\{\mathbf{v}_h \in \prod_{R=(t_{n-1},t_n)\times K \in \mathcal{R}_h} \mathbb{P}_{p_R} \otimes \mathbb{P}_{q_R}(K; \mathbb{R}^m) \subset \mathbb{P}(I_h \times \Omega_h; \mathbb{R}^m)\colon$$
$$\mathbf{v}_h(0) = \mathbf{0}\,,\ \mathbf{v}_{n,h}(t_{n-1}) = \Pi_{n,h} \mathbf{v}_{n-1,h}(t_{n-1}) \text{ for } n = 2, \dots, N\Big\} \subset \mathrm{H}^1(0, T; Y_h)\,.$$

By construction, we have $\partial_t V_h = W_h$ in I_h and $\dim V_h = \dim W_h$. Note that V_h includes homogeneous initial data. For inhomogeneous initial data $\mathbf{u}_0$ we define the affine space

$$V_h(\mathbf{u}_0) = \Big\{\mathbf{v}_h \in \prod_{R=(t_{n-1},t_n)\times K \in \mathcal{R}_h} \mathbb{P}_{p_R} \otimes \mathbb{P}_{q_R}(K; \mathbb{R}^m) \subset \mathbb{P}(I_h \times \Omega_h; \mathbb{R}^m)\colon$$
$$\mathbf{v}_h(0) = \Pi_{1,h} \mathbf{u}_0 \text{ for } t = 0\,, \tag{4.2}$$
$$\mathbf{v}_{n,h}(t_{n-1}) = \Pi_{n,h} \mathbf{v}_{n-1,h}(t_{n-1}) \text{ for } n = 2, \dots, N\Big\} \subset \mathrm{H}^1(0, T; Y_h)\,.$$

4.3 Inf-Sup Stability

Let

$$\Pi_h \colon \mathrm{L}_2(Q; \mathbb{R}^m) \longrightarrow W_h$$

be the projection defined by

$$\big(M_h \Pi_h \mathbf{v}, \mathbf{w}_h\big)_Q = \big(M_h \mathbf{v}, \mathbf{w}_h\big)_Q\,, \qquad \mathbf{v} \in \mathrm{L}_2(Q; \mathbb{R}^m)\,,\ \mathbf{w}_h \in W_h\,.$$

Note that $\Pi_h M_h \mathbf{v}_h = M_h \Pi_h \mathbf{v}_h$ and $\Pi_h A_h \mathbf{v}_h = A_h \Pi_h \mathbf{v}_h$ for $\mathbf{v}_h \in \mathrm{L}_2(0, T; Y_h)$.

The analysis of the discretization is based on the norms

$$\|\mathbf{w}_h\|_{W_h} = \sqrt{\big(M_h\mathbf{w}_h, \mathbf{w}_h\big)_Q}\,, \quad \|\mathbf{f}_h\|_{W_h^*} = \sqrt{\big(M_h^{-1}\mathbf{f}_h, \mathbf{f}_h\big)_Q}\,, \quad \mathbf{w}_h\,, \mathbf{f}_h \in \mathrm{L}_2(Q;\mathbb{R}^m)$$

and

$$\|\mathbf{v}_h\|_{V_h} = \sqrt{\|\mathbf{v}_h\|_{W_h}^2 + \|\Pi_h M_h^{-1} L_h \mathbf{v}_h\|_{W_h}^2}\,, \qquad \mathbf{v}_h \in \mathrm{H}^1(0,T;Y_h)\,. \tag{4.3}$$

Theorem 4.1 *The bilinear form* $b_h\colon \mathrm{H}^1(0,T;Y_h)\times \mathrm{L}_2(Q;\mathbb{R}^m) \longrightarrow \mathbb{R}$ *defined by* $b_h(\mathbf{v}_h,\mathbf{w}_h) = (L_h\mathbf{v}_h,\mathbf{w}_h)_Q$ *is inf-sup stable in* $V_h\times W_h$ *satisfying*

$$\sup_{\mathbf{w}_h\in W_h\setminus\{\mathbf{0}\}} \frac{b_h(\mathbf{v}_h,\mathbf{w}_h)}{\|\mathbf{w}_h\|_{W_h}} \ge \beta\,\|\mathbf{v}_h\|_{V_h}\,, \quad \mathbf{v}_h\in V_h \quad \textit{with } \beta = \frac{1}{\sqrt{4T^2+1}}\,.$$

Thus, for given $\mathbf{f}\in \mathrm{L}_2(Q;\mathbb{R}^m)$, *a unique solution* $\mathbf{u}_h\in V_h$ *of*

$$(L_h\mathbf{u}_h,\mathbf{w}_h)_Q = (\mathbf{f},\mathbf{w}_h)_Q\,, \qquad \mathbf{w}_h\in W_h, \tag{4.4}$$

exists satisfying the a priori bound $\|\mathbf{u}_h\|_{V_h} \le \beta^{-1}\|\Pi_h\mathbf{f}\|_{W_h^*}$.

The stability constant $\beta>0$ is the same as in the continuous case in Theorem 2.16. The proof of the inf-sup stability is based on the following estimates.

Lemma 4.2 *Let* $\lambda_{n,k}\in\mathbb{P}_k$, $k=0,1,2,\ldots$, *be the orthonormal Legendre polynomials in* $\mathrm{L}_2(t_{n-1},t_n)$. *Then, we have* $\big(t\partial_t\lambda_{n,k},\lambda_{n,k}\big)_{(t_{n-1},t_n)}\ge 0$.

Proof The orthonormal Legendre polynomials $\lambda_{n,k}$ with respect to $\big(\cdot,\cdot\big)_{(t_{n-1},t_n)}$ are given by scaling the orthogonal polynomials $\tilde\lambda_{n,k}$

$$\lambda_{n,k}(t) = c_{n,k}\tilde\lambda_{n,k}(t)\,,\quad \tilde\lambda_{n,k}(t) = \partial_t^k\big((t-t_{n-1})(t-t_n)\big)^k\,,\quad c_{n,k} = \|\tilde\lambda_{n,k}\|_{(t_{n-1},t_n)}^{-1}\,.$$

For $k=0$ we have $\partial_t\lambda_{n,0}=0$ and thus $\big(t\partial_t\lambda_{n,0},\lambda_{n,0}\big)_{(t_{n-1},t_n)}=0$. For $k\ge 1$ we have

$$\begin{aligned}
\big(t\partial_t\lambda_{n,k},\lambda_{n,k}\big)_{(t_{n-1},t_n)} &= \big(tc_{n,k}\partial_t^{k+1}((t-t_{n-1})(t-t_n))^k,\lambda_{n,k}\big)_{(t_{n-1},t_n)}\\
&= \big(tc_{n,k}\partial_t^{k+1}t^{2k},\lambda_{n,k}\big)_{(t_{n-1},t_n)}\\
&= \big(c_{n,k}k\partial_t^k t^{2k},\lambda_{n,k}\big)_{(t_{n-1},t_n)} = k\big(\lambda_{n,k},\lambda_{n,k}\big)_{(t_{n-1},t_n)} = k>0
\end{aligned}$$

using $t\,\partial_t^{k+1}t^{2k} = t\Big(2k(2k-1)\cdots(k+1)k\,t^{k-1}\Big) = k\,\partial_t^k t^{2k}$. □

Lemma 4.3 *Let* $\underline{X} = (X_{kn}) \in \mathbb{R}^{N\times N}_{\text{sym}}$ *be a symmetric and positive semidefinite matrix, and* $\underline{Y} = (Y_{kn}) \in \mathbb{R}^{N\times N}$ *be a positive semidefinite matrix. Then,*

$$\underline{X} : \underline{Y} = \sum_{k,m=1}^{N} X_{km} Y_{km} \geq 0 .$$

Proof Let $(\mu_n, \underline{w}_n)$, $n = 1, \dots, N$ be a complete eigensystem of $\underline{X}$ with $\mu_n \geq 0$ and $\underline{w}_n = (w_{nk})_{k=1,\dots,N} \in \mathbb{R}^N$, so that $\underline{X} = \sum_{n=1}^N \mu_n \underline{w}_n \underline{w}_n^\top$. Then, we have

$$\underline{X} : \underline{Y} = \sum_{k,m=1}^{N} X_{km} Y_{km} = \sum_{k,m,n=1}^{N} \mu_n w_{nk} w_{nm} Y_{km} = \sum_{n=1}^{N} \mu_n \underline{w}_n^\top \underline{Y}\, \underline{w}_n \geq 0 .$$

□

Lemma 4.4 *We have for* $\mathbf{v}_h \in V_h$

$$\|\mathbf{v}_h\|_{W_h} \leq 2T \, \|\Pi_h M_h^{-1} L_h \mathbf{v}_h\|_{W_h} , \qquad \mathbf{v}_h \in V_h . \tag{4.5}$$

This shows that Lemma 2.10 extends to the discrete estimate also with $C_L = 2T$.

Proof Set $p = \max\limits_{R\in\mathcal{R}_h} p_R$. For $\mathbf{v}_h \in V_h$ in every time slice (t_{n-1}, t_n) a representation

$$\mathbf{v}_{n,h}(t, \mathbf{x}) = \sum_{k=0}^{p} \lambda_{n,k}(t) \mathbf{v}_{n,k,h}(\mathbf{x}) , \qquad \mathbf{v}_{n,k,h} \in Y_{n,h} , \ (t, \mathbf{x}) \in (t_{n-1}, t_n) \times \Omega_h$$

exists with $\mathbf{v}_{n,k,h}(\mathbf{x}) = \mathbf{0}$ for $(t, \mathbf{x}) \in R = (t_{n-1}, t_n) \times K$ and $k > p_R$, so that

$$\Pi_h \mathbf{v}_{n,h}(t, \mathbf{x}) = \sum_{k=0}^{p_R-1} \lambda_{n,k}(t) \mathbf{v}_{n,k,h}(\mathbf{x}) = \sum_{k=0}^{p} \lambda_{n,k}(t) \hat{\mathbf{v}}_{n,k,h}(\mathbf{x}) , \qquad (t, \mathbf{x}) \in R$$

with $\hat{\mathbf{v}}_{n,k,h}(\mathbf{x}) = \mathbf{v}_{n,k,h}(\mathbf{x})$ for $k < p_R$ and $\hat{\mathbf{v}}_{n,k,h}(\mathbf{x}) = \mathbf{0}$ for $k \geq p_R$.

The proof of (4.5) relies on the application of Fubini's theorem

$$\int_0^T \int_0^t \phi(s) \mathrm{d}s \mathrm{d}t = \int_0^T d_T(t) \phi(t) \mathrm{d}t , \qquad \phi \in \mathrm{L}_1(0, T) \tag{4.6}$$

and on estimates with respect to the weighting function in time $d_T(t) = T - t$.

In the first step, we show

$$\big(M_h\partial_t\mathbf{v}_h, d_T\mathbf{v}_h\big)_Q \le \big(M_h\partial_t\mathbf{v}_h, d_T\Pi_h\mathbf{v}_h\big)_Q, \tag{4.7}$$

$$0 \le \big(\Pi_h A_h\mathbf{v}_h, d_T\Pi_h\mathbf{v}_h\big)_Q, \tag{4.8}$$

$$0 \le \big(\Pi_h D_h\mathbf{v}_h, d_T\Pi_h\mathbf{v}_h\big)_Q. \tag{4.9}$$

Since A_h and D_h are monotone, we obtain (4.8) and (4.9) from Lemma 4.3 applied to

$$\begin{aligned}
\big(\Pi_h A_h\mathbf{v}_h, d_T\Pi_h\mathbf{v}_h\big)_Q &= \sum_{n=1}^{N}\big(\Pi_h A_h\mathbf{v}_{n,h}, d_T\Pi_h\mathbf{v}_{n,h}\big)_{(t_{n-1},t_n)\times\Omega}\\
&= \sum_{n=1}^{N}\sum_{R=(t_{n-1},t_n)\times K}\sum_{k=0}^{p_R-1}\sum_{l=0}^{p_R-1}\big(\lambda_{n,k}, d_T\lambda_{n,l}\big)_{(t_{n-1},t_n)}\big(A_h\mathbf{v}_{n,k,h}, \mathbf{v}_{n,l,h}\big)_K\\
&= \sum_{n=1}^{N}\sum_{k=0}^{p}\sum_{l=0}^{p}\big(\lambda_{n,k}, d_T\lambda_{n,l}\big)_{(t_{n-1},t_n)}\big(A_h\hat{\mathbf{v}}_{n,k,h}, \hat{\mathbf{v}}_{n,l,h}\big)_\Omega \ge 0,
\end{aligned}$$

$$\begin{aligned}
&\big(\Pi_h D_h\mathbf{v}_h, d_T\Pi_h\mathbf{v}_h\big)_Q\\
&\quad= \sum_{n=1}^{N}\sum_{R=(t_{n-1},t_n)\times K}\sum_{k=0}^{p_R-1}\sum_{l=0}^{p_R-1}\big(\lambda_{n,k}, d_T\lambda_{n,l}\big)_{(t_{n-1},t_n)}\big(D_h\mathbf{v}_{n,k,h}, \mathbf{v}_{n,l,h}\big)_K\\
&\quad= \sum_{n=1}^{N}\sum_{k=0}^{p}\sum_{l=0}^{p}\big(\lambda_{n,k}, d_T\lambda_{n,l}\big)_{(t_{n-1},t_n)}\big(D_h\hat{\mathbf{v}}_{n,k,h}, \hat{\mathbf{v}}_{n,l,h}\big)_\Omega \ge 0.
\end{aligned}$$

For $k \ge 1$ we have $\big(d_T\partial_t\lambda_{n,k}, \lambda_{n,k}\big)_{(t_{n-1},t_n)} = -\big(t\partial_t\lambda_{n,k}, \lambda_{n,k}\big)_{(t_{n-1},t_n)} < 0$ by Lemma 4.2, which gives

$$\begin{aligned}
\big(d_T M_h\partial_t\mathbf{v}_h, \mathbf{v}_h - \Pi_h\mathbf{v}_h\big)_Q &= \sum_{n=1}^{N}\big(d_T M_h\partial_t\mathbf{v}_h, \mathbf{v}_h - \Pi_h\mathbf{v}_h\big)_{(t_{n-1},t_n)\times\Omega}\\
&= \sum_{n=1}^{N}\sum_{R=(t_{n-1},t_n)\times K}\sum_{k=0}^{p_R}\big(d_T\partial_t\lambda_{n,k}, \lambda_{n,p_R}\big)_{(t_{n-1},t_n)}\big(M_h\mathbf{v}_{n,k,h}, \mathbf{v}_{n,p_R,h}\big)_K\\
&= \sum_{n=1}^{N}\sum_{R=(t_{n-1},t_n)\times K}\big(d_T\partial_t\lambda_{n,p_R}, \lambda_{n,p_R}\big)_{(t_{n-1},t_n)}\big(M_h\mathbf{v}_{n,p_R,h}, \mathbf{v}_{n,p_R,h}\big)_K \le 0.
\end{aligned}$$

Thus we obtain (4.7) by

$$\begin{aligned}\big(M_h\partial_t\mathbf{v}_h, d_T\mathbf{v}_h\big)_Q &= \big(d_T M_h\partial_t\mathbf{v}_h, \mathbf{v}_h\big)_Q\\ &\le \big(d_T M_h\partial_t\mathbf{v}_h, \Pi_h\mathbf{v}_h\big)_Q = \big(M_h\partial_t\mathbf{v}_h, d_T\Pi_h\mathbf{v}_h\big)_Q\,.\end{aligned}$$

Finally, we show the assertion (4.5). We have for $k = 2, \dots, N$

$$\big\|\mathbf{v}_{k,h}(t_{k-1})\big\|_{Y_h} = \big\|\Pi_{k,h}\mathbf{v}_{k-1,h}(t_{k-1})\big\|_{Y_h} \le \big\|\mathbf{v}_{k-1,h}(t_{k-1})\big\|_{Y_h}\,,$$

so that for all $t \in (t_{n-1}, t_n)$ using $\mathbf{v}_h(0) = \mathbf{v}_{1,h}(0) = \mathbf{0}$

$$\begin{aligned}\big\|\mathbf{v}_h(t)\big\|_{Y_h}^2 &= \big\|\mathbf{v}_h(t)\big\|_{Y_h}^2 + \sum_{k=2}^{n}\Big(\|\Pi_{k,h}\mathbf{v}_{k-1,h}(t_{k-1})\|_{Y_h}^2 - \big\|\mathbf{v}_{k,h}(t_{k-1})\big\|_{Y_h}^2\Big) - \big\|\mathbf{v}_{1,h}(0)\big\|_{Y_h}^2\\ &\le \big\|\mathbf{v}_h(t)\big\|_{Y_h}^2 + \sum_{k=2}^{n}\Big(\|\mathbf{v}_{k-1,h}(t_{k-1})\|_{Y_h}^2 - \big\|\mathbf{v}_{k,h}(t_{k-1})\big\|_{Y_h}^2\Big) - \big\|\mathbf{v}_{1,h}(0)\big\|_{Y_h}^2\\ &= \big\|\mathbf{v}_h(t)\big\|_{Y_h}^2 - \big\|\mathbf{v}_{n,h}(t_{n-1})\big\|_{Y_h}^2 + \sum_{k=1}^{n-1}\Big(\big\|\mathbf{v}_{k,h}(t_k)\big\|_{Y_h}^2 - \big\|\mathbf{v}_{k,h}(t_{k-1})\big\|_{Y_h}^2\Big)\\ &= \int_{t_{n-1}}^{t}\partial_s\big(M_h\mathbf{v}_{n,h}(s), \mathbf{v}_{n,h}(s)\big)_\Omega\,\mathrm{d}s + \sum_{k=1}^{n-1}\int_{t_{k-1}}^{t_k}\partial_s\big(M_h\mathbf{v}_{n,h}(s), \mathbf{v}_{n,h}(s)\big)_\Omega\,\mathrm{d}s\\ &= 2\int_0^t\big(M_h\partial_s\mathbf{v}_h(s), \mathbf{v}_h(s)\big)_\Omega\,\mathrm{d}s\end{aligned}$$

and thus using (4.6), (4.7), (4.8), and (4.9) we obtain (4.5) by

$$\begin{aligned}\|\mathbf{v}_h\|_{W_h}^2 &= \int_0^T\big(M_h\mathbf{v}_h(t), \mathbf{v}_h(t)\big)_\Omega\mathrm{d}t \le 2\int_0^T\int_0^t\big(M_h\partial_s\mathbf{v}_h(s), \mathbf{v}_h(s)\big)_\Omega\mathrm{d}s\mathrm{d}t\\ &= 2\int_0^T d_T(t)\big(M_h\partial_t\mathbf{v}_h(t), \mathbf{v}_h(t)\big)_\Omega\mathrm{d}t = 2\big(M_h\partial_t\mathbf{v}_h, d_T\mathbf{v}_h\big)_Q\\ &\le 2\big(M_h\partial_t\mathbf{v}_h, d_T\Pi_h\mathbf{v}_h\big)_Q = 2\big(M_h\Pi_h\partial_t\mathbf{v}_h, d_T\Pi_h\mathbf{v}_h\big)_Q\\ &= 2\big(\Pi_h M_h\partial_t\mathbf{v}_h, d_T\Pi_h\mathbf{v}_h\big)_Q \le 2\big(\Pi_h L_h\mathbf{v}_h, d_T\Pi_h\mathbf{v}_h\big)_Q\\ &= 2\big(M_h^{-1}\Pi_h L_h\mathbf{v}_h, M_h d_T\Pi_h\mathbf{v}_h\big)_Q = 2\big(\Pi_h M_h^{-1}L_h\mathbf{v}_h, M_h d_T\Pi_h\mathbf{v}_h\big)_Q\\ &\le 2\,\|\Pi_h M_h^{-1}L_h\mathbf{v}_h\|_{W_h}\|d_T\Pi_h\mathbf{v}_h\|_{W_h} \le 2T\,\|\Pi_h M_h^{-1}L_h\mathbf{v}_h\|_{W_h}\|\mathbf{v}_h\|_{W_h}\,.\end{aligned}$$

□

Now we can prove Theorem 4.1.

Proof (Theorem 4.1) For $\mathbf{v}_h \in V_h \setminus \{\mathbf{0}\}$ we have

$$b_h(\mathbf{v}_h, \mathbf{w}_h) = \big(L_h \mathbf{v}_h, \mathbf{w}_h\big)_Q = \big(M_h^{-1} L_h \mathbf{v}_h, \mathbf{w}_h\big)_{W_h} = \big(\Pi_h M_h^{-1} L_h \mathbf{v}_h, \mathbf{w}_h\big)_{W_h},$$

and we test with $\mathbf{w}_h = \Pi_h M_h^{-1} L_h \mathbf{v}_h$, so that

$$\begin{aligned}\sup_{\mathbf{w}_h \in W_h \setminus \{\mathbf{0}\}} \frac{b_h(\mathbf{v}_h, \mathbf{w}_h)}{\|\mathbf{w}_h\|_{W_h}} \geq \frac{b_h(\mathbf{v}_h, \Pi_h M_h^{-1} L_h \mathbf{v}_h)}{\|\Pi_h M_h^{-1} L_h \mathbf{v}_h\|_{W_h}} &= \|\Pi_h M_h^{-1} L_h \mathbf{v}_h\|_{W_h} \\ &\geq (4T^2 + 1)^{-1/2} \|\mathbf{v}_h\|_{V_h}.\end{aligned}$$

using $\|\mathbf{v}_h\|_{V_h}^2 = \|\mathbf{v}_h\|_{W_h}^2 + \|\Pi_h M_h^{-1} L_h \mathbf{v}_h\|_{W_h}^2 \leq (4T^2 + 1)\|\Pi_h M_h^{-1} L_h \mathbf{v}_h\|_{W_h}^2$ inserting the estimate (4.5) in Lemma 4.4. □

4.4 Convergence for Strong Solutions

For the error estimate with respect to the norm in V_h we need to extend the norm $\|\cdot\|_{V_h}$ such that the error can be evaluated in this norm. For sufficiently smooth functions the operator A_h can be extended by (3.8), so that L_h and thus the norm in V_h is well-defined.

Theorem 4.5 *Let $\mathbf{u} \in V$ be the strong solution of $L\mathbf{u} = \mathbf{f}$, and let $\mathbf{u}_h \in V_h$ be the approximation solving* (4.4).

If the solution is sufficiently smooth, we obtain the a priori error estimate

$$\begin{aligned}\|\mathbf{u} - \mathbf{u}_h\|_{V_h} \leq C\big(\Delta t^p + \Delta x^q\big)\Big(\|\partial_t^{p+1}\mathbf{u}\|_Q + \|\mathrm{D}^{q+1}\mathbf{u}\|_Q\Big) \\ + \beta^{-1}\big\|M_h^{-1/2}(M_h - M)M^{-1/2}\big\|_\infty \|\partial_t \mathbf{u}\|_W \\ + \beta^{-1}\big\|M_h^{-1/2}(D_h - D)M^{-1/2}\big\|_\infty \|\mathbf{u}\|_W\end{aligned}$$

for Δt, Δx and $p, q \geq 1$ with $\Delta t \geq t_n - t_{n-1}$, $\Delta x \geq \mathrm{diam}(K)$, $p \leq p_R$ and $q \leq q_R$ (for all n, K, R), and with a constant $C > 0$ depending on $\beta = (4T^2+1)^{-1/2}$, on the material parameters in M, and on the mesh regularity.

Proof For the solution we assume the regularity $\mathbf{u} \in \mathrm{H}^{p+1}\big(0, T; \mathrm{L}_2(\Omega; \mathbb{R}^m)\big) \cap \mathrm{L}_2\big(0, T; \mathrm{H}^{q+1}(\Omega; \mathbb{R}^m)\big)$, hence there exists an interpolant $\mathbf{v}_h \in V_h$ such that

$$\|\mathbf{u} - \mathbf{v}_h\|_{V_h} \leq C\big(\Delta t^p + \Delta x^q\big)\Big(\|\partial_t^{p+1}\mathbf{u}\|_Q + \|\mathrm{D}^{q+1}\mathbf{u}\|_Q\Big). \tag{4.10}$$

Moreover, $A_h\mathbf{u}$ is well-defined and consistent satisfying (3.9). We have

$$\begin{aligned}
b_h(\mathbf{v}_h-\mathbf{u}_h,\mathbf{w}_h) &= b_h(\mathbf{v}_h,\mathbf{w}_h)-b_h(\mathbf{u}_h,\mathbf{w}_h) = b_h(\mathbf{v}_h,\mathbf{w}_h)-(\mathbf{f},\mathbf{w}_h)_Q\\
&= b_h(\mathbf{v}_h,\mathbf{w}_h)-b(\mathbf{u},\mathbf{w}_h)\\
&= (L_h\mathbf{v}_h,\mathbf{w}_h)_Q-(L\mathbf{u},\mathbf{w}_h)_Q\\
&= \big(L_h(\mathbf{v}_h-\mathbf{u}),\mathbf{w}_h\big)_Q-\big(L\mathbf{u},\mathbf{w}_h\big)_Q+\big(L_h\mathbf{u},\mathbf{w}_h\big)_Q\\
&= \big(L_h(\mathbf{v}_h-\mathbf{u}),\mathbf{w}_h\big)_Q-\big((M-M_h)\partial_t\mathbf{u},\mathbf{w}_h\big)_Q\\
&\qquad -\big((D-D_h)\mathbf{u},\mathbf{w}_h\big)_Q-\big((A-A_h)\mathbf{u},\mathbf{w}_h\big)_Q\\
&= \big(M_h\Pi_hM_h^{-1}L_h(\mathbf{v}_h-\mathbf{u}),\mathbf{w}_h\big)_Q\\
&\qquad -\big(M_hM_h^{-1}(M-M_h)\partial_t\mathbf{u},\mathbf{w}_h\big)_Q-\big(M_hM_h^{-1}(D-D_h)\mathbf{u},\mathbf{w}_h\big)_Q\\
&\le \Big(\big\|\Pi_hM_h^{-1}L_h(\mathbf{v}_h-\mathbf{u})\big\|_{W_h}+\big\|M_h^{-1}(M-M_h)\partial_t\mathbf{u}\big\|_{W_h}\\
&\qquad\qquad +\big\|M_h^{-1}(D-D_h)\mathbf{u}\big\|_{W_h}\Big)\|\mathbf{w}_h\|_{W_h}
\end{aligned}$$

and thus the assertion follows from

$$\begin{aligned}
\|\mathbf{u}-\mathbf{u}_h\|_{V_h} &\le \|\mathbf{u}-\mathbf{v}_h\|_{V_h}+\|\mathbf{v}_h-\mathbf{u}_h\|_{V_h}\\
&\le \|\mathbf{u}-\mathbf{v}_h\|_{V_h}+\beta^{-1}\sup_{\mathbf{w}_h\in W_h\setminus\{\mathbf{0}\}}\frac{b_h(\mathbf{v}_h-\mathbf{u}_h,\mathbf{w}_h)}{\|\mathbf{w}_h\|_{W_h}}\\
&\le \|\mathbf{u}-\mathbf{v}_h\|_{V_h}\\
&\qquad +\beta^{-1}\Big(\big\|\Pi_hM_h^{-1}L_h(\mathbf{v}_h-\mathbf{u})\big\|_{W_h}+\big\|M_h^{-1}(M-M_h)\partial_t\mathbf{u}\big\|_{W_h}\\
&\qquad\qquad +\big\|M_h^{-1}(D-D_h)\mathbf{u}\big\|_{W_h}\Big)\\
&\le (1+\beta^{-1})\|\mathbf{u}-\mathbf{v}_h\|_{V_h}+\beta^{-1}\big\|M_h^{-1/2}(M-M_h)\partial_t\mathbf{u}\big\|_Q\\
&\qquad +\beta^{-1}\big\|M_h^{-1/2}(D-D_h)\mathbf{u}\big\|_Q
\end{aligned}$$

by the interpolation estimate (4.10) and

$$\begin{aligned}
\big\|M_h^{-1/2}(M-M_h)\partial_t\mathbf{u}\big\|_Q &= \big\|M_h^{-1/2}(M-M_h)M^{-1/2}M^{1/2}\partial_t\mathbf{u}\big\|_Q\\
&\le \big\|M_h^{-1/2}(M-M_h)M^{-1/2}\big\|_\infty\|\partial_t\mathbf{u}\|_W\,.
\end{aligned}$$

□

Remark 4.6 The estimate is derived for homogeneous initial and boundary conditions. It transfers to the inhomogeneous case if initial and boundary data $\mathbf{u}_0$ and g_k can be extended to $\mathrm{H}(L, Q)$, i.e., if $\hat{\mathbf{u}} \in \mathrm{H}(L, Q)$ exists such that $\hat{\mathbf{u}}(0, x) = \mathbf{u}_0(x)$ for $x \in \Omega$ and $(\underline{A}_{\mathbf{n}}\hat{\mathbf{u}}(t, x))_k = g_k(t, x)$ and for $(t, x) \in (0, T) \times \Gamma_k$, $k = 1, \ldots, m$. Then, the approximation $\mathbf{u}_h \in V_h(\mathbf{u}_0)$ in the affine space (4.2) is computed by $b_h(\mathbf{u}_h, \mathbf{w}_h) = \langle \ell_h, \mathbf{w}_h \rangle$ for $\mathbf{w}_h \in W_h$, and the strong solution with inhomogeneous initial and boundary data is given by $\mathbf{u} = \tilde{\mathbf{u}} + \hat{\mathbf{u}} \in \mathrm{H}(L, Q)$, where $\tilde{\mathbf{u}} \in V$ solves $L\tilde{\mathbf{u}} = \mathbf{f} - L\hat{\mathbf{u}}$. Then, the result in Theorem 4.5 can be extended.

Remark 4.7 Since the norm (4.3) in $V_h + (\mathrm{H}^1(Q; \mathbb{R}^m) \cap V)$ is discrete in the derivatives, the topology in the space V with respect to this norm is equivalent to the topology in L_2 with mesh dependent bounds for the norm equivalence. Norm equivalence with respect to $\|\cdot\|_V$ is obtained in the limit: Let $(V_h)_{h\in\mathcal{H}}$ be a shape regular family of discrete spaces with $0 \in \overline{\mathcal{H}}$ such that $(V_h \cap V)_{h\in\mathcal{H}}$ is dense in V. Then, defining $\|\mathbf{v}\|_{V_\mathcal{H}} = \sup_{h\in\mathcal{H}} \|\mathbf{v}\|_{V_h}$ yields a norm, and for sufficiently smooth functions $\mathbf{v} \in \mathrm{H}^1(Q; \mathbb{R}^m) \cap V$ this norm is equivalent to $\|\cdot\|_V$.

4.5 Convergence for Weak Solutions

Qualitative convergence estimates with respect to the norm in $V \subset \mathrm{H}(L, Q)$ require additional regularity, so that these estimates do not apply to weak solutions with discontinuities or singularities. For weak solutions without additional regularity we only can derive asymptotic convergence. Here, this is shown for simplicity only for homogeneous boundary data. The given data are the right-hand side $\mathbf{f} \in \mathrm{L}_2(Q; \mathbb{R}^m)$ and the initial value $\mathbf{u}_0 \in Z$. We assume that Lemma 2.12 and dual consistency for A_h in Lemma 3.3 is satisfied.

In the first step, we show that the inf-sup stability of the Petrov–Galerkin method yields a uniform a priori bound for the approximation. We define the approximation of the initial value by $\Pi_{1,h}\mathbf{u}_0$; this is extended to the space-time cylinder by defining $\mathbf{u}_{0,h}(t) = (1 - t/T)\Pi_{n,h}\mathbf{u}_0$ for $t \in (t_{n-1}, t_n]$, $n = 1, \ldots, N$, so that $V_h(\mathbf{u}_0) = \mathbf{u}_{0,h} + V_h$.

Lemma 4.8 *The discrete solution $\mathbf{u}_h \in V_h(\mathbf{u}_0)$ of the variational space-time equation*

$$b_h(\mathbf{u}_h, \mathbf{w}_h) = \big(\mathbf{f}, \mathbf{w}_h\big)_Q, \qquad \mathbf{w}_h \in W_h$$

is bounded by

$$\|\mathbf{u}_h\|_{W_h} \le 2T\, \|M_h^{-1}\Pi_h\mathbf{f}\|_{W_h} + (1 + 2T)\, \|\mathbf{u}_{0,h}\|_{V_h} .$$

Proof For $\mathbf{v}_h = \mathbf{u}_h - \mathbf{u}_{0,h} \in V_h$ the estimate $\|\mathbf{v}_h\|_{W_h} \le 2T\|\Pi_h M_h^{-1} L_h \mathbf{v}_h\|_{W_h}$ in Lemma 4.4 together with

$$\begin{aligned}\big(\Pi_h M_h^{-1} L_h \mathbf{v}_h, \mathbf{w}_h\big)_{W_h} &= \big(M_h^{-1} L_h \mathbf{v}_h, \mathbf{w}_h\big)_{W_h} = \big(L_h \mathbf{v}_h, \mathbf{w}_h\big)_Q \\ &= b_h(\mathbf{u}_h, \mathbf{w}_h) - b_h(\mathbf{u}_{0,h}, \mathbf{w}_h) \\ &= \big(\mathbf{f}, \mathbf{w}_h\big)_Q - b_h(\mathbf{u}_{0,h}, \mathbf{w}_h) = \big(\Pi_h \mathbf{f}, \mathbf{w}_h\big)_Q - b_h(\mathbf{u}_{0,h}, \mathbf{w}_h)\end{aligned}$$

for $\mathbf{w}_h \in W_h$ yields by duality

$$\begin{aligned}\|\mathbf{v}_h\|_{W_h} &\le 2T\|\Pi_h M_h^{-1} L_h \mathbf{v}_h\|_{W_h} = 2T \sup_{\mathbf{w}_h \in W_h \setminus \{\mathbf{0}\}} \frac{\big(\Pi_h M_h^{-1} L_h \mathbf{v}_h, \mathbf{w}_h\big)_{W_h}}{\|\mathbf{w}_h\|_{W_h}} \\ &= 2T \sup_{\mathbf{w}_h \in W_h \setminus \{\mathbf{0}\}} \frac{\big(\Pi_h \mathbf{f}, \mathbf{w}_h\big)_Q - b_h(\mathbf{u}_{0,h}, \mathbf{w}_h)}{\|\mathbf{w}_h\|_{W_h}} \\ &\le 2T \Big(\|M_h^{-1} \Pi_h \mathbf{f}\|_{W_h} + \|\Pi_h M_h^{-1} L_h \mathbf{u}_{0,h}\|_{W_h}\Big),\end{aligned}$$

so that

$$\begin{aligned}\|\mathbf{u}_h\|_{W_h} &\le \|\mathbf{v}_h\|_{W_h} + \|\mathbf{u}_{0,h}\|_{W_h} \\ &\le 2T \Big(\|M_h^{-1} \Pi_h \mathbf{f}\|_{W_h} + \|\Pi_h M_h^{-1} L_h \mathbf{u}_{0,h}\|_{W_h}\Big) + \|\mathbf{u}_{0,h}\|_{W_h} \\ &\le 2T\ \|M_h^{-1} \Pi_h \mathbf{f}\|_{W_h} + (1 + 2T)\ \|\mathbf{u}_{0,h}\|_{V_h}\,.\end{aligned}$$

□

Next we show that the dual consistency of the DG operator implies dual consistency of the space-time method. For simplicity, we assume that the parameters in M and D are piecewise constant on all cells $K \in \mathcal{K}_h$, so that $M = M_h$ and $D = D_h$, which implies $\|\mathbf{z}_h\|_{W_h} = \|\mathbf{z}_h\|_W$ for $\mathbf{z}_h \in \mathrm{L}_2(0, T; Y_h)$.

Lemma 4.9 *We have*

$$b_h(\mathbf{v}_h, \mathbf{w}) = \big(\mathbf{v}_h, L^* \mathbf{w}\big)_Q - \big(M \Pi_{1,h} \mathbf{u}_0, \mathbf{w}\big)_\Omega\,, \quad \mathbf{v}_h \in V_h(\mathbf{u}_0)\,, \ \mathbf{w} \in W_h \cap V^*\,. \tag{4.11}$$

Proof We obtain for $\mathbf{v}_h \in V_h(\mathbf{u}_0) \subset \mathrm{H}^1(0, T; Y_h)$ and $\mathbf{w} \in W_h \cap V^*$

$$\begin{aligned}b_h(\mathbf{v}_h, \mathbf{w}) &= \big(M \partial_t \mathbf{v}_h, \mathbf{w}\big)_Q + \big(D \mathbf{v}_h, \mathbf{w}\big)_Q + \big(A_h \mathbf{v}_h, \mathbf{w}\big)_{Q_h} \\ &= \big(M \mathbf{v}_h(T), \mathbf{w}(T)\big)_\Omega - \big(M \mathbf{v}_h(0), \mathbf{w}(0)\big)_\Omega\end{aligned}$$

$$
\begin{aligned}
&- \big(\mathbf{v}_h, M\partial_t \mathbf{w}\big)_Q + \big(\mathbf{v}_h, D\mathbf{w}\big)_Q - \big(\mathbf{v}_h, A\mathbf{w}\big)_Q \\
&= \big(\mathbf{v}_h, L^*\mathbf{w}\big)_Q - \big(M\Pi_{1,h}\mathbf{u}_0, \mathbf{w}(0)\big)_\Omega
\end{aligned}
$$

using $M = M_h$, $D = D_h$, integration by parts for $\mathbf{v}_h, \mathbf{w} \in \mathrm{H}^1(0, T; Y_h)$, $\mathbf{w}(T) = \mathbf{0}$ in V^*, and the dual consistency of the DG operator A_h with upwind flux (see Lemmas 3.2 and 3.3 for acoustics). □

Let (V_h, W_h), $h \in \mathcal{H} \subset (0, h_0)$, be a dense family of nested discretizations with $V_h \subset V_{h'}$ and $W_h \subset W_{h'}$ for $h' < h$, $h, h' \in \mathcal{H}$ and $0 \in \overline{\mathcal{H}}$. We assume that the assumptions in this section are fulfilled for all discretizations, so that (V_h, W_h) is uniformly inf-sup stable by Theorem 4.1. We only consider the case $\mathbb{P}_1(Q_h; \mathbb{R}^m) \subset W_h$, so that $W_h \cap \mathrm{H}^1(Q; \mathbb{R}^m)$ includes the continuous linear elements and thus $\bigcup_{h\in\mathcal{H}}(V^* \cap W_h)$ is dense in V^*.

Theorem 4.10 *Assume $M = M_h$, $D = D_h$, and that $\|\mathbf{u}_{0,h}\|_{V_h} \le C$ is uniformly bounded for all $h \in \mathcal{H}$. Then, the discrete solutions $(\mathbf{u}_h)_{h\in\mathcal{H}}$ are weakly converging to the weak solution $\mathbf{u} \in W$ of the equation*

$$
(\mathbf{u}, L^*\mathbf{z})_Q = \big(\mathbf{f}, \mathbf{z}\big)_Q + \big(M\mathbf{u}_0, \mathbf{z}(0)\big)_\Omega , \qquad \mathbf{z} \in \mathcal{V}^* . \tag{4.12}
$$

Proof By Theorem 4.1 $\mathbf{u}_h - \mathbf{u}_{0,h}$ is uniformly bounded in V_h and thus, by Lemma 4.4, $(\mathbf{u}_h)_{h\in\mathcal{H}}$ is uniformly bounded in W, so that a subsequence $\mathcal{H}_0 \subset \mathcal{H}$ and a weak limit $\mathbf{u} \in W$ exists, i.e.,

$$
\lim_{h\in\mathcal{H}_0} \big(\mathbf{u}_h, \mathbf{w}\big)_W = \big(\mathbf{u}, \mathbf{w}\big)_W , \qquad \mathbf{w} \in W .
$$

The assumption that $(W_h \cap V^*)_{h\in\mathcal{H}}$ is dense in $V^* \supset \mathcal{V}^*$ implies that for all $\mathbf{z} \in \mathcal{V}^*$ there exists a sequence $(\mathbf{w}_h)_{h\in\mathcal{H}}$ with $\mathbf{w}_h \in W_h \cap V^*$ and $\lim_{h\in\mathcal{H}} \|\mathbf{w}_h - \mathbf{z}\|_{V^*} = 0$. This implies $\lim_{h\in\mathcal{H}} \|\mathbf{w}_h(0) - \mathbf{z}(0)\|_Y = 0$, cf. Remark 2.13. Using the weak convergence of $\mathbf{u}_h$, the strong convergence of $L^*\mathbf{w}_h$, and Lemma 4.9 yields

$$
\begin{aligned}
\big(\mathbf{u}, L^*\mathbf{z}\big)_Q &= \lim_{h\in\mathcal{H}_0} \big(\mathbf{u}_h, L^*\mathbf{z}\big)_Q = \lim_{h\in\mathcal{H}_0} \big(\mathbf{u}_h, L^*\mathbf{w}_h\big)_Q \\
&= \lim_{h\in\mathcal{H}_0} \Big(b_h(\mathbf{u}_h, \mathbf{w}_h) + \big(M\Pi_{1,h}\mathbf{u}_0, \mathbf{w}_h(0)\big)_\Omega\Big) \\
&= \lim_{h\in\mathcal{H}_0} \big(\mathbf{f}, \mathbf{w}_h\big)_Q + \big(M\mathbf{u}_0, \mathbf{z}(0)\big)_\Omega = \big(\mathbf{f}, \mathbf{z}\big)_Q + \big(M\mathbf{u}_0, \mathbf{z}(0)\big)_\Omega ,
\end{aligned}
$$

so that $\mathbf{u}$ is a weak solution of (4.12). Since the weak solution is unique by Theorem 2.8 and Lemma 2.12, this shows that the weak limit of all subsequences in $\mathcal{H}$ is the unique weak solution, so that the full sequence is convergent. □

Remark 4.11 Since we assume that $(\mathbf{u}_{0,h})_{h\in\mathcal{H}}$ is uniformly bounded in V_h, the initial value $\mathbf{u}_0$ extends to $\mathrm{H}(L, Q)$, and the weak solution is a strong solution.

4.6 Goal-Oriented Adaptivity

In order to find an efficient choice for the polynomial degrees (p_R, q_R), we introduce a dual-weighted residual error indicator with respect to a suitable goal functional. Its construction is based on a dual-primal error representation combined with a priori estimates constructed from an approximation of the dual solution. Note that this corresponds to a problem backward in time, so that the resulting error indicator only refines regions of the space-time domain which are relevant for the evaluation of the chosen goal functional.

Dual-Primal Error Bound

Let $E\colon W \longrightarrow \mathbb{R}$ be a linear error functional. Our goal is to estimate and then to reduce the error with respect to this functional. The dual solution $\mathbf{u}^* \in V^*$ is defined by

$$(\mathbf{w}, L^*\mathbf{u}^*)_Q = \langle E, \mathbf{w}\rangle\,, \qquad \mathbf{w} \in W\,.$$

For the local representation of E we define the pairing in $R \in \mathcal{R}_h$

$$\langle \mathbf{v}_R, \mathbf{w}_R\rangle_{\partial R} = (L\mathbf{v}_R, \mathbf{w}_R)_R - (\mathbf{v}_R, L^*\mathbf{w}_R)_R\,, \qquad \mathbf{v}_R \in \mathrm{H}(L, R)\,,\ \mathbf{w}_R \in \mathrm{H}(L^*, R)\,.$$

Lemma 4.12 *Let $\mathbf{u} \in V$ be the solution of $L\mathbf{u} = \mathbf{f}$, and let $\mathbf{u}_h \in V_h$ be the approximation solving* (4.4)*. Then, the error can be represented by*

$$\langle E, \mathbf{u} - \mathbf{u}_h\rangle = \sum_{R\in\mathcal{R}_h} \Big(\big(\mathbf{f} - L\mathbf{u}_h, \mathbf{u}^*\big)_R + \langle \mathbf{u}_h, \mathbf{u}^*\rangle_{\partial R}\Big)\,.$$

If the dual solution is sufficiently regular, the error is bounded for all $\mathbf{w}_h \in W_h$ by

$$\begin{aligned}
\big|\langle E, \mathbf{u} - \mathbf{u}_h\rangle\big| \le \sum_{n=1}^{N} \sum_{R=(t_{n-1},t_n)\times K} \Bigg(&\big\|\mathbf{f} - (M_h\partial_t + A + D_h)\mathbf{u}_h\big\|_R \big\|\mathbf{u}^* - \mathbf{w}_h\big\|_R \\
&+ \sum_{F\in\mathcal{F}_K} \big\|\underline{A}_{\mathbf{n}_K}\mathbf{u}_{h,R} - \underline{A}_{\mathbf{n}_K}^{\mathrm{upw}}\mathbf{u}_{n,h}\big\|_{(t_{n-1},t_n)\times F} \big\|\mathbf{u}^* - \mathbf{w}_h\big\|_{(t_{n-1},t_n)\times F}\Bigg) \\
+ \sum_{n=1}^{N-1} \big\|M_h\big(\mathbf{u}_{n,h}(t_n) - \Pi_{n+1,h}\mathbf{u}_{n,h}(t_n)\big)\big\|_\Omega &\big\|\mathbf{u}^*(t_n) - \mathbf{w}_{n+1,h}(t_n)\big\|_\Omega
\end{aligned}$$

$$+ \|M_h^{-1/2}(M - M_h)M^{-1/2}\|_\infty \sum_{n=1}^{N-1} \|\mathbf{u}_{n,h}(t_n) - \Pi_{n+1,h}\mathbf{u}_{n,h}(t_n)\|_Y \|\mathbf{u}^*(t_n)\|_Y$$

$$+ \|M_h^{-1/2}(M - M_h)M^{-1/2}\|_\infty \|\partial_t \mathbf{u}_h\|_W \|\mathbf{u}^*\|_W$$

$$+ \|M_h^{-1/2}(D - D_h)M^{-1/2}\|_\infty \|\mathbf{u}_h\|_W \|\mathbf{u}^*\|_W .$$

Proof We have by definition of $\mathbf{u}^*$

$$\begin{aligned}
\langle E, \mathbf{u} - \mathbf{u}_h\rangle &= (\mathbf{u} - \mathbf{u}_h, L^*\mathbf{u}^*)_Q = (\mathbf{u}, L^*\mathbf{u}^*)_Q - (\mathbf{u}_h, L^*\mathbf{u}^*)_Q \\
&= (L\mathbf{u}, \mathbf{u}^*)_Q - (\mathbf{u}_h, L^*\mathbf{u}^*)_Q = (\mathbf{f}, \mathbf{u}^*)_Q - \sum_{R\in\mathcal{R}_h} (\mathbf{u}_h, L^*\mathbf{u}^*)_R \\
&= (\mathbf{f}, \mathbf{u}^*)_Q - \sum_{R\in\mathcal{R}_h} \Big((L\mathbf{u}_h, \mathbf{u}^*)_R - \langle \mathbf{u}_h, \mathbf{u}^*\rangle_{\partial R}\Big) \\
&= \sum_{R\in\mathcal{R}_h} \Big(\big(\mathbf{f} - L\mathbf{u}_h, \mathbf{u}^*\big)_R + \langle \mathbf{u}_h, \mathbf{u}^*\rangle_{\partial R}\Big) .
\end{aligned}$$

Using $\mathbf{u}_h(0) = \mathbf{0}$ and $\mathbf{u}^*(T) = \mathbf{0}$, we obtain

$$\begin{aligned}
\sum_{n=1}^{N} \sum_{R=(t_{n-1},t_n)\times K} &(M\partial_t \mathbf{u}_{n,h}, \mathbf{u}^*)_R + (M\mathbf{u}_{n,h}, \partial_t \mathbf{u}^*)_R = \sum_{n=1}^{N} \int_{t_{n-1}}^{t_n} \partial_t (M\mathbf{u}_{n,h}, \mathbf{u}^*)_\Omega \,\mathrm{d}t \\
&= \sum_{n=1}^{N} \big(M\mathbf{u}_{n,h}(t_n), \mathbf{u}^*(t_n)\big)_\Omega - \big(M\mathbf{u}_{n,h}(t_{n-1}), \mathbf{u}^*(t_{n-1})\big)_\Omega \\
&= \sum_{n=1}^{N-1} \Big(M\big(\mathbf{u}_{n,h}(t_n) - \mathbf{u}_{n+1,h}(t_n)\big), \mathbf{u}^*(t_n)\Big)_\Omega \\
&= \sum_{n=1}^{N-1} \Big(M\big(\mathbf{u}_{n,h}(t_n) - \Pi_{n+1,h}\mathbf{u}_{n,h}(t_n)\big), \mathbf{u}^*(t_n)\Big)_\Omega
\end{aligned}$$

and in every time slice (t_{n-1}, t_n) we obtain, if the dual solution $\mathbf{u}^*$ is sufficiently smooth satisfying $\mathbf{u}^*|_{\partial Q_h} \in \mathrm{L}_2(\partial Q_h; \mathbb{R}^m)$, for the restriction to the space-time skeleton

$$\begin{aligned}
\sum_{K\in\mathcal{K}_h} &\big(A\mathbf{u}_{n,h}, \mathbf{u}^*\big)_{(t_{n-1},t_n)\times K} + \big(\mathbf{u}_{n,h}, A\mathbf{u}^*\big)_{(t_{n-1},t_n)\times K} \\
&= \sum_{K\in\mathcal{K}_h} \big(\underline{A}_{\mathbf{n}_K} \mathbf{u}_{n,h.K}, \mathbf{u}^*\big)_{(t_{n-1},t_n)\times\partial K}
\end{aligned}$$

$$= \sum_{K\in\mathcal{K}_h}\sum_{F\in\mathcal{F}_K}\big(\underline{A}_{\mathbf{n}_K}\mathbf{u}_{n,h.K},\mathbf{u}^*\big)_{(t_{n-1},t_n)\times F}$$
$$= \sum_{K\in\mathcal{K}_h}\sum_{F\in\mathcal{F}_K}\big(\underline{A}_{\mathbf{n}_K}\mathbf{u}_{n,h,K}-\underline{A}^{\mathrm{upw}}_{\mathbf{n}_K}\mathbf{u}_{n,h},\mathbf{u}^*\big)_{(t_{n-1},t_n)\times F},$$

where $\mathbf{u}_{n,h,K}$ is the extension of $\mathbf{u}_{n,h}|_K$ to $\overline{K}$. This gives, inserting (3.5),

$$\sum_{R\in\mathcal{R}_h}\langle\mathbf{u}_h,\mathbf{u}^*\rangle_{\partial R}=\sum_{n=1}^{N}\sum_{R=(t_{n-1},t_n)\times K}\big(L\mathbf{u}_{n,h},\mathbf{u}^*\big)_R-\big(\mathbf{u}_{n,h},L^*\mathbf{u}^*\big)_R$$
$$=\sum_{n=1}^{N}\sum_{R=(t_{n-1},t_n)\times K}\big(M\partial_t\mathbf{u}_{n,h},\mathbf{u}^*\big)_R+\big(M\mathbf{u}_{n,h},\partial_t\mathbf{u}^*\big)_R+\big(A\mathbf{u}_{n,h},\mathbf{u}^*\big)_R+\big(\mathbf{u}_{n,h},A\mathbf{u}^*\big)_R$$
$$=\sum_{n=1}^{N-1}\big(M\big(\mathbf{u}_{n,h}(t_n)-\Pi_{n+1,h}\mathbf{u}_{n,h}(t_n)\big),\mathbf{u}^*(t_n)\big)_\Omega$$
$$+\sum_{R=(t_{n-1},t_n)\times K}\sum_{F\in\mathcal{F}_K}\big(\underline{A}_{\mathbf{n}_K}\mathbf{u}_{h,R}-\underline{A}^{\mathrm{upw}}_{\mathbf{n}_K}\mathbf{u}_{n,h},\mathbf{u}^*\big)_{(t_{n-1},t_n)\times F},$$

where $\mathbf{u}_{h,R}$ is the extension of $\mathbf{u}_h|_R$ to $\overline{R}$.

For the discrete solution $\mathbf{u}_h\in V_h$ and any discrete test function $\mathbf{w}_h\in W_h$ we have

$$\big(\mathbf{f},\mathbf{w}_h\big)_Q=\big(L_h\mathbf{u}_h,\mathbf{w}_h\big)_Q=\big(M_h\partial_t\mathbf{u}_h,\mathbf{w}_h\big)_Q+\big(A_h\mathbf{u}_h,\mathbf{w}_h\big)_Q+\big(D_h\mathbf{u}_h,\mathbf{w}_h\big)_Q$$
$$=\big(M_h\partial_t\mathbf{u}_h,\mathbf{w}_h\big)_Q+\big(D_h\mathbf{u}_h,\mathbf{w}_h\big)_Q$$
$$+\sum_{n=1}^{N}\sum_{R=(t_{n-1},t_n)\times K}\Big(\big(A\mathbf{u}_h,\mathbf{w}_h\big)_R+\sum_{F\in\mathcal{F}_K}\big(\underline{A}^{\mathrm{upw}}_{\mathbf{n}_K}\mathbf{u}_{n,h}-\underline{A}_{\mathbf{n}_K}\mathbf{u}_{h,R},\mathbf{w}_{h,R}\big)_{(t_{n-1},t_n)\times F}\Big),$$

so that

$$0=\sum_{n=1}^{N}\sum_{R=(t_{n-1},t_n)\times K}\Big(\big(A\mathbf{u}_h-\mathbf{f},\mathbf{w}_h\big)_R+\big(M_h\partial_t\mathbf{u}_h,\mathbf{w}_h\big)_R+\big(D_h\mathbf{u}_h,\mathbf{w}_h\big)_R$$
$$+\sum_{F\in\mathcal{F}_K}\big(\underline{A}^{\mathrm{upw}}_{\mathbf{n}_K}\mathbf{u}_{n,h}-\underline{A}_{\mathbf{n}_K}\mathbf{u}_{h,R},\mathbf{w}_{h,R}\big)_{(t_{n-1},t_n)\times F}\Big).$$

Together, this gives

$$
\begin{aligned}
\langle E, \mathbf{u} - \mathbf{u}_h \rangle &= \sum_{R \in \mathcal{R}_h} \Big(\big(\mathbf{f} - L\mathbf{u}_h, \mathbf{u}^*\big)_R + \langle \mathbf{u}_h, \mathbf{u}^* \rangle_{\partial R} \Big) \\
&= \sum_{n=1}^{N} \sum_{R=(t_{n-1},t_n)\times K} \Bigg(\big(\mathbf{f} - (M_h \partial_t + A + D_h)\mathbf{u}_h, \mathbf{u}^*\big)_R \\
&\qquad\qquad - \big((M - M_h)\partial_t \mathbf{u}_h, \mathbf{u}^*\big)_R - \big((D - D_h)\mathbf{u}_h, \mathbf{u}^*\big)_R \\
&\qquad\qquad + \sum_{F \in \mathcal{F}_K} \big(\underline{A}_{\mathbf{n}_K} \mathbf{u}_{h,R} - \underline{A}_{\mathbf{n}_K}^{\mathrm{upw}} \mathbf{u}_{n,h}, \mathbf{u}^*\big)_{(t_{n-1},t_n)\times F} \Bigg) \\
&\quad + \sum_{n=1}^{N-1} \Big(M\big(\mathbf{u}_{n,h}(t_n) - \Pi_{n+1,h}\mathbf{u}_{n,h}(t_n)\big), \mathbf{u}^*(t_n) \Big)_\Omega \\
&= \sum_{n=1}^{N} \sum_{R=(t_{n-1},t_n)\times K} \Bigg(\big(\mathbf{f} - (M_h \partial_t + A + D_h)\mathbf{u}_h, \mathbf{u}^* - \mathbf{w}_h\big)_R \\
&\qquad\qquad + \sum_{F \in \mathcal{F}_K} \big(\underline{A}_{\mathbf{n}_K} \mathbf{u}_{h,R} - \underline{A}_{\mathbf{n}_K}^{\mathrm{upw}} \mathbf{u}_{n,h}, \mathbf{u}^* - \mathbf{w}_h\big)_{(t_{n-1},t_n)\times F} \Bigg) \\
&\quad + \sum_{n=1}^{N-1} \Big(M_h\big(\mathbf{u}_{n,h}(t_n) - \Pi_{n+1,h}\mathbf{u}_{n,h}(t_n)\big), \mathbf{u}^*(t_n) - \mathbf{w}_{n+1,h}(t_n) \Big)_\Omega \\
&\quad + \sum_{n=1}^{N-1} \Big((M - M_h)\big(\mathbf{u}_{n,h}(t_n) - \Pi_{n+1,h}\mathbf{u}_{n,h}(t_n)\big), \mathbf{u}^*(t_n) \Big)_\Omega \\
&\quad - \big((M - M_h)\partial_t \mathbf{u}_h, \mathbf{u}^*\big)_Q - \big((D - D_h)\mathbf{u}_h, \mathbf{u}^*\big)_Q \\
&\le \sum_{n=1}^{N} \sum_{R=(t_{n-1},t_n)\times K} \Bigg(\big\|\mathbf{f} - (M_h \partial_t + A + D_h)\mathbf{u}_h\big\|_R \big\|\mathbf{u}^* - \mathbf{w}_h\big\|_R \\
&\qquad + \sum_{F \in \mathcal{F}_K} \big\|\underline{A}_{\mathbf{n}_K} \mathbf{u}_{h,R} - \underline{A}_{\mathbf{n}_K}^{\mathrm{upw}} \mathbf{u}_{n,h}\big\|_{(t_{n-1},t_n)\times F} \big\|\mathbf{u}^* - \mathbf{w}_h\big\|_{(t_{n-1},t_n)\times F} \Bigg) \\
&\quad + \sum_{n=1}^{N-1} \big\| M_h\big(\mathbf{u}_{n,h}(t_n) - \Pi_{n+1,h}\mathbf{u}_{n,h}(t_n)\big)\big\|_\Omega \big\|\mathbf{u}^*(t_n) - \mathbf{w}_{n+1,h}(t_n)\big\|_\Omega
\end{aligned}
$$

$$+\sum_{n=1}^{N-1} \|M^{-1}(M-M_h)\big(\mathbf{u}_{n,h}(t_n)-\Pi_{n+1,h}\mathbf{u}_{n,h}(t_n)\big)\|_Y \|\mathbf{u}^*(t_n)\|_Y$$
$$+\big\|(M-M_h)\partial_t\mathbf{u}_h\big\|_{W^*}\big\|\mathbf{u}^*\big\|_W - \big\|(D-D_h)\mathbf{u}_h\big\|_{W^*}\big\|\mathbf{u}^*\big\|_W .$$

This yields the assertion. □

Dual-Primal Error Indicator

For the evaluation of the error bound the exact solution $\mathbf{u}^*$ of the dual problem is required, and for $\mathbf{w}_h$ an interpolation of $\mathbf{u}^*$ can be inserted. Then, the interpolation errors $\big\|\mathbf{u}^*-\mathbf{w}_h\big\|_R$ and $\big\|\mathbf{u}^*-\mathbf{w}_h\big\|_{(t_{n-1},t_n)\times F}$ can be estimated by the regularity of the dual solution.

Since $\mathbf{u}^* \in V^*$ cannot be computed exactly, it is approximated by $\mathbf{u}_h^* \in W_h$ solving the discrete dual solution

$$b_h(\mathbf{v}_h,\mathbf{u}_h^*) = \langle E,\mathbf{v}_h\rangle , \qquad \mathbf{v}_h \in V_h ,$$

and the regularity of the dual solution is estimated from the regularity of $\mathbf{u}_h^*$. Therefore, we compute the L_2 projection $\Pi_h^0 \colon \mathrm{L}_2(Q;\mathbb{R}^m) \longrightarrow \mathbb{P}_0(Q_h;\mathbb{R}^m)$ and the jump terms $[\Pi_h^0\mathbf{u}_h^*]_F$ with $[\mathbf{y}_h]_F = \mathbf{y}_{h,K_F} - \mathbf{y}_{h,K}$ on inner faces, $([\mathbf{y}_h]_F)_j = (\underline{A}_\mathbf{n}\mathbf{y}_h)_j$ for $F \subset \Gamma_j^*$, and $([\mathbf{y}_h]_F)_j = \mathbf{0}$ for $F \subset \partial\Omega \setminus \Gamma_j^*$. Then, the error indicator $\eta_h = \sum_{R\in\mathcal{R}_h}\eta_R$ for $R=(t_{n-1},t_n)\times K$ is defined by

$$\eta_R = \Big(\big\|(M_h\partial_t + A + D_h)\mathbf{u}_h - \mathbf{f}\big\|_R$$
$$+\big\|\mathbf{u}_{n,h}(t_{n-1}) - \Pi_{n,h}\mathbf{u}_{n-1,h}(t_{n-1})\|_K\Big)\, h_K^{1/2}\,\big\|[\Pi_h^0\mathbf{u}_h^*]_F\big\|_{(t_{n-1},t_n)\times\partial K}$$
$$+\big\|\big(\underline{A}_{\mathbf{n}_K} - \underline{A}_{\mathbf{n}_K}^{\mathrm{upw}}\big)\mathbf{u}_h\|_{(t_{n-1},t_n)\times\partial K}\big\|[\Pi_h^0\mathbf{u}_h^*]_F\big\|_{(t_{n-1},t_n)\times\partial K} .$$

Depending on threshold parameters $0 < \vartheta_0 < \vartheta_0 < 1$ this results in the following p-adaptive algorithm:

choose low order polynomial degrees on the initial mesh
while $\max_R(p_R) \le p_{\max}$ and $\max_R(q_R) \le q_{\max}$ **do**
  compute $\mathbf{u}_h$
  compute $\mathbf{u}_h^*$ and the projection $\Pi_h^0\mathbf{u}_h^*$
  compute η_R on every cell R
  if the estimated error η_h is small enough, then STOP
  mark space-time cell R for refinement if $\eta_R > \vartheta_1 \max_{R'}\eta_{R'}$
  and for derefinement if $\eta_R < \vartheta_0 \max_{R'}\eta_{R'}$
  increase/decrease polynomial degrees on marked cells
  redistribute cells on processes for better load balancing
end while

4.7 Reliable Error Estimation for Weak Solutions

Finally, we derive a posteriori estimates for weak solutions based on local conforming reconstructions. Here we consider the general case including inhomogeneous initial and boundary data, where initial data are included in the definition of the affine ansatz space (4.2), and the DG formulation for boundary data is derived in (3.6), see (3.13) for an example. For simplicity, we assume that the parameters in M and D are piecewise constant, so that $M = M_h$ and $D = D_h$.

For the data $\mathbf{f} \in \mathrm{L}_2(Q; \mathbb{R}^m)$, $\mathbf{u}_0 \in \mathrm{L}_2(\Omega; \mathbb{R}^m)$, $g_k \in \mathrm{L}_2((0, T) \times \Gamma_k)$ defining the linear functional ℓ by

$$\langle \ell, \mathbf{z} \rangle = (\mathbf{f}, \mathbf{z})_Q + \big(M\mathbf{u}_0, \mathbf{z}(0)\big)_\Omega - \sum_{k=1}^{m} \big(g_k, z_k\big)_{(0,T)\times\Gamma_k}, \qquad \mathbf{z} \in \mathcal{V}^*,$$

we select piecewise polynomial approximations $\mathbf{f}_h \in \mathbb{P}(Q_h; \mathbb{R}^m)$, $\mathbf{u}_{0,h} \in \mathbb{P}(\Omega_h; \mathbb{R}^m)$, and $g_{k,h} \in \mathbb{P}((0, T) \times \Gamma_k)$ defining the approximated linear functional ℓ_h by

$$\langle \ell_h, \mathbf{z}_h \rangle = (\mathbf{f}_h, \mathbf{z}_h)_Q + \big(M\mathbf{u}_{0,h}, \mathbf{z}_h(0)\big)_\Omega - \sum_{k=1}^{m} \big(g_{k,h}, z_{k,h}\big)_{(0,T)\times\Gamma_k}, \qquad \mathbf{z}_h \in \mathcal{V}^*.$$

We assume that ℓ is bounded by (2.12) so that a unique weak solution $\mathbf{u} \in W$ of

$$\big(\mathbf{u}, L^*\mathbf{z}\big)_Q = \langle \ell, \mathbf{z} \rangle, \qquad \mathbf{z} \in \mathcal{V}^*$$

exists by Theorem 2.8 and Corollary 2.17. For the approximation $\mathbf{u}_h \in V_h(\mathbf{u}_{0,h})$ solving

$$b_h(\mathbf{u}_h, \mathbf{w}_h) = \big(\mathbf{f}_h, \mathbf{w}_h\big)_Q - \big(\underline{A}_{\mathbf{n}}^{\mathrm{bnd}} \mathbf{g}_h, \mathbf{w}_h\big)_{(0,T)\times\partial\Omega}, \qquad \mathbf{w}_h \in W_h,$$

we now construct a conforming reconstruction in a continuous finite element space $V_h^{\mathrm{cf}} \subset \mathrm{H}(L, Q) \cap \mathbb{P}(Q_h; \mathbb{R}^m)$ as described in the following. Here, we set for the right-hand side $\mathbf{g}_h = (g_{k,h})_{k=1,\dots,m} \in \mathrm{L}_2((0, T) \times \partial\Omega; \mathbb{R}^m)$ with $g_{k,h} = 0$ on $\partial\Omega \setminus \Gamma_k$.

The reconstruction is defined on local patches associated to the corners of the space-time mesh. Therefore, let $C_K \subset \overline{K}$ be the corner points in space of the elements $K \in \mathcal{K}_h$ such that $\overline{K} = \operatorname{conv} C_K$, and define $C_h = \bigcup_{K\in\mathcal{K}_h} C_K$. For all $\mathbf{c} \in C_h$ we define $\mathcal{K}_{h,\mathbf{c}} = \big\{K \in \mathcal{K}_h : \mathbf{c} \in C_K\big\}$ and open subdomains $\omega_\mathbf{c} \subset \Omega$ with $\overline{\omega}_\mathbf{c} = \bigcup_{K\in\mathcal{K}_{h,\mathbf{c}}} \overline{K}$. This extends to space-time patches $Q_{0,\mathbf{c}} = (0, t_1) \times \omega_\mathbf{c}$, $Q_{n,\mathbf{c}} = (t_{n-1}, t_{n+1}) \times \omega_\mathbf{c}$ for $n = 1, \dots, N-1$, and $Q_{N,\mathbf{c}} = (t_{N-1}, T) \times \omega_\mathbf{c}$. Let $\psi_{n,\mathbf{c}} \in \mathrm{C}^0(\overline{Q}) \cap \mathbb{P}(Q_h)$ be a corresponding decomposition of $1 \equiv \sum_{n=0}^{N} \sum_{\mathbf{c}\in C_h} \psi_{n,\mathbf{c}}$ with $\operatorname{supp} \psi_{n,\mathbf{c}} = \overline{Q}_{n,\mathbf{c}}$.

On every patch we define discrete conforming local affine spaces

$$
\begin{aligned}
V_{n,\mathbf{c}}^{\mathrm{cf}}(\mathbf{u}_{0,h},\mathbf{g}_h) = \Big\{\mathbf{v}_h \in V_h^{\mathrm{cf}} \colon\ & \mathrm{supp}(\mathbf{v}_h) \subset \overline{Q}_{n,\mathbf{c}}\,,\\
& \mathbf{v}_h(0) = \psi_{n,\mathbf{c}}\mathbf{u}_{0,h} \text{ in } \Omega \text{ if } n = 0\,,\\
& \mathbf{v}_h(t_{n-1}) = \mathbf{0} \text{ in } \Omega \text{ if } n > 0\,,\\
& \mathbf{v}_h(t_{n+1}) = \mathbf{0} \text{ in } \Omega \text{ if } n < N\,,\\
& (\underline{A}_{\mathbf{n}}\mathbf{v}_h)_k = \psi_{n,\mathbf{c}} g_{k,h} \text{ on } (0,T)\times\Gamma_k\,,\ k = 1,\dots,m\,,\\
& \underline{A}_{\mathbf{n}}\mathbf{v}_h = \mathbf{0} \text{ on } (0,T)\times(\partial\omega_{\mathbf{c}}\setminus\partial\Omega)\Big\}\,.
\end{aligned}
$$

In the following we assume $V_{n,\mathbf{c}}^{\mathrm{cf}}(\mathbf{u}_{0,h},\mathbf{g}_h) \neq \emptyset$, which can be achieved by a suitable choice of the data approximation $\mathbf{u}_{0,h}$ and $\mathbf{g}_{k,h}$ depending on the reconstruction space V_h^{cf}.

Now, the local conforming reconstruction of the discrete solution $\mathbf{u}_h$ is defined by $\mathbf{u}_h^{\mathrm{cf}} = \sum_{n=0}^N \sum_{\mathbf{c}\in C_h} \mathbf{u}_{n,\mathbf{c}}^{\mathrm{cf}}$, where $\mathbf{u}_{n,\mathbf{c}}^{\mathrm{cf}} \in V_{n,\mathbf{c}}^{\mathrm{cf}}(\mathbf{u}_{0,h},\mathbf{g}_h)$ is the best approximation of $\psi_{n,\mathbf{c}}\mathbf{u}_h$ in the topology of W, i.e.,

$$
\big\|\psi_{n,\mathbf{c}}\mathbf{u}_h - \mathbf{u}_{n,\mathbf{c}}^{\mathrm{cf}}\big\|_W \le \big\|\psi_{n,\mathbf{c}}\mathbf{u}_h - \mathbf{v}_{n,\mathbf{c}}\big\|_W\,, \qquad \mathbf{v}_{n,\mathbf{c}} \in V_{n,\mathbf{c}}^{\mathrm{cf}}(\mathbf{u}_{0,h},\mathbf{g}_h)\,,
$$

so that $\mathbf{u}_{n,\mathbf{c}}^{\mathrm{cf}}$ is determined by a small local quadratic minimization problem with linear constraints.

Lemma 4.13 *The approximation error of the weak solution can be estimated by*

$$
\big\|\mathbf{u}-\mathbf{u}_h\big\|_W \le \big\|\mathbf{u}_h - \mathbf{u}_h^{\mathrm{cf}}\big\|_W + 2T\,\big\|L\mathbf{u}_h^{\mathrm{cf}} - \mathbf{f}_h\big\|_{W^*} + \beta^{-1} \sup_{\mathbf{z}\in\mathcal{V}^*\setminus\{\mathbf{0}\}} \frac{\langle \ell-\ell_h,\mathbf{z}\rangle}{\|\mathbf{z}\|_{V^*}}\,.
$$

Proof By construction we have for $\mathbf{u}_{n,\mathbf{c}}^{\mathrm{cf}} \in V_{n,\mathbf{c}}^{\mathrm{cf}}(\mathbf{u}_{0,h},\mathbf{g}_h)$

$$
\begin{aligned}
\mathbf{u}_h^{\mathrm{cf}}(0) &= \sum_{n=0}^N \sum_{\mathbf{c}\in C_h} \psi_{n,\mathbf{c}}\mathbf{u}_{0,h} = \mathbf{u}_{0,h} && \text{in } \Omega\,,\\
\big(\underline{A}_{\mathbf{n}}\mathbf{u}_h^{\mathrm{cf}}\big)_k &= \sum_{n=0}^N \sum_{\mathbf{c}\in C_h} \psi_{n,\mathbf{c}} g_{k,h} = g_{k,h} && \text{on } (0,T)\times\Gamma_k,\ k = 1,\dots,m\,,
\end{aligned}
$$

so that for all $\mathbf{z} \in \mathcal{V}^*$ integration by parts and the boundary conditions in $\mathcal{V}^*$ gives

$$
\begin{aligned}
\big(\mathbf{u}_h^{\mathrm{cf}}, L^*\mathbf{z}\big)_Q &= \big(L\mathbf{u}_h^{\mathrm{cf}},\mathbf{z}\big)_Q + \big(M\mathbf{u}_h^{\mathrm{cf}}(0),\mathbf{z}(0)\big)_\Omega - \big(\underline{A}_{\mathbf{n}}\mathbf{u}_h^{\mathrm{cf}},\mathbf{z}\big)_{(0,T)\times\partial\Omega}\\
&= \big(L\mathbf{u}_h^{\mathrm{cf}} - \mathbf{f}_h,\mathbf{z}\big)_Q + \langle \ell_h,\mathbf{z}\rangle\,.
\end{aligned}
$$

Since $\mathrm{L}_2(Q;\mathbb{R}^m) = M^{-1}L^*(V^*)$ and $\mathcal{V}^* \subset V^*$ is dense, we obtain by duality

$$
\begin{aligned}
\|\mathbf{u} - \mathbf{u}_h^{\mathrm{cf}}\|_W &= \sup_{\mathbf{v}\in \mathrm{L}_2(Q;\mathbb{R}^m)\setminus\{\mathbf{0}\}} \frac{\big(M(\mathbf{u}-\mathbf{u}_h^{\mathrm{cf}}),\mathbf{v}\big)_Q}{\|\mathbf{v}\|_W} = \sup_{\mathbf{z}\in\mathcal{V}^*:\,L^*\mathbf{z}\neq\mathbf{0}} \frac{\big(\mathbf{u}-\mathbf{u}_h^{\mathrm{cf}}, L^*\mathbf{z}\big)_Q}{\|M^{-1}L^*\mathbf{z}\|_W}\\
&= \sup_{\mathbf{z}\in\mathcal{V}^*:\,L^*\mathbf{z}\neq\mathbf{0}} \frac{\big(L\mathbf{u}_h^{\mathrm{cf}} - \mathbf{f}_h, \mathbf{z}\big)_Q + \langle \ell - \ell_h, \mathbf{z}\rangle}{\|L^*\mathbf{z}\|_{W^*}}\\
&\le \sup_{\mathbf{z}\in\mathcal{V}^*:\,L^*\mathbf{z}\neq\mathbf{0}} \frac{\big\|L\mathbf{u}_h^{\mathrm{cf}} - \mathbf{f}_h\big\|_{W^*}\big\|\mathbf{z}\big\|_W}{\|L^*\mathbf{z}\|_{W^*}} + \sup_{\mathbf{z}\in\mathcal{V}^*:\,L^*\mathbf{z}\neq\mathbf{0}} \frac{\langle \ell - \ell_h, \mathbf{z}\rangle}{\|L^*\mathbf{z}\|_{W^*}}\\
&\le 2T\,\big\|L\mathbf{u}_h^{\mathrm{cf}} - \mathbf{f}_h\big\|_{W^*} + \beta^{-1} \sup_{\mathbf{z}\in\mathcal{V}^*:\,L^*\mathbf{z}\neq\mathbf{0}} \frac{\langle \ell - \ell_h, \mathbf{z}\rangle}{\|\mathbf{z}\|_{V^*}}
\end{aligned}
$$

using the a priori estimate $\|\mathbf{z}\|_W \le 2T\,\|L^*\mathbf{z}\|_{W^*}$ from Remark 2.13 with $C_L = 2T$ and $\|\mathbf{z}\|_{V^*} \le \beta^{-1}\|L^*\mathbf{z}\|_{W^*}$ with $\beta^{-1} = \sqrt{1+4T^2}$ from Corollary 2.17, so that

$$
\|\mathbf{u} - \mathbf{u}_h\|_W \le \|\mathbf{u} - \mathbf{u}_h^{\mathrm{cf}}\|_W + \|\mathbf{u}_h^{\mathrm{cf}} - \mathbf{u}_h\|_W
$$

yields the assertion. □

This lemma shows that the corresponding error estimator with local contributions

$$
\eta_{n,K} = \Big(\sum_{\mathbf{c}\in\mathcal{C}_K} \eta_{n-1,\mathbf{c}}^2 + \eta_{n,\mathbf{c}}^2\Big)^{1/2},
$$

$$
\eta_{n,\mathbf{c}} = \Big(\big\|M^{1/2}(\psi_{n,\mathbf{c}}\mathbf{u}_h - \mathbf{u}_{n,\mathbf{c}}^{\mathrm{cf}})\big\|_{Q_{n,\mathbf{c}}}^2 + 2T\,\big\|M^{-1/2}(L\mathbf{u}_h^{\mathrm{cf}} - \mathbf{f}_h)\big\|_{Q_{n,\mathbf{c}}}^2\Big)^{1/2},
$$

is reliable up to the data approximation error, i.e.,

$$
\big\|\mathbf{u} - \mathbf{u}_h\big\|_W \le \Big(\sum_{n=0}^{N}\sum_{K\in\mathcal{K}_h} \eta_{n,K}^2\Big)^{1/2} + \beta^{-1}\sup_{\mathbf{z}\in\mathcal{V}^*\setminus\{\mathbf{0}\}} \frac{\langle \ell - \ell_h, \mathbf{z}\rangle}{\|\mathbf{z}\|_{V^*}}.
$$

Bibliographic Comments

This chapter is based on [48, 49], where also numerical results for the adaptive algorithm are presented. Further applications and several numerical applications are reported in [50, 71, 169].

The extension to estimates for weak solutions is based on the construction of a right-inverse as it is done in [62] for conforming Petrov–Galerkin approximations in reflexive Banach spaces.

The estimate for the Legendre polynomials can also be obtained recursively using [1, Lem. 8.5.3], see, e.g., [48, Lem. A.1].

The error estimation based on dual-weighted residuals transfers the approach in [10] to our space-time framework, and for the general concepts on error estimation by conforming reconstructions we refer to [63].

The results are closely related to the analysis of space-time discontinuous Galerkin methods for acoustics in [8, 98, 127]. Alternative concepts for space-time discretizations for wave equations are collected in [115]. See also the results in [6, 75] and more recently in [139, 160], and the references therein.

Part II

Local Wellposedness and Long-Time Behavior of Quasilinear Maxwell Equations

Roland Schnaubelt

The Maxwell system is the foundation of electro-magnetic theory. We develop the local wellposedness theory for the (non-autonomous) linear and the quasilinear Maxwell equations in the Sobolev space H^3, which is the natural state space for energy methods in this case. On $\mathbb{R}^3$ these results directly follow from the standard theory of symmetric hyperbolic systems, whereas it was shown only recently on domains. In the first chapter we present the theory on $\mathbb{R}^3$ in detail, using energy methods. We also treat the finite speed of propagation and, for the isotropic Maxwell system, a blow-up example in H^1. The second chapter is devoted to the problem with boundary conditions focusing on the half-space case. Here the main challenge is to establish regularity in normal direction at the boundary, employing the special structure of the Maxwell system. The general case is studied via localization, where severe additional difficulties arise, so that this step is only sketched. The last chapter then combines these results and methods with an observability-type estimate to show global existence and exponential convergence to 0 for small initial fields in the presence of a strictly positive conductivity.

5 Introduction and Local Wellposedness on $\mathbb{R}^3$

In this section we develop a local wellposedness theory for the quasilinear Maxwell equations on $\mathbb{R}^3$. Our approach is based on energy methods and a fixed-point argument, which make use of the linear system with time-depending coefficients. One has to work in Sobolev spaces H^s with $s > \frac{5}{2}$ in this context, where we take $s = 3$ for simplicity. Actually we treat general symmetric hyperbolic systems on $\mathbb{R}^3$. In the first subsection we introduce Maxwell equations and discuss some facts used throughout these notes. We then investigate the linear case, first in L^2 and then in H^3, also establishing the finite speed of propagation. Our main tools are energy estimates, duality arguments for existence in L^2, approximation by mollifiers for regularity and uniqueness, and finally a transformation from L^2 to H^3. The non-linear problem is solved by means of fixed-point arguments going back to Kato [106] at least, where the derivation of blow-up conditions in $W^{1,\infty}$ and the continuous dependence of data in H^3 requires significant additional effort. Finally, for the isotropic Maxwell system, we show the preservation of energy and construct a blow-up example in H^1.

The wellposedness results on $\mathbb{R}^3$ are due to Kato [105], but our proof differs from Kato's and instead uses energy methods from the theory of symmetric hyperbolic PDE, see [7, 11, 26, 123], for instance. This approach is well known, but we think that a detailed presentation in form of lecture notes is quite helpful. In particular, it gives us the opportunity to discuss in a simpler situation some core features of the domain case treated in the Chap. 6.

5.1 The Maxwell System

The Maxwell equations relate the *electric field* $\mathbf{E}(t, x) \in \mathbb{R}^3$, the (electric) *displacement field* $\mathbf{D}(t, x) \in \mathbb{R}^3$, the *magnetic field* $\mathbf{B}(t, x) \in \mathbb{R}^3$ and the *magnetizing field* $\mathbf{H}(t, x) \in \mathbb{R}^3$

W. Dörfler et al., *Wave Phenomena*, Oberwolfach Seminars 49,
https://doi.org/10.1007/978-3-031-05793-9_5

via the Maxwell–Ampère and Maxwell–Faraday laws

$$\partial_t \mathbf{D} = \operatorname{curl} \mathbf{H} - \mathbf{J}, \qquad \partial_t \mathbf{B} = -\operatorname{curl} \mathbf{E}, \qquad t \geq 0,\ x \in G, \tag{5.1}$$

where $G \subseteq \mathbb{R}^3$ is open and $\mathbf{J}(t,x) \in \mathbb{R}^3$ is the *current density*. (See e.g. [100] for the background in physics.) If $G \neq \mathbb{R}^3$ we have to add boundary conditions to (5.1) which are discussed in the next chapter. We use the standard differential expressions

$$\operatorname{curl} u = \nabla \times u = \begin{pmatrix} 0 & -\partial_3 & \partial_2 \\ \partial_3 & 0 & -\partial_1 \\ -\partial_2 & \partial_1 & 0 \end{pmatrix} \begin{pmatrix} u_1 \\ u_2 \\ u_3 \end{pmatrix}, \qquad \operatorname{div} u = \nabla \cdot u = \partial_1 u_1 + \partial_2 u_2 + \partial_3 u_3,$$

where the derivatives are interpreted in a weak sense if needed (see Sect. 5.2). Since $\operatorname{div} \operatorname{curl} = 0$, solutions to (5.1) fulfill Gauß' laws

$$\rho(t) := \operatorname{div} \mathbf{D}(t) = \operatorname{div} \mathbf{D}(0) - \int_0^t \operatorname{div} \mathbf{J}(s)\, \mathrm{d}s, \qquad \operatorname{div} \mathbf{B}(t) = \operatorname{div} \mathbf{B}(0), \quad t \geq 0. \tag{5.2}$$

The *electric charge density* ρ is thus determined by the initial charge and the current density. As there are no magnetic charges in physics, one often requires $\operatorname{div} \mathbf{B}(0) = 0$.

To complete the Maxwell system (5.1), we have to connect the fields via *material laws*. They involve the *polarization* $\mathbf{P} = \mathbf{D} - \varepsilon_0 \mathbf{E}$ and the *magnetization* $\mathbf{M} = \mathbf{B} - \mu_0 \mathbf{H}$ which describe the material response to the fields $\mathbf{E}$ and $\mathbf{B}$. Below we set $\varepsilon_0 = \mu_0 = 1$ for simplicity (thus destroying physical units). In these notes we use instantaneous constitutive relations, namely

$$(\mathbf{D}, \mathbf{B}) = \theta(x, \mathbf{E}, \mathbf{H}) = \theta(x, u) \quad \text{for regular } \theta : G \times \mathbb{R}^6 \to \mathbb{R}^6, \tag{5.3}$$

We choose $u = (\mathbf{E}, \mathbf{H})$ as state because this fits best to energy estimates. Other choices are possible since transformations like $\theta(x, \cdot)$ are typically invertible. Our main hypothesis will be that $\partial_u \theta(x,u) =: a_0(x,u)$ is symmetric and $a_0 \geq \eta I > 0$. Finally, the current is modelled as the sum

$$\mathbf{J} = \sigma(x, \mathbf{E}, \mathbf{H})\mathbf{E} + \mathbf{J}_0 \tag{5.4}$$

of a given external current density $\mathbf{J}_0 : \mathbb{R}_{\geq 0} \times G \to \mathbb{R}^3$ and a current induced via Ohm's law for a (possibly state-depending) *conductivity* $\sigma : G \times \mathbb{R}^6 \to \mathbb{R}^{3\times 3}$.

Example 5.1 A basic example in nonlinear optics is the (instantaneous) Kerr law

$$\mathbf{D} = \chi_1(x)\mathbf{E} + \chi_3(x)|\mathbf{E}|^2\, \mathbf{E}, \qquad \mathbf{H} = \mathbf{B},$$

for bounded functions $\chi_j : G \to \mathbb{R}$ with $\chi_1(x) \geq 2\eta > 0$ for all x, see e.g. [18] and also Example 5.20. It is isotropic; i.e., $\mathbf{D}(t, x)$ and $\mathbf{E}(t, x)$ are parallel. The Kerr law satisfies our assumption $a_0 = a_0^\top \geq \eta I$ for small $\mathbf{E}$ (and for all $\mathbf{E}$ if $\chi_3 \geq 0$). The latter also holds for the more general laws $\mathbf{D} = \chi_e(x)\mathbf{E} + \beta_e(x, |\mathbf{E}|^2)\mathbf{E}$ and $\mathbf{H} = \chi_m(x)\mathbf{B} + \beta_m(x, |\mathbf{B}|^2)\mathbf{B}$ for 3×3 matrices $\chi_j = \chi_j^\top \geq 2\eta I$ and smooth scalar β_j with $\beta_j(0) = 0$.

In physics material laws often also contain a time retardation, see [19] or [68]. Here we stick to the instantaneous case which stays within the PDE framework. (But we expect to tackle the Maxwell system with retardation with variants of our methods.)

It is often convenient to rewrite (5.1) with (5.3) and (5.4) as a quasilinear symmetric hyperbolic system. To this end, we first introduce the matrices

$$S_1 = \begin{pmatrix} 0 & 0 & 0 \\ 0 & 0 & -1 \\ 0 & 1 & 0 \end{pmatrix}, \qquad S_2 = \begin{pmatrix} 0 & 0 & 1 \\ 0 & 0 & 0 \\ -1 & 0 & 0 \end{pmatrix}, \qquad S_3 = \begin{pmatrix} 0 & -1 & 0 \\ 1 & 0 & 0 \\ 0 & 0 & 0 \end{pmatrix} \qquad \text{satisfying}$$

$$\operatorname{curl} = S_1\partial_1 + S_2\partial_2 + S_3\partial_3, \qquad a \times b = (a_1 S_1 + a_2 S_2 + a_3 S_3)b$$

for vectors $a, b \in \mathbb{R}^3$. We then define

$$A_j^{\mathrm{co}} = \begin{pmatrix} 0 & -S_j \\ S_j & 0 \end{pmatrix}, \qquad d = \begin{pmatrix} \sigma \\ 0 \end{pmatrix}, \qquad f = \begin{pmatrix} -\mathbf{J}_0 \\ 0 \end{pmatrix}, \qquad \partial_t = \partial_0 \tag{5.5}$$

for $j \in \{1, 2, 3\}$. Note that the matrices A_j^{co} are symmetric.

Then the Maxwell system (5.1) with material laws (5.3) and (5.4) becomes

$$L(u)u := a_0(u)\partial_t u + \sum\nolimits_{j=1}^{3} A_j^{\mathrm{co}} \partial_j u + d(u)u = f. \tag{5.6}$$

Our strategy to solve this problem goes (at least) back to Kato [106]. One freezes a function v from a suitable space E in the nonlinearities, setting $A_0 = a_0(v)$ and $D = d(v)$. One then solves the resulting non-autonomous linear problem $L(v)u = \sum_{j=0}^{3} A_j \partial_j u + Du = f$ in the space E. For small times $(0, T)$ one finds a fixed point of the map $v \mapsto u$ which then solves (5.6) and (5.1). The first linear step is more difficult; here it is crucial to control very well how the constants in the estimates depend on the coefficients. We carry out this program for $G = \mathbb{R}^3$ in the following sections.

5.2 The Linear Problem on $\mathbb{R}^3$ in L^2

Let $J = (0, T)$. We will solve the linear problem in the space $C(\overline{J}, L^2(\mathbb{R}^3, \mathbb{R}^6)) = C(\overline{J}, L^2_x)$ for coefficients and data subject to the assumptions

$$A_j = A_j^\top \in W^{1,\infty}_{t,x} = W^{1,\infty}(J \times \mathbb{R}^3, \mathbb{R}^{6\times 6}) \text{ for } j \in \{0,1,2,3\}, \quad A_0 = A_0^\top \geq \eta I > 0,$$
$$D \in L^\infty_{t,x} = L^\infty(J \times \mathbb{R}^3, \mathbb{R}^{6\times 6}), \quad u_0 \in L^2_x, \quad f \in L^2_{t,x} = L^2(J \times \mathbb{R}^3, \mathbb{R}^6). \tag{5.7}$$

(We often omit range spaces as $\mathbb{R}^6$ in the notation. We use the subscript t to indicate a function space over $t \in J$ or other time intervals, and x for a space over $x \in \mathbb{R}^3$ (or over $x \in U \subseteq \mathbb{R}^m$).) Compared to (5.6) we allow for D and f with non-zero 'magnetic' components, as needed in our analysis. We also deal with general (symmetric) x-depending coefficients A_1, A_2 and A_3, and thus with linear *symmetric hyperbolic systems*. Those occur in many applications, see [11, 106, 123], or Part I, and our reasoning would not differ much if we restricted to $A_j = A_j^{\text{co}}$. Moreover, when treating the Maxwell system on domains by localization arguments, one obtains x-depending coefficients. It is useful to see them first in an easier case.

Assuming (5.7), we look for a solution $u \in C(\overline{J}, L^2_x)$ of the system

$$Lu := \sum_{j=0}^{3} A_j \partial_j u + Du = f, \quad t \geq 0, \qquad u(0) = u_0, \tag{5.8}$$

with $\partial_0 = \partial_t$. Here the derivatives are understood in a weak sense. To explain this, we assume that the reader is familiar with Sobolev spaces $W^{k,p}(U) = W^{k,p}$ for an open subset U of $\mathbb{R}^m$, $k \in \mathbb{N}_0$, and $p \in [1, \infty]$. (See [2] or [17], for instance.) We mostly work with real scalars, endow $W^{k,p}$ with the (complete) norm $\|v\|_{k,p}^p = \sum_{0\leq|\alpha|\leq k} \|\partial^\alpha v\|_p^p$ (obvious modification for $p = \infty$), and write $H^k := W^{k,2}$ (which is a Hilbert space), $L^p = W^{0,p}$ and $\|v\|_p := \|v\|_{0,p}$. By $W_0^{k,p}(U)$ we denote the closure of test functions $C_c^\infty(U)$ in $W^{k,p}(U)$. If ∂U is compact and C^k (or Lipschitz if $k = 1$)), say, then $W_0^{k,p}$ is the closed subspace in $W^{k,p}$ of functions whose (weak) derivatives of order up to $k-1$ have trace 0. One can check that $W_0^{k,p}(\mathbb{R}^m) = W^{k,p}(\mathbb{R}^m)$.

Let $H^{-k}(U)$ be the dual space $H_0^k(U)^*$, where we restrict ourselves to $p = 2$ for simplicity. For $\varphi \in L^2(U)$, $j \in \{1, \ldots, m\}$ and $v \in H_0^1(U)$, we define the weak derivative $\partial_j \varphi \in H^{-1}(U)$ by setting

$$(\partial_j \varphi)(v) = \langle v, \partial_j \varphi \rangle_{H_0^1} := -\langle \partial_j v, \varphi \rangle_{L^2}.$$

(The brackets $\langle \cdot, \cdot \rangle_X$ designate the duality pairing between Banach spaces X and X^*.) Since $|\langle \partial_j v, \varphi \rangle| \leq \|v\|_{1,2} \|\varphi\|_2$, the linear map $\partial_j : L^2(U) \to H^{-1}(U)$ is bounded. Iteratively, one obtains bounded maps $\partial_j : H^{-k}(U) \to H^{-k-1}(U)$, and analogously

$\partial^\alpha : H^{-k}(U) \to H^{-k-|\alpha|}(U)$ for multi-indices $\alpha \in \mathbb{N}_0^m$ and $k \in \mathbb{N}_0$. The definitions imply that these derivatives commute.

For $a \in W^{1,\infty}(U)$ and $\varphi \in H^{-1}(U)$, we next define the map $a\varphi \in H^{-1}(U)$ by

$$(a\varphi)(v) = \langle v, a\varphi\rangle_{H_0^1} := \langle av, \varphi\rangle_{H_0^1}, \qquad v \in H_0^1(U).$$

Because of $\|av\|_{1,2} \lesssim \|a\|_{1,\infty} \|v\|_{1,2}$, we see as above that the multiplication operator $M_a : \varphi \mapsto a\varphi$ is bounded on $H^{-1}(U)$. (Here and below $A \lesssim_\alpha B$ stands for $A \le cB$ for a generic constant $c = c(\alpha)$ which is non-decreasing in each component of $\alpha \in \mathbb{R}^n$.) These facts easily extend to $\mathbb{R}^l$–valued functions.

We infer that $Lu \in H_{t,x}^{-1}$ if $u \in L_{t,x}^2$. If $Lu = f$ is contained in $L_{t,x}^2$, we obtain

$$\partial_t u = A_0^{-1} f - \sum\nolimits_{j=1}^3 A_0^{-1} A_j \partial_j u - A_0^{-1} Du \in L_t^2 H_x^{-1} := L^2(J, H_x^{-1}). \tag{5.9}$$

As u belongs to $H_t^1 H_x^{-1} \hookrightarrow C(\overline{J}, H_x^{-1})$, the initial condition in (5.8) is taken in H_x^{-1}.

We will first show the basic *energy* (or *apriori*) *estimate*. Here we use the temporal weights $e_{-\gamma}(t) = \mathrm{e}^{-\gamma t}$ for $\gamma \ge 0$ and $t \in J$ (or $t \in \mathbb{R}$) and the weighted spaces $L_\gamma^2 H_x^k$ of functions with (finite) norm $\|v\|_{L_\gamma^2 H_x^k} = \|e_{-\gamma} v\|_{L_t^2 H_x^k}$. On J these norms are equivalent to the unweighted case as $\|v\|_{L_\gamma^2 H_x^k} \le \|v\|_{L^2 H_x^k} \le \mathrm{e}^{\gamma T} \|v\|_{L_\gamma^2 H_x^k}$. Taking large γ in these norms, we can produce small constants in front of the contribution of f in the inequality below. This fact will be used to absorb error terms by the left-hand side, for instance. The estimate and the precise form of the constants is also crucial for the nonlinear problem. We write $\operatorname{div} A = \sum_{j=0}^3 \partial_j A_j$.

Lemma 5.2 *Assume that* (5.7) *is true and that* $u \in H^1(J \times \mathbb{R}^3)$ *solves* (5.8). *Let* $C := \frac{1}{2} \operatorname{div} A - D$, $\gamma \ge \gamma_0'(L) := \max\{1, 4\,\|C\|_\infty/\eta\}$, *and* $t \in \overline{J}$. *We then obtain*

$$\tfrac{\gamma\eta}{4} \|u\|^2_{L_\gamma^2((0,t),L_x^2)} + \tfrac{\eta}{2} \mathrm{e}^{-2\gamma t} \|u(t)\|^2_{L_x^2} \le \tfrac{1}{2} \|A_0(0)\|_\infty \|u_0\|^2_{L_x^2} + \tfrac{1}{2\gamma\eta} \|f\|^2_{L_\gamma^2((0,t),L_x^2)}.$$

Proof Set $v = e_{-\gamma} u$ and $g = e_{-\gamma} f$. We have $\gamma A_0 v + Lv = g$. Using the symmetry of A_j, we derive

$$\begin{aligned}
\langle g, v\rangle &= \gamma \langle A_0 v, v\rangle + \sum\nolimits_{j=0}^3 \langle A_j \partial_j v, v\rangle + \langle Dv, v\rangle \\
&= \gamma \langle A_0 v, v\rangle + \frac{1}{2} \sum\nolimits_{j=0}^3 \Big(\int_0^t \int_{\mathbb{R}^3} \partial_j (A_j v \cdot v)\, \mathrm{d}x\, \mathrm{d}s - \langle \partial_j A_j v, v\rangle \Big) + \langle Dv, v\rangle,
\end{aligned}$$

where we drop the subscript $L^2((0,t), L_x^2)$ of the brackets and denote the scalar product in $\mathbb{R}^6$ by a dot. Integration yields

$$\gamma \langle A_0 v, v\rangle + \tfrac{1}{2}\langle A_0(t)v(t), v(t)\rangle_{L_x^2} = \tfrac{1}{2}\langle A_0(0)v(0), v(0)\rangle_{L_x^2} + \langle Cv, v\rangle + \langle g, v\rangle.$$

We now replace $v = e_{-\gamma}u$, $g = e_{-\gamma}f$ as well as $u(0) = u_0$, and use (5.7) and $\|C\|_\infty \le \gamma\eta/4$. It follows

$$\begin{aligned}\gamma\eta\|u\|_{L_\gamma^2 L_x^2} &+ \tfrac{\eta}{2}\mathrm{e}^{-2\gamma t}\|u(t)\|_{L_x^2}^2 \\ &\le \tfrac{1}{2}\|A_0(0)\|_\infty \|u_0\|_{L_x^2}^2 + \|C\|_\infty \|u\|_{L_\gamma^2 L_x^2}^2 + \tfrac{\sqrt{\gamma\eta}}{\sqrt{\gamma\eta}}\|u\|_{L_\gamma^2 L_x^2}\|f\|_{L_\gamma^2 L_x^2} \\ &\le \tfrac{1}{2}\|A_0(0)\|_\infty \|u_0\|_{L_x^2}^2 + \big(\tfrac{\gamma\eta}{4} + \tfrac{\gamma\eta}{2}\big)\|u\|_{L_\gamma^2 L_x^2}^2 + \tfrac{1}{2\gamma\eta}\|f\|_{L_\gamma^2 L_x^2}^2,\end{aligned}$$

which implies the assertion. □

Below we use the above estimate for $\gamma \ge \gamma_0(r,\eta) := \max\{1, 12r/\eta\} \ge \gamma_0'(L)$ where $\|\partial_j A_j\|_\infty, \|D\|_\infty \le r$. For $\gamma = 0$ its proof yields the *energy equality*

$$\int_{\mathbb{R}^3} A_0(t)u(t)\cdot u(t)\,\mathrm{d}x = \int_{\mathbb{R}^3} A_0(0)u_0\cdot u_0\,\mathrm{d}x + 2\int_0^t\!\int_{\mathbb{R}^3}\big(C(s)u(s) + f(s)\big)\cdot u(s)\,\mathrm{d}x\,\mathrm{d}s. \tag{5.10}$$

In the term with $C = \frac{1}{2}\operatorname{div} A - D$ we have damping effects (if $D = D^\top \gneqq 0$) and extra errors terms coming from the t- or x-dependence of A_j.

Lemma 5.2 yields uniqueness of H^1–solutions to (5.8). However, we need uniqueness (and the energy estimate) for solutions in $C(\overline{J}, L_x^2)$. This fundamental gap can be closed by a crucial regularization argument based on *mollifiers*, see e.g. [17].

We set $g_\varepsilon(x) = \varepsilon^{-m}g(\varepsilon^{-1}x)$ for a function g on $\mathbb{R}^m$, $\varepsilon > 0$, and $x \in \mathbb{R}^m$. Take $0 \le \rho \in C_c^\infty(\mathbb{R}^m)$ with $\int \rho\,\mathrm{d}x = 1$, support $\operatorname{supp}\rho$ in the closed unit ball $\overline{B}(0,1)$, and $\rho(x) = \rho(-x)$ for $x \in \mathbb{R}^m$. Note that $\|\rho_\varepsilon\|_1 = 1$. For $\varepsilon > 0$ and $v \in L^1_{\mathrm{loc}}(\mathbb{R}^m)$, we set

$$R_\varepsilon v(x) = \rho_\varepsilon * v(x) = \int_{\mathbb{R}^m} \rho_\varepsilon(x-y)v(y)\,\mathrm{d}y, \quad x \in \mathbb{R}^m.$$

One can check that $R_\varepsilon v \in C^\infty(\mathbb{R}^m)$, $\operatorname{supp} R_\varepsilon v \subseteq \operatorname{supp} v + \overline{B}(0,\varepsilon)$, and $\partial^\alpha R_\varepsilon v = R_\varepsilon \partial^\alpha v$ for $v \in W^{|\alpha|,p}(\mathbb{R}^m)$. Young's inequality for convolutions yields $\|R_\varepsilon v\|_{k,p} \le \|v\|_{k,p}$ for $p \in [1,\infty]$ and $k \in \mathbb{N}_0$. Using this estimate, one derives that $R_\varepsilon v \to v$ in $W^{k,p}(\mathbb{R}^m)$ for $v \in W^{k,p}(\mathbb{R}^m)$ as $\varepsilon \to 0$ if $p < \infty$, since this limit is true for test functions v. Differentiating $\rho_\varepsilon(x-y)$ in x, one also obtains $\|R_\varepsilon v\|_{k,p} \lesssim_{\varepsilon,k} \|v\|_p$.

Finally, for $\varphi \in H^{-k}(\mathbb{R}^m)$, $v \in H^k(\mathbb{R}^m)$ and $k \in \mathbb{N}$, we put

$$(R_\varepsilon\varphi)(v) = \langle v, R_\varepsilon\varphi\rangle_{H^k} := \langle R_\varepsilon v, \varphi\rangle_{H^k} .$$

This definition is consistent with the symmetry $R_\varepsilon^* = R_\varepsilon$ on $L^2(\mathbb{R}^m)$ which follows from the symmetry of ρ and Fubini's theorem. By means of its properties in $H^k(\mathbb{R}^m)$, one can show that R_ε is contractive on $H^{-l}(\mathbb{R}^m)$ and that it maps this space into $H^k(\mathbb{R}^m)$ for all $l \in \mathbb{N}$. Moreover, it commutes with ∂^α.

Hence, the commutator $[R_\varepsilon, M_a] = R_\varepsilon M_a - M_a R_\varepsilon$ tends to 0 strongly in L^2_x if $a \in L^\infty_x$. It even gains a derivative if $a \in W^{1,\infty}_x$, which is crucial for our analysis.

Proposition 5.3 *Let $a \in W^{1,\infty}(\mathbb{R}^m)$, $u \in L^2(\mathbb{R}^m)$, $j \in \{1, \dots, m\}$, and $\varepsilon > 0$. Set $C_\varepsilon u := R_\varepsilon(a\partial_j u) - a\partial_j(R_\varepsilon u)$. Then there is a constant $c = c(\rho)$ such that*

$$\|C_\varepsilon u\|_2 \le c\|a\|_{1,\infty}\,\|u\|_2 \qquad \textit{and} \qquad C_\varepsilon u \to 0 \ \textit{ in } L^2_x \ \textit{ as } \varepsilon \to 0.$$

Proof Let $v \in H^1(\mathbb{R}^m)$. Using the above indicated facts, we compute

$$\langle v, C_\varepsilon u\rangle_{H^1} = \langle aR_\varepsilon v, \partial_j u\rangle_{H^1} - \langle av, R_\varepsilon\partial_j u\rangle_{H^1} = \langle \partial_j(R_\varepsilon(av) - aR_\varepsilon v), u\rangle_{L^2}.$$

We set $C'_\varepsilon v = \partial_j(R_\varepsilon(av) - aR_\varepsilon v)$ and R^j_ε for the convolution with $(|\partial_j\rho|)_\varepsilon$. For a.e. $x \in \mathbb{R}^m$, differentiation and $|x - y| \le \varepsilon$ yield

$$C'_\varepsilon v(x) = \int_{B(x,\varepsilon)} \varepsilon^{-m}(\partial_j\rho)(\varepsilon^{-1}(x-y))\,\varepsilon^{-1}(a(y)-a(x))v(y)\,\mathrm{d}y - \partial_j a(x)R_\varepsilon v(x),$$

$$|C'_\varepsilon v(x)| \le \|a\|_{1,\infty}\,(|R^j_\varepsilon v(x)| + |R_\varepsilon v(x)|).$$

(Recall that $W^{1,\infty}(\mathbb{R}^m)$ is isomorphic to the space of bounded Lipschitz functions, [17].) Young's inequality now implies the first assertion. The second one is true for u in the dense subspace $H^1(\mathbb{R}^m)$ and thus on $L^2(\mathbb{R}^m)$ by the uniform estimate. □

With this tool at hand we can extend Lemma 5.2 to all solutions of (5.8) in $C(\overline{J}, L^2_x)$.

Corollary 5.4 *Let* (5.7) *hold and $u \in C(\overline{J}, L^2_x)$ solve* (5.8). *Then the statement of Lemma 5.2 and* (5.10) *are also valid for u. Hence,* (5.8) *has at most one solution in $C(\overline{J}, L^2_x)$.*

Proof We note that $R_\varepsilon u$ belongs to $C(\overline{J}, H^k_x)$ for all $\varepsilon > 0$ and $k \in \mathbb{N}$. Moreover, $R_\varepsilon u$ tends to u in $C(\overline{J}, L^2_x)$ as $\varepsilon \to 0$ since $u(\overline{J})$ is compact and $R_\varepsilon \to I$ strongly in L^2_x. As $\|R_\varepsilon f(t)\|_2 \le \|f(t)\|_2$, dominated convergence also yields $R_\varepsilon f \to f$ in $L^2_{t,x}$. Using

$Lu = f$ and (5.9), we compute

$$LR_\varepsilon u = R_\varepsilon f + [D, R_\varepsilon]u + \sum\nolimits_{j=1}^{3} [A_j, R_\varepsilon]\partial_j u + [A_0, R_\varepsilon]\partial_t u \tag{5.11}$$

$$= R_\varepsilon f + [D, R_\varepsilon]u + [A_0, R_\varepsilon]A_0^{-1}(f - Du) + \sum_{j=1}^{3} \big([A_j, R_\varepsilon] - [A_0, R_\varepsilon]A_0^{-1}A_j\big)\partial_j u.$$

Proposition 5.3 shows that the right-hand side belongs to $L^2_{t,x}$ with uniform bounds. Hence, $R_\varepsilon u$ is also contained $H^1_t L^2_x$ by (5.9). Arguing as above, we further see that the commutator terms tend to 0 in $L^2_{t,x}$ and thus in $L^2_\gamma L^2_x$. Lemma 5.2 and (5.10) for $R_\varepsilon u$ now lead to the first assertion letting $\varepsilon \to 0$. The second one follows from linearity. □

Combining the energy estimate with a clever duality argument, one can also deduce the existence of a solution.

Theorem 5.5 *Let* (5.7) *be true. Then there is a unique function u in* $C(\overline{J}, L^2_x)$ *solving* (5.8). *It satisfies the estimate in Lemma 5.2 and* (5.10).

Proof (1) We need the (formal) adjoint $L^\circ = -\sum_{j=0}^{3} A_j\partial_j + D^\circ$ of L with $D^\circ = D^\top - \operatorname{div} A$. Let $V = \{v \in H^1(J \times \mathbb{R}^3, \mathbb{R}^6) \mid v(T) = 0\}$, $v \in V$, and $L^\circ v = h$. We introduce $\tilde{v}(t) = v(T - t)$ and $f(t) = h(T - t)$ for $t \in \overline{J}$ and the operator $\tilde{L}$ with coefficients $\tilde{A}_0(t) = A_0(T - t)$, $\tilde{A}_j(t) = -A_j(T - t)$ for $j \in \{1, 2, 3\}$ and $\tilde{D}(t) = D^\circ(T - t)$. Note that $\tilde{L}\tilde{v} = f$ and $\tilde{v}(0) = 0$. Applied to $\tilde{L}$, $\tilde{v}$ and $\gamma = \gamma_0(r, \eta)$ at time $T - t$, Lemma 5.2 yields the estimate

$$\|v(t)\|_2^2 = \|\tilde{v}(T-t)\|_2^2 \le \frac{2e^{2\gamma(T-t)}}{\eta \cdot 2\eta\gamma} \int_0^{T-t} e^{-2\gamma\tau} \|h(T - \tau)\|_2^2 \, d\tau \le \frac{e^{2\gamma T}}{\gamma\eta^2} \int_t^T \|h(s)\|_2^2 \, ds,$$

$$\|v\|_{L^2_{t,x}} \le \kappa\sqrt{T}\, \|L^\circ v\|_{L^2_{t,x}}, \qquad \kappa := \frac{1}{\eta\sqrt{\gamma}} e^{\gamma T}. \tag{5.12}$$

In particular, $L^\circ : V \to L^2(J \times \mathbb{R}^3)^6$ is injective. We can thus define the functional

$$\ell_0 : L^\circ V \to \mathbb{R}; \quad \ell_0(L^\circ v) = \langle v, f\rangle_{L^2_{t,x}} + \langle v(0), A_0(0)u_0\rangle_{L^2_x}.$$

The Cauchy–Schwarz inequality and estimate (5.12) imply

$$|\ell_0(L^\circ v)| \le \big(\|f\|_{L^2_{t,x}} + \|A_0(0)u_0\|_{L^2_x}\big)\,\kappa\big(\sqrt{T} + 1\big)\|L^\circ v\|_{L^2_{t,x}}.$$

By the Hahn–Banach theorem, ℓ_0 has an extension ℓ in $(L^2_{t,x})^*$ which can be represented by a function $u \in L^2(J, L^2_x)$ via

$$\begin{aligned}\langle v, f\rangle_{L^2_{t,x}} + \langle v(0), A_0(0)u_0\rangle_{L^2_x} &= \ell(L^\circ v) = \langle L^\circ v, u\rangle_{L^2_{t,x}} \\ &= \langle v, Du\rangle - \sum_{j=0}^{3} \int_0^T \int_{\mathbb{R}^3} \partial_j(A_j v) \cdot u \, dx \, dt \;(\forall\, v \in V).\end{aligned} \tag{5.13}$$

(2) To evaluate (5.13), we first take $v \in H_0^1(J \times \mathbb{R}^3)$. The definition of weak derivatives then leads to $\langle v, f\rangle_{L^2_{t,x}} = \langle v, Lu\rangle_{H_0^1}$; i.e., $Lu = f$ in $H^{-1}_{t,x}$. Hence, u belongs to $H^1_t H^{-1}_x$ because of (5.9) and $f \in L^2_{t,x}$. For $v \in V$, we can now integrate by parts the summand in (5.13) with $j = 0$ in H^{-1}_x; the others are treated as before. As $v(T) = 0$, it follows

$$\langle v, f\rangle_{L^2_{t,x}} + \langle v(0), A_0(0)u_0\rangle_{L^2_x} = \langle v, Lu\rangle_{H_0^1} + \langle A_0(0)v(0), u(0)\rangle_{L^2_x}.$$

Since $A_0(0)$ is symmetric and $Lu = f$, we have also shown that $u(0) = u_0$.

(3) We next use (5.11) for $w_{n,m} = R_{1/n}u - R_{1/m}u$. As in the proof of Corollary 5.4, Proposition 5.3 implies that $w_{n,m}$ is contained in $H^1_{t,x}$ and satisfies $Lw_{n,m} \to 0$ in $L^2_{t,x}$ and $w_{n,m}(0) \to 0$ in L^2_x as $n, m \to \infty$. So $(R_{1/n}u)$ is a Cauchy sequence in $C(\overline{J}, L^2_x)$ by Lemma 5.2, and it converges to u in $L^2_{t,x}$. Thus, u belongs to $C(\overline{J}, L^2_x)$. The other assertions were proven in Corollary 5.4. □

In the time-independent Maxwell case ($A_0 = A_0(x)$ and $A_j = A_j^{\text{co}}$) one can show a similar result if A_0 is only bounded and positive definite (even with boundary conditions), see e.g. Theorem 5.2.5 in [4] or §7.8 in [68]. In the non-autonomous case there are blow-up solutions even for the wave equation on $G = \mathbb{R}$ with Hölder continuous and x-independent coefficients, as shown in [28].

As indicated in Sect. 5.1 and described in the next example, the above result can easily be applied to the linear Maxwell system

$$\partial_t(\varepsilon \mathbf{E}) = \operatorname{curl} \mathbf{H} - \sigma \mathbf{E} - \mathbf{J}_0, \qquad \partial_t(\mu \mathbf{H}) = -\operatorname{curl} \mathbf{E}, \qquad t \ge 0,\ x \in \mathbb{R}^3, \tag{5.14}$$

which is (5.1) on $G = \mathbb{R}^3$ with the material laws $\mathbf{D} = \varepsilon(t,x)\mathbf{E}$ and $\mathbf{B} = \mu(t,x)\mathbf{H}$. We write $\mathbb{R}^{n\times n}_\eta$ for the space of real $n \times n$ matrices $M = M^\top \ge \eta I$.

Example 5.6 Let $\varepsilon, \mu \in W^{1,\infty}(J \times \mathbb{R}^3, \mathbb{R}^{3\times 3}_\eta)$ for some $\eta > 0$, $\sigma \in L^\infty(J \times \mathbb{R}^3, \mathbb{R}^{3\times 3})$, $\mathbf{E}_0, \mathbf{H}_0 \in L^2(\mathbb{R}^3, \mathbb{R}^3)$ and $\mathbf{J}_0 \in L^2(J \times \mathbb{R}^3, \mathbb{R}^3)$. As in (5.5), we set $A_0 = \operatorname{diag}(\varepsilon, \mu)$, $A_j = A_j^{\text{co}}$ for $j = \{1, 2, 3\}$, $D = \operatorname{diag}(\sigma + \partial_t \varepsilon, \partial_t \mu)$, $f = (-\mathbf{J}_0, 0)$, and $u_0 = (\mathbf{E}_0, \mathbf{H}_0)$. Theorem 5.5 then yields a unique solution $(\mathbf{E}, \mathbf{H}) \in C(\overline{J}, L^2_x)$ of (5.14) with $\mathbf{E}(0) = \mathbf{E}_0$

and $\mathbf{H}(0) = \mathbf{H}_0$. It satisfies the energy equality

$$\|\varepsilon(t)^{\frac{1}{2}}\mathbf{E}(t)\|_2^2 + \|\mu(t)^{\frac{1}{2}}\mathbf{H}(t)\|_2^2 = \|\varepsilon(0)^{\frac{1}{2}}\mathbf{E}_0\|_2^2 + \|\mu(0)^{\frac{1}{2}}\mathbf{H}_0\|_2^2 \\ - \int_0^t \int_{\mathbb{R}^3} ((2\sigma + \partial_t\varepsilon\,\mathbf{E} + 2\mathbf{J}_0) \cdot \mathbf{E} + \partial_t\mu\,\mathbf{H}\cdot\mathbf{H})\mathrm{d}x\,\mathrm{d}s.$$

One of the key features of hyperbolic systems is the finite propagation speed of their solutions. To see a simple example first, we look at the standard wave equation $\partial_{tt}u = c^2\partial_{xx}u$ on $\mathbb{R}$ for the wave speed $c > 0$ equipped with the initial conditions $u(0) = u_0$ and $\partial_t u(0) = v_0$. (As in Part I one can put this second-order equation in the above first-order framework for the new state $(\partial_t u, \partial_x u)$.) The solution of this wave problem is given by d'Alembert's formula

$$u(t,x) = \tfrac{1}{2}(u_0(x+ct) + u_0(x-ct)) + \frac{1}{2c}\int_{x-ct}^{x+ct} v_0(s)\,\mathrm{d}s, \qquad t \geq 0,\ x \in \mathbb{R}.$$

Hence, the solution at (x,t) only depends on the initial data on $[x-ct, x+ct]$; for instance, $u(t,x) = 0$ if u_0 and v_0 vanish on $[x-ct, x+ct]$. Conversely, the value of u_0 and v_0 at y influences u at most for (t,x) with $|x-y| \leq ct$; i.e., on a triangle with vertex $(y,0)$ and lateral sides of slope $\pm 1/c$. In this sense, c is the speed of propagation.

We extend these observations to the system (5.8), assuming (5.7). In the statement we use the backward 'light' cone

$$\Gamma(x_0, R, K) = \{(t,x) \in \mathbb{R}_{\geq 0} \times \mathbb{R}^3 \mid |x - x_0| < R - Kt\}.$$

It has the base $B(x_0, R)$ at $t = 0$ and the apex $(\frac{R}{K}, x_0)$. Set

$$k_0^2 = \|A_1\|_\infty^2 + \|A_2\|_\infty^2 + \|A_3\|_\infty^2$$

with the operator norm for $|\cdot|_2$ on $\mathbb{R}^{6\times 6}$. Note that $k_0 = \sqrt{3}$ in the Maxwell example.

Below we see (for $f = 0$) that u vanishes on $\Gamma(x_0, R, k_0/\eta)$ if $u_0 = 0$ on $B(x_0, R)$. Hence, if two initial functions u_0 and $\tilde{u}_0$ coincide on $B(x_0, R)$ then the corresponding solutions u and $\tilde{u}$ are equal on $\Gamma(x_0, R, k_0/\eta)$. In other words, the values of u_0 outside $B(x_0, R)$ influence $u(t)$ only off $\Gamma(x_0, R, k_0/\eta)$, that is, with maximal speed k_0/η. Our proof is based on energy estimates with an exponential weight, and the arguments are taken from §4.2.2 of [7].

Theorem 5.7 *Let* (5.7) *be true. Assume that* $u_0 = 0$ *on* $B(x_0, R)$ *and* $f = 0$ *on* $\Gamma(x_0, R, k_0/\eta)$ *for some* $R > 0$ *and* $x_0 \in \mathbb{R}^3$. *Then the solution* $u \in C(\overline{J}, L_x^2)$ *of* (5.8) *also vanishes on* $\Gamma(x_0, R, k_0/\eta)$.

Proof (1) Let $\delta, R > 0$ and $x_0 \in \mathbb{R}^3$ be given. There is a function $\psi \in C^\infty(\mathbb{R}^3)$ with $|\nabla\psi| \le \eta/k_0$ (for the euclidean norm) and

$$-2\delta + \eta k_0^{-1}(R - |x - x_0|) \le \psi(x) \le -\delta + \eta k_0^{-1}(R - |x - x_0|), \qquad x \in \mathbb{R}^3. \tag{5.15}$$

We construct ψ as in Theorem 6.1 of [157]. Take $\chi(s) = -\frac{3}{2}\delta + \eta k_0^{-1}(R - |s|)$ for $s \in \mathbb{R}$. This function is Lipschitz with constant η/k_0. The same is true for the mollified map $\chi_\varepsilon = R_\varepsilon \chi$ as $\nabla \chi_\varepsilon = R_\varepsilon \nabla \chi$. Also, χ_ε tends uniformly to χ as $\varepsilon \to 0$ since

$$|\chi_\varepsilon(s) - \chi(s)| \le \int_{\mathbb{R}} \varepsilon^{-1}\rho(\varepsilon^{-1}\tau)\,|\chi(s-\tau) - \chi(s)|\,\mathrm{d}\tau \le \eta k_0^{-1}\varepsilon \int_{\mathbb{R}} \rho(\sigma)\,|\sigma|\,\mathrm{d}\sigma.$$

We fix a small $\varepsilon > 0$ such that χ_ε satisfies (5.15) with s instead of $|x - x_0|$ and $5/3$ instead of 2. Then $\psi(x) = \chi_\varepsilon((\delta_0^2 + |x - x_0|^2)^{1/2})$ does the job, where $\delta_0 = k_0\delta(3\eta)^{-1}$.

Set $\phi(t, x) = \psi(x) - t$ and $u_\tau = \mathrm{e}^{\tau\phi}u$ for $\tau > 0$. Inequality (5.15) yields $\psi(x) \le -\delta + t$ if $|x - x_0| \ge R - k_0 t/\eta$ (i.e., $(t, x) \notin \Gamma(x_0, R, k_0/\eta)$), so that $\mathrm{e}^{\tau\phi} \le \mathrm{e}^{-\tau\delta} \le 1$ off $\Gamma(x_0, R, k_0/\eta)$ and $\mathrm{e}^{\tau\phi}$ is bounded on $J \times \mathbb{R}^3$. We further have $\nabla \mathrm{e}^{\tau\phi} = \tau\nabla\psi \mathrm{e}^{\tau\phi}$ and $\partial_t \mathrm{e}^{\tau\phi} = -\tau \mathrm{e}^{\tau\phi}$. As a result, u_τ is an element of $C(\overline{J}, L_x^2)$ and the right-hand side of

$$Lu_\tau = \mathrm{e}^{\tau\phi} f - \tau\Big(A_0 - \sum\nolimits_{j=1}^3 A_j \partial_j \psi\Big) u_\tau$$

belongs to $L^2_{t,x}$. The matrix in parentheses is denoted by M.

(2) For $\xi \in \mathbb{R}^6$ we have $M\xi \cdot \xi \ge (\eta - k_0|\nabla\psi|)|\xi|^2 \ge 0$. Set $C = \frac{1}{2}\operatorname{div} A - D$ and $\kappa = \|C\|_\infty$. By Theorem 5.5, the function u_τ satisfies the energy equality

$$\|A_0(t)^{\frac{1}{2}} u_\tau(t)\|_{L_x^2}^2 = \|A_0(0)^{\frac{1}{2}} u_\tau(0)\|_{L_x^2}^2 + 2\langle (C - \tau M)u_\tau + \mathrm{e}^{\tau\phi} f, u_\tau\rangle_{L^2_{t,x}}.$$

Using Cauchy–Schwarz, the above inequalities and Gronwall, we estimate

$$\eta\,\|u_\tau(t)\|_{L_x^2}^2 \le \|A_0(0)\|_\infty\,\|\mathrm{e}^{\tau\phi}u_0\|_{L_x^2}^2 + \|\mathrm{e}^{\tau\phi} f\|_{L^2_{t,x}}^2 + (2\kappa + 1)\int_0^t \|u_\tau(s)\|_{L_x^2}^2\,\mathrm{d}s,$$

$$\|\mathrm{e}^{\tau\phi}u(t)\|_{L_x^2}^2 \lesssim_T \|\mathrm{e}^{\tau\phi}u_0\|_{L_x^2}^2 + \|\mathrm{e}^{\tau\phi} f\|_{L^2_{t,x}}^2.$$

The right-hand side tends to 0 as $\tau \to \infty$ since u_0 and f vanish on $\Gamma(x_0, R, k_0/\eta)$ and $\mathrm{e}^{\tau\phi} \to 0$ uniformly off $\Gamma(x_0, R, k_0/\eta)$. Hence, $u(t)$ has to be 0 on $\{\phi > \delta\} = \{\psi > t + \delta\}$. By (5.15), this set includes points (t, x) with $|x - x_0| < R - k_0\eta^{-1}(t + 3\delta)$. Since $\delta > 0$ is arbitrary here, u equals 0 on $\Gamma(x_0, R, k_0/\eta)$. $\square$

5.3 The Linear Problem on $\mathbb{R}^3$ in H^3

As noted in Sect. 5.1, to solve the nonlinear problem (5.6) we will set $A_0 = a_0(v)$ for functions v having the same regularity as the desired solution u. Since A_0 has to be Lipschitz in Theorem 5.5, the same must be true for v. Working in H_x^k spaces, we thus need solutions in $L_t^\infty H_x^3 \cap W_t^{1,\infty} H_x^2$ at least. We want to reduce the problem in H_x^3 to that in L_x^2 by means of a transformation. (One could also perform the proof of Theorem 5.5 in H_x^3 instead of L_x^2, see e.g. [11] or [26], which would require more work in our context.)

To this end, we define the square root $\Lambda = (I - \Delta)^{1/2} = \mathcal{F}^{-1}(1 + |\xi|^2)^{1/2}\mathcal{F}$ of the shifted Laplacian on $L^2(\mathbb{R}^3)$, where $\mathcal{F}$ is the Fourier transform. Using standard properties of $\mathcal{F}$, one can check that Λ commutes with derivatives and that it can be extended, respectively restricted, to isomorphisms $H_x^k \to H_x^{k-1}$ for $k \in \mathbb{Z}$ with inverse given by $\Lambda^{-1} = (I - \Delta)^{-1/2} = \mathcal{F}^{-1}(1 + |\xi|^2)^{-1/2}\mathcal{F}$. Observe that $\Lambda = (I - \Delta)\Lambda^{-1}$ and that Λ^{-1} is a convolution operator with positive kernel, see Proposition 6.1.2 in [77]. Hence, Λ leaves invariant real-valued functions.

Our analysis relies on a commutator estimate for Λ^3 and $M_a : \varphi \mapsto a\varphi$ which gains a derivative. In Lemma A2 in [106] it is shown that

$$\|[\Lambda^3, M_a]\|_{\mathcal{B}(H^2(\mathbb{R}^3), L^2(\mathbb{R}^3))} \lesssim \|\nabla a\|_{H^2(\mathbb{R}^3)}. \tag{5.16}$$

Here the space dimension 3 is crucial; on $\mathbb{R}^m$ one obtains e.g. an analogous bound for $[\Lambda^k, M_a] : H_x^{k-1} \to L_x^2$ with $k > \frac{m}{2} + 1$. (Noninteger k are also allowed here.)

Guided by (5.16) and (5.7), we introduce the space

$$\tilde{F}^k(J) = \tilde{F}^k(T) = \{A \in W^{1,\infty}(J \times \mathbb{R}^3, \mathbb{R}^{6\times 6}) \mid \nabla_{t,x} A \in L_t^\infty H_x^{k-1}\}, \qquad k \in \mathbb{N},$$

for the coefficients, endowed with its natural norm. We will usually take $k = 3$. We use the same notation for vector- or scalar-valued functions of the same regularity. The subscript sym will refer to symmetric matrices and η to those with $A = A^\top \geq \eta I$ with $\eta > 0$. We state the hypotheses of the present section:

$$\begin{aligned} &A_0 \in \tilde{F}_\eta^3(J), \quad A_1, A_2, A_3 \in \tilde{F}_{\text{sym}}^3(J), \quad D \in \tilde{F}^3(J), \\ &u_0 \in H_x^3 = H^3(\mathbb{R}^3, \mathbb{R}^6), \quad f \in Z^3(J) = Z^3(T) := L^2(J, H_x^3) \cap H^1(J, H_x^2). \end{aligned} \tag{5.17}$$

Set $\|f\|_{Z_\gamma^3(J)}^2 = \|e_{-\gamma} f\|_{L_t^2 H_x^3}^2 + \|e_{-\gamma}\partial_t f\|_{L_t^2 H_x^2}^2$ for $\gamma \geq 0$. We also use the spaces $\hat{H}_x^k = \{v \in L^\infty(\mathbb{R}^3) \mid \nabla_x v \in H_x^{k-1}\}$ and $\tilde{G}^k(J) = \tilde{G}^k(T) = C(\overline{J}, H_x^k) \cap C^1(\overline{J}, H_x^{k-1})$ with their natural norms. (Such spaces will also be considered on other time intervals.)

On $\mathbb{R}^3$ we have the product estimates

$$\|vw\|_{H^j} \lesssim \|v\|_{H^k}\|w\|_{H^j} \qquad \text{for } v \in H_x^k,\ \ w \in H_x^j,\ \ k \geq \max\{j, 2\}.$$

Here one can replace H_x^k by $\hat{H}_x^k$, as well as H_x^j and H_x^k by $\tilde{G}^j(J)$ and $\tilde{G}^j(J)$ (or $\tilde{F}^j(J)$), or by $\tilde{F}^j(J)$ and $\tilde{F}^k(J)$. Also, if $A \in \hat{H}_\eta^k$ for $k \in \mathbb{N}$, then A^{-1} belongs to $\hat{H}_\eta^k$ with norm bounded by $c(\eta, k)(1+\|A\|_{\hat{H}_x^k})^{k-1}\|A\|_{\hat{H}_x^k}$. See Lemmas 2.1 and 2.3 in [159].

We sketch the proof of the first claim, using Sobolev embeddings such as $H^2 \hookrightarrow L^p$ for $p \in [2, \infty]$ and $H^1 \hookrightarrow L^q$ for $q \in [2, 6]$ on $\mathbb{R}^3$. By the product rule (and interpolative inequalities) we have to estimate $\partial_x^\beta v \partial_x^{\alpha-\beta} w$ for multi-indices $0 \leq \beta \leq \alpha$ with $|\alpha| = j$. Observe that $\partial_x^\beta v \in H_x^{k-|\beta|}$ and $\partial_x^{\alpha-\beta} w \in H_x^{|\beta|}$. This product can be estimated in L_x^2 as needed if $k - |\beta| \geq 2$ or $|\beta| \geq 2$ since then v respectively w is bounded. As $k \geq 2$, only the case $|\beta| = 1$ remains. Here $\partial_x^\beta v$ and $\partial_x^{\alpha-\beta} w$ belong to $H_x^1 \hookrightarrow L_x^4$ and thus the product to L_x^2. The listed variants are proved similarly.

We look for a solution $u \in \tilde{G}^3(J)$ of (5.8) assuming (5.17). The basic idea is to solve a modified problem for $w = \Lambda^3 u$ in $C(\overline{J}, L_x^2)$. Since the inequality (5.16) only improves space regularity, we first replace the equation $Lu = f$ by $\hat{L}u = \hat{f} := A_0^{-1} f$ where $\hat{L}$ has the coefficients $\hat{A}_0 = I$, $\hat{A}_j = A_0^{-1} A_j$ and $\hat{D} = A_0^{-1} D$. We then obtain

$$\hat{L}w = \Lambda^3 \hat{f} + \sum\nolimits_{j=1}^{3} [\hat{A}_j, \Lambda^3]\partial_j u + [\hat{D}, \Lambda^3]u,$$

$$Lw = A_0\Lambda^3 \hat{f} + \sum\nolimits_{j=1}^{3} A_0[\hat{A}_j, \Lambda^3]\partial_j u + A_0[\hat{D}, \Lambda^3]u =: g(f, u). \tag{5.18}$$

We now replace in g the unknown u by a given function $v \in C(\overline{J}, H_x^3)$. Theorem 5.5 gives a solution $w \in C(\overline{J}, L_x^2)$ of $Lw = g(f, v)$ with $w(0) = \Lambda^3 u_0$. The energy estimate from Lemma 5.2 (with a large γ) then implies that $\Phi : v \mapsto \Lambda^{-3} w$ is a strict contraction on $L_\gamma^\infty H_x^3$. This fact will lead to the desired regularity result. Let λ be the maximum of $\|\Lambda^k\|_{\mathcal{B}(H^k, L^2)}$ and $\|\Lambda^{-k}\|_{\mathcal{B}(L^2, H^k)}$ for $k \in \{2, 3\}$. It will be important in the fixed-point argument for the nonlinear problem that the constant c_0 in (5.19) only depends on r_0 (and η), but not on r.

Theorem 5.8 *Let* (5.17) *be true. Then there is a unique solution* $u \in C(\overline{J}, H_x^3) \cap C^1(\overline{J}, H_x^2)$ *of* (5.8)*. For* $t \in \overline{J}$ *and* $\gamma \geq \gamma_1(r, \eta) := \max\{\gamma_0(r, \eta), \sqrt{c_1}\}$ *we have*

$$\gamma \|u\|^2_{Z^3_\gamma(0,t)} + e^{-2\gamma t}(\|u(t)\|^2_{H_x^3} + \|\partial_t u(t)\|^2_{H_x^2})$$
$$\leq c_0(\|u_0\|^2_{H_x^3} + \|f(0)\|^2_{H_x^2}) + \tfrac{c_1}{\gamma}\|f\|^2_{Z^3_\gamma(0,t)}, \tag{5.19}$$

where $\|A_j(0)\|_{\hat{H}_x^2}, \|D(0)\|_{\hat{H}_x^2} \leq r_0$, $\|A_j\|_{\tilde{F}^3(J)}, \|D\|_{\tilde{F}^3(J)} \leq r$ *for* $j \in \{0, 1, 2, 3\}$*, and* $c_0 = c_0(r_0, \eta)$ *and* $c_1 = c_1(r, \eta)$ *are constants described in the proof.*

Proof (1) Take $v \in C(\overline{J}, H_x^3)$ and $\gamma \geq \gamma_0(r, \eta)$ from Lemma 5.2 and the text following it. Using the above product rules and (5.16) we see that the square of the norm in $L_\gamma^2 L_x^2$ of $g(f, v)$ from (5.18) is bounded by $c_1'(\|f\|_{L_\gamma^2 H_x^3}^2 + \|v\|_{L_\gamma^2 H_x^3}^2)$ for a constant $c_1' = c_1'(r, \eta)$. Theorem 5.5 yields a solution $w \in C(\overline{J}, L_x^2)$ of $Lw = g(f, v)$ and $w(0) = \Lambda^3 u_0 =: w_0$ which satisfies

$$\tfrac{\gamma\eta}{4}\|w\|_{L_\gamma^2 L_x^2}^2 + \tfrac{\eta}{2}\|w\|_{L_\gamma^\infty L_x^2}^2 \leq c_0'\|u_0\|_{H_x^3}^2 + \tfrac{c_1'}{2\gamma\eta}\big(\|f\|_{L_\gamma^2 H_x^3}^2 + \|v\|_{L_\gamma^2 H_x^3}^2\big) \tag{5.20}$$

with $c_0' = \frac{\lambda^2}{2}\|A_0(0)\|_\infty$. The map w also belongs to $C^1(\overline{J}, H_x^{-1})$ because of (5.9) and $f \in Z^3(J)$. Set $\Phi v = \Lambda^{-3} w \in \tilde{G}^3(J)$. Let $\overline{w}$ satisfy $L\overline{w} = g(f, \overline{v})$ and $\overline{w}(0) = w_0$ for some $\overline{v} \in C(\overline{J}, H_x^3)$. For $w - \overline{w}$ estimate (5.20) applies with $u_0 = 0$ and $f = 0$ so that

$$\|\Phi(v - \overline{v})\|_{L_\gamma^\infty H_x^3} = \|\Lambda^{-3}(w - \overline{w})\|_{L_\gamma^\infty H_x^3} \leq \frac{\lambda\sqrt{c_1' T}}{\sqrt{\gamma\eta}}\|v - \overline{v}\|_{L_\gamma^\infty H_x^3}.$$

Fixing a large $\gamma = \gamma(r, \eta, T)$, we obtain a fixed point u of Φ in $L_\gamma^\infty H_x^3$. It actually belongs to $\tilde{G}^3(J)$ and satisfies $u(0) = u_0$. Equation (5.18) implies that $Lu = f$. Uniqueness of solutions was already shown in Corollary 5.4.

(2) It remains to establish (5.19). We first insert $u = v$ and $w = \Lambda^3 u$ in (5.20) and take $\gamma \geq \max\Big\{\gamma_0(r, \eta), \frac{2\lambda\sqrt{c_1'}}{\eta}\Big\}$. Absorbing $\|u\|_{L_\gamma^2 H_x^3}^2$ by the left-hand side, we infer

$$\tfrac{\gamma\eta}{8}\|u\|_{L_\gamma^2 H_x^3}^2 + \tfrac{\eta}{2}\|u\|_{L_\gamma^\infty H_x^3}^2 \leq c_0'\lambda^2\|u_0\|_{H_x^3}^2 + \tfrac{c_1'\lambda^2}{2\gamma\eta}\|f\|_{L_\gamma^2 H_x^3}^2. \tag{5.21}$$

If we estimated $\partial_t u$ in H_x^2 by means of (5.9) and (5.21), we would obtain a constant depending on r in front of the norm of u_0. Instead we use that $\partial_t u \in C(\overline{J}, H_x^2)$ satisfies

$$L\partial_t u = \partial_t f - \partial_t D u - \sum\nolimits_{j=0}^{3} \partial_t A_j \partial_j u =: h,$$

$$\partial_t u(0) = A_0(0)^{-1} f(0) - A_0(0)^{-1} D(0) u_0 - \sum\nolimits_{j=1}^{3} A_0(0)^{-1} A_j(0)\partial_j u_0 =: v_0.$$

The above product rules yield

$$\|h(t)\|_{H_x^2} \leq \|\partial_t f(t)\|_{H_x^2} + \overline{c}(r)(\|u(t)\|_{H_x^3} + \|\partial_t u(t)\|_{H_x^2},$$

$$\|v_0\|_{H_x^2} \leq c(r_0, \eta)(\|f(0)\|_{H_x^2} + \|u_0\|_{H_x^3}).$$

The commutator $[M_a, \Lambda^2] = [M_a, -\Delta] : H_x^1 \to L_x^2$ is bounded if $a \in W_x^{1,\infty}$ and $D^2 a \in H_x^1 \hookrightarrow L_x^3$. Starting from $L\partial_t u = h$, as in (5.18) and (5.20) we thus deduce

$$\frac{\gamma\eta}{4}\|\partial_t u\|_{L^2_\gamma H^2_x}^2 + \frac{\eta}{2}\|\partial_t u\|_{L^\infty_\gamma H^2_x}^2$$
$$\le \hat{c}_0\lambda^2\big(\|u_0\|_{H^3_x}^2 + \|f(0)\|_{H^2_x}^2\big) + \frac{\hat{c}_1\lambda^2}{2\gamma\eta}\big(\|\partial_t f\|_{L^2_\gamma H^2_x}^2 + \|u\|_{L^2_\gamma H^3_x}^2 + \|\partial_t u\|_{L^2_\gamma H^2_x}^2\big)$$

for constants $\hat{c}_0 = \hat{c}_0(r_0, \eta)$ and $\hat{c}_1 = \hat{c}_1(r, \eta)$. Set $c_0 = 16\lambda^2\eta^{-1}(c_0' + \hat{c}_0)$ and $c_1 = \frac{8\lambda^2}{\eta^2}\max\{c_1', \hat{c}_1\}$. We add the above inequality to (5.21) and take $\gamma \ge \gamma_1(r, \eta) := \max\{\gamma_0(r, \eta), \sqrt{c_1}\}$. Estimate (5.19) follows after some calculations. □

In the above result we control more space than time derivatives. Under stronger assumptions on A_j, D and f, one can obtain analogous estimates on $\partial_t^2 u$ in H_x^1 and $\partial_t^3 u$ in L_x^2 by differentiating (5.8) in time, see (6.23). We discuss variants of the above theorem partly needed below.

Corollary 5.9 *Let A_j and D be as in Theorem 5.8, as well as $u_0 \in H_x^2$ and $f \in L^2(J, H_x^2)$. Then there is a unique solution $u \in C(\overline{J}, H_x^2) \cap C^1(\overline{J}, H_x^1)$ of (5.8). For $t \in \overline{J}$ and $\gamma \ge \tilde{\gamma}_1(r, \eta) := \max\{\gamma_0(r, \eta), \sqrt{\tilde{c}_1}\}$, we have*

$$\gamma\|u\|_{L^2_\gamma((0,t),H^2_x)}^2 + +\mathrm{e}^{-2\gamma t}\|u(t)\|_{H^2_x}^2 \le \tilde{c}_0\|u_0\|_{H^2_x}^2 + \frac{\tilde{c}_1}{\gamma}\|f\|_{L^2_\gamma((0,t),H^2_x)}^2$$

for constants $\tilde{c}_0 = \tilde{c}_0(r_0, \eta)$ and $\tilde{c}_1 = \tilde{c}_1(r, \eta)$. If $\partial_t f \in L^2(J, H_x^1)$ we also obtain

$$\gamma\|\partial_t u\|_{L^2_\gamma((0,t),H^1_x)}^2 + \mathrm{e}^{-2\gamma t}\|\partial_t u(t)\|_{H^1_x}^2 \le \tilde{c}_0(\|u_0\|_{H^2_x}^2 + \|f(0)\|_{H^1_x}^2) + \frac{\tilde{c}_1}{\gamma}\|f\|_{Z^2_\gamma(0,t)}^2$$

where $Z^k(J) := L^2(J, H_x^k) \cap H^1(J, H_x^{k-1})$ for $k \in \mathbb{N}$.

The result is shown as Theorem 5.8, replacing Λ^3 by Λ^2 in its proof up to (5.21) and Λ^2 by Λ afterwards. For the second part one also uses that the commutator $[M_a, \Lambda]$ is bounded on L_x^2 by Proposition 4.1.A in [162] if $a \in W_x^{1,\infty}$.

Remark 5.10 In Theorem 5.8 we have focused on the space H_x^3 needed for the quasilinear problem. Actually, one obtains a unique solution $u \in \tilde{G}^k(J)$ of (5.8) satisfying the analogue of (5.19) if $u_0 \in H_x^k$, $f \in Z^k(J)$, $A_j, D \in \tilde{F}^k(J)$, $A_j = A_j^\top$, $A_0 \ge \eta I$, and $k \in \mathbb{N}\setminus\{2\}$. For $k = 2$ one needs another assumption stated below. This can be shown as for $k = 3$, one only has to take care of estimates for products, inverse matrices and commutators.

Indeed, for $k > 3$ one can use the product and inversion results mentioned above and the higher-order version of (5.16) in [106]. For $k = 1$ (thus for coefficients in $W_{t,x}^{1,\infty}$) the needed product and inversion bounds are easy to check, and we have just seen that $[M_a, \Lambda]$ is bounded on L^2_x if $a \in W^{1,\infty}_x$. For $k = 2$ the second-order derivatives of A_j also have to belong to $L^\infty_t L^3_x$. Then the commutator $[M_a, \Lambda^2] = [M_a, -\Delta] : H^1_x \to L^2_x$ is bounded, and the extra condition is preserved by products and inverses.

Moreover, there is no problem to change the range space $\mathbb{R}^6$ to $\mathbb{R}^n$. Also other spatial domains $\mathbb{R}^m$ can be treated analogously, though one has to modify the assumptions on the coefficients in this case. Finally, invoking a bit more harmonic analysis one can also work in fractional Sobolev spaces H^s_x instead of H^k_x, see [105].

Remark 5.11 In (5.17) we have required that the derivatives of the coefficients belong to H^j_x spaces. So local singularities are allowed to some extent, but one enforces a certain decay at infinity which is an unnecessary restriction. Actually Theorem 5.8 remains valid if we replace the space $\hat{F}^3(J)$ by $\hat{F}^3_\infty(J) = \hat{F}^3(J) + W^{3,\infty}_{t,x}$, and $\hat{H}^2_x$ by $\hat{H}^2_\infty = \hat{H}^2_x + W^{2,\infty}_x$. (They have the norm of sums $X + Y$, namely $\|z\|_{X+Y} = \inf_{z=x+y} \|x\|_X + \|y\|_Y$.) To show this fact, we note that $[M_A, \Lambda^2] : H^2_x \to H^1_x$ is bounded uniformly in t if $A \in \hat{F}^3(J) + W^{3,\infty}_{t,x}$, and so the same is true for

$$[M_A, \Lambda^3] = [M_A, \Lambda]\Lambda^2 + \Lambda[M_A, \Lambda^2] : H^2_x \to L^2_x.$$

(Recall the boundedness of $[M_a, \Lambda]$ on L^2_x.) One can further show the appropriate bounds for products and inversions involving $\hat{F}^3(J) + W^{3,\infty}_{t,x}$ and $\hat{H}^2_x + W^{2,\infty}_x$, as well as $\tilde{G}^3(J)$. The analogue of Theorem 5.8 can now be proven as before.

As a preparation for Theorem 5.18 on the wellposedness of the nonlinear problem we show an approximation result for the coefficients.

Lemma 5.12 *Let $u_0 \in L^2_x$, $f \in L^2_{t,x}$, $n \in \mathbb{N} \cup \{\infty\}$, $j \in \{0, 1, 2, 3\}$, $A^n_j \in \hat{F}^3_\infty(J)$ be symmetric with $A^n_0 \geq \eta I$, and $D^n \in \hat{F}^3_\infty(J)$. Assume that $\|A^n_j\|_{W^{1,\infty}_{t,x}} \leq r$ and $\|D^n\|_{L^\infty_{t,x}} \leq r$, as well as $A^n_j \to A^\infty_j$ and $D^n \to D^\infty$ in $L^\infty_{t,x}$ as $n \to \infty$. Set $L_n = \sum_j A^n_j \partial_j + D^n$. We have functions $u_n \in C(\overline{J}, L^2_x)$ with $L_n u_n = f$ and $u_n(0) = u_0$. Then $u_n \to u_\infty$ in $C(\overline{J}, L^2_x)$ as $n \to \infty$.*

Proof For the given data there are functions $u_{0,m}$ in H^3_x and f_m in $Z^3(J)$ converging to u_0 and f in L^2_x and $L^2_{t,x}$, respectively, as $m \to \infty$. For these data Theorem 5.8 provides functions $u_{n,m} \in \tilde{G}^3(J)$ satisfying $L_n u_{n,m} = f_m$ and $u_{n,m}(0) = u_{0,m}$. Fixing $\gamma = \gamma_0(r, \eta)$ from Lemma 5.2, Corollary 5.4 then shows

$$\|u_n - u_{n,m}\|_{L^\infty_t L^2_x} \leq c\|u_n - u_{n,m}\|_{L^\infty_\gamma L^2_x} \leq c\big(\|u_0 - u_{0,m}\|^2_{L^2_x} + \|f - f_m\|^2_{L^2_{t,x}}\big).$$

with $c = c(r, \eta, T)$. The right-hand side tends to 0 as $m \to \infty$ uniformly for $n \in \mathbb{N}\cup\{\infty\}$. It is thus enough to take $u_0 \in H_x^3$, $f \in Z^3(J)$, and $u_n \in \tilde{G}^3(J)$. We have

$$L_n(u_n - u_\infty) = L_\infty u_\infty - L_n u_\infty = \sum\nolimits_{j=0}^{3} (A_j^\infty - A_j^n)\partial_j u_\infty + (D^\infty - D^n)u_\infty =: g_n$$

Since $u_\infty \in \tilde{G}^3(T)$, as above Lemma 5.2 yields

$$\|u_n - u_\infty\|_{L_t^\infty L_x^2} \le c(\gamma, T)\, \|g_n\|_{L_\gamma^\infty L_x^2} \longrightarrow 0, \qquad n \to \infty.$$

□

5.4 The Quasilinear Problem on $\mathbb{R}^3$

In this section we treat the nonlinear system

$$L(u)u := \sum\nolimits_{j=0}^{3} a_j(u)\partial_j u + d(u)u = f, \quad t \ge 0,\ x \in \mathbb{R}^3, \qquad u(0) = u_0, \tag{5.22}$$

under the assumptions

$$\begin{aligned} &a_j, d \in C^3(\mathbb{R}^3 \times \mathbb{R}^6, \mathbb{R}^{6\times 6}), \quad a_j = a_j^\top, \quad a_0 \ge \eta I, \ \ \eta \in (0,1], \\ &\forall r > 0: \sup_{|\xi|\le r} \max_{0\le|\alpha|\le 3} \|\partial_x^\alpha a_j(\cdot,\xi)\|_{L_x^\infty}, \|\partial_x^\alpha d(\cdot,\xi)\|_{L_x^\infty} < \infty, \quad j \in \{0,1,2,3\}, \\ &u_0 \in H_x^3, \quad \forall T > 0: f \in Z^3(T) = Z^3(J) = L^2(J, H_x^3) \cap H^1(J, H_x^2), \quad J = (0,T). \end{aligned} \tag{5.23}$$

One can also treat coefficients only defined for $(x, \xi) \in \mathbb{R}^3 \times U$ and an open subset $U \subseteq \mathbb{R}^6$, see Remark 5.19. This is already needed in the Kerr Example 5.1 if χ_3 is not non-negative. To simplify a bit, we focus on the case $U = \mathbb{R}^6$ in (5.23).

We look for solutions u of (5.22) in $C([0, T_+), H_x^3) \cap C^1([0, T_+), H_x^2)$ for a maximally chosen final time $T_+ \in (0, \infty]$. As indicated in the next section, solutions may blow up and so T_+ could be finite. The solutions will be constructed in a fixed-point argument on the space $\tilde{G}^{k-}(J) = L^\infty(J, H_x^k) \cap W^{1,\infty}(J, H_x^{k-1})$ endowed with its natural norm, where $k = 3$. The overall strategy of this section and many techniques are typical for quasilinear (or semilinear) evolution equations, though there are different (but related) approaches, see e.g. [7], [11], or [97].

We first state basic properties of substitution operators. (Recall Remark 5.11 concerning $\tilde{F}_\infty^3(J)$ and $\hat{H}_\infty^2$.) We set $E_\gamma = L_\gamma^\infty(J, H_x^2)$ for a moment.

Lemma 5.13 *Let a be as in* (5.23) *and* $\gamma \geq 0$.

(a) Let $v \in \tilde{G}^3(J)$ *with* $\|v\|_\infty \leq r$. *Then* $\|a(v)\|_{\tilde{F}^3_\infty(J)} \leq \kappa(r)(1 + \|v\|^3_{\tilde{G}^3(J)})$.
(b) Let $v, w \in L^\infty_t H^2_x$ *with norm* $\leq r$. *Then* $\|a(v) - a(w)\|_{E_\gamma} \leq \kappa(r)\,\|v - w\|_{E_\gamma}$. *Here we can also replace* $L^\infty_t H^2_x$ *and* E_γ *by* $\tilde{G}^2(J)$ *and* $\tilde{G}^2_\gamma(J)$, *respectively.*
(c) Let $v_0 \in H^2_x$ *with* $\|v_0\|_\infty \leq r_0$. *Then* $\|a(v_0)\|_{\hat{H}^2_\infty} \leq \kappa_0(r_0)(1 + \|v_0\|^2_{H^2_x})$.
(d) Let $v_0, w_0 \in H^2_x$ *with norm* $\leq r_0$. *Then* $\|a(v_0) - a(w_0)\|_{H^2_x} \leq \kappa_0(r_0)\|v_0 - w_0\|^2_{H^2_x}$.

We only sketch the proof. (See §7.1 in [157] or §2 in [158] for more details.) Take $\alpha \in \mathbb{N}_0^4$ with $1 \leq |\alpha| \leq 3$ and $\alpha_0 \in \{0, 1\}$. The latter refers to the time derivative. It is clear that the function $|(\partial^\beta a)(v)|$ is bounded by $c(r)$ for all $0 \leq |\beta| \leq 3$ where $\beta = (\beta_x, \beta_\xi) \in \mathbb{N}_0^3 \times \mathbb{N}_0^6$. Note that $\partial^\alpha a(v)$ is a linear combination of products of $(\partial^\beta a)(\cdot, v)$ and $j \in \{0, 1, 2, 3\}$ factors $\partial^{\gamma_i} v$ with $\beta_x + \gamma_1 + \cdots + \gamma_j = \alpha$. Since $v \in W^{1,\infty}_{t,x}$ by Sobolev's embedding, one can estimate $\partial^\alpha a(v)$ in $L^\infty_t L^2_x$ if $j \geq 1$ and in $L^\infty_{t,x}$ if $j = 0$, both by $c(r)(1 + \|v\|^3_{\tilde{G}^3(J)})$. For (b) we start from the formula

$$a(v) - a(w) = \int_0^1 \partial_\xi a(\cdot, v + s(w - v))\,(w - v)\,\mathrm{d}s$$

and proceed as above. Parts (c) and (d) are treated similarly.

As the space for the fixed-point argument we will use

$$E(R, T) := \{v \in \tilde{G}^{3-}(J) \mid \|v\|_{\tilde{G}^{3-}(J)} \leq R,\ v(0) = u_0.\}$$

for suitable $R \geq \|u_0\|_{H^3}$ and $T > 0$. This set is non-empty as it contains the constant function $t \mapsto v(t) = u_0$. It is crucial that $E(R, T)$ is complete for a metric involving only two derivatives, which can be shown by a standard application of the Banach–Alaoglu theorem. For this we recall that $L^\infty_t L^2_x$ is the dual space of $L^1_t L^2_x$, see Corollary 1.3.22 in [96]. (This is the reason to take L^∞ in time instead of C.)

Lemma 5.14 *The space* $E(R, T)$ *is complete with the metric* $\|u - v\|_{L^\infty_t H^2_x}$.

Proof Let (u_n) be Cauchy in $E(R, T)$ with this metric. Then (u_n) has a limit u in $C(\bar{J}, H^2_x)$. Take $\alpha \in \mathbb{N}_0^4$ with $\alpha_0 \leq 1$ and $0 \leq |\alpha| \leq 3$. Applying Banach–Alaoglu iteratively, we obtain a subsequence (also denoted by (u_n)) such that $\partial^\alpha u_n$ tends to a function v_α weak* in $L^\infty_t L^2_x$ which also satisfies $\sum_\alpha \|v_\alpha\|^2_{L^\infty_t L^2_x} \leq R^2$. It remains to check that $v_\alpha = \partial^\alpha u$. To this end, take $\varphi \in H^3_0(J \times \mathbb{R}^3)$. We compute

$$\langle \partial^\alpha \varphi, u\rangle = \lim_{n\to\infty} \langle \partial^\alpha \varphi, u_n\rangle = \lim_{n\to\infty} (-1)^{|\alpha|}\langle \varphi, \partial^\alpha u_n\rangle = (-1)^{|\alpha|}\langle \varphi, v_\alpha\rangle$$

in the duality pairing $L^1_t L^2_x \times L^\infty_t L^2_x$. There thus exists $\partial^\alpha u = v_\alpha$. □

In the next lemma we perform the core fixed-point argument.

Lemma 5.15 *Let* (5.23) *hold and* $\rho^2 \geq \|u_0\|_{H_x^3}^2 + \|f(0)\|_{H_x^2}^2 + \|f\|_{Z^3(1)}^2$. *Then there is a radius* $R = R(\rho) > \rho$ *given by* (5.24), *a time* $T_0 = T_0(\rho) \in (0, 1]$ *given by* (5.25), *and a unique solution* $u \in E(R, T_0)$ *of* (5.22).

Proof (1) Lemma 5.13 shows that $a_j(u_0)$ and $d(u_0)$ are bounded in $\hat{H}^2_\infty$ by some $\kappa_0(\rho)$. This yields a constant $c_0 = c_0(\rho) \geq 1$ in (5.19), in the setting of Remark 5.11. We define

$$R^2 = R(\rho)^2 = \mathrm{e} c_0(\rho)\rho^2 + 1 > \rho^2. \tag{5.24}$$

Take $v, w \in E(R, T)$ for some $T > 0$. Let $a \in \{a_0, a_1, a_2, a_3, d\}$ and $\gamma \geq 0$. By Lemma 5.13 and $H_x^2 \hookrightarrow L_x^\infty$ there is a constant $\kappa = \kappa(R)$ with

$$\|a(v)\|_{\tilde{F}^3_\infty(J)} \leq \kappa \qquad \text{and} \qquad \|a(v) - a(w)\|_{L^\infty_\gamma H^2_x} \leq \kappa \|v - w\|_{L^\infty_\gamma H^2_x}.$$

Let $c_1 = c_1(\kappa, \eta)$, $\tilde{c}_1 = \tilde{c}_1(\kappa, \eta)$, and $\gamma_1 = \max\{\gamma_1(\kappa, \eta), \tilde{\gamma}_1(\kappa, \eta)\}$ be given by Theorem 5.8 and Corollary 5.9. We fix

$$\gamma = \gamma(\rho) = \max\left\{\gamma_1, \mathrm{e} c_1 \rho^2, 2\mathrm{e}\tilde{c}_1 \overline{c}^2 \kappa^2 R^2\right\}, \qquad T_0 = T_0(\rho) = \min\{1, (2\gamma)^{-1}\}, \tag{5.25}$$

where the constant $\overline{c} > 0$ is introduced below.

(2) Theorem 5.8 gives a solution $u \in \tilde{G}^3(J_0)$ of $L(v)u = f$ and $u(0) = u_0$ satisfying

$$\|u(t)\|_{H_x^3}^2 + \|\partial_t u(t)\|_{H_x^2}^2 \leq \mathrm{e}^{2\gamma T_0}\big(c_0(\|u_0\|_{H_x^3}^2 + \|f(0)\|_{H_x^2}^2) + c_1\gamma^{-1}\|f\|_{Z^3(1)}^2\big) \leq R^2$$

for $t \in [0, T_0]$. So the map $\Phi : v \mapsto u =: \hat{v}$ leaves invariant $E(R, T_0)$. Observe that

$$L(v)(\hat{v} - \hat{w}) = (L(w) - L(v))\hat{w} = \sum\nolimits_{j=0}^{3}(a_j(w) - a_j(v))\partial_j \hat{w} + (d(w) - d(v))\hat{w}.$$

The right-hand side at time t is bounded in H_x^2 by $\overline{c}\kappa R\|v(t) - w(t)\|_{2,2}$ due to Lemma 5.13. Since $v(0) = w(0)$ and $T_0 \leq 1$, Corollary 5.9 then implies

$$\begin{aligned} \|\Phi(v) - \Phi(w)\|_{L_t^\infty H_x^2}^2 &\leq \mathrm{e}^{2\gamma T_0}\|\Phi(v) - \Phi(w)\|_{L_\gamma^\infty H_x^2}^2 \\ &\leq \mathrm{e}\tilde{c}_1\gamma^{-1}\overline{c}^2\kappa^2 R^2 T_0 \, \|v - w\|_{L_\gamma^\infty H_x^2}^2 \leq \tfrac{1}{2}\|v - w\|_{L_t^\infty H_x^2}^2. \end{aligned} \tag{5.26}$$

The assertion now follows from the contraction mapping principle. □

The above result yields uniqueness only in the ball $E(R, T_0)$, but the contraction estimate (5.26) itself will lead to a much more flexible uniqueness statement. Before showing it, we note that restrictions or translations of a solution $u \in \tilde{G}^3(J)$ to (5.22) satisfy (obvious) variants of (5.22). Let $u \in \tilde{G}^3(J)$ solve (5.22) and $v \in \tilde{G}^3(J')$ with $v(T) = u(T)$ solve it on $J' = (T, T')$. Then the concatenation w of u and v belongs to $\tilde{G}^3(0, T')$ and fulfills (5.22). (Use (5.22) to check $\partial_t w \in C([0, T'], H_x^2)$.)

Lemma 5.16 *Let* (5.23) *hold,* $\tilde{J} = (0, \tilde{T})$*,* $u \in \tilde{G}^3(J)$ *and* $\tilde{u} \in \tilde{G}^3(\tilde{J})$ *solve* (5.22) *on* J *and* $\tilde{J}$*, respectively. Then* $u = \tilde{u}$ *on* $J \cap \tilde{J} =: \hat{J}$.

Proof Let τ be the supremum of all $t \in [0, \sup \hat{J})$ for which $u = \tilde{u}$ on $[0, t]$. Note that $u(0) = u_0 = \tilde{u}(0)$. We suppose that $\tau < \sup \hat{J}$. Then $u = \tilde{u}$ on $[0, \tau]$ by continuity, and there exists a number $\overline{\delta} > 0$ with $J_{\overline{\delta}} := [\tau, \tau + \overline{\delta}] \subseteq \hat{J}$. Let $\overline{R}$ be the maximum of the norms of u and $\tilde{u}$ in $\tilde{G}^3(J_{\overline{\delta}})$. Fix γ as in (5.25) (with $\overline{\kappa} = \kappa(\overline{R})$ and $\rho = 0$) and take $\delta \in (0, \overline{\delta}]$. As in (5.26), Corollary 5.9 yields a constant $\overline{c}_1 = \tilde{c}_1(\overline{R}) > 0$ with

$$\|u - \tilde{u}\|^2_{L^\infty_\gamma(J_\delta, H^2_x)} \leq \mathrm{e}\overline{c}_1 \gamma^{-1} \overline{c}^2 \overline{\kappa}^2 \overline{R}^2 \delta \, \|u - \tilde{u}\|_{L^\infty_\gamma(J_\delta, H^2_x)}.$$

Choosing a sufficiently small $\delta > 0$, we infer $u = \tilde{u}$ on J_δ. This fact contradicts the definition of τ, so that $\tau = \sup \hat{J}$ as asserted. □

We now use the above results to define a *maximal solution* u to (5.22) assuming (5.23). The *maximal existence time* is given by

$$T_+ = T_+(u_0, f) := \sup\{T \geq 0 \,|\, \exists \text{ solution } u_T \in \hat{G}^3(T) \text{ of (5.22) on } [0, T]\} \in (0, \infty].$$

Lemma 5.15 shows $T_+(u_0, f) > T_0(\rho)$ as we can restart the problem at time $t_0 = T_0(\rho)$ with the initial value $u_T(T)$. Moreover, by Lemma 5.16 the solutions u_t and u_T coincide on $[0, t]$ for $0 < t < T < T_+$. Setting $u(t) = u_T(t)$ for such times thus yields a unique solution u of (5.22) on $[0, T_+)$ which belongs to $\tilde{G}^3(T)$ for each $T \in (0, T_+)$.

In the proof of our main result below, we need the following Moser-type estimates.

Lemma 5.17 *Let* $k \in \mathbb{N}$ *and* $\alpha, \beta \in \mathbb{N}_0^m$.

(a) For $v, w \in L^\infty(\mathbb{R}^m) \cap H^k(\mathbb{R}^m)$ *and* $|\alpha| + |\beta| = k$*, we have*

$$\|\partial^\alpha v \, \partial^\beta w\|_2 \leq c\|v\|_\infty \, \|w\|_{k,2} + \|v\|_{k,2} \, \|w\|_\infty.$$

(b) *For* $v, w \in W^{1,\infty}(\mathbb{R}^m) \cap H^k(\mathbb{R}^m)$ *with* $\partial^\alpha v, \partial^\beta w \in L^2(\mathbb{R}^m)$, $1 \le |\alpha| \le k$ *and* $|\alpha| + |\beta| = k+1$, *we have*

$$\|\partial^\alpha v \partial^\beta w\|_2 \le c\|\nabla v\|_\infty \sum_{j=1}^m \|\partial_j w\|_{k-1,2} + \|\nabla w\|_\infty \sum_{j=1}^m \|\partial_j v\|_{k-1,2}.$$

Proof We first recall the Gagliardo–Nirenberg inequality

$$\|\partial^\alpha \varphi\|_{2k/|\alpha|} \le c\,\|\varphi\|_\infty^{1-\frac{|\alpha|}{k}} \sum_{|\gamma|=k} \|\partial^\gamma \varphi\|_2^{\frac{|\alpha|}{k}}$$

for $\varphi \in L^\infty(\mathbb{R}^m)$ with $\partial^\gamma \varphi \in L^2(\mathbb{R}^m)$ for all $|\gamma| = k$, see [135].

Assertion (a) is clear if $|\alpha|$ is 0 or k. So let $k \ge 2$ and $1 \le |\alpha| \le k-1$. Note that $\frac{|\beta|}{k} = 1 - \frac{|\alpha|}{k}$. The inequalities of Hölder (with $\frac{1}{2} = \frac{|\alpha|}{2k} + \frac{|\beta|}{2k}$), Gagliardo–Nirenberg and Young yield

$$\begin{aligned}\|\partial^\alpha v \partial^\beta w\|_2 &\le \|\partial^\alpha v\|_{2k/|\alpha|}\,\|\partial^\beta w\|_{2k/|\beta|} \le c\|v\|_\infty^{1-\frac{|\alpha|}{k}}\,\|v\|_{k,2}^{\frac{|\alpha|}{k}}\,\|w\|_\infty^{1-\frac{|\beta|}{k}}\,\|w\|_{k,2}^{\frac{|\beta|}{k}}\\ &= (\|v\|_\infty\|w\|_{k,2})^{1-\frac{|\alpha|}{k}}\,(\|w\|_\infty\|v\|_{k,2})^{\frac{|\alpha|}{k}} \le c\|v\|_\infty\|w\|_{k,2} + \|v\|_{k,2}\|w\|_\infty.\end{aligned}$$

In part (b) we can assume that $k \ge 3$ and $2 \le |\alpha| \le k-1$. There are $i, j \in \{1, \dots, m\}$ with $\alpha = \alpha' + e_i$ and $\beta = \beta' + e_j$, where $|\alpha'| + |\beta'| = k-1$. From (a) we deduce

$$\|\partial^\alpha v \partial^\beta w\|_2 = \|\partial^{\alpha'} \partial_i v\, \partial^{\beta'} \partial_j w\|_2 \le c\|\partial_i v\|_\infty\, \|\partial_j w\|_{k-1,2} + \|\partial_i v\|_{k-1,2}\, \|\partial_j w\|_\infty$$

and thus statement (b). □

We state the core local wellposedness result for (5.22). Let $\mathcal{B}_T((u_0, f), r)$ be the closed ball in $H_x^3 \times Z^3(T)$ with center (u_0, f) and radius $r > 0$.

Theorem 5.18 *Let* (5.23) *hold and* $\rho^2 \ge \|u_0\|_{H_x^3}^2 + \|f(0)\|_{H_x^2}^2 + \|f\|_{Z^3(1)}^2$. *Then the following assertions are true.*

(a) *There is a unique solution* $u = \Psi(u_0, f)$ *of* (5.22) *on* $[0, T_+)$, *where* $T_+ = T_+(u_0, f) \in (T_0(\rho), \infty]$ *with* $T_0(\rho) > 0$ *from* (5.25) *and* $u \in \tilde{G}^3(T)$ *for all* $T \in (0, T_+)$.
(b) *Let* $T_+ < \infty$. *Then* $\lim_{t\to T_+} \|u(t)\|_{H_x^3} = \infty$ *and* $\limsup_{t\to T_+} \|u(t)\|_{W_x^{1,\infty}} = \infty$.
(c) *Let* $T \in [0, T_+)$. *Then there is a radius* $\delta > 0$ *such that for all* $(v_0, g) \in \mathcal{B}_T((u_0, f), \delta)$ *we have* $T_+(v_0, f) > T$ *and* $\Psi : \mathcal{B}_T((u_0, f), \delta) \to \tilde{G}^3(T)$ *is continuous. Moreover,* $\Psi : (\mathcal{B}_T((u_0, f), \delta), \|\cdot\|_{H_x^2 \times Z^2(T)}) \to \tilde{G}^2(T)$ *is Lipschitz.*

Proof (a)/(b) Above we have shown part (a). Let $T_+ < \infty$ and $u = \Psi(u_0, f)$.

(1) Suppose there are $t_n \to T_+$ with $r := \sup_n \|u(t_n)\|_{3,2} < \infty$. Let $T = T_+ + 1$. Set $\overline{\rho}^2 = r^2 + \|f\|^2_{Z^3(T)} + \sup_n \|f(t_n)\|^2_{2,2} < \infty$. Let $\overline{R} = R(\overline{\rho}) > 0$ and $\tau = T_0(\overline{\rho}) > 0$ be given by (5.24) and (5.25), respectively. Fix an index such that $t_N + \tau > T_+$. Lemma 5.15 and a time shift yield a solution $v \in \tilde{G}^3(t_N, t_N + \tau)$ of (5.22) with $v(t_N) = u(t_N)$. We thus obtain a solution on $[0, t_N + \tau]$. This fact contradicts the definition of T_+, and hence $\|u(t)\|_{3,2} \to \infty$ as $t \to T_+$.

(2) Next, set $\omega = \sup_{0 \le t < T_+} \|u(t)\|_{1,\infty}$ and suppose that $\omega < \infty$. Let $\alpha \in \mathbb{N}_0^3$ with $|\alpha| \le 3$. Using (5.9), we compute

$$\begin{aligned} L(u)\partial_x^\alpha u = \partial_x^\alpha f - \sum\nolimits_{0<\beta\le\alpha} \binom{\alpha}{\beta} \Big[\sum\nolimits_{j=1}^3 \partial_x^\beta a_j(u) \partial_x^{\alpha-\beta} \partial_j u + \partial_x^\beta d(u) \partial_x^{\alpha-\beta} u \\ + \partial_x^\beta a_0(u) \partial_x^{\alpha-\beta} \Big(a_0(u)^{-1} \Big(f - \sum\nolimits_{j=1}^3 a_j(u)\partial_j u - d(u)u \Big)\Big)\Big] \\ =: f_\alpha = \partial_x^\alpha f - g_\alpha. \end{aligned} \tag{5.27}$$

In view of the product rules and (the proof of) Lemma 5.13, the summands of f_α in the second line can be treated as the others (using Young's inequality for products of norms of f and u). Employing also Lemma 5.17 and $H_x^3 \hookrightarrow W_x^{1,\infty}$, we can estimate

$$\begin{aligned} \|f_\alpha(t)\|_2 &\le c(\omega)\Big(\|f(t)\|_{H_x^3} + 1 + \sum\nolimits_{k=1}^3 \sum_{|\gamma_i|\le 3, |\gamma_1|+\dots+|\gamma_k|\le 4} \|\partial_x^{\gamma_1} u \cdots \partial_x^{\gamma_k} u\|_2 \Big) \\ &\le c(\omega)\big(\|f(t)\|_{H_x^3} + 1 + (1+\omega^3)\|u(t)\|_{H_x^3} \big). \end{aligned}$$

Take $\gamma \ge \gamma_0(\omega)$ in Corollary 5.4. For $t \in [0, T_+)$, this corollary and the above estimate yield (with $J_t = (0, t)$)

$$\|\partial_x^\alpha u\|^2_{L^2_\gamma(J_t, L_x^2)} + \frac{2e^{-2\gamma t}}{\gamma} \|\partial_x^\alpha u(t)\|^2_{L_x^2} \le \frac{c(\omega)}{\eta\gamma} \|u_0\|^2_{H_x^3} + \frac{c(\omega)}{\eta^2\gamma^2} \big[\|f\|^2_{L^2_\gamma H_x^3} + 1 + \|u\|^2_{L^2_\gamma(J_t, H_x^3)} \big].$$

We now sum over $|\alpha| \le 3$ and fix a large γ to absorb the last summand. It turns out that $\|u(t)\|_{3,2}$ is bounded for $t < T_+$ contradicting step (1), and hence part (b) is shown.

(c) The proof of (c) is more demanding. We first fix some constants, and then show continuity of Ψ at (u_0, f) on an interval $[0, b]$ assuming that we have solutions with uniform bounds on $[0, b]$. Using this fact and Lemma 5.15, we then prove inductively that solutions on $[0, T]$ exist and satisfy such bounds if we start in a certain ball around (u_0, f). Finally, we replace (u_0, f) by different data in this ball to obtain the asserted continuity statements.

(1) Fix $T' \in (T, T_+)$ and set $J' = (0, T')$. We can extend functions g from $Z^3(T)$ to $Z^3(T')$ with norm bounded by $c_E \|g\|_{Z^3(T')}$ with $c_E \geq 1$. Let $c_S \geq 1$ be the norm of the embedding $C([0, T'], H_x^2) \hookrightarrow Z^3(T')$, $\tilde{\rho}^2 \geq \|u_0\|_{3,2}^2 + \|f\|_{Z^3(T')}^2 + \|f\|_{L^\infty(J', H_x^2)}^2$, $\delta_0 = \tilde{\rho}/c_E$, and $\tilde{r} \geq \max\{c_S \tilde{\rho}, \|u\|_{\tilde{G}^3(T')}\}$. Let $R \geq \tilde{r}$, $b \leq T'$, and v in $\tilde{G}^3(b)$ with norm less or equal R. Lemma 5.13 yields a constant $\overline{\kappa} = \overline{\kappa}(R)$ larger than the norms of $a_j(v)$ and $d(v)$ in $\hat{F}^3_\infty(b)$ and of $a_j(v)(0)$ and $d(v)(0)$ in $\hat{H}^2_\infty$.

(2) Take $b \in (0, T']$, $v_0 \in H_x^3$ and $g \in Z^3(T)$ such that $T_+(v_0, g) > b$. We write $v = \Psi(v_0, g) \in \tilde{G}^3(b)$. Let $R \geq \|v\|_{\tilde{G}^3(b)}$ with $R \geq \tilde{r}$. Observe that

$$L(u)(v-u) = g - f + (L(u) - L(v))v = g - f + \sum_{j=0}^{3} (a_j(u) - a_j(v))\partial_j v + (d(u) - d(v))v.$$

By Lemma 5.13 and Sobolev embeddings, the map $(L(u) - L(v))v$ belongs to $\tilde{G}^2_\gamma(b)$ with norm less than $c(\overline{\kappa}) R \|v - u\|_{\tilde{G}^2_\gamma(b)}$ for $\gamma \geq 0$. Corollary 5.9 yields

$$\|v - u\|_{\tilde{G}^2_\gamma(b)} \leq \tilde{c}(\overline{\kappa}, \eta, T')\big(\|u_0 - v_0\|_{H_x^2}^2 + \|f - g\|_{Z^2_\gamma(b)} + \gamma^{-1} R b \|v - u\|_{\tilde{G}^2_\gamma(b)}\big)$$

for $\gamma \geq \tilde{\gamma}_1(\overline{\kappa}, \eta) \geq 1$. Fixing a large $\overline{\gamma}_1 = \overline{\gamma}_1(\overline{\kappa}, R, T', \eta) \geq \tilde{\gamma}_1(\overline{\kappa}, \eta)$, we thus obtain

$$\|v - u\|_{\tilde{G}^2(b)} \leq \tilde{c}(\overline{\kappa}, R, T', \eta)\big(\|u_0 - v_0\|_{H_x^2}^2 + \|f - g\|_{Z^2(b)}\big). \tag{5.28}$$

(3) Estimate (5.28) is related to Lipschitz continuity of Ψ in $\tilde{G}^2$. The hard and core part of the proof is to check continuity of Ψ in $\tilde{G}^3$ at (u_0, f), assuming apriori bounds. So let $(u_{0,n}, f_n) \in \mathcal{B}_T((u_0, f), \tilde{\delta})$ tend to (u_0, f) on $H_x^3 \times Z^3(T)$ as $n \to \infty$, where $\tilde{\delta} > 0$. Hence, $f_n(0) \to f(0)$ in H_x^2 and $f_n \to f$ in $Z^3(T')$. Assume that $T_+(u_{0,n}, f_n) > b$ with $b \in (0, T']$ and that $u_n = \Psi(u_{0,n}, f_n)$ is bounded by some $R \geq \tilde{r}$ in $\tilde{G}^3(b)$ for all $n \in \mathbb{N}$. Then u_n tends to u in $\tilde{G}^2(b)$ as $n \to \infty$ by (5.28), and the coefficients $a_j(u_n)$ and $d(u_n)$ satisfy the estimates of step (1) with a uniform $\overline{\kappa} = \overline{\kappa}(R)$.

The main idea is to split the n-dependence of the coefficients and the data. Let $\alpha \in \mathbb{N}_0^3$ with $|\alpha| = 3$. As in (5.27) we write $L(u_n)\partial_x^\alpha u_n = \partial_x^\alpha f_n - g_{n,\alpha}$ and $L(u)\partial_x^\alpha u = \partial_x^\alpha f - g_\alpha$. Theorem 5.5 yields solutions $w_n, z_n \in C([0, b], L_x^2)$ of

$$\begin{aligned} L(u_n)w_n &= \partial_x^\alpha f - g_\alpha, \quad && w_n(0) = \partial_x^\alpha u_0, \\ L(u_n)z_n &= \partial_x^\alpha f_n - \partial_x^\alpha f + g_\alpha - g_{n,\alpha}, \quad && z_n(0) = \partial_x^\alpha u_{0,n} - \partial_x^\alpha u_0. \end{aligned}$$

By uniqueness, we have $w_n + z_n = \partial_x^\alpha u_n$ and hence

$$\partial_x^\alpha u_n - \partial_x^\alpha u = w_n - \partial_x^\alpha u + z_n.$$

Since $a_j(u_n) \to a_j(u)$ and $d(u_n) \to d(u)$ in $L^\infty_{t,x}$ as $n \to \infty$, Lemma 5.12 shows that $q_n = \|w_n - \partial_x^\alpha u\|_{L_t^\infty L_x^2}$ tends to 0. We thus have to prove that $z_n \to 0$ in $L_t^\infty L_x^2$.

Choose $\gamma = \overline{\gamma}_1(R)$ as in step (2). For $t \in [0, b]$, Corollary 5.4 then implies

$$\begin{aligned}\|\partial_x^\alpha(u_n(t) - u(t))\|_{L_x^2}^2 &\le 2q_n^2 + 2\|z_n(t)\|_{L_x^2}^2\\ &\le 2q_n^2 + c(R)\big(\|\partial_x^\alpha(u_{0,n} - u_0)\|_{L_x^2}^2 + \|\partial_x^\alpha(f_n - f)\|_{L_{t,x}^2}^2 + \|g_{n,\alpha} - g_\alpha\|_{L_{t,x}^2}^2\big).\end{aligned}$$

The estimation of $\|g_{n,\alpha} - g_\alpha\|$ is only sketched. Let $a \in \{a_j, a_0^{-1}, d\}$, $v \in \{u, u_n\}$, and $w \in \{u, u_n, f\}$. First, we look at summands of the type $\partial_x^\beta a(v(t))\partial_x^\gamma(u_n(t) - u(t))$ with $|\gamma| \le 4 - |\beta| \le 3$ and $|\beta| \le 3$. By the product rules and the bounds on the coefficients these terms are bounded in L_x^2 by $c(R)\|u_n(t) - u(t)\|_{3,2}$. Analogous summands with $f_n(t) - f(t)$ are treated similarly.

We next analyze terms like $W = \partial_x^\beta[a(u_n(t)) - a(u(t))]\,\partial_x^\gamma w(t)$. At first, we look at situations where we can estimate the first factor in L_x^2 by $u - u_n$ in $L_t^\infty H_x^2$ using Lemma 5.13. This works for $\beta = 0$ in all cases, for $|\beta| = 1$ if $|\gamma| \le 2$, and for $|\beta| = 2$ if $|\gamma| \le 1$; and it yields terms as in the first case. If this does not work (which implies $w \in \{u, u_n\}$), we have to compute $\partial_x^\beta(a(u_n) - a(u))$ as in the proof of Lemma 5.13. For these terms we define

$$h_n(t) = \sum_{a\in\{a_j,d,a_0^{-1}\}} \sum_{k=1}^{3}\sum_{l_i=1}^{9} \|(\partial_{l_k}\cdots\partial_{l_1}a)(u_n(t)) - (\partial_{l_k}\cdots\partial_{l_1}a)(u(t))\|_{L_x^\infty}.$$

The L_x^2-norm of W is bounded by linear combinations of $c(R)$ times

$$h_n(t)\|\partial_x^{\gamma_1}v(t)\cdots\partial_x^{\gamma_{m-1}}v(t)\partial_x^{\gamma_m}w(t)\|_{L_x^2} + \|\partial_x^{\gamma_1}v(t)\cdots\partial_x^{\gamma_{m-1}}\varphi_n(t)\partial_x^{\gamma_m}w(t)\|_{L_x^2},$$

where $\varphi_n = u_n - u$, $m \in \{1, 2, 3, 4\}$, $|\gamma_i| \le 3$, and $|\gamma_1| + \cdots + |\gamma_m| \le 4$. This sum can be estimated by $c(R)(h_n(t) + \|u_n(t) - u(t)\|_{3,2})$ due to Sobolev embeddings and the bounds on u and u_n. We have shown that

$$\begin{aligned}\|g_{n,\alpha} - g_\alpha\|_{L^2((0,t),L_x^2)}^2 \le c(R, T')\Big(&\|f_n - f\|_{L_t^2 H_x^2}^2 + \|u_n - u\|_{L_t^\infty H_x^2}^2 + \int_0^{T'} h_n(s)^2\,ds\\ &+ \int_0^t \sum_{|\gamma|=3}\|\partial_x^\gamma(u_n(s) - u(s))\|_{L_x^2}^2\,ds\Big).\end{aligned}$$

We write the last integrand as $\|\partial_x^3(u_n(s) - u(s))\|_2^2$. Observe that $h_n(s)$ tends to 0 as $n \to \infty$ since $u_n \to u$ in $L_{t,x}^\infty$ and that it is bounded uniformly in s and n. By dominated convergence $\int_0^{T'} h_n^2 \, ds$ tends to 0. Summing up, we conclude that

$$\|\partial_x^3(u_n(t) - u(t))\|_2^2 \le c(R, T')\varepsilon_n + c(R, T') \int_0^t \|\partial_x^3(u_n(s) - u(s))\|_2^2 \, ds$$

for a null sequence (ε_n). By Gronwall, $\partial_x^3(u_n - u)$ tends to 0 in $C([0, b], L_x^2)$ as $n \to \infty$, and so $u_n \to u$ in $C([0, b], H_x^3)$. Using (5.9) and Lemma 5.13, we infer $u_n \to u$ in $\tilde{G}^3(b)$.

(4) We now look for data to which we can apply steps (2) and (3). Let $(v_0, g) \in \mathcal{B}_T((u_0, f), \delta_0)$. We then obtain

$$\begin{aligned}
\|v_0\|_{H_x^3} &\le \|v_0 - u_0\|_{H_x^3} + \|u_0\|_{H_x^3} \le \delta_0 + \tilde{\rho} \le 2\tilde{\rho} \le 2\tilde{r}, \\
\|g\|_{Z^3(T')} &\le \|g - f\|_{Z^3(T')} + \|f\|_{Z^3(T')} \le c_E\delta_0 + \tilde{\rho} \le 2\tilde{\rho} \le 2\tilde{r}, \\
\|g\|_{L^\infty(J', H_x^2)} &\le c_S\|g\|_{Z^3(T')} \le 2c_S\tilde{\rho} \le 2\tilde{r}.
\end{aligned}$$

Lemma 5.15 thus yields a time $\tau = \tau(\tilde{r})$ and a solution $v \in \tilde{G}^3(\tau)$ of (5.22) with data v_0 and g, where $\|v\|_{\tilde{G}^3(\tau)} \le \tilde{R} = \tilde{R}(\tilde{r})$ and $\tilde{R} > 2\tilde{r}$. By parts (a) and (b), we have $v = \Psi(v_0, g)$ and $T_+(v_0, g) > \tau$. Fix $N \in \mathbb{N}$ with $(N-1)\tau \le T < N\tau$, set $t_k = k\tau$ for $k \in \{0, 1, \dots, N-1\}$ and $t_N = \min\{T', N\tau\}$.

Steps (2) and (3) show that (5.28) is true on $[0, \tau]$ for such v with a constant $\tilde{c} = \tilde{c}(\tilde{r})$ and that $\Psi : \mathcal{B}_T((u_0, f), \delta_0) \to \tilde{G}^3(\tau)$ is continuous at (u_0, f). We can thus find a radius $\delta_1 \in (0, \delta_0]$ such that $\|v - u\|_{\tilde{G}^3(\tau)} \le \tilde{r}$, and hence $\|v\|_{\tilde{G}^3(\tau)} \le 2\tilde{r}$, for all $(v_0, g) \in \mathcal{B}_T((u_0, f), \delta_1)$.

(5) We iterate the above argument. Assume that for some $k \in \{1, \dots, N-1\}$ and $\delta_k \in (0, \delta_0]$, we have $T_+(v_0, g) > t_k$ and $\|v - u\|_{\tilde{G}^3(t_k)} \le \tilde{r}$ for all $(v_0, g) \in \mathcal{B}_T((u_0, f), \delta_k)$ and the map $\Psi : \mathcal{B}_T((u_0, f), \delta_k) \to \tilde{G}^3(t_k)$ is continuous at (u_0, f). It follows $\|v\|_{\tilde{G}^3(t_k)} \le 2\tilde{r}$. Since $\|v(t_k)\|_{3,2} \le 2\tilde{r}$, step (4) and a time shift provide a solution $\tilde{v} \in \tilde{G}^3([t_k, t_{k+1}])$ of (5.22) with $\tilde{v}(t_k) = v(t_k)$ and norm less or equal $\tilde{R}$. We can thus extend v to a solution in $\tilde{G}^3([0, t_{k+1}])$ bounded by $\tilde{R}$ and so $T_+(v_0, g) > t_{k+1}$. Because of this bound, steps (2) and (3) imply (5.28) on $[0, t_{k+1}]$ with $\tilde{c} = \tilde{c}(\tilde{r})$ for all $(v_0, g) \in \mathcal{B}_T((u_0, f), \delta_k)$ and the continuity of $\Psi : \mathcal{B}_T((u_0, f), \delta_k) \to \tilde{G}^3(t_{k+1})$ at (u_0, f). Using the latter property, we find a radius $\delta_{k+1} \in (0, \delta_k]$ such that $\|v - u\|_{\tilde{G}^3(t_{k+1})} \le \tilde{r}$ for $v = \Psi(v_0, g)$ and all $(v_0, g) \in \mathcal{B}_T((u_0, f), \delta_{k+1})$, and hence $\|v\|_{\tilde{G}^3(t_{k+1})} \le 2\tilde{r}$.

Induction yields a radius $\delta = \delta_N$ such that for all $(v_0, g) \in \mathcal{B}_T((u_0, f), \delta)$ we have $T_+(v_0, g) > T$, the continuity of $\Psi : \mathcal{B}_T((u_0, f), \delta) \to \tilde{G}^3(T)$ at (u_0, f), and $\|\Psi(v_0, g)\|_{\tilde{G}^3(T)} \le 2\tilde{r}$. Moreover, (5.28) holds on $[0, T]$ for such $v = \Psi(v_0, g)$ and u.

(6) Finally, we take any $(v_0, g), (w_0, h) \in \mathcal{B}_T((u_0, f), \delta)$ with corresponding solutions v and w. Replacing u by w in step (2), we then obtain the last assertion in (c). Also step (3) can be repeated on $[0, T]$ for data converging to (w_0, h) in $\mathcal{B}_T((u_0, f), \delta)$. □

Observe that Theorem 5.7 yields finite speed of propagation for a solution $u \in \tilde{G}^3(T)$ of (5.22), setting $A_j = a_j(u)$ and $D = d(u)$. We comment on variants of Theorem 5.18.

Remark 5.19 One can easily extend Theorem 5.18 to negative times (e.g., by time reversion). Moreover, in (5.23) one can replace the domain $\mathbb{R}^3 \times \mathbb{R}^6$ of a_j and d by $\mathbb{R}^3 \times U$ for an open $U \subseteq \mathbb{R}^6$, restricting ξ in the supremum not to each closed ball $\overline{B}(0, r) \subseteq \mathbb{R}^6$ but to each compact subset of U. One further has to require that the closure K_0 of $u_0(\mathbb{R}^3)$ is contained in U, and the solution u has to take values in U. Theorem 5.18 is then valid with one modification. In part (b) now $T_+ < \infty$ implies that $\limsup_{t<T_+} \|u(t)\|_{W_x^{1,\infty}} = \infty$ or that $u(t)$ leaves any compact subset of U as $t \to T_+$.

The proofs are very similar in this more general case. In the fixed-point argument one chooses a bounded open set V with $K_0 \subseteq V \subseteq \overline{V} \subseteq U$. Let $d > 0$ be the distance between V and ∂U. In $E(R, T)$ one then also includes the condition that $\|v(t) - u_0\|_\infty \le d/2$ for all $t \in [0, T]$ which is preserved by limits in $L_t^\infty H_x^2$. Other steps in the reasoning are modified accordingly. Compare Theorem 3.3 of [158].

As explained in Sect. 5.1, one can easily apply Theorem 5.18 to the Maxwell system (5.1) with material laws (5.3) and (5.4). We state the needed assumptions in a situation motivated by nonlinear optics.

Example 5.20 Let $\theta(x, \mathbf{E}, \mathbf{H}) = (\varepsilon_{\mathrm{lin}}(x)\mathbf{E} + \varepsilon_{\mathrm{nl}}(x, \mathbf{E})\mathbf{E}, \mu_{\mathrm{lin}}(x)\mathbf{H})$ and $\mathbf{J} = \sigma(x, \mathbf{E})\mathbf{E} + \mathbf{J}_0$ in (5.3) and (5.4). Here we assume that $\varepsilon_{\mathrm{lin}}, \mu_{\mathrm{lin}} \in C_b^3(\mathbb{R}^3, \mathbb{R}^{3\times3}_{\mathrm{sym}})$ and $\sigma \in C^3(\mathbb{R}^6, \mathbb{R}^{3\times3})$ satisfy $\varepsilon_{\mathrm{lin}}, \mu_{\mathrm{lin}} \ge 2\eta I > 0$ and $\sup_{|\xi|\le r} \|\partial_x^\alpha \sigma(\cdot, \xi)\|_{L_x^\infty} < \infty$ for all $r \ge 0$ and $0 \le |\alpha| \le 3$, respectively. (The subscript b means that the functions and all occurring derivatives are bounded.) In Example 5.1 we had seen rather general isotropic nonlinear terms which fit to (5.23). A typical anisotropic example is furnished by

$$\varepsilon_{\mathrm{nl}}(x, \mathbf{E}) = \Big(\sum\nolimits_{j,k=1}^{3} \chi_i^{jkl}(x)\mathbf{E}_j\mathbf{E}_k\Big)_{il}$$

for scalar coefficients $\chi_i^{jkl} \in C_b^3(\mathbb{R}^3)$, cf. [19]. Because of the triple sum in $\varepsilon_{\mathrm{nl}}(x, \mathbf{E})\mathbf{E}$, the tensor $(\chi_i^{jkl})_{i,j,k,l}$ has to be symmetric in $\{j, k, l\}$. For (5.23) we also require symmetry in $\{i, l\}$, i.e., we can only prescribe χ_i^{jkl} for, say, $1 \le i \le j \le l \le 3$. For $|\mathbf{E}| < r$ and a suitable $r \in (0, \infty]$ and all $x, \mathbf{H} \in \mathbb{R}^3$ we then obtain $\partial_{(\mathbf{E},\mathbf{H})}(x, \mathbf{E}, \mathbf{H}) \ge \eta I$. Rewriting the system as in (5.6), we see that hypothesis (5.23) (modified as in Remark 5.19 if $r < \infty$) is fulfilled. For initial fields in H_x^3 with $|\mathbf{E}_0| < r/2$ and a current density $\mathbf{J}_0 \in Z^3(T)$ for all

$T > 0$, Theorem 5.18 and Remark 5.19 thus provide wellposedness in H_x^3 of the Maxwell system (5.1) with the above material laws.

5.5 Energy and Blow-Up

In the preceding sections we have worked with the linear energy estimate which contains error terms caused by the time derivative of coefficients. (The space derivatives in C of (5.10) disappear in the Maxwell case.) These error terms have led to the H_x^3 setting, which is quite inconvenient. The time dependence arises since we freeze a function in the nonlinearities of (5.22). One may wonder whether this is really necessary and whether it is not better to solve (5.22) based on a nonlinear energy identity. Actually, this can be done in the semilinear case where $\mathbf{D} = \varepsilon(x)\mathbf{E}$, $\mathbf{B} = \mu(x)\mathbf{H}$, and $\mathbf{J} = \sigma(x, \mathbf{E})\mathbf{E}$ under appropriate conditions on σ, cf. [56]. Below we see that this does not seem to work in the quasilinear case.

In this section we first establish an energy equality in the quasilinear case, without conductivity and for isotropic nonlinearities

$$\mathbf{D} = \varepsilon_{\text{lin}}\mathbf{E} + \beta_e(\cdot, |\mathbf{E}|^2)\mathbf{E}, \qquad \mathbf{B} = \mu_{\text{lin}}\mathbf{H} + \beta_m(\cdot, |\mathbf{H}|^2)\mathbf{H}, \tag{5.29}$$

Here $\varepsilon_{\text{lin}}, \mu_{\text{lin}} \in L^\infty(\mathbb{R}^3, \mathbb{R}^{3\times 3}_\eta)$ for some $\eta > 0$ and the functions $\beta_e, \beta_m : \mathbb{R}^3 \times \mathbb{R}_{\geq 0} \to \mathbb{R}$ are C^1, bounded in $x \in \mathbb{R}^3$ and increasing in $s \in \mathbb{R}_{\geq 0}$. We set $u = (\mathbf{E}, \mathbf{H})$ and

$$A_0 = \begin{pmatrix} \varepsilon_{\text{lin}} & 0 \\ 0 & \mu_{\text{lin}} \end{pmatrix}, \qquad \beta(|u|^2) = \begin{pmatrix} \beta_e(\cdot, |\mathbf{E}|^2) I_{3\times 3} & 0 \\ 0 & \beta_m(\cdot, |\mathbf{H}|^2) I_{3\times 3} \end{pmatrix},$$

$$M = \begin{pmatrix} 0 & \operatorname{curl} \\ -\operatorname{curl} & 0 \end{pmatrix} = -\sum\nolimits_{j=1}^3 S_j \partial_j, \qquad \mathcal{D}(M) = H(\operatorname{curl}) \times H(\operatorname{curl}),$$

where $H(\operatorname{curl}) = H(\operatorname{curl}, U) = \{v \in L^2(U, \mathbb{R}^3) \mid \operatorname{curl} v \in L^2(U, \mathbb{R}^3)\}$. The operator M is skew-adjoint in $L^2(\mathbb{R}^3, \mathbb{R}^6)$. Maxwell equations (5.1) here become

$$\partial_t[A_0 u(t) + \beta(|u(t)|^2)u(t)] = Mu(t), \quad t \geq 0, \qquad u(0) = u_0 = (\mathbf{E}_0, \mathbf{H}_0). \tag{5.30}$$

Omitting the argument x in the notation, we further define

$$b_j(s) = \int_0^s \beta_j(r)\,\mathrm{d}r, \qquad h_j(s) = s\beta_j(s) - \tfrac{1}{2}b_j(s).$$

Note that $h_j(s) \geq s\beta_j(s)/2$ since β_j increases and of $h'_j(s) = \beta_j(s)/2 + s\beta'_j(s)$, where $\beta'_j = \partial_2\beta_j$. We now introduce the 'energy' for $u = (u_1, u_2)$ by

$$\mathcal{E}(u) = \int_{\mathbb{R}^3} \left[\tfrac{1}{2} A_0 u \cdot u + h_1(|u_1|^2) + h_2(|u_2|^2)\right] \mathrm{d}x$$

Observe that $\mathcal{E}(u) \geq \frac{\eta}{2}\|u\|_2^2$ if $\beta_j \geq 0$. In the Kerr case $\varepsilon_{\text{lin}} = \mu_{\text{lin}} = 1$, $\beta_e(x, s) = \chi_3(x)s$ and $\beta_m = 0$, we obtain

$$\mathcal{E}_{\text{Kerr}}(\mathbf{E}, \mathbf{H}) = \int_{\mathbb{R}^3} \left[\tfrac{1}{2}|\mathbf{E}(t)|^2 + \tfrac{3}{4}\chi_3|\mathbf{E}(t)|^4 + \tfrac{1}{2}|\mathbf{H}(t)|^2\right] \mathrm{d}x.$$

Let $u \in \tilde{G}^1(T)$ solve (5.30). The energy equality $\mathcal{E}(u(t)) = \mathcal{E}(u_0)$, $t \in [0, T]$, follows by

$$\begin{aligned} \tfrac{\mathrm{d}}{\mathrm{d}t}\mathcal{E}(u) &= \int_{\mathbb{R}^3} \left[u \cdot \partial_t(A_0 u) + \beta(|u|^2)u \cdot \partial_t u + 2|u|^2\beta'(|u|^2)u \cdot \partial_t u\right] \mathrm{d}x \\ &= \int_{\mathbb{R}^3} \partial_t\left[A_0 u + \beta(|u|^2)u\right] \cdot u \,\mathrm{d}x = \int_{\mathbb{R}^3} Mu \cdot u \,\mathrm{d}x = 0. \end{aligned} \tag{5.31}$$

In the Kerr case (with $\chi_3 \geq 0$) we can thus bound powers of p-norms of solutions. This is not enough control to pass to a weak limit in the nonlinearity when performing an approximation argument (which would typically produce a global solution). One would need an estimate involving derivatives. Such estimates are not known, and the next result on blow-up indicates that they do not hold.

We first stress that it is well known that the gradient of a solution to (5.30) may blow up in sup-norm in finite time, see [123]. However in the semilinear case one relies on estimates in $H(\text{curl})$, so we are interested blow-up in this space (or at least in H^1). Below we give such an example on a domain with (unphysical) periodic boundary conditions, taken from [34]. (See this paper for a weaker result on $\mathbb{R}^3$.) We work in the following more specific setting given by $\mathbf{D} = (1+\alpha(|\mathbf{E}|))\mathbf{E}$ and $\mathbf{B} = \mathbf{H}$. We set $a(s) = (1+\alpha(|s|))s$ for $s \in \mathbb{R}$ and assume

$$\begin{aligned} &a \in C^2(\mathbb{R}, \mathbb{R}), \quad \exists\, s_- < 0 < s_0 < s_+ : a' > 0 \text{ on } S := (s_-, s_+), \\ &q : S \to \mathbb{R};\ q(s) = \frac{a''(s)}{2a'(s)^{3/2}}, \text{ has a global maximum at } s = s_0, \\ &q \text{ is } C^1 \text{ near } s_0, \quad q(s) > 0 \text{ for } 0 < s \leq s_0. \end{aligned} \tag{5.32}$$

Let $\gamma > 2$ and $\alpha_0 > 0$. A simple example for (5.32) is furnished by any C^2-extension of $a : [0, s_+] \to \mathbb{R}$; $a(s) = s + \alpha_0 s^\gamma$, which is strictly growing on (s_-, s_+) for some

$s_- < 0 < s_0 < s_+$ with

$$s_0 = \Big(\frac{2(\gamma-2)}{\alpha_0\gamma(\gamma+1)}\Big)^{\frac{1}{\gamma-1}}$$

in this case. We stress that the behavior of a for large s is arbitrary here.

Theorem 5.21 *Assume that* (5.32) *is true. Then there are numbers* $M, T > 0$ *and a divergence-free map* $(\mathbf{E}, \mathbf{H}) \in C^1([0, T) \times [-M, M]^3)$ *which solves* (5.1) *on* $(-M, M)^3$ *with periodic boundary conditions and the above material laws and which satisfies*

$$\|\operatorname{curl} \mathbf{E}(t)\|_{L^2_x} \to \infty \qquad \text{as } t \to T.$$

We look for a solution of the form

$$(\mathbf{E}(t, x), \mathbf{B}(t, x)) = (u(t, x_2), 0, 0, 0, 0, v(t, x_2)).$$

for $x \in (-M, M)^3$ and $t \in [0, T)$. Observe that such $\mathbf{E}$ and $\mathbf{B}$ are divergence-free. If u and v have support in $[0, T) \times (-M, M)$, then $\mathbf{E}$ and $\mathbf{B}$ fulfill the periodic boundary condition. Moreover, $(\mathbf{E}, \mathbf{B}) \in C^1$ satisfy (5.1) on $(-M, M)^3$ with the above material laws if and only if $(u, v) \in C^1$ solve

$$\partial_t a(u) = \partial_x v, \qquad \partial_t v = \partial_x u, \qquad (u(0), v(0)) = (u_0, v_0),$$

for $t \in [0, T)$ and $x \in \mathbb{R}$. This system can be rewritten as

$$\partial_t \begin{pmatrix} u \\ v \end{pmatrix} + A(u, v)\partial_x \begin{pmatrix} u \\ v \end{pmatrix} = 0 \qquad \text{with } A(u, v) = \begin{pmatrix} 0 & -a'(u)^{-1} \\ -1 & 0 \end{pmatrix} \tag{5.33}$$

on $\mathbb{R}$. Here we assume that u takes values in S from (5.32). Since also $\partial_x u = \operatorname{curl} \mathbf{E}$, the theorem thus follows from the next one-dimensional result.

The following proof uses a standard construction from Section 1.4 of [123]. However, it requires a rather detailed analysis to find a class of initial values for which we get the blow-up of $\partial_x u$ in L^2 instead of L^∞.

Proposition 5.22 *Assume that* (5.32) *is true. Then there exist initial data* $(u_0, v_0) \in C^1_c(\mathbb{R}, \mathbb{R}^2)$ *and a* C^1*-solution* (u, v) *to* (5.33) *on* $[0, T) \times \mathbb{R}$ *for some* $T \in (0, \infty)$ *which is compactly supported and which satisfies* $\|\partial_x u(t, \cdot)\|_{L^2(\mathbb{R})} \to \infty$ *as* $t \to T^-$.

Proof (1) For $(u, v) \in S \times \mathbb{R}$, the matrix $A(u, v)$ has the eigenvalues and eigenvectors

$$\lambda_{1,2}(u, v) = \pm a'(u)^{-\frac{1}{2}}, \qquad w_{1,2}(u, v) = (\mp 1, a'(u)^{\frac{1}{2}}).$$

(Recall $S = (s_-, s_+)$, s_0 and q from (5.32).) These observations are a special case of the analysis in Section 3 of [3]. In the following we take $\lambda = \lambda_1$ and $w = w_1$ and drop the index 1. Fix $(\xi, \zeta) \in (s_0, s_+) \times \mathbb{R}$ such that

$$q(s) > 0 \qquad \text{for} \quad 0 < s \leq \xi.$$

Observe that the interval $\xi - S = (\xi - s_+, \xi - s_-)$ contains $[0, \xi]$. The C^2-function $\phi : \xi - S \to S \times \mathbb{R}$

$$\phi_1(s) = \xi - s, \qquad \phi_2(s) = \zeta + \int_0^s a'(\xi - \tau)^{1/2}\, d\tau,$$

solves the ordinary differential equation

$$\phi'(s) = w(\phi(s)), \quad s \in \xi - S, \qquad \phi(0) = (\xi, \zeta).$$

For later use, we note the identities

$$\nabla\lambda(\phi(s)) \cdot \phi'(s) = \nabla\lambda(\phi(s)) \cdot w(\phi(s)) = q(\xi - s), \qquad s \in \xi - S. \tag{5.34}$$

Let $\sigma_0 : \mathbb{R} \to [0, \xi]$ be C^1 and equal to ξ outside a compact set. There is a unique C^1-solution σ of the scalar partial differential equation

$$\begin{aligned} \partial_t \sigma(t, x) + \lambda(\phi(\sigma(t, x)))\partial_x \sigma(t, x) &= 0, \qquad t \geq 0,\ x \in \mathbb{R}, \\ \sigma(0, x) &= \sigma_0(x), \qquad x \in \mathbb{R}, \end{aligned} \tag{5.35}$$

on a bounded time interval $[0, \bar{t})$, where σ takes values in $\xi - S$. See e.g. Theorems 2.1 and 2.2 of [123]. We now define

$$\begin{pmatrix} u(t, x) \\ v(t, x) \end{pmatrix} = \phi(\sigma(t, x)).$$

It is easy to check that (u, v) is a C^1-solution of (5.33) on $[0, \bar{t}) \times \mathbb{R}$. We observe that

$$\partial_x u = \phi_1'(\sigma)\partial_x \sigma = -\partial_x \sigma. \tag{5.36}$$

(2) The method of characteristics yields the implicit formula

$$\begin{aligned}\sigma(t,x) &= \sigma_0(x - t\lambda(\phi(\sigma(t,x)))) = \sigma_0(y(t,x)),\\ y(t,x) &:= x - t\lambda(\phi(\sigma(t,x))) = x - ta'(\xi - \sigma(t,x))^{-1/2},\end{aligned} \tag{5.37}$$

for the solution of (5.35) as long as

$$\begin{aligned}&1+t\nabla\lambda(\phi(\sigma(t,x)))\cdot w(\phi(\sigma(t,x)))\sigma_0'(x - t\lambda(\phi(\sigma(t,x))))\\ &\quad = 1 + t\sigma_0'(x - t\lambda(\phi(\sigma(t,x))))q(\xi - \sigma(t,x)) \;\; > 0,\end{aligned} \tag{5.38}$$

see (5.34). Hence, σ is bounded. We now set

$$\gamma(t) := \inf_{x\in\mathbb{R}} \sigma_0'(y(t,x))q(\xi - \sigma(t,x)) \quad \text{for} \quad t \in [0,\bar{t}).$$

Let $t_0 \geq 0$ be the supremum of $t \in [0,\bar{t})$ such that $\tau\gamma(\tau) > -1$ for all $\tau \in [0,t]$. In the following, we take $t \in [0,t_0)$ so that the inequality (5.38) is valid for all $x \in \mathbb{R}$. Equations (5.37) then imply

$$\partial_x\sigma(t,x) = \sigma_0'(x - t\lambda(\phi(\sigma(t,x))))\Big(1 - tq(\xi - \sigma(t,x))\partial_x\sigma(t,x)\Big),$$

$$\partial_x\sigma(t,x) = \frac{\sigma_0'(y(t,x))}{1 + tq(\xi-\sigma(t,x))\sigma_0'(y(t,x))}.$$

In particular, $\partial_x\sigma$ is bounded on $[0, t_0-\delta]\times\mathbb{R}$ for each $\delta \in (0,t_0]$. The blow–up condition in Theorem 2.2 Annex of [123] (a variant of Theorem 5.18) thus yields $\bar{t} = t_0$. From formula (5.37) we further deduce $\partial_x\sigma(t,x) = \sigma_0'(y(t,x))\partial_x y(t,x)$ and therefore

$$\partial_x y(t,x) = \frac{1}{1 + tq(\xi - \sigma(t,x))\sigma_0'(y(t,x))} > 0. \tag{5.39}$$

(In the case $\sigma_0'(y(t,x)) = 0$ the identity $\partial_x y(t,x) = 1 > 0$ follows from (5.37).) Using also (5.37), we see that the map $x \mapsto y(t,x)$ is a bijection from $\mathbb{R}$ to $\mathbb{R}$. This fact and (5.37) lead to the equation

$$\gamma(t) = \inf_{z\in\mathbb{R}} \sigma_0'(z)q(\xi - \sigma_0(z)) =: \gamma_0.$$

(3) We now fix a C^1-function $\sigma_0 : \mathbb{R} \to [0, \xi]$ which is equal to ξ outside some compact set and satisfies

$$\sigma_0(0) = \xi - s_0, \qquad \sigma_0'(0) = \min_{z \in \mathbb{R}} \sigma_0'(z) < 0.$$

In view of (5.32), we can determine

$$\gamma_0 = \sigma_0'(0) q(s_0) \quad \text{and} \quad t_0 = -\frac{1}{\gamma_0}. \tag{5.40}$$

Substituting $z = y(t, x)$ and using (5.39), we infer from (5.37) the identities

$$\begin{aligned}
\|\partial_x \sigma(t, \cdot)\|_2^2 &= \int_{\mathbb{R}} |\partial_x \sigma(t, x)|^2 \, \mathrm{d}x = \int_{\mathbb{R}} |\sigma_0'(y(t, x)) \partial_x y(t, x)|^2 \, \mathrm{d}x \\
&= \int_{\mathbb{R}} \frac{|\sigma_0'(z)|^2}{1 + tq(\xi - \sigma_0(z)) \, \sigma_0'(z)} \, \mathrm{d}z.
\end{aligned}$$

Since q has a global maximum at s_0 while σ_0' has a global minimum at 0, we obtain the expansions

$$q(s) = q(s_0) - o_+(s - s_0), \qquad \sigma_0'(z) = \sigma_0'(0) + o_+(z), \qquad \sigma_0(z) = \xi - s_0 + O(z),$$

where $o_+(z)$ denotes any nonnegative function with the property $o_+(z)/z \to 0$ as $z \to 0$. Hence, (5.40) yields

$$\begin{aligned}
1 + tq(\xi - \sigma_0(z)) \, \sigma_0'(z) &= 1 + t\gamma_0 + t[q(s_0) o_+(z) + o_+(z) \, |\sigma_0'(0)| - o_+(z)^2] \\
&= 1 + t\gamma_0 + t o_+(z)
\end{aligned}$$

for small $|z|$. Fix a number $\delta_0 > 0$ such that the above identity is true and $|\sigma_0'(z)|^2 \geq \frac{1}{2} |\sigma_0'(0)|^2 =: c_0$ if $|z| \leq \delta_0$. For each $\epsilon > 0$ there exists a radius $\delta \in (0, \delta_0)$ with $0 \leq o_+(z) \leq \epsilon\delta$ for $z \in (-\delta, \delta)$. We can then estimate

$$\|\partial_x \sigma(t, \cdot)\|_2^2 \geq \int_{-\delta}^{\delta} \frac{|\sigma_0'(z)|^2}{1 + t\gamma_0 + t o_+(z)} \, \mathrm{d}z \geq \int_{-\delta}^{\delta} \frac{c_0}{1 + t\gamma_0 + t\epsilon\delta} \, \mathrm{d}z = \frac{2 c_0 \delta}{1 + t\gamma_0 + t\epsilon\delta}.$$

Because of $t_0 = -1/\gamma_0 =: T$ in (5.40), it follows

$$\liminf_{t \to T^-} \|\partial_x \sigma(t, \cdot)\|_2^2 \geq \frac{2 c_0}{T \epsilon}.$$

Since $\epsilon > 0$ is arbitrary, Eq. (5.36) finally implies that

$$\liminf_{t \to T^-} \|\partial_x u(t, \cdot)\|_2^2 = \liminf_{t \to T^-} \|\partial_x \sigma(t, \cdot)\|_2^2 = +\infty.$$

(4) Note that $\sigma(t, x) = \sigma_0(y(t, x)) = \xi$ if $|y|$ is large enough. This fact holds for some $x_0 > 0$ and all $t \in [0, T)$ and $|x| \geq x_0$ because of (5.37) and the strict positivity of a' on $[0, \xi]$. So $u = \xi - \sigma$ has compact support. Fixing

$$\zeta = -\int_0^\xi a'(\xi - \tau)^{1/2} \, \mathrm{d}\tau,$$

also the function

$$v = \zeta + \int_0^\sigma a'(\xi - \tau)^{1/2} \, \mathrm{d}\tau$$

has compact support. □

Local Wellposedness on a Domain 6

In this chapter we extend the results from the previous one to linear and quasilinear Maxwell systems on a spatial domain G, endowed with boundary conditions. The general theory of symmetric hyperbolic systems is much more sophisticated in this case. It uses Sobolev spaces of higher order and with weights encoding a loss of derivatives in normal direction, see [79] or [155]. Fortunately the Maxwell equations have a special structure which allows us to derive analogous theorems as on $\mathbb{R}^3$ using a similar approach. However, already in the half-space case $G = \mathbb{R}^3_+ := \{x \in \mathbb{R}^3 | x_3 > 0\}$ many new difficulties arise, which we describe and solve below (sketching or omitting some technical steps). The general case is treated via localization arguments and thus reduced to hyperbolic problems on $\mathbb{R}^3_+$. They still resemble the Maxwell system, but the resulting coefficients A_j, $j \in \{1, 2, 3\}$, are far more complicated than A_j^{co}. Here we can only indicate how one deals with the new situation. In a first section we start with a derivation of the boundary conditions and a discussion of the relevant trace operator and the compatibility conditions.

6.1 The Maxwell System on a Domain

We continue to study the Maxwell equations

$$\partial_t \mathbf{D} = \operatorname{curl} \mathbf{H} - \mathbf{J}, \qquad \partial_t \mathbf{B} = -\operatorname{curl} \mathbf{E}, \qquad t \geq 0,\ x \in G, \tag{6.1}$$

for $t \geq 0$ and $x \in G$, where $G \subseteq \mathbb{R}^3$ is open and bounded with a smooth boundary or $G = \mathbb{R}^3_+ = \mathbb{R}^2 \times \mathbb{R}_+$. As before, we can define solutions to these equations in $C(\overline{J}, L^2_x)$. Observe that the solutions still satisfy Gauß' laws (5.2). Below we will equip the system again with the material laws (5.3) and (5.4), or their linear variants. However, the derivation of the boundary conditions is independent of these laws.

W. Dörfler et al., *Wave Phenomena*, Oberwolfach Seminars 49,
https://doi.org/10.1007/978-3-031-05793-9_6

We first establish the interface conditions for (6.1), arguing a bit informal. Let Σ be a surface in G, which is given by a chart $\varphi : U \to V$ with $\varphi(\Sigma) = V_0 \times \{0\}$. Set $\psi = \varphi^{-1}$ and $U_\pm = \psi(V \cap \mathbb{R}^3_\pm)$ with $\mathbb{R}^3_- = \mathbb{R}^2 \times \mathbb{R}_-$. We equip Σ with the unit normal n_Σ pointing into U_+, whereas n and $\mathrm{n}_\pm$ are the outer unit normal of U and $U_\pm$, respectively. Moreover, let $S \subseteq V_0$ be a line segment with direction p and $a > 0$ such that $Q = S \times [-a, a] \subseteq V$. Let ∂Q be oriented counter-clockwise and choose the normal ν to Q with $\det[\nu, p, e_3] > 0$. The surface $\Gamma = \psi(S \times [-a, a]) \subseteq U$ shall carry the induced orientation; i.e., its boundary (with a parametrization $\gamma = \gamma(\theta)$) winds positively around the unit normal n_Γ of Γ. Note that n_Γ is perpendicular to n_Σ. Let $\Gamma_\pm = \psi(Q \cap \mathbb{R}^3_\pm)$ be oriented accordingly. For a function f on $U_\pm$ we denote its trace on Σ by $f_\pm$ (assuming that it exists) and its jump across Σ by $[f] = f_+ - f_-$.

It is better to start from the more fundamental integral versions of the Maxwell equations and the Gauß' laws (5.2), namely

$$\begin{aligned} &\int_{\partial\Gamma} \mathbf{H} \cdot \mathrm{ds} = \int_\Gamma (\partial_t \mathbf{D} + \mathbf{J}) \cdot \mathrm{n}_\Gamma \, \mathrm{d}\sigma, \qquad \int_{\partial\Gamma} \mathbf{E} \cdot \mathrm{ds} = -\int_\Gamma \partial_t \mathbf{B} \cdot \mathrm{n}_\Gamma \, \mathrm{d}\sigma, \\ &\int_{\partial U} \mathbf{D} \cdot \mathrm{n} \, \mathrm{d}\sigma = \int_U \rho \, \mathrm{d}x, \qquad \int_{\partial U} \mathbf{B} \cdot \mathrm{n} \, \mathrm{d}\sigma = 0, \end{aligned} \tag{6.2}$$

where we require that these traces and integrals exist. (Here $\mathbf{H} \cdot \mathrm{ds} = (\mathbf{H} \circ \gamma) \cdot \gamma' \, \mathrm{d}\theta$ and σ is the surface measure.) If the fields belong to $H^1(U)$, say, these equations follow from (6.1) and (5.2) by means of Stokes' and Gauß' theorems. To show the converse implication, after applying Stokes and Gauß again, one divides the integrals by the volume of U and Γ, respectively, and lets them tend to 0. (Note that n_Γ can be any unit vector in $\mathbb{R}^3$ if one varies Σ and Γ.)

Let $\rho_\pm = \rho|_{U_\pm}$ and $\mathbf{J}_\pm = \mathbf{J}|_{U_\pm}$. We also allow for surface charges ρ_Σ and surface currents $\mathbf{J}_\Sigma$ concentrated on Σ, where $\mathbf{J}_\Sigma \perp \mathrm{n}_\Sigma$. Let $\mathbf{D}$ be regular on $U_\pm$ so that the jump $[\mathbf{D} \cdot \mathrm{n}_\Sigma] = [\mathbf{D}] \cdot \mathrm{n}_\Sigma$ is integrable on Σ. We then infer from (6.2) on U that

$$\begin{aligned} \int_{U_+} \rho_+ \, \mathrm{d}x + \int_{U_-} \rho_- \, \mathrm{d}x + \int_\Sigma \rho_\Sigma \, \mathrm{d}\sigma &= \int_U \rho \, \mathrm{d}x = \int_{\partial U} \mathbf{D} \cdot \mathrm{n} \, \mathrm{d}\sigma \\ &= \int_{\partial U_+} \mathbf{D} \cdot \mathrm{n}_+ \, \mathrm{d}\sigma + \int_{\partial U_-} \mathbf{D} \cdot \mathrm{n}_- \, \mathrm{d}\sigma + \int_\Sigma [\mathbf{D} \cdot \mathrm{n}_\Sigma] \, \mathrm{d}\sigma. \end{aligned}$$

By (6.2) on $U_\pm$, the first two terms on both sides cancel. We can replace V_0 by subsets V_0'. Dividing by the area of Σ and shrinking V_0', we see that $\rho_\Sigma = [\mathbf{D} \cdot \mathrm{n}_\Sigma]$ on Σ. In the same way one shows that $[\mathbf{B} \cdot \mathrm{n}_\Sigma] = 0$. Similarly, (6.2) also yields

$$\begin{aligned} &\int_{\partial\Gamma_+} \mathbf{H} \cdot \mathrm{ds} + \int_{\partial\Gamma_-} \mathbf{H} \cdot \mathrm{ds} - \int_{\Gamma\cap\Sigma} [\mathbf{H}] \cdot \mathrm{ds} = \int_{\partial\Gamma} \mathbf{H} \cdot \mathrm{ds} \\ &= \int_{\Gamma_+} (\partial_t \mathbf{D} + \mathbf{J}) \cdot \mathrm{n}_{\Gamma_+} \, \mathrm{d}\sigma + \int_{\Gamma_-} (\partial_t \mathbf{D} + \mathbf{J}) \cdot \mathrm{n}_{\Gamma_-} \, \mathrm{d}\sigma + \int_{\Gamma\cap\Sigma} \mathbf{J}_\Sigma \cdot \mathrm{n}_\Gamma \, \mathrm{d}s \end{aligned}$$

for $ds = |\gamma'|\,d\theta$. Since $\tau \mapsto \mathrm{n}_\Sigma \times \tau$ rotates tangent vectors of Σ counter-clockwise by $\pi/2$, we deduce as above that $-\mathbf{J}_\Sigma \cdot \mathrm{n}_\Gamma = [\mathbf{H}] \cdot (\mathrm{n}_\Sigma \times \mathrm{n}_\Gamma) = \mathrm{n}_\Gamma \cdot [\mathbf{H} \times \mathrm{n}_\Sigma]$, and hence $[\mathbf{H} \times \mathrm{n}_\Sigma] = -\mathbf{J}_\Sigma$ as n_Γ is an arbitrary tangent vector of Σ. Analogously one shows that $[\mathbf{E} \times \mathrm{n}_\Sigma] = 0$. We summarize the *interface conditions*

$$[\mathbf{E} \times \mathrm{n}_\Sigma] = 0, \qquad [\mathbf{D} \cdot \mathrm{n}_\Sigma] = \rho_\Sigma, \qquad [\mathbf{B} \cdot \mathrm{n}_\Sigma] = 0, \qquad [\mathbf{H} \times \mathrm{n}_\Sigma] = -\mathbf{J}_\Sigma \tag{6.3}$$

for fields which are sufficiently regular on $U_\pm$. (Cf. §I.4.2.4 in [35] or §1.7 in [68].)

As for Gauß' laws (5.2), the equations for $\mathbf{D}$ and $\mathbf{B}$ are redundant for solutions to (6.1) satisfying the interface conditions for $\mathbf{E}$ and $\mathbf{H}$. More precisely, one has $[\mathbf{B}(t) \cdot \mathrm{n}_\Sigma] = [\mathbf{B}(0) \cdot \mathrm{n}_\Sigma]$, and $\rho_\Sigma(t)$ can be computed in terms of $[\mathbf{D}(0) \cdot \mathrm{n}_\Sigma] = \rho_\Sigma(0)$, $\mathbf{J}$ and $\mathbf{J}_\Sigma$. See Lemma 8.1 in [153] and also §I.4.2.4 in [35] or our Lemma 6.1.

Arguably the basic set-up for the Maxwell system is $\mathbb{R}^3$ endowed with different material laws on subsets G and $\mathbb{R}^3 \setminus \overline{G}$ (e.g., having vacuum $\mathbf{D} = \mathbf{E}$ and $\mathbf{H} = \mathbf{B}$ on $\mathbb{R}^3 \setminus \overline{G}$) and equipped with initial conditions and interface conditions for $\mathbf{E}$ and $\mathbf{H}$ on $\Sigma = \partial G$. In fact, one can extend the local wellposedness theory discussed in this chapter to this setting, see [153]. Here we treat the simpler situation that the traces at ∂G of the fields on $\mathbb{R}^3 \setminus \overline{G}$ are assumed to be 0. For the electric fields, this is reasonable in the case of a perfect conductor on $\mathbb{R}^3 \setminus \overline{G}$ which refers to the limit case of infinite conductivity σ so that $\mathbf{E} = \frac{1}{\sigma}\mathbf{J} = 0$ in Ohm's law, see §I.4.2.4+6 in [35] or §7.12 in [68]. In this setting we will derive a local wellposedness theory with the boundary condition of a *perfect conductor*

$$\mathbf{E} \times \mathrm{n} = 0 \qquad \text{on } \partial G. \tag{6.4}$$

It is usually combined with the condition

$$\mathbf{B} \cdot \mathrm{n} = 0 \qquad \text{on } \partial G, \tag{6.5}$$

which however turns out to be true if it holds at time 0 by Lemma 6.1.

Before we continue, we have to explain the meaning of the above equations for functions $\mathbf{E}, \mathbf{B} \in C(\overline{J}, L^2(G, \mathbb{R}^3))$ solving the Maxwell system. To this aim, we first recall several known results about traces, see [2] and [36], for instance. Let $U \subseteq \mathbb{R}^m$ be an open subset with a Lipschitz boundary given by local graphs which yield a covering of U by finitely many charts $\varphi_j : U_j \to V_j$ with inverses ψ_j and parametrizations $F_j = \psi_j|_{\{y_m=0\}}$. Let n be its outer unit normal. For $s \geq 0$ we have the fractional Sobolev spaces $H^s(\mathbb{R}^m)$ consisting of $v \in L^2(\mathbb{R}^m)$ such that $|\xi|^s \mathcal{F} v$ belongs to $L^2(\mathbb{R}^m)$ (where $|\xi|^s$ stands for the map $\xi \mapsto |\xi|^s$). They are endowed with the norm given $\|v\|_s^2 = \|v\|_2^2 + \||\xi|^s \mathcal{F} v\|_2^2$. Their dual spaces are denoted by $H^{-s}(\mathbb{R}^m)$. The space $H^s(U)$ contains the restrictions $v|_U$ for $v \in H^s(\mathbb{R}^m)$. For an open subset $\Gamma \subseteq \partial U$ and $s \in (0, 1)$, we define

$$H^s(\Gamma) = \big\{ v \in L^2(\Gamma, \sigma) \,\big|\, v \circ F_j \in H^s(\varphi_j(\Gamma \cap V_j)), \ \ \forall j \big\}.$$

Again we let $H^{-s}(\Gamma)$ be the dual space. If ∂U is C^k, one can take here $s \in [0, k]$.

It is known that the trace operator $\mathrm{tr} : v \mapsto v|_{\partial U}$ (defined on $H^s(U) \cap C(\overline{U})$) extends to a continuous and surjective map from $H^1(U)$ to $H^{\frac{1}{2}}(\partial U)$. Its kernel is $H_0^1(U)$. Let $H(\mathrm{div}, U) = H(\mathrm{div}) = \{v \in L^2(U)^m \mid \mathrm{div}\, v \in L^2(U)\}$. Then the normal trace $\mathrm{tr_{no}} : v \mapsto (v \cdot \mathrm{n})|_{\partial U}$ defined on $H(\mathrm{div}) \cap C(\overline{U})$ extends to a continuous and surjective map from $H(\mathrm{div})$ to $H^{-\frac{1}{2}}(\partial U)$ with kernel $H_0(\mathrm{div})$, which is the closure of test functions in $H(\mathrm{div})$. One also has the divergence theorem

$$\int_U v \cdot \nabla\phi \,\mathrm{d}x = -\int_U \phi \,\mathrm{div}\, v \,\mathrm{d}x + \langle \mathrm{tr}\,\phi, \mathrm{tr_{no}}\, v\rangle_{H^{1/2}(\partial U)} \tag{6.6}$$

for $v \in H(\mathrm{div})$ and $\phi \in H^1(U)$. (Actually, one defines $\mathrm{tr_{no}}$ via continuous extension using (6.6) for $v \in H^1(U)$. The hard part is to show the density of $H^1(U)$ in $H(\mathrm{div})$.) Analogously, if $m = 3$ the tangential trace $\mathrm{tr_{ta}} : v \mapsto (v \times \mathrm{n})|_{\partial U}$ can be extended from $H(\mathrm{curl}) \cap C(\overline{U})$ to a continuous map from $H(\mathrm{curl})$ to $H^{-\frac{1}{2}}(\partial U)$ with kernel $H_0(\mathrm{curl})$, i.e, the closure of test functions in $H(\mathrm{curl})$. Moreover,

$$\int_U v \cdot \mathrm{curl}\,\phi \,\mathrm{d}x = \int_U \mathrm{curl}\, v \cdot \phi \,\mathrm{d}x + \langle \mathrm{tr}\,\phi, \mathrm{tr_{ta}}\, v\rangle_{H^{1/2}(\partial U)} \tag{6.7}$$

for $v \in H(\mathrm{curl})$ and $\phi \in H^1(U, \mathbb{R}^3)$. See Theorem 2.4 in [24] for a description of the range of $\mathrm{tr_{ta}}$ if $\partial U \in C^2$, say.

These results justify the boundary condition (6.5) in view of (5.2), but not yet (6.4) since we only require $u = (\mathbf{E}, \mathbf{H}) \in C(\overline{J}, L_x^2)$. We first fix our assumptions for the linear problem $Lu = \sum_{j=0}^3 A_j \partial_j u + Du = f$ on G as

$$\begin{aligned}
&A_0 \in W_{t,x}^{1,\infty} = W^{1,\infty}(J \times G, \mathbb{R}^{6\times 6}), \quad A_0 = A_0^\top \geq \eta I > 0, \quad J = (0, T),\\
&A_j = A_j^{\mathrm{co}} \text{ for } j \in \{1, 2, 3\}, \quad D \in L_{t,x}^\infty = L^\infty(J \times G, \mathbb{R}^{6\times 6}),\\
&u_0 \in L^2(G, \mathbb{R}^6) = L_x^2, \quad f \in L_{t,x}^2 = L^2(J \times G, \mathbb{R}^6).
\end{aligned} \tag{6.8}$$

(The matrices A_j^{co} were defined in (5.5).) Let $Lu = f$ for $u \in C(\overline{J}, L_x^2)$. We thus obtain

$$\sum_{j=0}^3 \partial_j (A_j u) = f - Du + \partial_t A_0\, u \in L_{t,x}^2.$$

By a natural extension of the above results, the function $\sum_{j=0}^3 \mathrm{N}_j A_j u$ has a trace in $H^{-1/2}(\partial(J \times G))$, where N is the outer unit normal of $J \times G$. Restricting to the subset $\{0\} \times G$ with $\mathrm{N} = -e_0$ this gives a meaning to the initial condition $u(0) = u_0$ as $A_0(0)$ is Lipschitz and invertible. On the lateral boundary $J \times \partial G$ with $\mathrm{N} = (0, \mathrm{n})$, by means of the comments before (5.5) we infer that $(-\mathrm{n} \times u^2, \mathrm{n} \times u^1)$ has a trace in $H^{-1/2}(J \times \partial G)$, and so (6.4) for $\mathbf{E} = u^1$ is well defined. We denote the latter trace also by $\mathrm{tr_{ta}}$. See §2.1

in [157] for a detailed exposition in which several basic properties are shown that are used below without further notice.

We now check that condition (6.5) is preserved for H^1-solutions of the Maxwell system with (6.4).

Lemma 6.1 *Let* $\mathbf{B}, \mathbf{E} \in C^1(\overline{J}, L^2_x) \cap C(\overline{J}, H^1_x)$ *satisfy* $\partial_t \mathbf{B} = -\operatorname{curl} \mathbf{E}$ *and* $\operatorname{tr_{ta}} \mathbf{E} = 0$. *Then* $\operatorname{tr_{no}} \mathbf{B}(t) = \operatorname{tr_{no}} \mathbf{B}(0)$ *for all* $t \in \overline{J}$.

Proof Let $t \in \overline{J}$ and $\varphi \in H^2(G)$. The assumption implies $\operatorname{div} \partial_t \mathbf{B} = 0$ so that $\operatorname{tr_{no}} \partial_t \mathbf{B}(t)$ exists in $H^{-\frac{1}{2}}(\partial G)$, and the same is true for $\operatorname{curl} \mathbf{E}(t)$. Using also (6.6) and (6.7), we thus obtain

$$\begin{aligned}
\partial_t \langle \mathbf{B}(t) \cdot \mathrm{n}, \varphi \rangle_{L^2(\partial G)} &= \langle \operatorname{tr_{no}}(\partial_t \mathbf{B})(t), \varphi \rangle_{H^{-1/2}(\partial G)} = \langle -\operatorname{tr_{no}} \operatorname{curl} \mathbf{E}(t), \varphi \rangle_{H^{-1/2}(\partial G)} \\
&= -\int_G \operatorname{div} \operatorname{curl} \mathbf{E}(t)\, \varphi \,\mathrm{d}x - \int_G \operatorname{curl} \mathbf{E}(t) \cdot \nabla \varphi \,\mathrm{d}x \\
&= -\int_G \mathbf{E}(t) \cdot \operatorname{curl} \nabla \varphi \,\mathrm{d}x + \langle \operatorname{tr_{ta}} \mathbf{E}(t), \nabla \varphi \rangle_{H^{-1/2}(\partial G)} = 0
\end{aligned}$$

omitting tr in front of φ. The result follows by density. □

6.2 The Linear Problem on $\mathbb{R}^3_+$ in L^2

We treat the linear Maxwell equations (6.1) on $G = \mathbb{R}^3_+$ with the boundary condition (6.4) of a perfect conductor for $\mathrm{n} = -e_3$. As in Example 5.6, we rewrite them as the symmetric hyperbolic system

$$\begin{aligned}
Lu = \sum\nolimits_{j=0}^{3} A_j \partial_j u + Du = f, \quad & t \geq 0,\ x \in \mathbb{R}^3_+, \\
Bu = -\mathbf{E} \times e_3 = 0, \quad & t \geq 0,\ x \in \partial\mathbb{R}^3_+, \\
u(0) = u_0, \quad & x \in \mathbb{R}^3_+
\end{aligned} \tag{6.9}$$

assuming hypothesis (6.8) for $G = \mathbb{R}^3_+$. We first look for a solution $u = (\mathbf{E}, \mathbf{H}) \in C(\overline{J}, L^2_x)$, using the notation of (6.8) for $G = \mathbb{R}^3_+$, proceeding as in [11], [26] or [147], for instance. The trace operator B is defined as in the previous section and will be identified with the matrix $B^{\mathrm{co}} = (S_3\ 0) \in \mathbb{R}^{3\times 6}$, where the matrices $S_j \in \mathbb{R}^{3\times 3}$ were introduced before (5.5). We proceed similar to Sect. 5.2 starting with an energy estimate. Later it turns out to be important that we require a bit less than H^1 in the lemma. We set $v_{\mathrm{ta}} = (v_1, v_2, v_4, v_5)$ for the tangential components of a function $v : \mathbb{R}^3_+ \to \mathbb{R}^6$, and $v_{\mathrm{no}} = (v_3, v_6)$ for the normal ones.

Lemma 6.2 *Assume that* (6.8) *is true for* $G = \mathbb{R}^3_+$ *and that* $u \in C(\overline{J}, L^2_x)$ *solves* (6.9) *and has derivatives* $\partial_j u$ *for* $j \in \{0, 1, 2\}$ *and* $\partial_3 u_{\mathrm{ta}}$ *in* $L^2_{t,x}$. *Let* $C^+ := \frac{1}{2}\partial_t A_0 - D$, $\gamma \geq \gamma_0^{+\prime}(L) := \max\{1, 4\,\|C^+\|_\infty/\eta\}$, $t \in \overline{J}$, *and* $L^2_\gamma = L^2_\gamma(0, t)$. *We then obtain*

$$\tfrac{\gamma\eta}{4}\,\|u\|^2_{L^2_\gamma L^2_x} + \tfrac{\eta}{2}\mathrm{e}^{-2\gamma t}\,\|u(t)\|^2_{L^2_x} \leq \tfrac{1}{2}\|A_0(0)\|_\infty\,\|u_0\|^2_{L^2_x} + \tfrac{1}{2\gamma\eta}\,\|f\|^2_{L^2_\gamma L^2_x}\,. \tag{6.10}$$

Proof Let $v = e_{-\gamma}u$ and $g = e_{-\gamma}f$. By assumption, $\partial_j u$ for $j \in \{0, 1, 2\}$ and $\partial_3 A_3^{\mathrm{co}} u$ belong to $L^2_{t,x}$, and hence u_{ta} has a trace on $\{x_3 = 0\}$ in $L^2(J \times \mathbb{R}^2)$. It is 0 for u_1 and u_2 by the boundary condition. As in Lemma 5.2, the equation $\gamma A_0 v + Lv = g$ yields

$$\langle g, v\rangle = \gamma\langle A_0 v, v\rangle + \sum\nolimits_{j=0}^{3}\langle A_j\partial_j v, v\rangle + \langle Dv, v\rangle$$

for the scalar products in $L^2((0, t), L^2_x)$. For $j \in \{1, 2, 3\}$ the summand with A_j is equal to $\int \frac{1}{2}\partial_j(A_j v\cdot v)\,\mathrm{d}(s, x)$ since A_j is constant and symmetric. The integral in x_j then vanishes by the above properties and by $A_3^{\mathrm{co}} u \cdot u = (u_5 u_1, -u_4 u_2, 0, -u_2 u_4, u_1 u_5, 0)$ has trace 0 on $\{x_3 = 0\}$. For $j = 0$, one obtains $2A_0\partial_t v \cdot v = \partial_t(A_0 v \cdot v) - \partial_t A_0 v \cdot v$. Integrating in t, we derive

$$\gamma\langle A_0 v, v\rangle + \frac{1}{2}\int_{\mathbb{R}^3_+} A_0(t)v(t)\cdot v(t)\,\mathrm{d}x = \frac{1}{2}\int_{\mathbb{R}^3_+} A_0(0)u_0\cdot u_0\,\mathrm{d}x + \langle C^+ v, v\rangle + \langle g, v\rangle.$$

The assertion now follows as in Lemma 5.2. □

We use (6.10) only for $\gamma \geq \gamma_0^+(r, \eta) := \max\{1, 6r/\eta\} \geq \gamma_0^{+\prime}(L)$ assuming that $\|\partial_t A_0\|_\infty, \|D\|_\infty \leq r$. As in (5.10) the above proof yields the energy equality

$$\int_{\mathbb{R}^3_+} A_0(t)u(t)\cdot u(t)\,\mathrm{d}x = \int_{\mathbb{R}^3_+} A_0(0)u_0\cdot u_0\,\mathrm{d}x + 2\int_0^t\int_{\mathbb{R}^3_+}\big(C^+(s)u(s) + f(s)\big)\cdot u(t)\,\mathrm{d}x\,\mathrm{d}s. \tag{6.11}$$

For the existence result we need analogous estimates for the (formal) adjoint

$$L^\circ = -\sum\nolimits_{j=0}^{3} A_j\partial_j + D^T - \partial_t A_0$$

in backward time and on the time interval $\mathbb{R}$. To this aim, we extend the coefficients A_0 constantly and D by 0 to $t \in \mathbb{R}$.

Corollary 6.3 *Assume that* (6.8) *is true for* $G = \mathbb{R}^3_+$ *with* $\|\partial_t A_0\|_\infty, \|D\|_\infty \leq r$. *Extend* A_0 *constantly and* D *by 0 to* $t \in \mathbb{R}$. *Let* $\gamma \geq \gamma_0^+(r, \eta)$.

(a) *Let $v \in C(\overline{J}, L^2_x)$ with $\partial_j v, \partial_3 v_{\mathrm{ta}} \in L^2(J, L^2_x)$ for $j \in \{0,1,2\}$ satisfy $L^\circ v = f$, $Bv = 0$ and $v(T) = v_0$. For the weight $\tilde{e}_\gamma(t) = \mathrm{e}^{\gamma(t-T)}$, $t \in \overline{T}$, and $L^2_{t,x} = L^2((t,T), L^2_x)$ we obtain*

$$\tfrac{\gamma\eta}{4} \|\tilde{e}_\gamma v\|^2_{L^2_{t,x}} + \tfrac{\eta}{2} \mathrm{e}^{2\gamma(t-T)} \|v(t)\|^2_{L^2_x} \le \tfrac{1}{2} \|A_0(T)\|_\infty \|v_0\|^2_{L^2_x} + \tfrac{1}{2\gamma\eta} \|\tilde{e}_\gamma f\|^2_{L^2_{t,x}}. \tag{6.12}$$

(b) *Let $h, v \in L^2_{-\gamma}(\mathbb{R}, L^2_x)$ with $\partial_j v, \partial_3 v_{\mathrm{ta}} \in L^2_{-\gamma}(\mathbb{R}, L^2_x)$ for $j \in \{0,1,2\}$ satisfy $L^\circ v = h$ and $Bv = 0$. We then have*

$$\tfrac{\gamma\eta}{4} \|v\|^2_{L^2_{-\gamma}(\mathbb{R}, L^2_x)} \le \tfrac{1}{2\gamma\eta} \|h\|^2_{L^2_{-\gamma}(\mathbb{R}, L^2_x)}. \tag{6.13}$$

The same estimate holds if we replace $-\gamma$ by γ and L° by L.

Proof Assertion (a) can be reduced to Lemma 6.2 as in Theorem 5.5. For (b), we first show the addendum. For $t \in \mathbb{R}$, as in Lemma 6.2 and with $L^2_{\gamma,t} = L^2_\gamma(-\infty, t)$ we derive

$$\tfrac{\gamma\eta}{4} \|v\|^2_{L^2_{\gamma,t} L^2_x} + \tfrac{\eta}{2} \mathrm{e}^{-2\gamma t} \|v(t)\|^2_{L^2_x} \le \tfrac{1}{2\gamma\eta} \|h\|^2_{L^2_{\gamma,t} L^2_x} \le \tfrac{1}{2\gamma\eta} \|h\|^2_{L^2_\gamma L^2_x}. \tag{6.14}$$

On the left we can drop the second summand and then let $t \to \infty$ using Fatou's lemma. Transforming $t \mapsto -t$ as in Theorem 5.5, estimate (6.13) follows from (6.14). □

Also in the present setting the duality argument from Theorem 5.5 provides a solution of (6.9) in $L^2_{t,x}$. However, the regularization argument does not work anymore in x_3. To obtain uniqueness and a continuous solution satisfying the energy estimate, we first pass to a problem with $t \in \mathbb{R}$ so that we can use regularization in t instead.

Proposition 6.4 *Assume that (6.8) is true for $G = \mathbb{R}^3_+$ with $\|\partial_t A_0\|_\infty, \|D\|_\infty \le r$.*

(a) *Then we have a solution $u \in L^2(J, L^2_x)$ of (6.9).*
(b) *Let $\gamma \ge \gamma_0^+(r, \eta)$ and $\tilde{f} \in L^2_\gamma(\mathbb{R}, L^2_x)$. Then there is a function $u \in L^2_\gamma(\mathbb{R}, L^2_x) \cap C(\mathbb{R}, L^2_x)$ satisfying $Lu = \tilde{f}$ and $Bu = 0$. Let $\tilde{f}$ also have support in $\mathbb{R}_{\ge 0}$. Then u solves (6.9) on J with $u_0 = 0$ and $\tilde{f}$, and it fulfills (6.10) and (6.11).*

Proof (a) We proceed as in Theorem 5.5 and define for $v \in V := \{v \in H^1(J \times \mathbb{R}^3_+) \mid Bv = 0,\ v(T) = 0\}$ the functional

$$\ell_0 : L^\circ V \to \mathbb{R}; \quad \ell_0(L^\circ v) = \langle v, f \rangle_{L^2_{t,x}} + \langle v(0), A_0(0) u_0 \rangle_{L^2_x}.$$

Estimate (6.12) with $\gamma = \gamma_0^+(r,\eta)$ shows that ℓ_0 is well defined and that

$$|\ell_0(L^\circ v)| \le \|f\|_{L^2_{t,x}} \|v\|_{L^2_{t,x}} + \|A_0(0)u_0\|_{L^2_x} \|v(0)\|_{L^2_x} \le c\|L^\circ v\|_{L^2_{t,x}}.$$

As a result, ℓ_0 can be extended to a functional on $L^2_{t,x}$ which in turn is represented by a function $u \in L^2(J, L^2_x)$ satisfying $\ell_0(L^\circ v) = \langle L^\circ v, u\rangle_{L^2_{t,x}}$ for all $v \in V$; i.e.,

$$\langle v, f\rangle_{L^2_{t,x}} + \langle v(0), A_0(0)u_0\rangle_{L^2_x} \tag{6.15}$$

$$= \langle v, Du\rangle_{L^2_{t,x}} - \sum_{j=0}^{2} \int_0^T \int_{\mathbb{R}^3_+} \partial_j(A_j v)\cdot u \,\mathrm{d}x\,\mathrm{d}t - \int_0^T \int_{\mathbb{R}^3_+} \partial_3 v \cdot A_3^{\mathrm{co}} u \,\mathrm{d}x\,\mathrm{d}t.$$

First, for $v \in H_0^1(J\times\mathbb{R}^3_+)$ this formula yields

$$\langle v, f\rangle_{L^2_{t,x}} = \langle v, Du\rangle_{L^2_{t,x}} + \sum\nolimits_{j=0}^{3} \langle v, A_j\partial_j u\rangle_{H_0^1(J\times\mathbb{R}^3_+)} = \langle v, Lu\rangle_{H_0^1(J\times\mathbb{R}^3_+)},$$

so that $Lu = f$ in $H^{-1}_{t,x}$. Since $f \in L^2_{t,x}$, from (5.9) we deduce that $\partial_t u$ belongs to $L^2_t H^{-1}_x$ and from $Lu = f$ that $\partial_3 u_{\mathrm{ta}}$ is contained in $L^2_{x_3}(\mathbb{R}_+, H^{-1}(J\times\mathbb{R}^2))$.

In a second step, we take $v = \phi v_0$ for some $v_0 \in C_c^\infty(\mathbb{R}^3_+)$ and $\phi \in C^1([0,T])$ with $\phi(0) = 1$ and $\phi(T) = 0$. Equation (6.15) now implies

$$\langle v, Lu\rangle_{L^2_{t,x}} + \langle v_0, A_0(0)u_0\rangle_{L^2_x}$$

$$= \langle v, Du\rangle_{L^2_{t,x}} + \sum_{j=0}^{3} \int_0^T \langle v(s), A_j(s)\partial_j u(s)\rangle_{H_0^1(\mathbb{R}^3_+)} \,\mathrm{d}s + \langle v_0, A_0(0)u(0)\rangle_{L^2_x}.$$

As $C_c^\infty(\mathbb{R}^3_+)$ is dense in $H_0^1(\mathbb{R}^3_+)$, it follows $A_0(0)u_0 = A_0(0)u(0)$ in H_x^{-1}, and so $u(0) = u_0$.

Finally, let $v \in C_c^\infty(J\times\mathbb{R}^2\times\mathbb{R}_{\ge 0})$. Identity (6.15) then leads to

$$\langle v, Lu\rangle_{L^2_{t,x}} = \langle v, Du\rangle_{L^2_{t,x}} + \sum\nolimits_{j=0}^{2} \langle v, A_j\partial_j u\rangle_{H_0^1(J\times\mathbb{R}^2, L^2(\mathbb{R}_+))} - \int_0^T \int_{\mathbb{R}^3_+} \partial_3 v\cdot A_3^{\mathrm{co}} u \,\mathrm{d}x\,\mathrm{d}t.$$

We now choose v such that only $v_5 \neq 0$. Write $\Gamma = \partial\mathbb{R}^3_+ = \{x_3 = 0\}$. Combined with the identity

$$-\int_0^\infty \langle \partial_3 v_5, u_1\rangle_{L^2(J\times\mathbb{R}^2)} \,\mathrm{d}x_3 = \int_0^\infty \langle v_5, \partial_3 u_1\rangle_{H_0^1(J\times\mathbb{R}^2)} \,\mathrm{d}x_3 + \langle \mathrm{tr}_\Gamma\, v_5, \mathrm{tr}_\Gamma\, u_1\rangle_{H_0^1(J\times\mathbb{R}^2)}$$

the above equation in display yields $\langle v, Lu\rangle = \langle v, Lu\rangle + \langle \mathrm{tr}_\Gamma\, v_5, \mathrm{tr}_\Gamma\, u_1\rangle$ so that the last term is equal to 0. Again by density we conclude $\mathrm{tr}_\Gamma\, u_1 = 0$ in $H^{-1}(J\times\mathbb{R}^2)$, and similarly $\mathrm{tr}_\Gamma\, u_2 = 0$. Therefore $u \in L^2(J, L^2_x)$ solves (6.9).

(b) (1) Let $\tilde{f} \in L^2_\gamma(\mathbb{R}, L^2_x)$ for a fixed $\gamma \geq \gamma_0^+(r, \eta)$. We proceed as above on the time interval $\mathbb{R}$, setting $V = \{v \in H^1_{-\gamma}(\mathbb{R} \times \mathbb{R}^3_+) \mid Bv = 0\}$ and $\ell_0(L^\circ v) = \langle v, \tilde{f} \rangle_{L^2_{t,x}}$ for $v \in V$. We note that L^2_γ is the dual of $L^2_{-\gamma}$ via the L^2-scalar product. Estimate (6.13) then implies that ℓ_0 is welldefined and bounded. As in part (a) we can then represent ℓ_0 by a function $u \in L^2_\gamma(\mathbb{R}, L^2_x)$ and show that $Lu = \tilde{f}$ and $Bu = 0$.

(2) Let $R_{1/n}$ be a mollifier in (t, x_1, x_2) for $n \in \mathbb{N}$. Then $R_{1/n}u$ is an element of $H^1_\gamma(\mathbb{R} \times \mathbb{R}^2, L^2(\mathbb{R}_+))$, and it satisfies $BR_{1/n}u = 0$ and

$$LR_{\frac{1}{n}}u = R_{\frac{1}{n}}\tilde{f} + [A_0, R_{\frac{1}{n}}]\partial_t u + [D, R_{\frac{1}{n}}]u =: \tilde{f}_n. \tag{6.16}$$

By Proposition 5.3 and dominated convergence, the functions $\tilde{f}_n$ tend to $\tilde{f}$ in $L^2_\gamma(\mathbb{R}, L^2_x)$ as $n \to \infty$. Moreover,

$$\partial_3 A_3^{\text{co}} R_{\frac{1}{n}}u = \tilde{f}_n - \sum\nolimits_{j=0}^{2} A_j \partial_j R_{\frac{1}{n}}u - DR_{\frac{1}{n}}u \tag{6.17}$$

is contained in $L^2_\gamma L^2_x$. So $(R_{1/n}u)_n$ is Cauchy in $C_{b,\gamma}L^2_x \cap L^2_\gamma L^2_x$ by (6.14) and (6.16). Since $R_{1/n}u \to u$ in $L^2_\gamma L^2_x$, we conclude that u belongs to $C(\mathbb{R}, L^2_x)$ and fulfills (6.14).

(3) Let $\tilde{f}$ have support in $\mathbb{R}_{\geq 0}$. Using (6.14) for u, we estimate

$$\int_{-\infty}^{0} \|u(s)\|^2_{L^2_x}\, \mathrm{d}s \leq \int_{-\infty}^{0} \mathrm{e}^{-2\gamma s} \|u(s)\|^2_{L^2_x}\, \mathrm{d}s \leq \frac{2}{\gamma^2\eta^2}\|\tilde{f}\|^2_{L^2_\gamma(\mathbb{R}, L^2_x)} \leq \frac{2}{\gamma^2\eta^2}\|\tilde{f}\|^2_{L^2_{t,x}}.$$

Letting $\gamma \to \infty$, we infer that u vanishes for $t \leq 0$. Inequality (6.14) then shows (6.10) with $u_0 = 0$. Moreover, the functions $R_{1/n}u$ from step (2) satisfy the energy equality (6.11) with $u_0 = 0$, and thus also u by approximation. □

The above proof also yields uniqueness of solutions to (6.9) in $L^2_{t,x}$.

Corollary 6.5 *Assume that* (6.8) *is true for* $G = \mathbb{R}^3_+$. *Let* $u, v \in L^2(J, L^2_x)$ *solve* (6.9). *Then* $u = v$.

Proof The function $w = u - v \in L^2_{t,x}$ solves (6.9) with $u_0 = 0$ and $f = 0$. Extend w by 0 to $\mathbb{R}$. We then have $w \in H^1((0, T), H^1_{-x})$ and $Lw = 0$ on $(-\infty, T)$. Take times $0 < t_0 < t_1 < T$ and a function $\theta \in C^\infty(\mathbb{R})$ being 1 on $(-\infty, t_0]$ and 0 on $[t_1, \infty)$. The map $\tilde{w} = \theta w$ has support in $\overline{J}$ and satisfies $L\tilde{w} = \theta' A_0 w =: g \in L^2_t L^2_x$, where $\operatorname{supp} g \subseteq [t_0, t_1]$. As in parts (2) and (3) of the proof of Proposition 6.4 (b), we then check that $w = 0$ on $[0, t_0]$ (using the weight $\mathrm{e}^{-\gamma(t-t_0)}$ and replacing the time 0 by t_0). Here $t_0 < T$ is arbitrary. □

As a final preliminary step we show the desired result if $f = 0$. Recall that we have extended A_0 and D to $\mathbb{R}$.

Lemma 6.6 *Assume that* (6.8) *is true and* $f = 0$. *Then there is a unique solution* $u \in C(\overline{J}, L^2_x)$ *of* (6.9), *and it satisfies* (6.10) *and* (6.11) *with* $f = 0$.

Proof Proposition 6.4 provides a solution $u \in L^2_{t,x}$ on $(0, T+1)$.

(1) First, let u_0 be 0 outside a compact set in $\mathbb{R}^3_+$. Then extend it by 0 to $\mathbb{R}^3$. Theorem 5.5 (and a backward version) yield a solution $\tilde{u} \in C(\mathbb{R}, L^2_x)$ of $L\tilde{u} = 0$ with $\tilde{u}(0) = u_0$. There is a time $\tau > 0$ such that $\tilde{u}(t)$ is supported in $\mathbb{R}^3_+$ for all $t \in [-\tau, \tau]$ due to the finite speed of propagation, see Theorem 5.7. So the restriction v of $\tilde{u}$ to $[-\tau, \tau] \times \mathbb{R}^3_+$ solves (6.9) with $f = 0$. Corollary 6.3 shows that $u = v$ on $[0, \tau]$ and hence $u \in C([0, \tau], L^2_x)$. We extend u by $\tilde{u}|_{\mathbb{R}^3_+}$ continuously to $t < 0$.

As in the proof of Proposition 6.4, we set $u_n = R_{1/n} u$ on $[0, T]$ for the mollifier in $(t, x_1, x_2) = (t, x')$. Then the functions $u_n \in C(\overline{J}, L^2_x)$ tend to u in $L^2(J, L^2_x)$ and satisfy $Bu_n = 0$, $Lu_n =: f_n \to 0$ in $L^2(J, L^2_x)$ and (6.10) on J. (See (6.16) and (6.17) and use Lemma 6.2.) Moreover, we have

$$u_n(0, x) = \int_{-1/n}^{1/n} \int_{B(x', 1/n)} \rho_{1/n}(-s, x' - y') u(s, y', x_3) \, \mathrm{d}y' \, \mathrm{d}s$$

which tends to u_0 in L^2_x by the time continuity of u. For $\gamma = \gamma_0^+$ and $t \in \overline{J}$, estimate (6.10) then yields

$$\|u_n(t) - u_m(t)\|^2_{L^2_x} \le c\big(\|u_n(0) - u_m(0)\|^2_{L^2_x} + \|f_n - f_m\|^2_{L^2_{t,x}}\big) \longrightarrow 0$$

for $n, m \to \infty$ so that (u_n) is Cauchy in $C(\overline{J}, L^2_x)$. As a result, u belongs to this space and fulfills (6.10) and (6.11) with $f = 0$.

(2) Let $u_0 \in L^2_x$. Let $u_{0,n} = \mathbb{1}_{K_n} u_0$ for compact sets $K_n \subseteq \mathbb{R}^3_+$ with $\bigcup_{n \in \mathbb{N}} K_n = \mathbb{R}^3_+$. Step (1) provides a function $u_n \in C(\overline{J}, L^2_x)$ with $Lu_n = 0$, $Bu_n = 0$ and $u_n(0) = u_{0,n}$ which satisfies (6.10) and (6.11). This estimate then implies that (u_n) is Cauchy in $C(\overline{J}, L^2_x)$, and hence the limit u has the asserted properties. □

We now obtain the basic linear well-posedness result in L^2_x. (The additional factor 2 could be avoided using $R_{1/n}$ as above.)

Theorem 6.7 *Assume that* (6.8) *is true for* $G = \mathbb{R}^3_+$ *with* $\|\partial_t A_0\|_\infty, \|D\|_\infty \le r$. *Then there is a unique solution* $u \in C(\overline{J}, L^2_x)$ *of* (6.9), *and it satisfies* (6.11) *as well as* (6.10) *with a factor 2 on the right-hand side for* $\gamma \ge \gamma_0^+(r, \eta)$.

Proof Uniqueness was shown in Corollary 6.3. Proposition 6.4 and Lemma 6.6 provide functions $v, w \in C(\overline{J}, L^2_x)$ satisfying $Lv = f$, $Bv = 0$, $v(0) = 0$, as well as $Lw = 0$, $Bw = 0$, $w(0) = u_0$. Then $u = v + w \in C(\overline{J}, L^2_x)$ solves (6.9). Since v and w fulfill (6.10) and (6.11) for the respective data, the last assertion also follows. □

6.3 The Linear Problem on $\mathbb{R}^3_+$ in H^3

On $G = \mathbb{R}^3$ we have reduced the wellposedness of the linear problem in H^3 to that in L^2 by means of the transformation $v \mapsto (I - \Delta)^{3/2}v$. For the Maxwell system on domains such a procedure does not seem to work anymore because of the boundary condition (6.4). (See [107] for cases where one can proceed in such a way also in the presence of (simpler) boundary conditions.) Instead we will first derive apriori estimates for H^3-solutions and then show by regularization arguments that the L^2-solution of Theorem 6.7 is actually an H^3-solution if the data satisfy natural assumptions.

In our reasoning we will mix space and time regularity so that we need the same number of derivatives in space and in time. We thus look for solutions in $G^3(J \times \mathbb{R}^3_+)$ where we set

$$G^m(J \times G) = \bigcap\nolimits_{k=0}^{m} C^k(\overline{J}, H^{m-k}(G, \mathbb{R}^6)),$$

$$G^{m-}(J \times G) = \bigcap\nolimits_{k=0}^{m} W^{k,\infty}(J, H^{m-k}(G, \mathbb{R}^6))$$

for $m \in \mathbb{N}_0$ and an open bounded set $G \subseteq \mathbb{R}^3$ with smooth boundary or $G \in \{\mathbb{R}^3_+, \mathbb{R}^3\}$. These spaces are endowed with their canonical norms. Sometimes we only write $G^m(T)$ if G is clear from the context and $J = (0, T)$. For the coefficients we use

$$F^m(J \times G) = \big\{A \in W^{1,\infty}(J \times G, \mathbb{R}^{6\times 6}) \,\big|\, \forall \alpha \in \mathbb{N}_0^4 \text{ s.t. } 1 \le |\alpha| \le m : \partial^\alpha A \in L^\infty(J, L^2_x)\big\},$$

These spaces are endowed with their natural norms, and the same symbols also denote spaces with different range spaces. The spaces $\hat{H}^m_x = \hat{H}^m(G)$ are defined as on $\mathbb{R}^3$ after (5.17). As before, the subscript 'sym' means that the functions take values in symmetric matrices, 'η' that they are bounded from below by ηI in addition, and 'γ' refers to norms with weight $\mathrm{e}^{-\gamma t}$. To obtain solutions in G^3, we strengthen hypotheses (6.8) to

$$\begin{aligned}
&A_0, D \in F^3(J \times G, \mathbb{R}^{6\times 6}), \quad A_0 = A_0^\top \ge \eta I > 0, \quad J = (0, T), \qquad (6.18)\\
&A_j = A_j^{\mathrm{co}} \text{ for } j \in \{1, 2, 3\}, \quad u_0 \in H^k(G, \mathbb{R}^6) = H^k_x, \quad f \in H^k(J \times G, \mathbb{R}^6) = H^k_{t,x},\\
&\|A_0\|_{F^3}, \|D\|_{F^3} \le r, \quad \|A_0(0)\|_{\hat{H}^2_x}, \|D(0)\|_{\hat{H}^2_x} \|\partial_t^l A_0(0)\|_{H^{2-l}_x}, \|\partial_t^l D(0)\|_{H^{2-l}_x} \le r_0
\end{aligned}$$

for all $l \in \{1, 2\}$, some $k \in \{1, 2, 3\}$, and constants $r \ge r_0 \ge 1$. We note that the product and inversion rules stated after (5.17) for $G = \mathbb{R}^3$ remain true on the present spatial domain G and for G^3, since G admits an extension operator and the additional time derivatives can be treated similarly, see §2 of [159] or §2.2 of [157].

If one has a solution $u \in C(\overline{J}, H^1_x)$ of (6.9), the initial value $u(0) = u_0$ must satisfy the boundary condition $Bu_0 = 0$ by continuity. If u even belongs to G^3, also $u^1 = \partial_t u(0)$ and $u^2 = \partial_{tt} u(0)$ have to fulfill $Bu^j = 0$, where we put $u^0 = u_0$. In view of (6.9) and (6.18), the following (linear) *compatibility conditions* (of order 3) are thus necessary for the existence of a solution $u \in G^3$.

$$Bu^j = 0 \text{ for } j \in \{0,1,2\}, \quad u^1 := A_0(0)^{-1}\Big[f(0) - D(0)u_0 - \sum\nolimits_{j=1}^{3} A_j\partial_j u_0\Big], \tag{6.19}$$
$$u^2 := A_0(0)^{-1}\Big[\partial_t f(0) - \partial_t D(0)u_0 - D(0)u^1 - \partial_t A_0(0)u^1 - \sum\nolimits_{j=1}^{3} A_j\partial_j u^1\Big].$$

The function u^3 is defined analogously applying ∂_{tt} to $Lu = f$. Assuming (6.18), the product and inversion rules easily yield

$$\|u^j\|_{H_x^{k-j}} \le c(r_0,\eta)\big(\|u_0\|_{H_x^k} + \|f(0)\|_{H_x^{k-1}} + \dots + \|\partial_t^{j-1} f(0)\|_{H_x^{k-j}}\big) \tag{6.20}$$

for $k \in \{1,2,3\}$ and $j \in \{0,\dots,k\}$. See Lemma 2.3 in [159] or Lemma 2.33 of [157].

We start with the apriori estimates for the time and tangential derivatives. We write $H^k_{\mathrm{ta}}(J \times G)$ for functions $g \in L^2_{t,x}$ with $\partial^\alpha g \in L^2_{t,x}$ for all $\alpha = (\alpha_0,\dots,\alpha_3) \in \mathbb{N}_0^4$ with $|\alpha| \le k$ and $\alpha_3 = 0$. Analogously, we define $G^k_{\mathrm{ta},\gamma}$ and $H^k_{\mathrm{ta}}(G) = H^k_{\mathrm{ta},x}$.

Lemma 6.8 *Let* (6.18) *be true for* $G = \mathbb{R}^3_+$ *and some* $k \in \{1,2,3\}$, *where we only assume* $f \in H^3_{\mathrm{ta}}(J \times G)$, *and let* $u \in G^k(J \times G)$ *solve* (6.9). *Then there exist constants* $\tilde\gamma_k^+ = \tilde\gamma_k^+(r,\eta) \ge \gamma_0^+$, $c_k^+ = c_k^+(r,\eta)$ *and* $c_{k,0}^+ = c_{k,0}^+(r_0,\eta)$ *such that* u *satisfies*

$$\|u\|^2_{G^k_{\mathrm{ta},\gamma}} + \gamma\|u\|^2_{H^k_{\mathrm{ta},\gamma}} \le c_{k,0}^+\big[\|u_0\|^2_{H_x^k} + \|f(0)\|^2_{H_x^{k-1}} + \dots + \|\partial_t^{k-1} f(0)\|^2_{L_x^2}\big]$$
$$+ \frac{c_k^+}{\gamma}\big(\|f\|^2_{H^k_{\mathrm{ta},\gamma}} + \|u\|^2_{G^k_\gamma}\big)$$

for all $\gamma \ge \tilde\gamma_k^+$.

Proof Take $\alpha \in \mathbb{N}_0^4$ with $|\alpha| \le k$ and $\alpha_3 = 0$ and apply ∂^α to (6.9). We then have $B\partial^\alpha u = 0$, $\partial^\alpha u(0) = \partial^{(\alpha_1,\alpha_2)} u_{\alpha_0}$ and

$$L\partial^\alpha u = \partial^\alpha f - \sum_{0<\beta\le\alpha} \tbinom{\alpha}{\beta}\big(\partial^\beta A_0 \partial^{\alpha-\beta}\partial_t u + \partial^\beta D\partial^{\alpha-\beta} u\big). \tag{6.21}$$

Combined with (6.20) and the product rules, the energy inequality (6.10) applied to the above equation yields the assertion. □

The extra term involving u on the right-hand side will be absorbed below. The above argument fails for the normal derivative ∂_3 since ∂_3 destroys the boundary condition $Bu =$

0. However, the equation (6.9) directly allows to bound $\partial_3 u_{\mathrm{ta}}$ in terms of u and $\partial_j u$ for $j \in \{0, 1, 2\}$. For instance, the first line of (6.9) yields

$$\partial_3 \mathbf{H}_2 = \partial_2 \mathbf{H}_3 - (A_0 \partial_t u)_1 - (Du)_1 + f_1. \tag{6.22}$$

Here we mix space and time regularity which forces us to use the solution space G^3. The remaining derivatives $\partial_3 u_3$ and $\partial_3 u_6$ (and higher-order analogues) can be treated using div curl $= 0$. We stress that we do not employ boundary conditions in these two steps.

Proposition 6.9 *Let* (6.18) *be true for* $G = \mathbb{R}^3_+$ *and some* $T' > 0$ *and* $k \in \{1, 2, 3\}$. *Let* $T \in (0, T']$ *and* $u \in G^k(J \times G)$ *solve* (6.9). *Then there exist constants* $\gamma_k^+ = \gamma_k^+(r, \eta, T') \geq \tilde{\gamma}_k^+$, $C_k^+ = C_k^+(r, \eta, T')$ *and* $C_{k,0}^+ = C_{k,0}^+(r_0, \eta)$ *such that*

$$\begin{aligned} \|u\|_{G_\gamma^k}^2 &\leq (C_{k,0}^+ + TC_k^+) \mathrm{e}^{kC_1^+ T} \big(\|u_0\|_{H_x^k}^2 + \|f(0)\|_{H_x^2}^2 + \cdots + \|\partial_t^{k-1} f(0)\|_{L_x^2}^2 \big) \\ &\quad + \tfrac{C_k^+}{\gamma} \mathrm{e}^{kC_1^+ T} \|f\|_{H_\gamma^k}^2 \qquad \textit{for all } \gamma \geq \gamma_k^+. \end{aligned} \tag{6.23}$$

Proof (1) Let $k = 1$ and $t \in [0, T]$. To carry out the argument indicated above, we first note that $\|A_0(t)\|_{L_x^\infty} \leq r_0 + rt$ since $A_0(t) = A_0(0) + \int_0^t \partial_t A_0(s)\,\mathrm{d}s$ and analogously for D and f. So (6.9) and Lemma 6.8 yield

$$\begin{aligned} \|\partial_3 u_{\mathrm{ta}}\|_{G_\gamma^0(t)} &\leq c(r_0 + Tr) \|(u, \partial_{t,x_1,x_2} u)\|_{G_\gamma^0(t)} + \|f(0)\|_{L_x^2} \\ &\quad + \sup_{\tau \in [0,t]} \int_0^\tau \mathrm{e}^{-\gamma(\tau - s)} \mathrm{e}^{-\gamma s} \|\partial_t f(s)\|_{L_x^2}\,\mathrm{d}s, \\ &\leq c(r_0 + Tr) \|(u, \partial_{t,x_1,x_2} u)\|_{G_\gamma^0(t)} + \|f(0)\|_{L_x^2} + \tfrac{c}{\sqrt{\gamma}} \|\partial_t f\|_{L_\gamma^2((0,t), L_x^2)} \\ &\leq (c_0(r_0, \eta) + Tc(r, \eta)) \big(\|u_0\|_{H_x^1} + \|f(0)\|_{L_x^2} \big) + \tfrac{c(r,\eta,T')}{\sqrt{\gamma}} (\|f\|_{H_{\gamma,\mathrm{ta}}^1(t)} + \|u\|_{G_\gamma^1(t)}), \end{aligned} \tag{6.24}$$

using also Hölder's inequality. We next treat $\partial_3 u_{\mathrm{no}}$ with $u_{\mathrm{no}} = (u_3, u_6)$. To simplify notation, we assume in this proof that $D = 0$ and $A_0 = \operatorname{diag}(a^e, a^m)$, where a^j maps into $\mathbb{R}_\eta^{3\times 3}$, and we write $f = (f^e, f^m)$. Equation (6.9) then leads to

$$\begin{aligned} \partial_t (a^e \nabla_x \mathbf{E}) &= \partial_t a^e \nabla_x \mathbf{E} + a^e \nabla_x (a_e^{-1} \operatorname{curl} \mathbf{H} + a_e^{-1} f^e) \\ &= \partial_t a^e \nabla_x \mathbf{E} + a^e \nabla_x a_e^{-1} (\operatorname{curl} \mathbf{H} + f^e) + \nabla_x f^e + \nabla_x \operatorname{curl} \mathbf{H} \end{aligned}$$

in H_x^{-1}. The first two summands in the last line are denoted by Λ and are bounded pointwise by $c(r)(|\nabla_x u| + |f|)$. Observe that the trace $\mathrm{sp}(\nabla_x \operatorname{curl} \mathbf{H})$ is equal to $\operatorname{div}\operatorname{curl}\mathbf{H} = 0$. Integrating in time, we thus obtain

$$a_{33}^e(t)\partial_3\mathbf{E}_3(t) = \mathrm{sp}(a^e(0)\nabla\mathbf{E}_0) - \sum_{(j,k)\neq(3,3)} a_{jk}^e(t)\partial_k\mathbf{E}_j(t) + \int_0^t \mathrm{sp}(\nabla f^e + \Lambda)\,\mathrm{d}s. \tag{6.25}$$

Let $\partial' u = (u, \partial_1 u, \partial_2 u, \partial_3 u_{\mathrm{ta}})$. Since $a_{33}^e \geq \eta$, as in (6.24) we derive

$$\begin{aligned}\|\partial_3\mathbf{E}_3(t)\|_{L_x^2} &\leq c(r_0,\eta)\|u_0\|_{H_x^1} + c\tfrac{r_0+rt}{\eta}\|\partial' u(t)\|_{L_x^2} + c(r,\eta)\int_0^t \|(f(s),u(s))\|_{H_x^1}\,\mathrm{d}s\\ &\leq c(r_0)\|u_0\|_{H_x^1} + (c_0(r)+tc(r))\mathrm{e}^{\gamma t}\|\partial' u(t)\|_{G_\gamma^0(t)} + \frac{c(r)}{\sqrt{\gamma}}\mathrm{e}^{\gamma t}\|f\|_{H_\gamma^1(t)}\\ &\quad + c_1(r)\int_0^t \|\partial_3 u_{\mathrm{no}}(s)\|_{L_x^2}\,\mathrm{d}s,\end{aligned} \tag{6.26}$$

dropping the dependence on η in the constants. We now multiply this inequality by $\mathrm{e}^{-\gamma t}$, and add the analogous one for $\partial_3\mathbf{H}_3$ as well as Lemma 6.8 and (6.24). It follows

$$\begin{aligned}\mathrm{e}^{-\gamma t}\|(u(t),\partial_{t,x}u(t))\|_{L_x^2} &\leq (c(r_0)+Tc(r))\big(\|u_0\|_{H_x^1} + \|f(0)\|_{L_x^2}\big) + \tfrac{c(r,T')}{\sqrt{\gamma}}\|f\|_{H_\gamma^1(t)}\\ &\quad + \tfrac{c(r,T')}{\sqrt{\gamma}}\|u\|_{G_\gamma^1(t)} + c(r)\int_0^t \mathrm{e}^{-\gamma s}\|\partial_3 u_{\mathrm{no}}(s)\|_{L_x^2}\,\mathrm{d}s.\end{aligned} \tag{6.27}$$

Gronwall's inequality yields (6.23) for $k = 1$, if we take sufficiently large γ to absorb the norm of u on the right-hand side.

(2) Let $k = 3$, the variant for $k = 2$ is shown by a modification of the proof.

(a) We first take $\alpha \in \mathbb{N}_0^4$ with $|\alpha| \leq 2$ and $\alpha_3 = 0$. Equation (6.21) and our product rules show that $L\partial^\alpha u = f_\alpha$ with $\|f_\alpha\|_{H^1_{\gamma,\mathrm{ta}}} \leq c(r)\,(\|f\|_{H^3_{\gamma,\mathrm{ta}}} + \|u\|_{H^3_\gamma})$. Using also (6.20), we can bound $\partial^\alpha u(0)$ in H_x^1 by

$$c(r_0)(\|u_0\|_{H_x^3} + \|f(0)\|_{H_x^2}^2 + \|\partial_t f(0)\|_{H_x^1} + \|\partial_{tt} f(0)\|_{L_x^2}) =: c(r_0)\kappa(u_0,f).$$

Lemma 6.8 and (6.24) thus yield

$$\|u\|_{G^3_{\mathrm{ta},\gamma}} + \|\partial_3\partial^\alpha u_{\mathrm{ta}}\|_{G_\gamma^0(t)} \leq (c(r_0)+Tc(r))\kappa(u_0,f) + \tfrac{c(r,T')}{\sqrt{\gamma}}\big(\|f\|_{H^3_{\gamma,\mathrm{ta}}} + \|u\|_{G^3_\gamma}\big).$$

We next employ (6.26) with $\partial^\alpha u$ and f_α instead of u and f. Using also the above estimates, we derive (6.27) for $\partial^\alpha u$ and f. Note that we still work with L and so the constant $c_1(r)$ in (6.26) is unchanged. Gronwall's inquality thus implies (6.23) for $k = 3$ up to the term $c(r,T')\gamma^{-1}\mathrm{e}^{c_1(r)t}\|u\|_{G_\gamma^3}^2$ on the right-hand side, if restrict ourselves to derivatives of u and f with $\alpha_3 \leq 1$.

(b) In a next step, in (a) we choose α with $|\alpha| \le 2$ and $\alpha_3 = 1$. Proceeding as in step a) and using it, we first obtain

$$\|u\|_{G^3_{\mathrm{ta},\gamma}(t)} + \|\partial_3 u\|_{G^2_{\mathrm{ta},\gamma}(t)} + \|\partial_3^2 u_{\mathrm{ta}}\|_{G^1_{\mathrm{ta},\gamma}(t)}$$
$$\le (c(r_0)+Tc(r))\mathrm{e}^{c_1(r)t}\kappa(u_0, f) + \tfrac{c(r,T')}{\sqrt{\gamma}}\big[\|\partial_3 f\|_{H^2_{\gamma,\mathrm{ta}}} + \mathrm{e}^{c_1(r)t}(\|f\|_{H^3_{\gamma,\mathrm{ta}}} + \|u\|_{G^3_\gamma})\big].$$

In (6.26) we now insert $\partial_3\partial_{\mathrm{ta}}u$ and $\partial_{3j} f$ with $j \in \{0, 1, 2\}$. As in (6.27), we derive from the above estimate and Gronwall's inquality the assertion up to the error term for all derivatives except ∂_3^3 and $k = 2$ in the exponent. The step for ∂_3^3 can be performed analogously. Finally we absorb $c(r, T')\gamma^{-1}\|u\|^2_{G^3_\gamma}$ by the left-hand side, choosing large $\gamma \ge \gamma_3^+(r, \eta, T')$. □

Using the above estimate, we now show in several steps that the solution $u \in G^0(J)$ of (6.9) given by Theorem 6.7 actually belongs to $G^3(J)$ if (6.18) is true with $k = 3$. For $k = 1$, we first show that $u \in C^1(\overline{J}, L^2_x)$ by solving an equation formally satisfied by $\partial_t u$. We need the compatibility condition for time regularity in this step. The tangential regularity is then derived by means of mollifiers as in Proposition 6.4. The normal regularity finally follows from (6.9) and (6.25).

Lemma 6.10 *Let* (6.18) *be true for* $G = \mathbb{R}^3_+$ *and* $k = 1$, *and let* $u \in G^0(J \times G)$ *solve* (6.9). *Assume that* $Bu_0 = 0$. *Then* u *belongs to* $C^1(\overline{J}, L^2_x)$.

Proof (1) Define $u^1 \in L^2_x$ as in (6.19). We look for a solution $v \in C(\overline{J}, L^2_x)$ of

$$\begin{aligned} L'v &:= \sum\nolimits_{j=0}^{3} A_j\partial_j v + (D + \partial_t A_0)v = \partial_t f - \partial_t D\Big(u_0 + V(t)\Big), \quad t \ge 0,\ x \in \mathbb{R}^3_+,\\ Bv &= 0, \quad t \ge 0,\ x \in \partial\mathbb{R}^3_+, \\ v(0) &= u^1, \quad x \in \mathbb{R}^3_+, \end{aligned} \tag{6.28}$$

where we have set $V(t) = \int_0^t v(s)\,\mathrm{d}s$. If already knew that u belonged to $C^1(\overline{J}, L^2_x)$, then $v = \partial_t u$ would satisfy (6.28). However, we can solve this problem directly using a simple fixed-point argument. Indeed, take $w \in G^0(J)$ and replace v by w on the right-hand side of the evolution equation in (6.28). Theorem 6.7 then yields a solution $v \in G^0(J)$ of the resulting problem. For $\gamma \ge \gamma_0^+$ and $c = c(r, \eta)$, we further obtain

$$\|v - \overline{v}\|^2_{G^0_\gamma} \le \tfrac{c}{\gamma}\,\|W - \overline{W}\|^2_{L^2_\gamma L^2_x} \le \tfrac{cT^2}{2\gamma^2}\,\|w - \overline{w}\|^2_{G^0_\gamma},$$

where $\overline{v} \in G^0(J)$ solves (6.28) for $\overline{w} \in G^0(J)$ instead of v on the right. Fixing a large γ, we obtain a unique fixed point $v \in G^0(J)$ solving (6.28).

(2) The function $w := u_0 + V \in C^1(\overline{J}, L_x^2)$ satisfies $w(0) = u_0$ and $Bw = 0$ due to the compatibility condition $Bu_0 = 0$. Observe that $\partial_t v \in L_t^2 H_x^{-1}$ by (6.28) and hence

$$A_0(t)v(t) = A_0(0)u^1 + \int_0^t \big(\partial_t A_0(s)v(s) + A_0(s)\partial_t v(s)\big)\, ds$$

in H_x^{-1} for $t \in [0, T]$. Similarly, we have

$$D(t)w(t) = D(0)u_0 + \int_0^t \big(\partial_t D(s)w(s) + D(s)v(s)\big)\, ds.$$

These identities, (6.28) and (6.19) imply

$$\begin{aligned} Lw(t) &= (A_0 v)(t) + \sum\nolimits_{j=1}^3 A_j^{\mathrm{co}} \partial_j (u_0 + V(t)) + D(t)w(t) \\ &= A_0(0)u^1 + \sum\nolimits_{j=1}^3 A_j^{\mathrm{co}} \partial_j u_0 + D(0)u_0 + \int_0^t \big(L'v(s) + \partial_t D(s)w(s)\big)\, ds \\ &= f(0) + \int_0^t \partial_t f(s)\, ds = f(t). \end{aligned}$$

The uniqueness statement of Theorem 6.7 now yields $u = w \in C^1(\overline{J}, L_x^2)$. □

Lemma 6.11 *Let* (6.18) *be true for* $G = \mathbb{R}_+^3$ *and* $k = 1$*, and let* $u \in C^1(\overline{J}, L_x^2)$ *solve* (6.9). *Then* u *belongs to* $C(\overline{J}, H^1_{\mathrm{ta},x})$.

Proof Let $R_{1/n}$ be the mollifier with respect to (x_1, x_2) and set $u_n = R_{1/n}u$ for $n \in \mathbb{N}$. This function belongs to $C(\overline{J}, H^1_{\mathrm{ta},x}) \cap C^1(\overline{J}, L_x^2)$ and tends to u in $C(\overline{J}, L_x^2)$ by the properties of u. As in (6.16), we have $BR_{1/n}u = 0$ and

$$Lu_n = R_{\frac{1}{n}} f + [A_0, R_{\frac{1}{n}}]\partial_t u + [D, R_{\frac{1}{n}}]u =: f_n.$$

Hence, $\partial_3 A_3^{\mathrm{co}} u_n$ is contained in $L^2_{t,x}$. We can thus apply ∂_j for $j \in \{1, 2\}$ to $Lu_n = f_n$ resulting in

$$L\partial_j u_n = R_{\frac{1}{n}} \partial_j f + [\partial_j A_0, R_{\frac{1}{n}}]\partial_t u + [\partial_j D, R_{\frac{1}{n}}]u - \partial_j A_0 \partial_t u_n - \partial_j D u_n =: g_n.$$

The regularity of u and the data implies that g_n tends to $\partial_j f - \partial_j A_0 \partial_t u - \partial_j D u$ in $L^2_{t,x}$ and $\partial_j u_n(0)$ to $\partial_j u_0$ in L_x^2 as $n \to \infty$. Hence, $(\partial_j u_n)_n$ is Cauchy in $C(\overline{J}, L_x^2)$ by (6.10). This means that (u_n) converges to u in $C(\overline{J}, H^1_{\mathrm{ta},x})$. □

The next lemma on normal regularity does not involve boundary conditions. If $Bu = 0$, then u satisfies its regularity assumptions thanks to the previous lemmas.

Lemma 6.12 *Let* (6.18) *be true for* $G = \mathbb{R}^3_+$ *and* $k = 1$, *and let* $u \in C^1(\overline{J}, L^2_x) \cap C(\overline{J}, H^1_{\mathrm{ta},x})$ *solve* $Lu = f$. *Then* u *belongs to* $G^1(J)$.

Proof Observe that $\partial_3 u_{\mathrm{ta}}$ is contained in $C(\overline{J}, L^2_x)$ by (6.22) and the assumptions. We want to regularize u in x_3 direction and then use (6.26) to pass to the limit. To this aim, we simplify a bit and restrict ourselves to the special case $A_0 = \mathrm{diag}(a^e, a^m)$ as in the proof of Proposition 6.9. More importantly, for technical reasons we shift the functions on $\mathbb{R}^3_+$ downwards by $S_\delta v(x) = v(x', x_3 + \delta)$ for $x = (x', x_3) \in \mathbb{R}^3_+$ and $\delta > 0$. This destroys the boundary condition, which is not needed fortunately. We then extend the function $S_\delta v$ by 0 to $\mathbb{R}^3$ and apply the mollifier $R_{1/n}$ in x_3. Afterwards we will restrict to $x \in \mathbb{R}^3_+$ again. This allows us to justify the calculations below, see the proof of Lemma 4.1 in [159] for the details.

We write $u_\delta = S_\delta u$, $f_\delta = S_\delta f$, and L_δ for the operator with shifted coefficients $S_\delta A_0$ and $S_\delta D$. One has $L_\delta u_\delta = f_\delta$. We compute

$$L_\delta R_{\frac{1}{n}} u_\delta = R_{\frac{1}{n}} f_\delta + [S_\delta A_0, R_{\frac{1}{n}}]\partial_t u_\delta + [S_\delta D, R_{\frac{1}{n}}]u_\delta =: g_{\delta,n}$$

for $n \in \mathbb{N}$ and $\delta > 0$ with $\frac{1}{n} < \delta$. Then $R_{1/n} u_\delta$ belongs to $G^1(J \times \mathbb{R}^3_+)$, $g_{\delta,n}$ tends to f_δ in $H^1_{t,x}$ as $n \to \infty$, and $u^0_{\delta,n} := R_{1/n} S_\delta u_0$ to $S_\delta u_0$ in H^1_x by our assumptions. We can now derive the analogue of formula (6.25) as before. As in (6.26) it follows

$$\begin{aligned}\|\partial_3 (R_{\frac{1}{n}} - R_{\frac{1}{m}}) S_\delta u_{\mathrm{no}}(t)\|_{L^2_x} \le c\Big[&\|u^0_{\delta,n} - u^0_{\delta,m}\|_{H^1_x} + \|(R_{\frac{1}{n}} - R_{\frac{1}{m}})\partial' S_\delta u\|_{G^0(t)} \\ &+ \|g_{\delta,n} - g_{\delta,m}\|_{H^1_{t,x}}\Big] + c\int_0^t \|\partial_3 (R_{\frac{1}{n}} - R_{\frac{1}{m}}) S_\delta u_{\mathrm{no}}\|_{L^2_x}\,\mathrm{d}s,\end{aligned}$$

where $u_{\mathrm{no}} = (u_3, u_6)$ and $\partial' u = (\partial_1 u, \partial_2 u, \partial_3 u_{\mathrm{ta}})$. In view of the above comments, the terms in brackets tend to 0 as $n, m \to \infty$, and hence the same is true for the left-hand side due to Gronwall's inequality.

As $(R_{1/n} S_\delta u_{\mathrm{no}})_n$ has the limit $S_\delta u_{\mathrm{no}}$ in G^0, we infer that $\partial_3 S_\delta u_{\mathrm{no}} = S_\delta \partial_3 u_{\mathrm{no}}$ is an element of $G^0(J \times \mathbb{R}^3_+)$. The strong continuity of $(S_\delta)_\delta$ on $L^2(\mathbb{R}^3_+)$ then implies that also $\partial_3 u_{\mathrm{no}}$ belongs to $G^0(J \times \mathbb{R}^3_+)$, so that u is an element of G^1. □

We can now show the linear wellposedness result in $H^3(\mathbb{R}^3_+)$.

Theorem 6.13 *Let* (6.18) *be true for* $G = \mathbb{R}^3_+$ *and* $k = 3$. *Assume that the compatibility conditions* (6.19) *hold. Then there is a unique solution* $u \in G^3(J \times \mathbb{R}^3_+)$ *of* (6.9). *It satisfies* (6.23).

Proof (1) Theorem 6.7 provides a unique solution u in $G^0(J)$. If we can prove that u belongs to $G^3(J)$, then it satisfies (6.23) by Proposition 6.9. Lemmas 6.10, 6.11 and 6.12 already show that u is an element of G^1.

(2) For the iteration steps, we also assume that $\partial_t A_0 \in F^3(J)$. Let $\tilde{L}$ be the operator with $\tilde{D} = D + \partial_t A_0 \in F^3(J)$ instead of D. We then have $\partial_t u \in G^0(J)$, $B\partial_t u = 0$, and

$$\tilde{L}\partial_t u = \partial_t f - \partial_t Du = \tilde{f} \in H^1_{t,x}.$$

Since $\partial_t u(0) = u^1 \in H^1_x$ and $Bu^1 = 0$ by (6.20) and (6.19), step (1) yields $\partial_t u \in G^1$.

Due to this regularity, $L\partial_j u = \partial_j f - \partial_j A_0 \partial_t u - \partial_j Du =: f_j$ belongs to $H^1_{t,x}$, and we have $\partial_j u(0) = \partial_j u_0 \in H^1_x$ for $j \in \{1, 2, 3\}$. If $j \neq 3$ also the boundary conditions $B\partial_j u = 0$ and $B = \partial_j u_0$ are preserved. Lemmas 6.11 and 6.12 thus show that $\partial_j u$ is an element of $C(\overline{J}, H^1_x)$ for $j = \{1, 2\}$. In particular, $\partial_3 u$ is contained in $C^1(\overline{J}, L^2_x) \cap C(\overline{J}, H^1_{\mathrm{ta},x})$, and hence in $G^1(J)$ due to Lemma 6.12 and $f_3, \partial_3 u_0 \in H^1$. Therefore u belongs to $G^2(J)$.

(3) To show $u \in G^3(J)$, we proceed similarly as in step 2), writing $\hat{L}$ for the operator with D replaced by $\hat{D} = D + 2\partial_t A_0 \in F^3(J)$. Because of step (2) and the assumption on A_0, the function $\partial_{tt} u \in G^0(J)$ satisfies

$$\hat{L}\partial_{tt} u = \partial_{tt} f - \partial_{tt} A_0 \partial_t u - \partial_{tt} Du - 2\partial_t D\partial_t u \in H^1_{t,x}$$

and $B\partial_{tt} u = 0$. Starting from $\tilde{L}\partial_t u = \tilde{f}$ and $\partial_t u(0) = u^1$, we compute

$$\partial_{tt} u(0) = A_0(0)^{-1}\Big[\partial_t f(0) - \partial_t D(0)u_0 - D(0)u^1 - \partial_t A_0(0)u^1 - \sum\nolimits_{j=1}^{3} A_j \partial_j u^1\Big] = u^2,$$

see (6.19). Using also (6.20), we infer $u^2 \in H^1_x$ and $Bu^2 = 0$. So $\partial_{tt} u$ belongs to $G^1(J)$ by step (1).

We next look at $\partial_{jt} u \in G^0(J)$ for $j \in \{1, 2, 3\}$. By the above established properties of u and the regularity of A_0, the map

$$\tilde{L}\partial_{jt} u = \partial_{jt} f - \partial_j A_0 \partial_{tt} u - \partial_{jt} A_0 \partial_t u - \partial_{jt} Du - \partial_t D\partial_j u - \partial_j D\partial_t u$$

is an element of $H^1_{t,x}$ and $\partial_{jt} u(0) = \partial_j u^1$ of H^1_x. Because of $B\partial_{jt} u = 0$ and $B\partial_j u^1 = 0$ if $j \neq 3$, step (1) yields $\partial_{jt} u \in G^1(J)$ in this case. As in step (2), $\partial_t u$ thus belongs to $C(\overline{J}, H^2_x)$ by Lemma 6.12.

Finally, we treat $\partial_{jk} u \in G^0(J)$ for $j, k \in \{1, 2, 3\}$. Again the right-hand side

$$L\partial_{jk} u = \partial_{jk} f - \partial_{jk} A_0 \partial_t u - \partial_j A_0 \partial_{kt} u - \partial_k A_0 \partial_{jt} u - \partial_{jk} Du - \partial_j D\partial_k u - \partial_k D\partial_j u$$

is contained in $H^1_{t,x}$ and $\partial_{jk} u(0) = \partial_{jk} u_0$ in H^1_x. For $j, k \leq 2$, we deduce that $\partial_{jk} u \in G^1(J)$ again from step (1). Hence, $\partial_{j3} u$ belongs to $C^1(\overline{J}, L^2_x) \cap C(\overline{J}, H^1_{\mathrm{ta},x})$ and thus to $G^1(J)$ by Lemma 6.12. The remaining property $\partial_{33} u \in C(\overline{J}, H^1_x)$ is shown analogously.

(4) We still have to remove the extra assumption $\partial_t A_0 \in F^3(J)$. To this aim, one has to regularize A_0 and to approximate u_0 in H^3_x so that the compatibility conditions (6.19) remain valid. This technical step is omitted, see Lemma 4.8 in [159]. □

As in Remarks 5.10 and 5.11, we list variants of the above theorem, which can be shown analogously.

Remark 6.14 Theorem 6.13 remains valid if we replace in (6.18) the differentiation order 3 by $m \in \mathbb{N}$ throughout, and impose corresponding variants of the compatibility conditions (6.19). If $m = 2$, the second-order derivatives of A_0 also have to belong to $L_t^\infty L_x^3$. On the other hand, in (6.8) we can replace $F^3 = F^3(J \times \mathbb{R}^3_+)$ by $\hat{F}^3_\infty = \hat{F}^3 + W^{3,\infty}_{t,x}$ and $\hat{H}^2_x = \hat{H}^2_x(\mathbb{R}^3_+)$ by $\hat{H}^2_\infty = \hat{H}^2_x + W^{2,\infty}$. We use this notation below also for other domains G.

6.4 The Quasilinear Problem on $\mathbb{R}^3_+$

We now treat the nonlinear problem

$$\begin{aligned} a_0(u)\partial_t u + \sum\nolimits_{j=1}^3 A_j^{\mathrm{co}}\partial_j u + d(u)u &= f, \quad t \geq 0,\ x \in G, \\ Bu = \mathbf{E} \times \mathrm{n} &= 0, \quad t \geq 0,\ x \in \partial G, \\ u(0) &= u_0, \quad x \in G. \end{aligned} \tag{6.29}$$

on $G = \mathbb{R}^3_+$ under the hypothesis

$$\begin{aligned} &a_0, d \in C^3(G \times \mathbb{R}^6, \mathbb{R}^{6\times 6}), \qquad a_0 = a_0^\top \geq \eta I, \\ &\forall r > 0: \sup_{|\xi|\leq r} \max_{0\leq|\alpha|\leq 3} \|\partial_x^\alpha a_0(\cdot,\xi)\|_{L_x^\infty}, \|\partial_x^\alpha d(\cdot,\xi)\|_{L_x^\infty} < \infty, \\ &u_0 \in H^3(G, \mathbb{R}^6), \qquad \forall T > 0: f \in H^3((0,T)\times G, \mathbb{R}^6) = H^3_{t,x}(T), \\ &\rho^2 \geq \|u_0\|^2_{H_x^3} + \|f\|^2_{H^3_{t,x}} + \|f(0)\|^2_{H_x^2} + \|\partial_t f(0)\|^2_{H_x^1} + \|\partial_t^2 f(0)\|^2_{L_x^2}. \end{aligned} \tag{6.30}$$

We state a version of Lemma 5.13 on G in the framework of G^3 and F^3. The proof is similar and thus omitted, see §2 of [158] or §7.1 in [157].

Lemma 6.15 *Let a be as in* (6.30) *and $\gamma \geq 0$.*

(a) Let $v \in G^3(J)$ with $\|v\|_\infty \leq r$. Then $\|a(v)\|_{\tilde{F}^m_\infty(J)} \leq \kappa(r)(1 + \|v\|^3_{G^3(J)})$.
(b) Let $v, w \in G^2(J)$ with norm $\leq r$. Then $\|a(v) - a(w)\|_{G^2_\gamma(J)} \leq \kappa(r)\,\|v - w\|_{G^2_\gamma(J)}$.
(c) Let $v_0 \in H_x^2$ with $\|v_0\|_\infty \leq r_0$. Then $\|a(v_0)\|_{\hat{H}^2_\infty} \leq \kappa_0(r_0)(1 + \|v_0\|^2_{H_x^2})$.
(d) Let $v_0, w_0 \in H_x^2$ with norm $\leq r_0$. Then $\|a(v_0) - a(w_0)\|_{H_x^2} \leq \kappa_0(r_0)\|v_0 - w_0\|^2_{H_x^2}$.

We need a nonlinear variant of the compatibility conditions (6.19), derived similarly:

$$\begin{aligned} &Bu^j = 0 \text{ for } j \in \{0,1,2\}, \quad u^1 := a_0(u_0)^{-1}\Big(f(0) - d(u_0)u_0 - \sum\nolimits_{j=1}^3 A_j\partial_j u_0\Big),\\ &u^2 := a_0(u_0)^{-1}\Big(\partial_t f(0) - \partial_\xi a_0(u_0)[u^1, u^1] - d(u_0)u^1 - \partial_\xi d(u_0)[u^1, u_0] \qquad (6.31)\\ &\qquad - \sum\nolimits_{j=1}^3 A_j\partial_j u^1\Big), \end{aligned}$$

where $u^0 := u_0$ and the derivatives in $\xi \in \mathbb{R}^6$ act on the vectors in brackets bilinearly. We set $u^j =: S_j(u_0, f, a_0, d)$. Lemma 5.13 (and analogous versions for $\partial_\xi a_0$ and $\partial_\xi d$) imply that S_j satisfy estimates as in (6.20) and related Lipschitz bounds, see Lemma 2.4 in [158] or Lemma 7.7 in [157].

We now state the local wellposedness result on $G = \mathbb{R}^3_+$, using the data manifold

$$\mathcal{D}_{T,a_0,d}((u_0, f), r) = \big\{(u_0, f) \in H^3_x \times H^3_{t,x}(T) \,\big|\, \|u_0\|^2_{H^3_x} + \|f\|^2_{H^3_x} \le r^2, \quad (6.31) \text{ holds}\big\}.$$

It is endowed with the metric of $H^3_x \times H^3_{t,x}(T)$ unless something else is stated.

Theorem 6.16 *Let* (6.30) *and* (6.31) *hold with* $G = \mathbb{R}^3_+$. *The following assertions are true.*

(a) *There is a maximal existence time* $T_+ = T_+(u_0, f) \in (T_0(\rho), \infty]$ *and a unique (maximal) solution* $u = \Psi(u_0, f) \in G^3([0, T_+))$ *of* (6.29). *(For* $T_0(\rho)$ *see the proof.)*
(b) *Let* $T_+ < \infty$. *Then* $\lim_{t \to T_+} \|u(t)\|_{H^3_x} = \infty$ *and* $\sup_{t<T_+} \|u(t)\|_{W^{1,\infty}_x} = \infty$.
(c) *Let* $T \in [0, T_+)$. *Then there is a radius* $\delta > 0$ *such that for all* (v_0, g) *in* $\mathcal{D}_{T,a_0,d}((u_0, f), \delta)$ *we have* $T_+(v_0, f) > T$ *and* $\Psi : \mathcal{D}_T((u_0, f), \delta) \to G^3(T)$ *is continuous. Moreover,* $\Psi : (\mathcal{D}_T((u_0, f), \delta), \|\cdot\|_{H^2_x \times H^2_{t,x}(T)}) \to G^2(T)$ *is Lipschitz.*

Proof The arguments are only sketched, see [159] for full proofs in a more general setting. As a fixed-point space we employ

$$E(R, T) := \big\{v \in G^{3-}(J) \,\big|\, \|v\|_{G^{3-}(J)} \le R, \; \partial_j^t v(0) = S_j(u_0, f, a_0, d) \text{ for } j \in \{0, 1, 2\}\big\}.$$

One can check as before that this set is complete for the metric induced by the norm of $G^2(T)$. (Compared to Lemma 5.14, we need the time derivatives in the norm because of the new initial conditions.) It is now more difficult to show that $E(R, T)$ is non-empty for sufficiently large R, see Lemma 2.6 in [159]. For $v \in E(R, T)$ one sets $A_0 = a_0(v)$ and $D = d(v)$ with a corresponding linear operator $L(v)$. Observe that the nonlinear compatibility conditions (6.31) for a_0 and d coincide with the linear ones (6.19) for A_0 and D because of the initial conditions in $E(R, T)$.

We then choose R, γ and T_0 depending on ρ as in Lemma 5.15. Here we have also to control first and second time derivatives of A_0 and D at $t = 0$ since these bounds enter the higher-order energy estimate (6.23) via (6.18). Theorem 6.13 (and Remark 6.14) now yield a unique solution $u = \Phi(v) \in G^3(T_0)$ of $L(v) = f$ with $Bu = 0$ and $u(0) = u_0$. The constants are arranged so that $\|u\|_{G^3} \leq R$ and Φ is strictly contractive. The compatibility conditions of this equation also imply that u belongs to $E(R, T_0)$. So we have solved (6.29) on $[0, T_0]$.

The proofs of Lemma 5.16 as well as of assertion (a) and the first part of (b) of Theorem 5.18 follow a general pattern so that these arguments and statements can easily be extended to the present situation. The second part of assertion (b) is more involved. One proceeds as in the proof of Theorem 5.18 and estimates $\partial_x^\alpha u$ in terms of the data and $\omega = \sup_{t<T_+} \|u(t)\|_{W_x^{1,\infty}}$. This works as before if $\alpha_3 = 0$. Using the resulting inequality in this case, one then follows the iteration steps of the proof of Theorem 6.13. The analogous difficulty occurs in step (3) of the proof of Theorem 5.18 (c), the other steps do not change much. □

We discuss variants of the above theorem in Remark 6.20 in greater generality.

6.5 The Main Wellposedness Result

We now treat the Maxwell system on an open and bounded set $G \subseteq \mathbb{R}^3$ with a smooth boundary. Again we first look at the linear problem

$$\begin{aligned} Lu = \sum\nolimits_{j=0}^{3} A_j \partial_j u + Du &= f, \quad t \geq 0,\ x \in G, \\ Bu = \mathbf{E} \times \mathrm{n} &= 0, \quad t \geq 0,\ x \in \partial G, \\ u(0) &= u_0, \quad x \in G, \end{aligned} \tag{6.32}$$

assuming hypothesis (6.18), and the nonlinear system (6.29) under assumption (6.30). We further assume the compatibility conditions (6.19) respectively (6.31) on G are satisfied.

Below we state the wellposedness theorems on the spatial domain G. We cannot give full proofs (since they are too lengthy and technical), but rather explain main features and differences to the case $G = \mathbb{R}^3_+$. We start with the localization procedure which is the core point.

Localization

In principle, we follow a standard localization procedure. One covers ∂G by finitely many charts $\varphi_i : U_i \to V_i$ and adds another open set U_0 with $\overline{U_0} \subseteq G$ so that $U_0, U_1, \ldots, U_N$ cover $\overline{G}$. Let $\varphi_0 : U_0 \to U_0$ be the identity, $\{\theta_0, \cdots, \theta_N\}$ be a smooth partition of unity for this cover and $V_i^+ = \{x \in V_i \mid x_3 > 0\}$ be the range $\varphi_i(U_i \cap G)$ for $i \geq 1$, where we put $\bar{V}_i^+ = \{x \in V_i \mid x_3 \geq 0\}$. Set $\psi_i = \varphi_i^{-1}$ and $\Phi_i : L^2(U_i) \to L^2(V_i)$; $\Phi_i u = u \circ \psi_i$, with inverse $\Phi_i^{-1} v = v \circ \varphi_i$.

First, let u solve (6.18). One looks at the transformed function $\Phi_i(\theta_i u) \in L^2(J \times V_i^+)$. After extension by 0, the map $\Phi_0(\theta_0 u)$ solves the original problem (5.8) on $\mathbb{R}^3$. For $i \geq 1$ and $v \in L^2(J \times V_i^+)$, we compute

$$\begin{aligned}
\tilde{L}^i v &:= \Phi_i L \Phi_i^{-1} v = \Phi_i\Big(A_0(\partial_t v) \circ \varphi_i + \sum\nolimits_{j=1}^3 A_j^{\mathrm{co}} \partial_j (v \circ \varphi_i) + Dv \circ \varphi_i\Big) \\
&= \Phi_i A_0 \partial_t v + \Phi_i D v + \sum\nolimits_{j=1}^3 A_j^{\mathrm{co}} \Phi_i \Big(\sum\nolimits_{k=1}^3 (\partial_k v) \circ \varphi_i \, \partial_j \varphi_{i,k}\Big) \\
&= \Phi_i A_0 \partial_t v + \Phi_i D + v \sum\nolimits_{k=1}^3 \Big(\sum\nolimits_{j=1}^3 A_j^{\mathrm{co}} \Phi_i (\partial_j \varphi_{i,k})\Big) \partial_k v \\
&=: \sum\nolimits_{k=0}^3 \tilde{A}_k^i \partial_k v + \tilde{D}^i v,
\end{aligned} \tag{6.33}$$

where $\varphi_{i,k}$ is the k-th component of φ_i. Note that $\tilde{A}_k^i$ is symmetric and $\tilde{A}_0^i \geq \eta I$.

One can check that there is a constant $\tau > 0$ and for each $i \geq 1$ an index $k(i) \in \{1, 2, 3\}$ such that $|\partial_{k(i)} \varphi_{i,3}| \geq \tau$ on U_i, see Lemma 5.1 in [157]. To simplify, we only look at the case $k(i) = 3$ and $\partial_3 \varphi_{i,3} \geq \tau$. Since $U_i \cap \partial G$ equals $U_i^0 := \{x \in U_i \mid \varphi_{i,3}(x) = 0\}$, the vector $\nabla \varphi_{i,3}(x)$ is orthogonal to ∂U_i^0 at x and hence given by $\nabla \varphi_{i,3}(x) = -\kappa_i(x) \mathrm{n}(x)$ for the smooth function $\kappa_i = -\mathrm{n} \cdot \nabla \varphi_{i,3}$. On U_i^0 the boundary condition $Bu = \mathbf{E} \times \mathrm{n} = 0$ thus is equivalent to $\kappa_i B u = \kappa_i \mathbf{E} \times \mathrm{n} = 0$. Using this reformulation, the transformation then yields the new boundary condition

$$\tilde{B}^i v := \Phi_i(\kappa_i B \Phi_i^{-1} v) = \Phi_i(\kappa_i B) v$$

on $\bar{V}_i^+$. The coefficients $\tilde{A}_j^i$, $\tilde{D}^i$, and $\tilde{B}^i$ are extended to $\mathbb{R}_+^3$ or its closure keeping their properties. (This extension is omitted below, cf. Chapter 5 of [157].)

The matrices $\tilde{A}_j^i$ for $j \in \{1, 2, 3\}$ and $\tilde{B}^i$ now depend (smoothly) on the space variable x. Moreover, $\tilde{A}_j^i$ has twelve instead of four non-zero entries, whereas $\tilde{B}^i$ still has just two of them. For instance, we have

$$\tilde{A}_3^i = \begin{pmatrix} 0 & -\tilde{S}_3^i \\ \tilde{S}_3^i & 0 \end{pmatrix}, \qquad \tilde{S}_3^i = \begin{pmatrix} 0 & -\Phi_i \partial_3 \varphi_{i,3} & \Phi_i \partial_2 \varphi_{i,3} \\ \Phi_i \partial_3 \varphi_{i,3} & 0 & -\Phi_i \partial_1 \varphi_{i,3} \\ -\Phi_i \partial_2 \varphi_{i,3} & \Phi_i \partial_1 \varphi_{i,3} & 0 \end{pmatrix}.$$

In the calculations of this chapter, these changes lead to plenty of additional commutators, which are partly hard to control and which make the iteration arguments much more complicated. Even worse is the change from A_3^{co} to $\tilde{A}_3^i$ since the form of A_3^{co} plays a crucial role in the above treatment of normal regularity. It is not clear how to extend the corresponding arguments to the transformed operator.

Instead, one passes to the function $v^i = R_i^{-1} \Phi_i(\theta_i u)$ for invertible matrices $R_i(x) = \mathrm{diag}(\hat{R}_i(x), \hat{R}_i(x))$ that are defined using φ_i. Let L^i be the operator on $\mathbb{R}_+^3$ with coefficients $A_j^i = R_i^\top \tilde{A}_j^i R_i$ and $D^i = R_i^\top \tilde{D}^i R_i - \sum_{j=1}^3 A_j^i \partial_j R_i^{-1} R_i$ and set $B^i = \hat{R}_i^\top \tilde{B}^i R_i$ as well as

$v_0^i = R_i^{-1}\Phi_i(\theta_i u_0)$. We then infer $v^i(0) = v_0^i$, $B^i v^i = 0$, and

$$\begin{aligned}
L^i v^i &= \sum\nolimits_{j=0}^{3} A_j^i \partial_j (R_i^{-1}\Phi_i(\theta_i u)) + R_i^\top \tilde{D}^i \Phi_i(\theta_i u) - \sum\nolimits_{j=1}^{3} A_j^i \partial_j R_i^{-1}\Phi_i(\theta_i u) \\
&= \sum\nolimits_{j=0}^{3} R_i^\top \tilde{A}_j^i \partial_j \Phi_i(\theta_i u) + R_i^\top \tilde{D}^i \Phi_i(\theta_i u) = R_i^\top \tilde{L}^i \Phi_i(\theta_i u) \\
&= R_i^\top \Phi_i(L(\theta_i u)) = R_i^\top \Phi_i(\theta_i f) + R_i^\top \Phi_i\Big(\sum\nolimits_{j=1}^{3} A_j^{\mathrm{co}} \partial_j \theta_i u\Big) =: f^i(f, u).
\end{aligned}$$

We now choose R_3^i so that $A_3^i = A_3^{\mathrm{co}}$ and $B^i = B^{\mathrm{co}}$, namely

$$R_i = \operatorname{diag}(\hat{R}_i, \hat{R}_i) \quad \text{with} \qquad \hat{R}_i = \frac{1}{\sqrt{\Phi_i \partial_3 \varphi_{i,3}}} \begin{pmatrix} 1 & 0 & \Phi_i \partial_1 \varphi_{i,3} \\ 0 & 1 & \Phi_i \partial_2 \varphi_{i,3} \\ 0 & 0 & \Phi_i \partial_3 \varphi_{i,3} \end{pmatrix}.$$

A computation shows $\hat{R}_i^\top \tilde{S}_3^i \hat{R}_i = S_3$, and hence $R_i^\top \tilde{A}_3^i R_i = A_3^{\mathrm{co}}$. We write $B = (B_1 \; 0_{3\times 3})$ and recall that $B_1 \mathbf{E} = \mathbf{E} \times \mathrm{n} = -\sum_j \mathrm{n}_j S_j \mathbf{E}$. It follows

$$\Phi_i(\kappa_i B_1) = -\sum\nolimits_{j=1}^{3} \Phi_i(\kappa_i \mathrm{n}_j) S_j = \sum\nolimits_{j=1}^{3} \Phi_i(\partial_j \varphi_{i,3}) S_j = \tilde{S}_3^i.$$

Therefore we obtain $B^i = \hat{R}_i^\top \tilde{B}_i R_i = B^{\mathrm{co}}$.

We only sketch the remaining steps, see Chapter 5 of [157] for details. One can check that the new coefficients and data satisfy hypothesis (6.18) with A_1^{co} and A_2^{co} replaced by $A_1^i, A_2^i \in F_{\mathrm{sym}}^3$ and the compatibility conditions (6.19) on $G = \mathbb{R}_+^3$. Moreover, the relevant norms of the transformed functions on $\mathbb{R}_+^3$ are bounded by a constant $c(G)$ times the same norms of the corresponding functions on G.

So the apriori estimates of Theorems 6.13 and 5.8 for v^i on $\mathbb{R}_+^3$ respectively $\theta_0 u$ on $\mathbb{R}^3$ yield analogous inequalities for u on G. This also shows uniqueness of solutions. To construct a solution, one solves the transformed problems on V_i^+ and V_0 and glues the solutions together. For this, we need another set of cut-off functions $\sigma_i \in C_c^\infty(U_i)$ which are 1 on the support of θ_i. We have included the original solution u into f^i to compensate for error terms with θ_i when deriving the transformed system. This forces us to set up a fixed-point argument on the space of functions v in $G^3(J \times G)$ which satisfy $\partial_t^j v(0) = u^j$ for $j = \{0, 1, 2\}$ and u^j from (6.19) given by the data. The fixed point then solves the problem.

Main Wellposedness Results

Besides Theorem 5.8 on $\mathbb{R}^3$, the above reasoning requires a variant of Theorem 6.13 on $\mathbb{R}_+^3$ for the modified coefficients A_1^i and A_2^i, where A_0^i and D^i have the same properties as before and $A_3 = A_3^{\mathrm{co}}$ and $B = B^{\mathrm{co}}$ do not change. This modification has quite unexpected consequences, as many estimates and iteration arguments become much more involved since the additional commutators intertwine our three steps (time, tangential and normal regularity) to a larger extent. However, with some effort these problems can be solved.

We first state the result in $C(\overline{J}, L_x^2)$ which follows by localization from Theorem 6.7. The result is surely older, but it is also a special case of the more general Theorem 1.4 of [53], which is actually devoted to boundary regularity. (The validity of (6.11) in this case is derived in Lemma 4.2 of [116] based on [159].)

Theorem 6.17 *Assume that* (6.8) *is true for* G *with* $\|\partial_t A_0\|_\infty, \|D\|_\infty \le r$. *Then there is a unique solution* $u \in C(\overline{J}, L_x^2)$ *of* (6.32)*, and it satisfies* (6.11) *on* G *as well as* (6.10) *with a factor 2 on the right-hand side.*

From Theorem 1.1 of [159] we take the linear wellposedness theorem in $G^3(J \times G)$.

Theorem 6.18 *Let* (6.18) *be true for* $k = 3$ *and the compatibility conditions* (6.19) *hold. Then there is a unique solution* $u \in G^3(J \times G)$ *of* (6.32)*. It satisfies* (6.23) *on* G.

Remark 6.14 remains valid on G (after replacing $\mathbb{R}^3_+$ by G). The first part of the proof of the quasilinear result on G is close to that of Theorem 6.16 on $\mathbb{R}^3_+$, now based on Theorem 6.18. In the proof of the blow-up condition in $W_x^{1,\infty}$ and of the continuous dependence on data, on $\mathbb{R}^3_+$ we have repeated the derivation of the apriori estimates on a nonlinear level. These steps have been omitted in the above proof of Theorem 6.16. On G the procedure is even more involved since it requires to perform the localization procedure also for the nonlinear problem. So we skip these arguments, too. The following local wellposedness result is proved in Theorem 5.3 of [158]. We use the notation from Theorem 6.16.

Theorem 6.19 *Let* (6.30) *and the compatibility conditions* (6.31) *hold. Then the following assertions are true.*

(a) There is a maximal existence time $T_+ = T_+(u_0, f) \in (T_0(\rho), \infty]$ *and a unique (maximal) solution* $u = \Psi(u_0, f) \in G^3([0, T_+))$ *of* (6.29).
(b) Let $T_+ < \infty$. *Then* $\lim_{t\to T_+} \|u(t)\|_{H_x^3} = \infty$ *and* $\sup_{t<T_+} \|u(t)\|_{W_x^{1,\infty}} = \infty$.
(c) Let $T \in [0, T_+)$. *Then there is a radius* $\delta > 0$ *such that for all* (v_0, g) *in* $\mathcal{D}_{T,a_0,d}((u_0, f), \delta)$ *we have* $T_+(v_0, f) > T$ *and* $\Psi : \mathcal{D}_T((u_0, f), \delta) \to G^3(T)$ *is continuous. Moreover,* $\Psi : (\mathcal{D}_T((u_0, f), \delta), \|\cdot\|_{H_x^2 \times H_{t,x}^2(T)}) \to G^2(T)$ *is Lipschitz.*

Example 5.20 directly carries over from $\mathbb{R}^3$ to G. In Theorem 5.3 of [158] more general results were shown which we sketch below, cf. Remark 5.19.

Remark 6.20

(a) In [158] one allows for unbounded domains G having a 'uniformly smooth' boundary (e.g. a compact, smooth one). In a similar way as above, the result in [158] deals with solutions on an interval (T_-, T_+) containing 0.
(b) One also obtains solutions in G^m for data in H^m or C^m with $m \geq 3$ and $m \in \mathbb{N}$, assuming higher-order compatibility conditions.
(c) One can admit nonlinearities a_0 and d taking values in an open subset $U \subseteq \mathbb{R}^6$. The necessary modifications are like those described in Remark 5.19.
(d) Also boundary data $Bu = g$ from the space $\bigcap_{j=0}^{m} H^j(J, H^{m+\frac{1}{2}-j}(G))$ can be included. The corresponding linear result in $G^0(J)$ is taken from [53] where it is assumed that the coefficients are constant outside a compact set. This leads to a restriction on a_0 and d in [158]: They have to converge if $|x| \to \infty$ (if G is unbounded).
(e) Finite speed of propagation on G is shown for the linear system in [157, 158].

The local wellposedness theory for general hyperbolic systems would require much more regularity for the above theorem, see [79] or [155]. In [153] we show results analogous to Theorem 6.19 for corresponding interface problems, and in [152] for so-called absorbing boundary conditions. For the latter case, in [143] an existence result was established (without uniqueness or continuous dependence on data).

Exponential Decay Caused by Conductivity 7

In this chapter we use the wellposedness Theorem 6.19 to show global existence and exponential decay to 0 for small initial data in the presence of a strictly positive conductivity σ. The result is taken from [116]. Its proof is based on a standard procedure for quasilinear problems, going back to [131] at least. Besides local wellposedness, it uses the energy estimates for $\partial_t^k u$ with $k \in \{0, 1, 2, 3\}$ including the dissipation terms $\|\sigma^{1/2}\partial_t^k \mathbf{E}(t)\|_{L_x^2}^2$. One further needs an observability-type estimate for the time-differentiated linear problem (inspired by [54] in our case) to control the norms of $\partial_t^k u$ in H_x^{3-k} by the dissipation terms, globally in time. This can only be done up to error terms which are small, but only in the stronger topology of G^3. Astonishingly, a variant of the apriori estimates from Chap. 6 allows us to bound space by time derivatives, again globally in time. We closely follow [116] in our presentation.

7.1 Introduction and Theorem on Decay

We study the special case of the Maxwell system (5.1) given by

$$\begin{aligned}
\partial_t\big(\varepsilon(\mathbf{E})\mathbf{E}\big) &= \operatorname{curl}\mathbf{H} - \sigma\mathbf{E}, && t \geq 0,\ x \in G,\\
\partial_t\big(\mu(\mathbf{H})\mathbf{H}\big) &= -\operatorname{curl} E, && t \geq 0,\ x \in G, \qquad (7.1)\\
\operatorname{tr}_{\mathrm{ta}}\mathbf{E} = \mathbf{E}\times \mathrm{n} &= 0, && t \geq 0,\ x \in \Gamma,\\
\mathbf{E}(0) = \mathbf{E}_0, \quad \mathbf{H}(0) &= \mathbf{H}_0, && x \in G,
\end{aligned}$$

W. Dörfler et al., *Wave Phenomena*, Oberwolfach Seminars 49,
https://doi.org/10.1007/978-3-031-05793-9_7

on an open, bounded, simply connected domain $G \subseteq \mathbb{R}^3$ with smooth boundary $\partial G = \Gamma$. As before, we also use the equivalent version

$$\begin{aligned} \varepsilon^{\mathrm{d}}(\mathbf{E})\partial_t \mathbf{E} &= \operatorname{curl} \mathbf{H} - \sigma \mathbf{E}, && t \geq 0,\ x \in G, \\ \mu^{\mathrm{d}}(\mathbf{H})\partial_t \mathbf{H} &= -\operatorname{curl} \mathbf{E}, && t \geq 0,\ x \in G, \\ \operatorname{tr}_{\mathrm{ta}} \mathbf{E} = \mathbf{E} \times \mathrm{n} &= 0, && t \geq 0,\ x \in \Gamma, \\ \mathbf{E}(0) = \mathbf{E}_0, &\quad \mathbf{H}(0) = \mathbf{H}_0, && x \in G, \end{aligned} \tag{7.2}$$

for energy estimates, with the differentiated coefficients

$$\varepsilon^{\mathrm{d}}(\cdot, \xi) := \varepsilon(\cdot, \xi) + \Big(\sum_{l=1}^{3} \partial_{\xi_k} \varepsilon_{jl}(\cdot, \xi)\xi_l\Big)_{jk}, \qquad \mu^{\mathrm{d}} \text{ analogously defined.}$$

We modify our assumptions (6.30) and impose the hypothesis

$$\begin{aligned} &\varepsilon, \mu, \varepsilon^{\mathrm{d}}, \mu^{\mathrm{d}}, \sigma \in C^3(\overline{G} \times \mathbb{R}^3, \mathbb{R}^{3\times 3}_{\mathrm{sym}}), \\ &\sigma \geq \eta I > 0, \quad \varepsilon(\cdot, 0), \mu(\cdot, 0), \varepsilon^{\mathrm{d}}(\cdot, 0), \mu^{\mathrm{d}}(\cdot, 0) \geq 2\eta I \quad \text{on } G, \end{aligned} \tag{7.3}$$

thus assuming that σ is uniformly positive definite. The material laws in Example 5.20 on G fulfill the above conditions for ε and μ. By continuity, we can fix a constant $\kappa > 0$ such that

$$\varepsilon(\cdot, \xi), \mu(\cdot, \xi), \varepsilon^{\mathrm{d}}(\cdot, \xi), \mu^{\mathrm{d}}(\cdot, \xi) \geq \eta \qquad \text{if } |\xi| \leq 2\kappa. \tag{7.4}$$

The initial fields shall also satisfy the magnetic divergence and boundary conditions now. Together with the simple connectedness of G, these conditions exclude non-zero $\mathbf{H}_0$ in the kernel of curl, cf. Proposition 7.8, which would produce a constant-in-time solution $(\mathbf{E}, \mathbf{H}) = (0, \mathbf{H}_0)$ of our system (7.1). Let C_S be the norm of the Sobolev embedding $H^2(G) \hookrightarrow L^\infty(G)$. We assume

$$\begin{aligned} &\mathbf{E}_0, \mathbf{H}_0 \in H^3(G, \mathbb{R}^3), \quad \|\mathbf{E}_0\|^2_{H^3_x} + \|H_0\|^2_{H^3_x} \leq r^2, \quad \text{where: } r \leq \kappa / C_S, \\ &\operatorname{div}(\mu(\mathbf{H}_0)\mathbf{H}_0) = 0, \quad \operatorname{tr}_{\mathrm{no}}(\mu(\mathbf{H}_0)\mathbf{H}_0) = 0, \quad \operatorname{tr}_{\mathrm{ta}} \mathbf{E}_0 = \operatorname{tr}_{\mathrm{ta}} \mathbf{E}^1 = \operatorname{tr}_{\mathrm{ta}} \mathbf{E}^2 = 0, \\ &\mathbf{E}^1 := \varepsilon^{\mathrm{d}}(\mathbf{E}_0)^{-1}[\operatorname{curl} \mathbf{H}_0 - \sigma \mathbf{E}_0], \qquad \mathbf{H}^1 := -\mu^{\mathrm{d}}(\mathbf{H}_0)^{-1} \operatorname{curl} \mathbf{E}_0, \\ &\mathbf{E}^2 := \varepsilon^{\mathrm{d}}(\mathbf{E}_0)^{-1}\big[\operatorname{curl} \mathbf{H}^1 - \sigma \mathbf{E}^1 - (\nabla_E \varepsilon^{\mathrm{d}}(\mathbf{E}_0)\mathbf{E}^1) \cdot \mathbf{E}^1\big]. \end{aligned} \tag{7.5}$$

Note that the initial data are bounded by κ. In view of (7.4), Theorem 6.19 and Remark 6.20 provide a unique local solution $u = (\mathbf{E}, \mathbf{H}) \in G^3(J_+)$ of (7.1) with a maximal existence time $T_+ = T_+(\mathbf{E}_0, \mathbf{H}_0)$ and $J_+ = [0, T_+)$. Moreover, (5.2) and Lemma 6.1 show

$$\operatorname{div}\big(\mu(\mathbf{H}(t))\mathbf{H}(t)\big) = 0, \qquad \operatorname{tr}_{\mathrm{no}}(\mu(\mathbf{H}(t))\mathbf{H}(t)) = 0, \tag{7.6}$$

$$\operatorname{div}\big(\varepsilon(\mathbf{E}(t))\mathbf{E}(t)\big) = \operatorname{div}\big(\varepsilon(\mathbf{E}_0)\mathbf{E}_0\big) - \int_0^t \operatorname{div}\big(\sigma\mathbf{E}(s)\big)\,\mathrm{d}s, \tag{7.7}$$

for $t \in J_+$. We state our decay result for small data, which is Theorem 2.2 of [116].

Theorem 7.1 *Let* (7.3) *and* (7.5) *hold. Then there exist a radius* $r > 0$ *in* (7.5) *and numbers* $M, \omega > 0$ *such that* $T_+(\mathbf{E}_0, \mathbf{H}_0) = \infty$ *and*

$$\max_{k\in\{0,1,2,3\}} \|\partial_t^k(\mathbf{E}(t), \mathbf{H}(t))\|_{H_x^{3-k}} \le M\mathrm{e}^{-\omega t} \qquad \textit{for all } t \ge 0.$$

The theorem is proved at the end of Sect. 7.2. In [145] we prove such a result for boundary damping $\mathbf{H} \times \mathrm{n} + (\zeta(\mathbf{E} \times \mathrm{n})) \times \mathrm{n} = 0$ with $\zeta(x) \ge \eta$ on ∂G, where G is strictly starlike. These theorems are the first decay results for quasilinear Maxwell systems on domains. On $\mathbb{R}^3$ one has global existence for small data and certain material laws exploiting dispersive estimates, see §11.6 of [146] (with polynomial decay), [122] or [156]. In [9] convergence to equilibria is shown for a class of hyperbolic systems with damping on $\mathbb{R}^m$ (not including the Maxwell system).

There are some decay results for linear Maxwell systems with conductivity. In [56, 114, 133] and [142], for instance, isotropic constitutive relations and (semilinear) strictly positive conductivity were considered, whereas matrix-valued coefficients were investigated only recently in [54], see also [52] for related results on boundary observability. Partially positive conductivities were treated in [134] in some cases, as well as in [142] for constant $\varepsilon, \mu > 0$ and in [54] without decay rates.

We discuss the background of the proof which employs time-differentiated versions of (7.1). For the sake of brevity, we set

$$\widehat{\varepsilon}_k = \begin{cases} \varepsilon(\mathbf{E}), & k = 0, \\ \varepsilon^{\mathrm{d}}(\mathbf{E}), & k \in \{1, 2, 3\}, \end{cases} \qquad \widehat{\mu}_k = \begin{cases} \mu(\mathbf{H}), & k = 0, \\ \mu^{\mathrm{d}}(\mathbf{H}), & k \in \{1, 2, 3\}. \end{cases} \tag{7.8}$$

For $k \in \{0, 1, 2, 3\}$, we then obtain the system

$$\begin{aligned} \partial_t(\widehat{\varepsilon}_k\partial_t^k\mathbf{E}) &= \operatorname{curl}\partial_t^k\mathbf{H} - \sigma\partial_t^k\mathbf{E} - \partial_t f_k, && t \in J_+,\ x \in G, \\ \partial_t(\widehat{\mu}_k\partial_t^k\mathbf{H}) &= -\operatorname{curl}\partial_t^k\mathbf{E} - \partial_t g_k, && t \in J_+,\ x \in G, \\ \operatorname{tr}_{\mathrm{ta}}\partial_t^k\mathbf{E} = 0, \quad \operatorname{tr}_{\mathrm{no}}&(\widehat{\mu}_k\partial_t^k\mathbf{H}) = -\operatorname{tr}_{\mathrm{no}} g_k, && t \in J_+,\ x \in \Gamma, \end{aligned} \tag{7.9}$$

with the commutator terms

$$
\begin{aligned}
&f_0 = f_1 = 0, \quad f_2 = \partial_t \varepsilon^{\mathrm{d}}(\mathbf{E})\, \partial_t \mathbf{E}, \quad f_3 = \partial_t^2 \varepsilon^{\mathrm{d}}(\mathbf{E})\, \partial_t \mathbf{E} + 2\partial_t \varepsilon^{\mathrm{d}}(\mathbf{E})\, \partial_t^2 \mathbf{E}, \\
&g_0 = g_1 = 0, \quad g_2 = \partial_t \mu^{\mathrm{d}}(\mathbf{H})\, \partial_t \mathbf{H}, \quad g_3 = \partial_t^2 \mu^{\mathrm{d}}(\mathbf{H})\, \partial_t \mathbf{H} + 2\partial_t \mu^{\mathrm{d}}(\mathbf{H})\, \partial_t^2 \mathbf{H}.
\end{aligned}
\tag{7.10}
$$

Note that the electric boundary condition remains unchanged. The magnetic one is well-defined in $H^{-1/2}(\Gamma)$ by the comments in Sect. 6.1 and the divergence relations

$$
\operatorname{div}(\mu^{\mathrm{d}}(\mathbf{H})\partial_t^k \mathbf{H}) = -\operatorname{div} g_k, \qquad \operatorname{div}(\varepsilon^{\mathrm{d}}(\mathbf{E})\partial_t^k \mathbf{E}) = -\operatorname{div}(\sigma \partial_t^{k-1}\mathbf{E} + f_k) \tag{7.11}
$$

which follow from (7.6) and from (7.7) for $k \in \{1, 2, 3\}$. Estimate (7.16) below shows that all maps $\partial_t f_k$, $\partial_t g_k$, $\operatorname{div} f_k$, and $\operatorname{div} g_k$ belong to $L^\infty(J, L^2(G))$ for $T < T_+$.

For the energy estimate, it is useful to consider the equivalent version of (7.9)

$$
\begin{aligned}
\varepsilon^{\mathrm{d}}(\mathbf{E})\, \partial_t \partial_t^k \mathbf{E} &= \operatorname{curl} \partial_t^k \mathbf{H} - \sigma \partial_t^k \mathbf{E} - \tilde{f}_k, && t \in J_+,\ x \in G, \\
\mu^{\mathrm{d}}(\mathbf{H})\, \partial_t \partial_t^k \mathbf{H} &= -\operatorname{curl} \partial_t^k \mathbf{E} - \tilde{g}_k, && t \in J_+,\ x \in G, \\
\operatorname{tr_{ta}} \partial_t^k \mathbf{E} &= 0, && t \in J_+,\ x \in \Gamma,
\end{aligned}
\tag{7.12}
$$

for $k \in \{0, 1, 2, 3\}$ which is based on (7.2) and has the new commutator terms

$$
\tilde{f}_k = \sum_{j=1}^{k} \binom{k}{j} \partial_t^j \varepsilon^{\mathrm{d}}(\mathbf{E})\, \partial_t^{k+1-j}\mathbf{E}, \qquad \tilde{g}_k = \sum_{j=1}^{k} \binom{k}{j} \partial_t^j \mu^{\mathrm{d}}(\mathbf{H})\, \partial_t^{k+1-j}\mathbf{H},
$$

where we put $\tilde{f}_0 = \tilde{g}_0 = 0$. We further introduce the quantities

$$
\begin{aligned}
e_k(t) &= \tfrac{1}{2} \max_{0 \le j \le k} \big(\|\widehat{\varepsilon}_k^{1/2} \partial_t^j \mathbf{E}(t)\|_{L_x^2}^2 + \|\widehat{\mu}_k^{1/2} \partial_t^j \mathbf{H}(t)\|_{L_x^2}^2 \big), && e = e_3, \\
d_k(t) &= \max_{0 \le j \le k} \|\sigma^{1/2} \partial_t^j \mathbf{E}(t)\|_{L_x^2}^2, \qquad d = d_3, \\
z_k(t) &= \max_{0 \le j \le k} \big(\|\partial_t^j \mathbf{E}(t)\|_{H_x^{k-j}}^2 + \|\partial_t^j \mathbf{H}(t)\|_{H_x^{k-j}}^2 \big), && z = z_3,
\end{aligned}
\tag{7.13}
$$

for $k \in \{0, 1, 2, 3\}$ and $t \in J_+$. The choice of weights simplifies some estimates below. Here e_k is related to energy and d_k to dissipation. We stress that d_k only contains the electric field and that d_k and e_k only involve time derivatives, in contrast to z_k.

To control the norms of $(\mathbf{E}, \mathbf{H})$ and the above quantities, we set $\delta_0 = \min\{1, \kappa/C_S\}$ and take $\delta \in (0, \delta_0]$, to be fixed in the proof of Theorem 7.1. Theorem 6.19 then yields a radius $r(\delta) \in (0, \delta]$ such that for all $r \in (0, r(\delta)]$ and $(\mathbf{E}_0, \mathbf{H}_0)$ as in (7.5) we have $T_+ > 1$ and $z(t) \le \delta^2$ for $t \in [0, 1]$. Given such $(\mathbf{E}_0, \mathbf{H}_0)$, we now introduce the time

$$
T_*(\mathbf{E}_0, \mathbf{H}_0) = T_* = \sup\big\{T \in [1, T_+) \,\big|\, \forall t \in [0, T] : z(t) \le \delta^2\big\}. \tag{7.14}
$$

The blow-up condition in Theorem 6.19 implies that $T_+ > T_*$ and hence

$$z(T_*) = \max_{0\le k\le 3} \left(\|\partial_t^k \mathbf{E}(T_*)\|^2_{H_x^{3-k}} + \|\partial_t^k \mathbf{H}(T_*)\|^2_{H_x^{3-k}}\right) = \delta^2 \qquad (\text{if } T_* < \infty) \tag{7.15}$$

by continuity. We will *suppose* that $T_* < \infty$. For sufficiently small $\delta > 0$ (and thus $r > 0$), below we then show that $z(T_*) < \delta^2$. This contradiction to (7.15) then establishes $T_* = \infty$. The exponential decay in Theorem 7.1 will be a by-product of this argument, see the end of the next section.

In the following we always look at solutions with data $(\mathbf{E}_0, \mathbf{H}_0)$ as in (7.5) for some $r \in (0, r(\delta)]$ and a corresponding solution $u = (\mathbf{E}, \mathbf{H})$ of (7.1) on $J_* = [0, T_*)$, which thus satisfies $z(t) \le \delta^2 \le 1$ for all $t \in J_*$. The constants c, c_k, C or C_k below do not depend on $s, t \in J_*$, T_*, $\delta \in (0, \delta_0]$, $r \in (0, r(\delta_0)]$, or $(\mathbf{E}_0, \mathbf{H}_0)$ satisfying (7.5).

Using the product rules, Lemma 6.15 and formula (7.14), one can estimate the above commutator terms by

$$\begin{aligned}
&\|\widehat{\varepsilon}_k(t)\|_{L_x^\infty}, \|\widehat{\mu}_k(t)\|_{L_x^\infty}, \|\widehat{\varepsilon}_k^{-1}(t)\|_{L_x^\infty}, \|\widehat{\mu}_k^{-1}(t)\|_{L_x^\infty} \le c,\\
&\qquad \|\partial^\alpha \widehat{\varepsilon}_j(t)\|_{L_x^2}, \|\partial^\alpha \widehat{\mu}_j(t)\|_{L_x^2} \le c(z_k^{1/2}(t) + \delta_{\alpha_0=0}),\\
&\max_{k\in\{2,3\}, j\in\{0,1\}} \left(\|\partial_t^j f_k(t)\|_{H_x^{4-j-k}} + \|\partial_t^j g_k(t)\|_{H_x^{4-j-k}}\right) \le cz(t),\\
&\|f_2(t)\|_{L_x^2}, \|g_2(t)\|_{L_x^2}, \|f_3(t)\|_{L_x^2}, \|g_3(t)\|_{L_x^2} \le ce_2^{1/2}(t),\\
&\qquad \|\tilde{f}_k(t)\|_{H_x^{3-k}}, \|\tilde{g}_k(t)\|_{H_x^{3-k}} \le cz(t)
\end{aligned} \tag{7.16}$$

for $j, k \in \{0, 1, 2, 3\}$, $\alpha \in \mathbb{N}_0^4$ with $|\alpha| = k > 0$, $t \in J_*$, where we set $\delta_{\alpha_0=0} = 1$ if $\alpha_0 = 0$ and $\delta_{\alpha_0=0} = 0$ if $\alpha_0 > 0$. The second summand in the second line of (7.16) arises if all derivatives in ∂^α are applied to the x-variable of ε or μ.

7.2 Energy and Observability-Type Inequalities

We establish an energy inequality for $\partial_t^k u$ involving dissipation. The error term $z^{3/2}$ is caused by various commutators with $\varepsilon(\mathbf{E})$ and $\mu(\mathbf{H})$.

Proposition 7.2 *We assume the conditions of Theorem 7.1 except for the simple connectedness of G. For $0 \le s \le t < T_*$ and $k \in \{0, 1, 2, 3\}$, we obtain the inequality*

$$e_k(t) + \int_s^t d_k(\tau)\,\mathrm{d}\tau \le e_k(s) + c_1 \int_s^t z^{3/2}(\tau)\,\mathrm{d}\tau. \tag{7.17}$$

We first give the direct proof for the case $k = 0$. Since $(\mathbf{E}, \mathbf{H})$ even belongs to G^3, the system (7.1) and the integration by parts formula (6.7) yield

$$\begin{aligned}
&\frac{\mathrm{d}}{\mathrm{d}t} \frac{1}{2} \int_G \big(\varepsilon(\mathbf{E}(t))\mathbf{E}(t) \cdot \mathbf{E}(t) + \mu(\mathbf{H}(t))\mathbf{H}(t) \cdot \mathbf{H}(t)\big)\,\mathrm{d}x \\
&= \frac{1}{2} \int_G \Big[\partial_t(\varepsilon(\mathbf{E})\mathbf{E}) \cdot \mathbf{E} + \varepsilon(\mathbf{E})\mathbf{E} \cdot \big[\varepsilon(\mathbf{E})^{-1}\partial_t(\varepsilon(\mathbf{E})\mathbf{E})\big] + \varepsilon(\mathbf{E})\mathbf{E} \cdot \big[\partial_t \varepsilon(\mathbf{E})^{-1} \varepsilon(\mathbf{E})\mathbf{E}\big] \\
&\quad + \partial_t(\mu(\mathbf{H})\mathbf{H}) \cdot \mathbf{H} + \mu(\mathbf{H})\mathbf{H} \cdot \big[\mu(\mathbf{H})^{-1}\partial_t(\mu(\mathbf{H})\mathbf{H})\big] + \mu(\mathbf{H})\mathbf{H} \cdot \big[\partial_t \mu(\mathbf{H})^{-1} \mu(\mathbf{H})\mathbf{H}\big]\Big]\mathrm{d}x \\
&= \int_G \big[\operatorname{curl}\mathbf{H} \cdot \mathbf{E} - \sigma\mathbf{E} \cdot \mathbf{E} - \operatorname{curl}\mathbf{E} \cdot \mathbf{H} - \tfrac{1}{2}\partial_t \varepsilon(\mathbf{E})\,\mathbf{E} \cdot \mathbf{E} - \tfrac{1}{2}\partial_t \mu(\mathbf{H})\,\mathbf{H} \cdot \mathbf{H}\big]\,\mathrm{d}x \\
&= -\int_G \big[\sigma\mathbf{E} \cdot \mathbf{E} + \tfrac{1}{2}\partial_t \varepsilon(\mathbf{E})\,\mathbf{E} \cdot \mathbf{E} + \tfrac{1}{2}\partial_t \mu(\mathbf{H})\,\mathbf{H} \cdot \mathbf{H}\big]\,\mathrm{d}x.
\end{aligned}$$

We thus obtain the energy equality

$$e_0(t) + \int_s^t d_0(\tau)\,\mathrm{d}\tau = e_0(t) - \frac{1}{2} \int_s^t \int_G \big(\partial_t \varepsilon(\mathbf{E})\,\mathbf{E} \cdot \mathbf{E} + \partial_t \mu(\mathbf{H})\,\mathbf{H} \cdot \mathbf{H}\big)\,\mathrm{d}x\,\mathrm{d}\tau.$$

Combined with estimate (7.16), we derive (7.17) for the case $k = 0$.

For $k \in \{1, 2, 3\}$ in Proposition 7.2, we have different coefficients in the energy e_k defined in (7.13) and more error terms. In this case, (7.17) follows from Lemma 7.3 below, the system (7.12) and the estimates (7.16). This lemma provides an energy identity in a more general situation to be encountered later.

Take coefficients $a, b \in W^{1,\infty}(J \times G, \mathbb{R}_\eta^{3+3})$ for some $T, \eta > 0$ and data $v_0, w_0 \in L_x^2$, $\varphi, \psi \in L_{t,x}^2$, and $\omega \in L^2(J, H^{1/2}(\Gamma))^3$ with $\mathrm{n} \cdot \omega = 0$. Theorem 1.4 of [53] yields a unique solution $(v, w) \in G^0(J)$ with $\operatorname{tr_{ta}}(v, w) \in L^2(J, H^{-1/2}(\Gamma))^6$ of the linear system

$$\begin{aligned}
a\partial_t v &= \operatorname{curl} w - \sigma v + \varphi, && t \in J,\ x \in G, \\
b\partial_t v &= -\operatorname{curl} v + \psi, && t \in J,\ x \in G, \\
\operatorname{tr_{ta}} v &= \omega, && t \in J,\ x \in \Gamma, \\
v(0) &= v_0, \quad w(0) = w_0, && x \in G.
\end{aligned}$$

(Theorem 6.17 deals with the case $\omega = 0$ without the regularity of $\operatorname{tr_{ta}} w$.) For $\omega = 0$ and $G = \mathbb{R}_+^3$, the next lemma is a part of Theorem 6.7. In the present form it follows from Theorem 6.18 by approximation arguments omitted here, see Lemma 4.2 in [116].

Lemma 7.3 *Under the assumptions above, for* $0 \le s \le t \le T$ *we have*

$$\frac{1}{2}\int_G \big(a(t)v(t)\cdot v(t) + b(t)w(t)\cdot w(t)\big)\mathrm{d}x + \int_s^t \int_G \sigma v \cdot v \,\mathrm{d}x\,\mathrm{d}\tau$$
$$= \frac{1}{2}\int_G \big(a(0)v_0\cdot v_0 + b(0)w_0\cdot w_0\big)\mathrm{d}x + \int_s^t \int_\Gamma \omega \cdot \mathrm{tr}_{\mathrm{ta}}\, w \,\mathrm{d}x\,\mathrm{d}\tau$$
$$+ \int_s^t \int_G \big(\tfrac{1}{2}\partial_t a\, v\cdot v + \tfrac{1}{2}\partial_t b\, w\cdot w + \varphi\cdot v + \psi\cdot w\big)\mathrm{d}x\,\mathrm{d}\tau.$$

In the next proposition we control the energy by the dissipation, i.e., $\partial_t^k \mathbf{E}$ by $\partial_t^k \mathbf{H}$. Following [54], our approach is based on a Helmholtz decomposition. We use the following spaces on G, where Γ_j are the components of Γ and N denotes the kernel of div and curl as maps from L_x^2 to H_x^{-1}:

$$\mathrm{N}_0(\mathrm{curl}) = \{v \in \mathrm{N}(\mathrm{curl}) \mid \mathrm{tr}_{\mathrm{ta}}\, v = 0\}, \qquad \mathrm{N}_0(\mathrm{div}) = \{v \in \mathrm{N}(\mathrm{div}) \mid \mathrm{tr}_{\mathrm{no}}\, v = 0\},$$
$$\mathrm{N}^\Gamma(\mathrm{div}) = \big\{v \in \mathrm{N}(\mathrm{div}) \,\big|\, \forall j : \textstyle\int_{\Gamma_j} \mathrm{tr}_{\mathrm{no}}\, v \,\mathrm{d}x = 0\big\}, \qquad \mathcal{N} = \mathrm{N}(\mathrm{div}) \cap \mathrm{N}_0(\mathrm{curl}),$$
$$H^1_{\mathrm{ta}0}(G) = \{v \in H^1(G)^3 \mid \mathrm{tr}_{\mathrm{ta}}\, v = 0\} = H(\mathrm{div}) \cap H_0(\mathrm{curl}).$$

The last identity is shown in Theorem XI.1.3 of [36]. The first three spaces are endowed with the L^2-norm, and we use the H^1-norm for $H^1_{\mathrm{ta}0}(G)$ and other subspaces of H^1_x. We list some consequences of Theorems 2.8–2.10' from [24] for simply connected G (see also §IX.1 in [36]), used several times below:

$$\mathrm{curl} : H^1_{\mathrm{ta}0}(G) \cap \mathrm{N}^\Gamma(\mathrm{div}) \to \mathrm{N}_0(\mathrm{div}) \quad \text{is invertible,} \tag{7.18}$$
$$\mathrm{curl} : H^1(G)^3 \cap \mathrm{N}_0(\mathrm{div}) \to \mathrm{N}^\Gamma(\mathrm{div}) \quad \text{is invertible,} \tag{7.19}$$
$$\mathrm{N}_0(\mathrm{curl}) = \nabla H^1_0(G) + \mathcal{N}, \quad L^2(G)^3 = \mathrm{N}^\Gamma(\mathrm{div}) + \nabla H^1_0(G) + \mathcal{N} \tag{7.20}$$

with orthogonal sums in L_x^2. We now establish the desired Helmholtz decomposition. Our result is a variant of Proposition 2 in [54] where the case of time-independent ε and μ and less regular solutions was treated.

Lemma 7.4 *Let the assumptions of Theorem 7.1 be satisfied and let* $(\mathbf{E}, \mathbf{H})$ *solve* (7.1). *Then there exist functions* w *in* $C^3\big(J_+, H^1_{\mathrm{ta}0}(G) \cap \mathrm{N}^\Gamma(\mathrm{div})\big) \cap C^4(J_+, L^2(G))^3$, p *in* $C^3(J_+, H^1_0(G))$ *and* h *in* $C^3(J_+, \mathcal{N})$ *with*

$$\partial_t^k \mathbf{E} = -\partial_t^{k+1} w + \nabla \partial_t^k p + \partial_t^k h, \qquad \widehat{\mu}_k \partial_t^k \mathbf{H} = \mathrm{curl}\, \partial_t^k w - g_k \tag{7.21}$$

for $k \in \{0, 1, 2, 3\}$, *cf.* (7.8) *and* (7.10), *where the sum for* $\partial_t^k \mathbf{E}$ *is orthogonal in* L_x^2.

Proof Let $t \in J_+$. Equation (7.6) implies that the function $\mu(\mathbf{H}(t))\mathbf{H}(t)$ is contained in $\mathrm{N}_0(\mathrm{div})$. Since G is simply connected, (7.18) then yields a vector field $w(t)$ in $H^1_{\mathrm{ta}0}(G) \cap \mathrm{N}^\Gamma(\mathrm{div})$ satisfying $\operatorname{curl} w(t) = \mu(\mathbf{H}(t))\mathbf{H}(t)$. Moreover, the map w belongs to $C^3(J_+, H^1_{\mathrm{ta}0}(G) \cap \mathrm{N}^\Gamma(\mathrm{div}))$ because of $(\mathbf{E}, \mathbf{H}) \in G^3$ and (7.18). Differentiating $\operatorname{curl} w = \mu(\mathbf{H})\mathbf{H}$ in t, we deduce

$$\operatorname{curl} \partial_t^k w = \partial_t^k(\mu(\mathbf{H})\mathbf{H}) = \mu^{\mathrm{d}}(\mathbf{H})\partial_t^k\mathbf{H} + g_k$$

for $k \in \{1, 2, 3\}$ which shows the second part of (7.21). Comparing this relation for $k = 1$ with (7.2), we infer $\operatorname{curl}(\mathbf{E} + \partial_t w) = 0$. Moreover, $\mathbf{E} + \partial_t w$ belongs to the kernel of $\mathrm{tr}_{\mathrm{ta}}$. From (7.20) we obtain functions $p(t) \in H^1_0(G)$ and $h(t) \in \mathcal{N}$ such that

$$\mathbf{E}(t) = -\partial_t w(t) + \nabla p(t) + h(t)$$

for $t \in J_+$ with orthogonal sums. This fact and $(\mathbf{E}, \mathbf{H}) \in G^3$ imply the remaining regularity assertions. We can now differentiate the above identity in t, proving (7.21). □

We can now show the desired observability-type estimate. Let us explain this name. For solutions of (7.1) with $\sigma = 0$, $\varepsilon = \varepsilon(x)$ and $\mu = \mu(x)$, Lemma 7.3 shows the energy equality $e_0(t) = e_0(0)$ for $t \geq 0$. Take $\sigma = 1$ in the definition of d_0. Then the next inequality can still be shown with modified constants and $z = 0$, implying $(t - 2c_3')e_0(0) \leq c_2' \int_0^t \|\mathbf{E}(\tau)\|^2_{L^2_x} \,\mathrm{d}\tau$. Hence, the initial fields can be determined by observing the electric field alone until $t > 2c_3'$.

Proposition 7.5 *Let the conditions of Theorem 7.1 be satisfied. For* $0 \leq s \leq t < T_*$ *and* $k \in \{0, 1, 2, 3\}$*, we can estimate*

$$\int_s^t e_k(\tau)\,\mathrm{d}\tau \leq c_2 \int_s^t d_k(\tau)\,\mathrm{d}\tau + c_3(e_k(t) + e_k(s)) + c_4 \int_s^t z^{3/2}(\tau)\,\mathrm{d}\tau.$$

Proof Let $k \in \{0, 1, 2, 3\}$. To simplify, we take $s = 0$. Equality (7.21) yields

$$\int_{G_t} \widehat{\mu}_k \partial_t^k \mathbf{H} \cdot \partial_t^k \mathbf{H} \,\mathrm{d}(x, \tau) = \int_{G_t} \operatorname{curl} \partial_t^k w \cdot \partial_t^k \mathbf{H} \,\mathrm{d}(x, \tau) - \int_{G_t} g_k \cdot \partial_t^k \mathbf{H} \,\mathrm{d}(x, \tau), \qquad (7.22)$$

where $G_t = G \times (0,t)$. Using $\partial_t^k w \in C(J_+, H^1_{\mathrm{ta0}}(G))$ by Lemma 7.4, we apply (6.7), insert the first line of the system (7.9), and integrate by parts in t. It follows

$$\begin{aligned}\int_{G_t} \operatorname{curl} \partial_t^k w \cdot \partial_t^k \mathbf{H}\,\mathrm{d}(x,\tau) &= \langle \partial_t^k w, \operatorname{curl} \partial_t^k \mathbf{H}\rangle_{L^2(0,t),H_0(\mathrm{curl}))} \\ &= \langle \partial_t^k w, \partial_t(\widehat{\varepsilon}_k \partial_t^k \mathbf{E})\rangle_{L^2((0,t),H_0(\mathrm{curl}))} + \int_{G_t} \partial_t^k w \cdot (\sigma \partial_t^k \mathbf{E} + \partial_t f_k)\,\mathrm{d}(x,\tau) \\ &= \int_G \partial_t^k w(t) \cdot \widehat{\varepsilon}_k(t) \partial_t^k \mathbf{E}(t)\,\mathrm{d}x - \int_G \partial_t^k w(0) \cdot \widehat{\varepsilon}_k(0) \partial_t^k \mathbf{E}(0)\,\mathrm{d}x \\ &\quad - \int_{G_t} \partial_t^{k+1} w \cdot \widehat{\varepsilon}_k \partial_t^k \mathbf{E}\,\mathrm{d}(x,\tau) + \int_{G_t} \partial_t^k w \cdot (\sigma \partial_t^k \mathbf{E} + \partial_t f_k)\,\mathrm{d}(x,\tau).\end{aligned} \tag{7.23}$$

Since $\partial_t^k w(t) \in H^1_{\mathrm{ta0}}(G)^3 \cap \mathrm{N}^\Gamma(\mathrm{div})$, formula (7.18) yields the Poincaré-type estimate $\|\partial_t^k w(\tau)\|_{L^2_x} \le c\|\operatorname{curl} \partial_t^k w(\tau)\|_{L^2_x}$. From (7.21) and (7.16), we then infer the bound

$$\|\partial_t^k w(\tau)\|_{L^2_x} \le c\| \operatorname{curl} \partial_t^k w(\tau)\|_{L^2_x} = c\|\hat{\mu}_k \partial_t^k \mathbf{H}(\tau) + g_k(\tau)\|_{L^2_x} \le c e_k^{1/2}(\tau). \tag{7.24}$$

The orthogonality in the first part of (7.21) implies $\|\partial_t^{k+1} w(\tau)\|_{L^2_x} \le \|\partial_t^k \mathbf{E}(\tau)\|_{L^2_x}$. For any $\theta > 0$, these inequalities along with (7.23) and (7.16) lead to the estimate

$$\begin{aligned}\Big|\int_{G_t} \operatorname{curl} \partial_t^k w \cdot \partial_t^k \mathbf{H}\,\mathrm{d}(x,\tau)\Big| &\le c(e_k(t) + e_k(0)) + c\int_{G_t} |\partial_t^k \mathbf{E}|^2\,\mathrm{d}(x,\tau) \\ &\quad + \theta \int_{G_t} |\partial_t^k w|^2 \mathrm{d}(x,\tau) + c_\theta \int_{G_t} |\partial_t^k \mathbf{E}|^2\,\mathrm{d}(x,\tau) + c\int_0^t z^{\frac{3}{2}}(\tau)\,\mathrm{d}\tau.\end{aligned} \tag{7.25}$$

As in (7.24), we further compute

$$\begin{aligned}\int_{G_t} |\partial_t^k w|^2\,\mathrm{d}(x,\tau) &\le c\int_{G_t} \operatorname{curl} \partial_t^k w \cdot \widehat{\mu}_k^{-1} \operatorname{curl} \partial_t^k w\,\mathrm{d}(x,\tau) \\ &= c\int_{G_t} \operatorname{curl} \partial_t^k w \cdot (\partial_t^k \mathbf{H} + \widehat{\mu}_k^{-1} g_k)\,\mathrm{d}(x,\tau) \\ &\le c\Big|\int_{G_t} \operatorname{curl} \partial_t^k w \cdot \partial_t^k \mathbf{H}\,\mathrm{d}(x,\tau)\Big| + c\int_0^t z^{\frac{3}{2}}(\tau)\,\mathrm{d}\tau.\end{aligned}$$

Fixing a small number $\theta > 0$, the term with $|\partial_t^k w|^2$ in Eq. (7.25) can now be absorbed by the left-hand side and by the integral of $z^{3/2}$. So we arrive at

$$\Big|\int_{G_t} \operatorname{curl} \partial_t^k w \cdot \partial_t^k \mathbf{H}\,\mathrm{d}(x,\tau)\Big| \le c(e_k(t) + e_k(0)) + c\int_0^t d_k(\tau)\,\mathrm{d}\tau + c\int_0^t z^{\frac{3}{2}}(\tau)\,\mathrm{d}\tau,$$

also using that $d_k(t)$ is equivalent to $\max_{j\le k} \|\partial_t^j \mathbf{E}(t)\|_{L^2_x}^2$. This fact, Eq. (7.22), the last inequality, and the estimates (7.16) yield the claim. □

Combining Propositions 7.2 and 7.5, we arrive at the following energy bound.

Corollary 7.6 *Under the conditions of Theorem 7.1, we have the inequality*

$$e_k(t) + \int_s^t e_k(s)ds \le C_1 e_k(s) + C_2 \int_s^t z^{3/2}(\tau)\,d\tau$$

for $0 \le s \le t < T_*$ *and* $k \in \{0, 1, 2, 3\}$.

Proof We multiply the inequality in Proposition 7.5 by $\alpha = \min\{c_2^{-1}, (2c_3)^{-1}\}$ and add it to (7.17), obtaining

$$e_k(t) + 2\alpha \int_s^t e_k(\tau)\,d\tau \le 3e_k(s) + 2(c_1 + \alpha c_4) \int_s^t z^{3/2}(\tau)\,d\tau.$$ □

For $z = 0$, from Corollary 7.6 one could easily infer exponential decay by a standard argument, see below. The extra term can be made small since $z^{1/2}(\tau) \le \delta$ for $\tau < T_*$ by (7.14). However, z involves space derivatives so that it cannot be absorbed by e that does not contain them. This gap is closed by the next surprising result proved in the next section. It then allows us to show Theorem 7.1.

Proposition 7.7 *We impose the conditions of Theorem 7.1 with the exception of the simple connectedness of G. Then the solutions* $(\mathbf{E}, \mathbf{H})$ *to (7.1) satisfy*

$$z(t) + \int_s^t z(\tau)\,d\tau \le c_5\big(z(s) + e(t) + z^2(t)\big) + c_6 \int_s^t \big(e(\tau) + z^{3/2}(\tau)\big)\,d\tau$$

for all $0 \le s \le t < T_*$.

Proof (Of Theorem 7.1) Proposition 7.7 and Corollary 7.6 show that

$$z(t) + \int_s^t z(\tau)\,d\tau \le (c_5 + C_1(c_5 + c_6))z(s) + c_5 z^2(t) + (c_6 + C_2(c_5 + c_6)) \int_s^t z^{3/2}(\tau)\,d\tau.$$

Fixing a sufficiently small radius $\delta \in (0, \delta_0]$, we can now absorb the superlinear terms involving z^2 and $z^{3/2}$ by the left-hand side and hence obtain

$$z(t) + \int_s^t z(\tau)\,d\tau \le Cz(s), \qquad \text{for all } 0 \le s \le t < T_*$$

and some constant $C > 0$. Since then $z(\tau) \geq C^{-1}z(t)$, we infer that

$$(1 + (t - s)C^{-1})z(t) \leq Cz(s). \tag{7.26}$$

The differentiated Maxwell system (7.12) and the bounds from (7.16) yield

$$z(0) \leq c_0 \|(\mathbf{E}_0, \mathbf{H}_0)\|_{H^3}^2 \leq c_0 r^2$$

for a constant $c_0 > 0$. We now fix the radius

$$r := \min\left\{r(\delta), \frac{\delta}{\sqrt{2c_0C}}\right\},$$

where $r(\delta)$ was introduced before (7.14).

We suppose that $T_* < \infty$, yielding $z(T_*) = \delta^2$ by (7.15). Because of (7.26), the number $z(t)$ is bounded by $Cz(0) \leq \delta^2/2$ for $t < T_*$ and by continuity also for $t = T_*$. This contradiction shows that $T_* = \infty$ and hence $T_+ = \infty$.

In particular, (7.26) is true for all $t \geq s \geq 0$. Fixing the time $T > 0$ with $C^2/(C+T) = 1/2$, we derive $z(nT) \leq \frac{1}{2}z((n-1)T)$ for $n \in \mathbb{N}$ and then $z(nT) \leq 2^{-n}z(0)$ by induction. With (7.26) one then obtains the asserted exponential decay. □

7.3 Time Regularity Controls Space Regularity

We first collect some preparations for the proof of Proposition 7.7. One can bound the H^1-norm of a field v by its norms in $H(\mathrm{curl}) \cap H(\mathrm{div})$ and the $H^{1/2}$-norm of $\mathrm{tr_{ta}}\, v$ or $\mathrm{tr_{no}}\, v$, see Corollary XI.1.1 of [36]. We need a version of this result with regular, matrix-valued coefficients a (which does not directly follow from the case $a = I$ unless a is scalar). It is stated in Remark 4 of [54] with a brief indication of a proof. We present a (different) proof inspired by Lemma 4.5.5 of [32].

Proposition 7.8 *Let $a \in W^{1,\infty}(G, \mathbb{R}^{3\times 3}_\eta)$ for some $\eta > 0$ and let $v \in H(\mathrm{curl})$ fulfill $\mathrm{div}(av) \in L^2(G)$ and $\mathrm{tr_{no}}(av) \in H^{1/2}(\Gamma)$. Then v belongs to $H^1(G)^3$ and satisfies*

$$\|v\|_{H^1_x} \leq c\big(\|v\|_{H(\mathrm{curl})} + \|\mathrm{div}(av)\|_{L^2_x} + \|\mathrm{tr_{no}}(av)\|_{H^{1/2}(\Gamma)}\big) =: c\kappa(v).$$

Proof There exists a finite partition of unity $\{\chi_i\}_i$ on $\overline{G}$ such that the support of each χ_i is contained in a simply connected subset of $\overline{G}$ with a connected smooth boundary. Since each χ_i is scalar, we obtain the estimate

$$\|\chi_i v\|_{L^2_x} + \|\mathrm{curl}(\chi_i v)\|_{L^2_x} + \|\mathrm{div}(a\chi_i v)\|_{L^2_x} + \|\mathrm{tr_{no}}(a\chi_i v)\|_{H^{1/2}(\Gamma)} \leq c\kappa(v).$$

We can thus assume that Γ is connected (and G simply connected). In this case, $\operatorname{curl} v$ belongs to $\mathrm{N}^{\Gamma}(\mathrm{div})$ and so (7.19) yields a vector field $w \in H^1(G) \cap \mathrm{N}_0(\mathrm{div})$ with $\operatorname{curl} v = \operatorname{curl} w$ and $\|w\|_{H^1_x} \le c\|\operatorname{curl} v\|_{L^2_x}$. As the difference $v - w$ belongs to N(curl), it is represented by $v - w = \nabla\varphi$ for a function $\varphi \in H^1(G)$ by Proposition IX.1.2 of [36]. Here we can assume that $\int_G \varphi \, \mathrm{d}x$ and so $\|\varphi\|_2 \lesssim \|\nabla\varphi\|_2 \lesssim \|v\|_2 + \|w\|_2$ by Poincaré's inequality. We further have

$$\operatorname{div}(a\nabla\varphi) = \operatorname{div}(av) - \operatorname{div}(aw) \in L^2(G),$$

$$\operatorname{tr}_{\mathrm{no}}(a\nabla\varphi) = \operatorname{tr}_{\mathrm{no}}(av) - \operatorname{tr}_{\mathrm{no}}(aw) \in H^{1/2}(\Gamma),$$

because of the assumptions and $w \in H^1(G)$. Due to the uniform ellipticity, φ is thus an element of $H^2(G)$ satisfying

$$\|\varphi\|_{H^2_x} \le c\big(\|v\|_{L^2_x} + \|\operatorname{div}(av)\|_{L^2_x} + \|\operatorname{tr}_{\mathrm{no}}(av)\|_{H^{1/2}(\Gamma)} + \|w\|_{H^1_x}\big) \le c\kappa(v).$$

The assertion now follows from the equation $v = w + \nabla\varphi$. □

In the proof of Proposition 7.7, we want to avoid the localization procedure since we need global-in-time estimates. This can be done using a new coordinate system near $\Gamma = \partial G$. (Possibly, one could derive the apriori estimates in Sect. 6.3 in a similar way; but for the regularization this is not clear because of the mollifier arguments.)

For a fixed distance $\rho > 0$, on the collar $\Gamma_\rho = \{x \in \overline{G} \mid \operatorname{dist}(x, \Gamma) < \rho\}$, we can find smooth functions $\tau^1, \tau^2, \mathrm{n} : \Gamma_\rho \to \mathbb{R}^3$ such that the vectors $\{\tau^1(x), \tau^2(x), \mathrm{n}(x)\}$ form an orthonormal basis of $\mathbb{R}^3$ for each point $x \in \Gamma_\rho$ and n extends the outer unit normal at Γ. Hence, τ^1 and τ^2 span the tangential planes at Γ. For $\xi, \zeta \in \{\tau^1, \tau^2, \mathrm{n}\}$, $v \in \mathbb{R}^3$ and $a \in \mathbb{R}^{3\times 3}$, we set

$$\partial_\xi = \sum\nolimits_j \xi_j \partial_j, \quad v_\xi = v \cdot \xi, \quad v^\xi = v_\xi \xi, \quad v^\tau = v_{\tau^1}\tau^1 + v_{\tau^2}\tau^2, \quad a_{\xi\zeta} = \xi^\top a \zeta.$$

We state several calculus formulas needed below, where it is always assumed that the functions involved are sufficiently regular. We can switch between the derivatives of the coefficient v_ξ and the component v^ξ up to a zero-order term since

$$\partial_\zeta v^\xi = \partial_\zeta v_\xi \xi + v_\xi \partial_\zeta \xi.$$

The commutator of tangential derivatives and traces

$$\partial_\tau \operatorname{tr}_{\mathrm{ta}} v = \partial_\tau (v \times \mathrm{n}) = \operatorname{tr}_{\mathrm{ta}} \partial_\tau v + v \times \partial_\tau \mathrm{n} \qquad \text{on } \Gamma$$

is also of lower order. Similarly, the directional derivatives commute

$$\partial_\xi \partial_\zeta v = \sum_{j,k} \xi_j \partial_j (\zeta_k \partial_k v) = \partial_\zeta \partial_\xi v + \sum_{j,k} \xi_j \partial_j \zeta_k \, \partial_k v - \zeta_k \partial_k \xi_j \, \partial_j v$$

up to a first-order operator with bounded coefficients.

The gradient of a scalar function φ is expanded as

$$\nabla \varphi = \sum_\xi \xi \cdot \nabla \varphi \, \xi = \sum_\xi \xi \partial_\xi \varphi,$$

so that $\partial_j = \sum_\xi \xi_j \partial_\xi$ for $j \in \{1, 2, 3\}$. Because of the formulas before (5.5) we have

$$\mathrm{curl} = \sum_j S_j \partial_j = \sum_{j,\xi} S_j \xi_j \partial_\xi =: \sum_\xi S(\xi) \partial_\xi .$$

Since the kernel of $S(\mathrm{n})$ is spanned by n, we can write $S(\mathrm{n})v = S(\mathrm{n})v^\tau$, and the restriction of $S(\mathrm{n})$ to $\mathrm{span}\{\tau^1, \tau^2\}$ has an inverse $R(\mathrm{n})$.

We now provide the tools that allow us transfer to the arguments of Proposition 6.9 from $\mathbb{R}^+_3$ to the present setting. We first isolate the normal derivative of the tangential components of v in the equation $\mathrm{curl}\, v = f$. By the above expansion

$$\mathrm{curl}\, v = J(\mathrm{n})(\partial_\mathrm{n} v)^\tau + J(\tau^1)\partial_{\tau^1} v + J(\tau^2)\partial_{\tau^2} v,$$

we obtain

$$\partial_\mathrm{n} v^\tau = \sum_i (\partial_\mathrm{n} \tau^i \, v_{\tau^i} + \tau^i \partial_\mathrm{n} \tau^i \cdot v) + R(\mathrm{n}) \Big(f - \sum_i J(\tau^i) \partial_{\tau^i} v \Big) \tag{7.27}$$

where the first sum only contains zero-order terms.

In order to recover the normal derivative of the normal component of v, we resort to the divergence operator. The divergence of a vector field v can be expressed as

$$\mathrm{div}\, v = \sum_j \partial_j \sum_\xi v_\xi \xi_j = \sum_\xi \big(\partial_\xi v_\xi + \mathrm{div}(\xi) v_\xi \big).$$

Letting $\varphi = \mathrm{div}(av)$ for a matrix-valued function a, we derive

$$\begin{aligned}
\mathrm{div}(av) &= \sum_{\xi,\zeta} \partial_\xi (\xi^\top a \zeta v_\zeta) + \sum_\xi \mathrm{div}(\xi)\, \xi^\top av \\
&= \sum_{\xi,\zeta} (a_{\xi\zeta} \partial_\xi v_\zeta + \partial_\xi a_{\xi\zeta} v_\zeta) + \sum_\xi \mathrm{div}(\xi)\, \xi^\top av, \\
a_{\mathrm{nn}} \partial_\mathrm{n} v_\mathrm{n} &= \varphi - \sum_{(\xi,\zeta) \neq (\mathrm{n},\mathrm{n})} a_{\xi\zeta} \partial_\xi v_\zeta - \sum_{\xi,\zeta} \partial_\xi a_{\xi\zeta} v_\zeta - \sum_\xi \mathrm{div}(\xi)\, \xi^\top av \\
&=: \varphi - D(a)v,
\end{aligned} \tag{7.28}$$

where $D(a)v$ contains all tangential derivatives and normal derivatives of tangential components of v plus zero-order terms. Next, let $a \in W^{1,\infty}(J \times G, \mathbb{R}^{3\times 3}_{\mathrm{sym}})$ be positive definite, $v \in C^1(\overline{J}, H^1_x)$, and $\psi \in L^2_{t,x}$. In view of (7.7), we look at the equation

$$\operatorname{div}\big(a(t)v(t)\big) = \operatorname{div}\big(a(0)u(0)\big) - \int_0^t \big(\operatorname{div}(\sigma u(s)) + \psi(s)\big)\,\mathrm{d}s \tag{7.29}$$

for $0 \le t \le T$. We set $\gamma = \sigma_{\mathrm{nn}}/a_{\mathrm{nn}}$ and $\Gamma(t,s) = \exp(-\int_s^t \gamma(\tau)\,\mathrm{d}\tau)$. Equations (7.28) and (7.29) yield

$$\begin{aligned} a_{\mathrm{nn}}(t)\partial_{\mathrm{n}} v_{\mathrm{n}}(t) &= \operatorname{div}\big(a(0)v(0)\big) - D\big(a(t)\big)v(t) \\ &\quad - \int_0^t \big(\gamma(s)a_{\mathrm{nn}}(s)\partial_{\mathrm{n}} v_{\mathrm{n}}(s) + D\big(\sigma\big)v(s) + \psi(s)\big)\,\mathrm{d}s, \end{aligned}$$

cf. (6.25). Differentiating with respect to t and solving the resulting ODE, we obtain

$$\begin{aligned} &a_{\mathrm{nn}}(t)\partial_{\mathrm{n}} v_{\mathrm{n}}(t) \\ &= \Gamma(t,0)a_{\mathrm{nn}}(0)\partial_{\mathrm{n}} n_{\mathrm{n}}(0) - \int_0^t \Gamma(t,s)\big(D(\sigma)v(s) + \psi(s) + \partial_s\big(D(a(s))v(s)\big)\big)\,\mathrm{d}s \\ &= \Gamma(t,0)\operatorname{div}(a(0)v(0)) - D(a(t))v(t) \\ &\quad + \int_0^t \Gamma(t,s)\big(\gamma(s)D\big(a(s)\big)v(s) - D(\sigma)v(s) - \psi(s)\big)\,\mathrm{d}s. \end{aligned} \tag{7.30}$$

Before tackling the (quite demanding) proof of Proposition 7.7, we describe our reasoning. We have to bound $\partial_t^k \mathbf{E}$ and $\partial_t^k \mathbf{H}$ in H_x^{3-k} for $k \in \{0,1,2\}$ by the L_x^2-norms of $\partial_t^j \mathbf{E}$ and $\partial_t^j \mathbf{H}$ for $j \in \{0,1,2,3\}$.

The H_x^1-norm of $\partial_t^k \mathbf{H}$ with $k \in \{0,1,2\}$ can easily be estimated by means of the curl–div estimates from Proposition 7.8 since we control curl, divergence and normal trace of $\partial_t^k \mathbf{H}$ via the time differentiated Maxwell system (7.9) and (7.11). Aiming at higher space regularity, we can apply the above strategy to tangential derivatives of $\partial_t^k \mathbf{H}$ only, whereas normal derivatives destroy the boundary conditions in (7.9). Here we proceed as in Proposition 6.9: The tangential components of normal derivatives are read off the differentiated Maxwell system using the expansion (7.27) of the curl-operator, while the normal components are bounded employing the divergence condition (7.11) and formula (7.28). In these arguments we have to restrict ourselves to fields localized near the boundary. The localized fields in the interior can be controlled more easily since the boundary conditions become trivial for them.

The electric fields $\mathbf{E}$ have less favorable divergence properties because of the conductivity term in (7.9). Instead of Proposition 7.8, we thus employ the energy bound of the system (7.32) derived by differentiating the Maxwell equations in time and tangential directions. The normal derivatives are again treated by the curl-div-strategy indicated in the previous paragraph. However, to handle the extra divergence term in (7.11) caused by the conductivity, we need the more sophisticated divergence formula (7.30) which relies on an ODE derived from (7.11).

This program is carried out by iteration on the space regularity. In each step one has to start with the magnetic fields in order to use their better properties when estimating the electric ones.

***Proof** (Of Proposition 7.7)* Let $(\mathbf{E}, \mathbf{H})$ be a solution of (7.1) on $J_* = [0, T_*)$ satisfying $z(t) \le \delta^2$ and Eqs. (7.6) and (7.7). Take $k \in \{0, 1, 2\}$ and $0 \le t < T_*$, where we let $s = 0$ for simplicity. To localize the fields, we choose smooth scalar functions χ and $1 - \chi =: \vartheta$ on $\overline{G}$ having compact support in $G \setminus \Gamma_{\rho/2}$ and Γ_ρ, respectively. The proof is divided into several steps following the above outline.

(1) *Estimate of* $\partial_t^k \mathbf{H}$ *in* H_x^1. The time differentiated Maxwell system (7.9) and (7.11) combined with estimates (7.16) yield

$$\begin{aligned} \left\| \operatorname{curl} \partial_t^k \mathbf{H}(t) \right\|_{L_x^2} &\le c e_{k+1}^{1/2}(t) + c z(t) \delta_{k2}, \\ \left\| \operatorname{div}(\widehat{\mu}_k \partial_t^k \mathbf{H}(t)) \right\|_{L_x^2} &\le c z(t) \delta_{k2}, \\ \left\| \operatorname{tr}_{\mathrm{no}}(\widehat{\mu}_k \partial_t^k \mathbf{H}(t)) \right\|_{H^{1/2}(\Gamma)} &\le c z(t) \delta_{k2}, \end{aligned}$$

where $\delta_{k2} = 1$ for $k = 2$ and $\delta_{k2} = 0$ for $k \in \{0, 1\}$. Proposition 7.8 thus implies

$$\begin{aligned} \left\| \partial_t^k \mathbf{H}(t) \right\|_{H_x^1}^2 &\le c e_{k+1}(t) + c z^2(t) \delta_{k2}, \\ \int_0^t \left\| \partial_t^k \mathbf{H}(s) \right\|_{H_x^1}^2 \mathrm{d}s &\le c \int_0^t (e_{k+1}(s) + z^2(s) \delta_{k2}) \, \mathrm{d}s. \end{aligned} \tag{7.31}$$

We stress the core fact that the inhomogeneities in (7.9) and (7.11) are quadratic in $(\mathbf{E}, \mathbf{H})$ and can thus be bounded by z via (7.16).

(2) *Estimates in the interior for* $\mathbf{E}$ *and* $\mathbf{H}$. We look at the localized fields $\partial_t^k(\chi \mathbf{E})$ and $\partial_t^k(\chi \mathbf{H})$ whose support supp χ is strictly separated from the boundary. Hence, their spatial derivatives satisfy the boundary conditions of the Maxwell system so that we can treat the electric fields via energy bounds and the magnetic ones via the curl-div estimates.

(a) Let $\alpha \in \mathbb{N}_0^3$ with $|\alpha| \le 3-k$. We apply $\partial_x^\alpha \chi$ to the Maxwell system (7.12), deriving the equations

$$\begin{aligned}
\varepsilon^{\mathrm{d}}(\mathbf{E})\, \partial_t \partial_x^\alpha \partial_t^k (\chi \mathbf{E}) &= \operatorname{curl} \partial_x^\alpha \partial_t^k (\chi \mathbf{H}) - \sigma \partial_x^\alpha \partial_t^k (\chi \mathbf{E}) + \partial_x^\alpha ([\chi, \operatorname{curl}] \partial_t^k \mathbf{H}) \\
&\quad - \sum_{0 \le \beta < \alpha} \binom{\alpha}{\beta} \partial_x^{\alpha-\beta} (\sigma + \varepsilon^{\mathrm{d}}(\mathbf{E}))\, \partial_x^\beta \partial_t^k (\chi \mathbf{E}) - \partial_x^\alpha (\chi \tilde{f}_k), \\
\mu^{\mathrm{d}}(\mathbf{H})\, \partial_t \partial_x^\alpha \partial_t^k (\chi \mathbf{H}) &= - \operatorname{curl} \partial_x^\alpha \partial_t^k (\chi \mathbf{E}) - \partial_x^\alpha ([\chi, \operatorname{curl}] \partial_t^k \mathbf{E}) - \partial_x^\alpha (\chi \tilde{g}_k) \qquad (7.32) \\
&\quad - \sum_{0 \le \beta < \alpha} \binom{\alpha}{\beta} \partial_x^{\alpha-\beta} \mu^{\mathrm{d}}(\mathbf{H})\, \partial_x^\beta \partial_t^k (\chi \mathbf{H}), \\
\operatorname{tr}_{\mathrm{ta}} \partial_x^\alpha \partial_t^k (\chi \mathbf{E}) &= 0, \qquad \operatorname{tr}_{\mathrm{no}} \partial_x^\alpha \partial_t^k (\chi \mathbf{H}) = 0.
\end{aligned}$$

Note that the commutator $m := [\chi, \operatorname{curl}]$ is merely a multiplication operator. Lemma 7.3 and the inequalities (7.16) thus yield

$$\begin{aligned}
&\|\partial_x^\alpha \partial_t^k (\chi \mathbf{E})(t)\|_{L_x^2}^2 + \int_0^t \|\partial_x^\alpha \partial_t^k (\chi \mathbf{E})(s)\|_{L_x^2}^2 \, \mathrm{d}s \\
&\le c z(0) + c \int_0^t \big(z^{3/2}(s) + \|\partial_t^k (\chi \mathbf{E}(s))\|_{H_x^{|\alpha|-1}}^2 + \|\partial_t^k (\chi \mathbf{H}(s))\|_{H_x^{|\alpha|-1}}^2 \big) \, \mathrm{d}s \\
&\quad + c \int_{G_t} \big(\partial_x^\alpha (m \partial_t^k \mathbf{H}) \cdot \partial_x^\alpha \partial_t^k (\chi \mathbf{E})) - \partial_x^\alpha (m \partial_t^k \mathbf{E}) \cdot \partial_x^\alpha \partial_t^k (\chi \mathbf{H}) \big) \, \mathrm{d}(x, s),
\end{aligned}$$

where $G_t = (0, t) \times G$. The former part of the last line can be estimated by

$$\frac{1}{4} \int_0^t \big\|\partial_x^\alpha \partial_t^k (\chi \mathbf{E})(s)\big\|_{L_x^2}^2 \, \mathrm{d}s + c \int_0^t \big\|\tilde{\chi} \partial_t^k \mathbf{H}(s)\big\|_{H_x^{|\alpha|}}^2 \, \mathrm{d}s$$

with another cut-off function $\tilde{\chi} \in C_c^\infty(G \setminus \Gamma_{\rho/2})$ that is equal to 1 on $\operatorname{supp} \chi$. The first summand is absorbed by the left-hand side, while the second one only involves $\mathbf{H}$ and can be treated separately. The latter part of the integral on G_t is similarly bounded by

$$\theta \int_0^t \big\|\partial_t^k \mathbf{E}(s)\big\|_{H_x^{|\alpha|}}^2 \, \mathrm{d}s + c(\theta) \int_0^t \big\|\tilde{\chi} \partial_t^k \mathbf{H}(s)\big\|_{H_x^{|\alpha|}}^2 \, \mathrm{d}s$$

for an arbitrary (small) $\theta > 0$. It follows

$$\begin{aligned}
&\big\|\partial_x^\alpha \partial_t^k(\chi\mathbf{E})(t)\big\|_{L_x^2}^2 + \int_0^t \big\|\partial_x^\alpha \partial_t^k(\chi\mathbf{E})(s)\big\|_{L_x^2}^2 \,\mathrm{d}s \\
&\qquad \le cz(0) + c\int_0^t (z^{3/2}(\tau) + \big\|\partial_t^k(\chi\mathbf{E}(s)\big\|_{H_x^{|\alpha|-1}}^2)\,\mathrm{d}s \\
&\qquad\qquad + \theta\int_0^t \big\|\partial_t^k\mathbf{E}(s)\big\|_{H_x^{|\alpha|}}^2\,\mathrm{d}s + c(\theta)\int_0^t \big\|\tilde{\chi}\partial_t^k\mathbf{H}(s)\big\|_{H_x^{|\alpha|}}^2\,\mathrm{d}s.
\end{aligned} \tag{7.33}$$

(b) To treat $\mathbf{H}$, we only need the case $|\alpha| \le 2-k$. Equations (7.6) and (7.11) yield

$$\begin{aligned}
&\operatorname{div}\big(\widehat{\mu}_k\partial_x^\alpha\partial_t^k(\chi\mathbf{H})\big) \\
&= \partial_x^\alpha([\operatorname{div},\chi]\widehat{\mu}_k\partial_t^k\mathbf{H}) - \sum_{0\le\beta<\alpha}\binom{\alpha}{\beta}\operatorname{div}\big(\partial_x^{\alpha-\beta}\widehat{\mu}_k\,\partial_x^\beta(\partial_t^k(\chi\mathbf{H})\big) - \partial_x^\alpha(\chi\operatorname{div} g_k).
\end{aligned} \tag{7.34}$$

Recalling formulas (7.32) and (7.16), we deduce

$$\begin{aligned}
&\big\|\operatorname{curl}\partial_x^\alpha\partial_t^k(\chi\mathbf{H}(t))\big\|_{L_x^2} + \big\|\widehat{\mu}_k\operatorname{div}\partial_x^\alpha\partial_t^k(\chi\mathbf{H}(t))\big\|_{L_x^2} \\
&\qquad \le c\Big(z(t) + \big\|\partial_t^k\tilde{\chi}\mathbf{H}(t)\big\|_{H_x^{|\alpha|}} + \big\|\partial_t^{k+1}(\chi\mathbf{E}(t))\big\|_{H_x^{|\alpha|}} + \big\|\partial_t^k(\chi\mathbf{E}(t))\big\|_{H_x^{|\alpha|}}\Big)
\end{aligned}$$

Proposition 7.8 now implies the inequalities

$$\big\|\partial_t^k\chi\mathbf{H}(t)\big\|_{H_x^{|\alpha|+1}}^2 \le c\big[z^2(t) + \big\|\partial_t^k\tilde{\chi}\mathbf{H}(t)\big\|_{H_x^{|\alpha|}}^2 + \max_{j\le k+1}\|\partial_t^j(\chi\mathbf{E}(t))\|_{H_x^{|\alpha|}}^2\big], \tag{7.35}$$

$$\int_0^t\big\|\partial_t^k\chi\mathbf{H}(s)\big\|_{H_x^{|\alpha|+1}}^2\mathrm{d}s \le c\int_0^t\big[z^2(s) + \big\|\partial_t^k\tilde{\chi}\mathbf{H}(s)\big\|_{H_x^{|\alpha|}}^2 + \max_{j\le k+1}\big\|\partial_t^j(\chi\mathbf{E}(s))\big\|_{H_x^{|\alpha|}}^2\big]\mathrm{d}s.$$

Here, we can replace χ by $\tilde{\chi}$ from inequality (7.33) and $\tilde{\chi}$ by a function $\breve{\chi} \in C_c^\infty(G\setminus\Gamma_{\rho/2})$ which is equal to 1 on supp $\tilde{\chi}$.

We set $y_j(t) = \max_{0\le k\le 3-j}\big\|\partial_t^k\chi(\mathbf{E}(t),\mathbf{H}(t))\big\|_{H_x^j}^2$. The estimates (7.31), (7.33) and (7.35) iteratively imply

$$\begin{aligned}
y_j(t) + \int_0^t y_j(s)\,\mathrm{d}s &\le cz(0) + c\big(e(t) + z^2(t)\big) + c(\theta)\int_0^t (e(s) + z^{3/2}(s))\,\mathrm{d}s \\
&\qquad + \theta\max_{1\le l\le j}\max_{0\le k\le 3-l}\int_0^t \|\partial_t^k\mathbf{E}(s)\|_{H_x^l}^2\,\mathrm{d}s
\end{aligned} \tag{7.36}$$

for any $\theta > 0$ and $j \in \{1, 2, 3\}$.

(3) *Boundary-collar estimate of* $\partial_t^k \mathbf{E}$ *in* H_x^1. We write $\vartheta = 1 - \chi$ and $\partial_\tau = (\partial_{\tau^1}, \partial_{\tau^2})$. Let $\alpha \in \mathbb{N}_0^2$ with $0 < |\alpha| \le 3 - k$. (For the later use, also higher-order space derivatives are treated.)

(a) We localize the system near the boundary by including the cut-off ϑ into Eqs. (7.12), and then apply ∂_τ^α to the resulting system. The localized tangential-time derivatives of $(\mathbf{E}, \mathbf{H})$ thus satisfy

$$\begin{aligned}
\varepsilon^{\mathrm{d}}(\mathbf{E})\, \partial_t \partial_\tau^\alpha \partial_t^k(\vartheta \mathbf{E}) &= \operatorname{curl} \partial_\tau^\alpha \partial_t^k(\vartheta \mathbf{H}) - \sigma\, \partial_\tau^\alpha \partial_t^k(\vartheta \mathbf{E}) + [\partial_\tau^\alpha, \operatorname{curl}]\partial_t^k(\vartheta \mathbf{H}) \\
&\quad + \partial_\tau^\alpha([\vartheta, \operatorname{curl}]\partial_t^k \mathbf{H}) - \sum_{0 \le \beta < \alpha} \binom{\alpha}{\beta} \partial_\tau^{\alpha-\beta}(\sigma + \varepsilon^{\mathrm{d}}(\mathbf{E}))\, \partial_\tau^\beta \partial_t^k(\vartheta \mathbf{E}) - \partial_\tau^\alpha(\vartheta \tilde{f}_k), \\
\mu^{\mathrm{d}}(\mathbf{H})\, \partial_t \partial_\tau^\alpha \partial_t^k(\vartheta \mathbf{H}) &= -\operatorname{curl} \partial_\tau^\alpha \partial_t^k(\vartheta \mathbf{E}) - \partial_\tau^\alpha([\vartheta, \operatorname{curl}]\partial_t^k \mathbf{E}) - [\partial_\tau^\alpha, \operatorname{curl}]\partial_t^k(\vartheta \mathbf{E}) \\
&\quad - \sum_{0 \le \beta < \alpha} \binom{\alpha}{\beta} \partial_\tau^{\alpha-\beta} \mu^{\mathrm{d}}(\mathbf{H})\, \partial_\tau^\beta \partial_t^k(\vartheta \mathbf{H}) - \partial_\tau^\alpha(\vartheta \tilde{g}_k), \\
\operatorname{tr}_{\mathrm{ta}} \partial_\tau^\alpha \partial_t^k(\vartheta \mathbf{E}) &= [\partial_\tau^\alpha, \operatorname{tr}_{\mathrm{ta}}]\partial_t^k(\vartheta \mathbf{E}) =: \omega.
\end{aligned} \tag{7.37}$$

The commutators $[\partial_\tau^\alpha, \operatorname{curl}]$ are differential operators of order $|\alpha|$ with bounded coefficients, whereas $[\partial_\tau^\alpha, \operatorname{tr}_{\mathrm{ta}}]$ is of order $|\alpha| - 1$ on the boundary and hence a bounded operator from $H^{|\alpha|-1/2}(\Gamma)$ to $H^{1/2}(\Gamma)$. We now use the energy identity in Lemma 7.3 with $a = \varepsilon^{\mathrm{d}}(\mathbf{E})$, $b = \mu^{\mathrm{d}}(\mathbf{H})$, $v = \partial_\tau^\alpha \partial_t^k(\vartheta \mathbf{E})$, and $w = \partial_\tau^\alpha \partial_t^k(\vartheta \mathbf{H})$. The commutator terms, the sums, and the summands with $\tilde{f}_k$ and $\tilde{g}_k$ yield the inhomogeneities φ and ψ, respectively. From Lemma 7.3 we deduce the inequality

$$\begin{aligned}
&\|\partial_\tau^\alpha \partial_t^k(\vartheta \mathbf{E})(t)\|_{L_x^2}^2 + \int_0^t \|\partial_\tau^\alpha \partial_t^k(\vartheta \mathbf{E})(s)\|_{L_x^2}^2 \,\mathrm{d}s \\
&\quad \le cz(0) + c \int_{G_t} \big(|\partial_t a v \cdot v| + |\partial_t b w \cdot w| + |\varphi \cdot v| + |\psi \cdot w|\big)\, \mathrm{d}(s, x) \\
&\qquad + c \int_{\Gamma_t} |\omega \cdot \operatorname{tr}_{\mathrm{ta}} w|\, \mathrm{d}(s, x).
\end{aligned}$$

Several terms on the right-hand side are super-quadratic in $(\mathbf{E}, \mathbf{H})$ and can be bounded by $cz^{3/2}$ due to (7.16). The quadratic ones need more care. The summands in $\varphi \cdot v$ and $\psi \cdot w$ containing the commutators are less or equal to

$$\theta \int_0^t \|\partial_t^k \mathbf{E}(s)\|_{H_x^{|\alpha|}}^2 \,\mathrm{d}s + c(\theta) \int_0^t \|\tilde{\vartheta}\, \partial_t^k \mathbf{H}(s)\|_{H_x^{|\alpha|}}^2 \,\mathrm{d}s$$

with any (small) constant $\theta > 0$ and a cut-off $\tilde{\vartheta} \in C_c^\infty(\Gamma_\rho)$ being equal to 1 on $\operatorname{supp} \vartheta$. The boundary integral is estimated by the same expression, where we use the dual paring

$H^{1/2}(\Gamma) \times H^{-1/2}(\Gamma)$ and that ∂_{τ^i} belongs to $\mathcal{B}(H^{1/2}(\Gamma), H^{-1/2}(\Gamma))$. The sums over β give rise to the terms

$$\frac{1}{4}\int_0^t \left\|\partial_\tau^\alpha(\partial_t^k \vartheta \mathbf{E}(s))\right\|_{L_x^2}^2 \, ds + c\int_0^t \left\|\vartheta \partial_t^k \mathbf{E}(s)\right\|_{H_x^{|\alpha|-1}}^2 \, ds + c\int_0^t \left\|\vartheta \partial_t^k \mathbf{H}(s)\right\|_{H_x^{|\alpha|}}^2 \, ds$$

plus super-quadratic terms. We thus arrive at

$$\begin{aligned} &\left\|\partial_\tau^\alpha \partial_t^k(\vartheta \mathbf{E})(t)\right\|_{L_x^2}^2 + \int_0^t \left\|\partial_\tau^\alpha \partial_t^k(\vartheta \mathbf{E})(s)\right\|_{L_x^2}^2 \, ds \\ &\quad \le cz(0) + c(\theta)\int_0^t \left(\left\|\vartheta \partial_t^k \mathbf{E}(s)\right\|_{H_x^{|\alpha|-1}}^2 + \left\|\tilde{\vartheta} \partial_t^k \mathbf{H}(s)\right\|_{H_x^{|\alpha|}}^2\right) ds \\ &\qquad + \theta\int_0^t \left\|\partial_t^k \mathbf{E}(s)\right\|_{H_x^{|\alpha|}}^2 \, ds + c\int_0^t z^{3/2}(s)\, ds. \end{aligned} \tag{7.38}$$

(b) To finalize the H_x^1-estimate for $\mathbf{E}$, we must control the normal derivatives. As in Proposition 6.9, we first treat their tangential component using the second equation in (7.12). Combined with formula (7.27) and estimate (7.16) it implies

$$\left\|\partial_{\mathrm{n}}(\partial_t^k(\vartheta \mathbf{E}(t))^\tau\right\|_{L_x^2}^2 \le c\big(e_{k+1}(t) + z^2(t) + \left\|\partial_\tau \partial_t^k(\vartheta \mathbf{E}(t))\right\|_{L_x^2}^2\big). \tag{7.39}$$

For the normal component we use the div-relations, where we also consider higher tangential derivatives for later use. We first look at the case $k \in \{1, 2\}$ and apply $\partial_\tau^\alpha \vartheta$ to Eq. (7.11) with $|\alpha| \le 2 - k$. It follows

$$\begin{aligned} \operatorname{div}\big(\varepsilon^{\mathrm{d}}(\mathbf{E})\partial_\tau^\alpha \partial_t^k(\vartheta \mathbf{E})\big) &= -D(\varepsilon^{\mathrm{d}}(\mathbf{E}), \alpha)\partial_t^k \mathbf{E} - \operatorname{div}(\sigma \partial_\tau^\alpha(\vartheta \partial_t^{k-1}\mathbf{E})) \\ &\quad - D(\sigma, \alpha)\partial_t^{k-1}\mathbf{E} - \partial_\tau^\alpha(\vartheta \operatorname{div} f_k\big). \end{aligned} \tag{7.40}$$

Here we abbreviate the commutator terms

$$D(a, \alpha)v := \partial_\tau^\alpha\big([\vartheta, \operatorname{div}](av)\big) + [\partial_\tau^\alpha, \operatorname{div}](\vartheta a v) + \sum_{0 \le \beta < \alpha} \binom{\alpha}{\beta} \operatorname{div}\big(\partial_\tau^{\alpha-\beta} a\, \partial_\tau^\beta(\vartheta v)\big)$$

for a matrix-valued function a and a vector function v. Observe that $D(a, \alpha)$ is a differential operator of order $|\alpha|$ and that $|D(a, 0)v| \le c|v|$. Below we treat the equality (7.40) by means of formula (7.28). For $k = 0$, the divergence equation contains a time integral and initial data which are handled using identity (7.30). To avoid terms which grow linearly in time, we have to derive another equation from (7.1), namely,

$$\partial_t(\varepsilon(\mathbf{E})\partial_\tau^\alpha(\vartheta\mathbf{E})) = \operatorname{curl}\partial_\tau^\alpha(\vartheta\mathbf{H}) - \sigma\partial_\tau^\alpha(\vartheta\mathbf{E}) - [\operatorname{curl},\partial_\tau^\alpha](\vartheta\mathbf{H}) - \partial_\tau^\alpha([\operatorname{curl},\vartheta]\mathbf{H})$$
$$- \sum_{0\le\beta<\alpha}\binom{\alpha}{\beta}\partial_\tau^{\alpha-\beta}(\sigma+\varepsilon(\mathbf{E}))\,\partial_\tau^\beta(\vartheta\mathbf{E}).$$

Writing h for the sum of the three errors terms, we infer the divergence relation

$$\operatorname{div}\big(\varepsilon(\mathbf{E}(t))\partial_\tau^\alpha(\vartheta\mathbf{E}(t))\big) = \operatorname{div}\big(\varepsilon(\mathbf{E}_0)\partial_\tau^\alpha(\vartheta\mathbf{E}_0)\big) \tag{7.41}$$
$$- \int_0^t \big(\operatorname{div}(\sigma\partial_\tau^\alpha(\vartheta\mathbf{E}(s))) + \operatorname{div} h(s)\big)\,\mathrm{d}s.$$

(c) To control $\partial_\mathrm{n} E_\mathrm{n}$, we use Eq. (7.41) with $\alpha = 0$ and identity (7.30), where we put $a = \varepsilon(\mathbf{E})$, $v = \vartheta E$, and $\psi = \operatorname{div} h$. The function $\gamma = \sigma_{\mathrm{nn}}/a_{\mathrm{nn}}$ is bounded from below by $\gamma_0 = c\eta > 0$. We then get the estimate

$$\big\|\partial_\mathrm{n}(\vartheta\mathbf{E}(t))_\mathrm{n}\big\|_{L_x^2}^2 \le c\mathrm{e}^{-\gamma_0 t} z(0) + c\big[\|\mathbf{E}(t)\|_{L_x^2}^2 + \|\partial_\tau(\vartheta\mathbf{E}(t))\|_{L_x^2}^2 + \|\partial_\mathrm{n}(\vartheta\mathbf{E}(t))^\tau\|_{L_x^2}^2\big]$$
$$+ c\int_0^t \mathrm{e}^{-\gamma_0(t-s)}\Big[\|\mathbf{E}(s)\|_{L_x^2}^2 + \|\partial_\tau(\vartheta\mathbf{E}(s))\|_{L^2}^2 + \|\partial_\mathrm{n}(\vartheta\mathbf{E}(s))^\tau\|_{L_x^2}^2 + \|\mathbf{H}(s)\|_{H_x^1}^2\Big]\mathrm{d}s.$$

This bound together with Eqs. (7.38), (7.39) and (7.31) implies

$$\big\|\partial_\mathrm{n}(\vartheta\mathbf{E}(t))_\mathrm{n}\big\|_{L_x^2}^2 + \int_0^t \big\|\partial_\mathrm{n}(\vartheta\mathbf{E}(s))_\mathrm{n}\big\|_{L_x^2}^2\,\mathrm{d}s \tag{7.42}$$
$$\le c\big(z(0) + e(t) + z^2(t)\big) + \theta\int_0^t \|\mathbf{E}(s)\|_{H_x^1}^2\,\mathrm{d}s + c(\theta)\int_0^t \big(e(s) + z^{3/2}(s)\big)\,\mathrm{d}s,$$

where the (small) number $\theta > 0$ comes from (7.38). Combining (7.38), (7.39), (7.42) and (7.31), we conclude

$$\big\|\vartheta\mathbf{E}(t)\big\|_{H_x^1}^2 + \int_0^t \big\|\vartheta\mathbf{E}(s)\big\|_{H_x^1}^2\,\mathrm{d}s$$
$$\le c\big(z(0) + e(t) + z^2(t)\big) + \theta\int_0^t \|\mathbf{E}(s)\|_{H_x^1}^2\,\mathrm{d}s + c(\theta)\int_0^t \big(e(s) + z^{3/2}(s)\big)\,\mathrm{d}s.$$

For $k \in \{1, 2\}$, we proceed similarly using Eq. (7.40) with $\alpha = 0$ and formula (7.28) for the normal component. Here the term $\|\partial_t^{k-1}\vartheta\mathbf{E}(t)\|_{H_x^1}^2$ appears on the right-hand side, which can be treated iteratively. We thus show the inequality

$$\begin{aligned}
&\|\partial_t^k \vartheta \mathbf{E}(t)\|_{H_x^1}^2 + \int_0^t \|\partial_t^k \vartheta \mathbf{E}(s)\|_{H_x^1}^2 \,\mathrm{d}s \qquad (7.43)\\
&\quad \le c\big(z(0) + e(t) + z^2(t)\big) + \theta \int_0^t \|\partial_t^k \mathbf{E}(s)\|_{H_x^1}^2 \,\mathrm{d}s + c(\theta) \int_0^t \big(e(s) + z^{3/2}(s)\big)\,\mathrm{d}s
\end{aligned}$$

for $k \in \{0, 1, 2\}$. Both in (7.43) and (7.33) for $|\alpha| = 1$, we now choose a sufficiently small $\theta > 0$. Together with (7.31) for $k \in \{0, 1, 2\}$, we derive the first-order bound

$$\begin{aligned}
&\|\partial_t^k (\mathbf{E}(t), \mathbf{H}(t))\|_{H_x^1}^2 + \int_0^t \|\partial_t^k (\mathbf{E}(s), \mathbf{H}(s))\|_{H_x^1}^2 \,\mathrm{d}s \qquad (7.44)\\
&\qquad \le c\big(z(0) + e(t) + z^2(t)\big) + c \int_0^t \big(e(s) + z^{3/2}(s)\big)\,\mathrm{d}s.
\end{aligned}$$

(4) *Estimate in* H_x^2. While the bound of $\mathbf{H}$ in H_x^1 was entirely based on the curl-div-estimates of Proposition 7.8, this is only partly possible in H_x^2 or H_x^3 since normal derivatives violate the boundary conditions. We thus have to employ the curl-div strategy of step (3) also for $\mathbf{H}$. Let $k \in \{0, 1\}$.

(a) We first control tangential space-time derivatives of $\mathbf{H}$ in H_x^1 by means of Proposition 7.8, which yields

$$\|v\|_{H_x^1} \le c\big(\|v\|_{H(\mathrm{curl})} + \|\operatorname{div}(\widehat{\mu}_k v)\|_{L_x^2} + \|\operatorname{tr}_{\mathrm{no}}(\widehat{\mu}_k v)\|_{H^{1/2}(\Gamma)}\big) \qquad (7.45)$$

for $v = \partial_\tau \partial_t^k \vartheta \mathbf{H}$. The curl-term appears in the first equation in (7.37) with $|\alpha| = 1$. From Eqs. (7.9), (7.6) and (7.11) we further deduce

$$\begin{aligned}
\operatorname{tr}_{\mathrm{no}}\big(\widehat{\mu}_k v\big) &= [\operatorname{tr}_{\mathrm{no}}, \partial_\tau](\partial_t^k \vartheta \mathbf{H}) - \operatorname{tr}_{\mathrm{no}}\big(\partial_\tau \widehat{\mu}_k\, \partial_t^k(\vartheta \mathbf{H})\big),\\
\operatorname{div}\big(\widehat{\mu}_k v\big) &= \partial_\tau([\operatorname{div}, \vartheta]\widehat{\mu}_k \partial_t^k \mathbf{H}) - [\partial_\tau, \operatorname{div}](\widehat{\mu}_k \partial_t^k(\vartheta \mathbf{H})) - \operatorname{div}(\partial_\tau \widehat{\mu}_k\, \partial_t^k(\vartheta \mathbf{H})).
\end{aligned}$$

The commutator $[\partial_\tau, \operatorname{div}]$ is of order one and the others are of order zero. By means of (7.16), we then estimate

$$\begin{aligned}
\|\operatorname{div}\big(\widehat{\mu}_k \partial_\tau \partial_t^k(\vartheta \mathbf{H}(t))\big)\|_{L_x^2} &\le c\|\partial_t^k \mathbf{H}(t)\|_{H_x^1},\\
\|\operatorname{curl}\big(\partial_\tau \partial_t^k \vartheta \mathbf{H}(t)\big)\|_{L_x^2} &\le c\big(\|\partial_t^{k+1}\mathbf{E}(t)\|_{H_x^1} + \|\partial_t^k(\mathbf{E}(t), \mathbf{H}(t))\|_{H_x^1} + z(t)\big),\\
\|\operatorname{tr}_{\mathrm{no}}\big(\widehat{\mu}_k \partial_\tau \partial_t^k(\vartheta \mathbf{H}(t))\big)\|_{H^{1/2}(\Gamma)} &\le c\|\partial_t^k \mathbf{H}(t)\|_{H_x^1}. \qquad (7.46)
\end{aligned}$$

Since $k+1 \le 2$, inequalities (7.44), (7.45) and (7.46) now imply

$$\|\partial_\tau \partial_t^k(\vartheta \mathbf{H}(t))\|_{H_x^1}^2 + \int_0^t \|\partial_\tau \partial_t^k(\vartheta \mathbf{H}(s))\|_{H_x^1}^2 \,\mathrm{d}s \tag{7.47}$$
$$\le c\big(z(0)+e(t)+z^2(t)\big) + c\int_0^t \big(e(s)+z^{3/2}(s)\big)\,\mathrm{d}s.$$

(b) To treat $\partial_{\mathrm{n}}\partial_t^k \mathbf{H}$ in H_x^1, we first solve in the first equation of (7.37) with $\alpha = 0$ for the tangential component $\partial_{\mathrm{n}}(\partial_t^k \vartheta \mathbf{H}(t))^\tau$ using formula (7.27). It follows

$$\|\partial_{\mathrm{n}}\partial_t^k(\vartheta \mathbf{H}(t))^\tau\|_{H_x^1} \le c\big(\|\partial_\tau \partial_t^k(\vartheta \mathbf{H}(t))\|_{H_x^1} + \|\partial_t^k(\mathbf{E}(t),\mathbf{H}(t))\|_{H_x^1} + \|\partial_t^{k+1}\mathbf{E}(t)\|_{H_x^1}\big).$$

Equations (7.44) and (7.47) thus allow us to bound the tangential component by

$$\|\partial_{\mathrm{n}}\partial_t^k(\vartheta \mathbf{H}(t))^\tau\|_{H_x^1}^2 + \int_0^t \|\partial_{\mathrm{n}}\partial_t^k(\vartheta \mathbf{H}(s))^\tau\|_{H_x^1}^2 \,\mathrm{d}s \tag{7.48}$$
$$\le c\big(z(0)+e(t)+z^2(t)\big) + c\int_0^t \big(e(s)+z^{3/2}(s)\big)\,\mathrm{d}s.$$

As to the normal component, we apply identity (7.28) to the divergence equation (7.34) with $\alpha = 0$ and ϑ instead of χ. The H_x^1-norm of $\partial_{\mathrm{n}}\partial_t^k(\vartheta \mathbf{H}(t))_{\mathrm{n}}$ is thus controlled by that of $\partial_t^k \vartheta \mathbf{H}(t)$, $\partial_\tau \partial_t^k(\vartheta \mathbf{H}(t))$, and $\partial_{\mathrm{n}}\partial_t^k(\vartheta \mathbf{H}(t))^\tau$. Formulas (7.44), (7.47), and (7.48) then yield

$$\|\partial_{\mathrm{n}}\partial_t^k(\vartheta \mathbf{H}(t))_{\mathrm{n}}\|_{H_x^1}^2 + \int_0^t \|\partial_{\mathrm{n}}\partial_t^k(\vartheta \mathbf{H}(s))_{\mathrm{n}}\|_{H_x^1}^2 \,\mathrm{d}s \tag{7.49}$$
$$\le c\big(z(0)+e(t)+z^2(t)\big) + c\int_0^t \big(e(s)+z^{3/2}(s)\big)\,\mathrm{d}s.$$

Collecting the inequalities (7.47), (7.48), (7.49), (7.35) and (7.44), we arrive at the H_x^2-estimate for the fields $\mathbf{H}$ and $\partial_t \mathbf{H}$

$$\|\partial_t^k \mathbf{H}(t)\|_{H_x^2}^2 + \int_0^t \|\partial_t^k \mathbf{H}(s)\|_{H_x^2}^2 \,\mathrm{d}s \tag{7.50}$$
$$\le c\big(z(0)+e(t)+z^2(t)\big) + c\int_0^t \big(e(s)+z^{3/2}(s)\big)\,\mathrm{d}s.$$

(c) We now turn our attention to **E**. Let $|\alpha| = 2$. The L^2_x-norm of the tangential derivative $\partial^\alpha_\tau(\vartheta\partial_t^k\mathbf{E})$ is already controlled via inequalities (7.38), (7.44), and (7.50) up to the term

$$\theta\int_0^t \|\partial_t^k\mathbf{E}(s)\|^2_{H^2_x}\,\mathrm{d}s.$$

The second equation in (7.37) with $|\alpha| = 1$ and formula (7.27) lead to the estimate

$$\begin{aligned}\|\partial_{\mathrm{n}}\big[\partial_\tau\partial_t^k(\vartheta\mathbf{E}(t))\big]^\tau\|_{L^2_x} \le c\big(&\|\partial_\tau^2\partial_t^k(\vartheta\mathbf{E}(t))\|_{L^2_x} + \|\partial_t^k(\mathbf{E}(t),\mathbf{H}(t))\|_{H^1_x}\\ &+ \|\partial_t^{k+1}\mathbf{E}(t)\|_{H^1_x} + z(t)\big).\end{aligned}$$

Combined with the tangential bound and the H^1_x-result (7.44), we obtain

$$\begin{aligned}&\|\partial_{\mathrm{n}}\big(\partial_\tau\partial_t^k(\vartheta\mathbf{E}(t)\big)^\tau\|^2_{L^2_x} + \|\partial_\tau^\alpha\partial_t^k(\vartheta\mathbf{E}(t))\|^2_{L^2_x}\\ &\quad + \int_0^t\Big(\|\partial_{\mathrm{n}}\big(\partial_\tau\partial_t^k(\vartheta\mathbf{E}(s))\big)^\tau\|^2_{L^2} + \|\partial_\tau^\alpha\partial_t^k(\vartheta\mathbf{E}(s))\|^2_{L^2_x}\Big)\mathrm{d}s \qquad (7.51)\\ &\le c(z(0)+e(t)+z^2(t)) + \theta\int_0^t \|\partial_t^k\mathbf{E}(s)\|^2_{H^2_x}\,\mathrm{d}s + c(\theta)\int_0^t\big(e(s)+z^{3/2}(s)\big)\,\mathrm{d}s.\end{aligned}$$

(d) For the normal component and $k = 0$, we look at the divergence relation (7.41) with $|\alpha| = 1$. As in (7.42), we deduce from (7.30) the estimate

$$\begin{aligned}&\|\partial_{\mathrm{n}}\big(\partial_\tau(\vartheta\mathbf{E}(t))\big)_{\mathrm{n}}\|^2_{L^2_x} + \int_0^t\|\partial_{\mathrm{n}}\big(\partial_\tau(\vartheta\mathbf{E}(s))\big)_{\mathrm{n}}\|^2_{L^2_x}\,\mathrm{d}s \qquad (7.52)\\ &\quad\le c(z(0)+e(t)+z^2(t)) + \theta\int_0^t\|\mathbf{E}(s)\|^2_{H^2_x}\,\mathrm{d}s + c(\theta)\int_0^t\big(e(s)+z^{3/2}(s)\big)\,\mathrm{d}s.\end{aligned}$$

The two above inequalities imply

$$\begin{aligned}&\|\partial_\tau(\vartheta\mathbf{E}(t))\|^2_{H^1_x} + \int_0^t\|\partial_\tau(\vartheta\mathbf{E}(s))\|^2_{H^1_x}\,\mathrm{d}s \qquad (7.53)\\ &\quad\le c\big(z(0)+e(t)+z^2(t)\big) + \theta\int_0^t\|\mathbf{E}(s)\|^2_{H^2_x}\,\mathrm{d}s + c(\theta)\int_0^t\big(e(s)+z^{3/2}(s)\big)\,\mathrm{d}s.\end{aligned}$$

To treat the case $k = 1$, we start from the divergence equation (7.40) with $|\alpha| = 1$ and use formula (7.28). Employing also estimates (7.51), (7.53) and (7.16), we get

$$\|\partial_{\mathrm{n}}\big(\partial_\tau(\vartheta\partial_t\mathbf{E}(t))\big)_{\mathrm{n}}\|_{L_x^2}^2 + \int_0^t \|\partial_{\mathrm{n}}\big(\partial_\tau(\vartheta\partial_t\mathbf{E}(s))\big)_{\mathrm{n}}\|_{L_x^2}^2\,\mathrm{d}s \tag{7.54}$$

$$\le c(z(0) + e(t) + z^2(t)) + \theta \int_0^t \big(\|\mathbf{E}(s)\|_{H_x^2}^2 + \|\partial_t\mathbf{E}(s)\|_{H_x^2}^2\big)\,\mathrm{d}s$$

$$+ c(\theta)\int_0^t \big(e(s) + z^{3/2}(s)\big)\,\mathrm{d}s.$$

Together with inequality (7.51), this relation leads to

$$\|\partial_\tau\partial_t(\vartheta\mathbf{E}(t))\|_{H_x^1}^2 + \int_0^t \|\partial_\tau\partial_t(\vartheta\mathbf{E}(s))\|_{H_x^1}^2\,\mathrm{d}s \tag{7.55}$$

$$\le c(z(0) + e(t) + z^2(t)) + \theta \int_0^t \big(\|\mathbf{E}(s)\|_{H_x^2}^2 + \|\partial_t\mathbf{E}(s)\|_{H_x^2}^2\big)\,\mathrm{d}s$$

$$+ c(\theta)\int_0^t \big(e(s) + z^{3/2}(s)\big)\,\mathrm{d}s.$$

(e) It remains to control the term $\partial_{\mathrm{n}}^2(\partial_t^k\vartheta\mathbf{E})$. We first replace the derivative ∂_τ^α by ∂_{n} in system (7.37). The resulting second equation, the curl-formula (7.27) and estimates (7.16) imply

$$\|\partial_{\mathrm{n}}\big(\partial_{\mathrm{n}}\partial_t^k(\vartheta\mathbf{E}(t))\big)^\tau\|_{L_x^2} \le c\Big(\big\|\partial_\tau\partial_{\mathrm{n}}\partial_t^k(\vartheta\mathbf{E}(t))\big\|_{L_x^2} + \max_{j\le 2}\big\|\partial_t^j(\mathbf{E}(t),\mathbf{H}(t))\big\|_{H_x^1} + z(t)\Big).$$

The right-hand side can be estimated via inequalities (7.44) and (7.55).

For the normal component, we employ the modifications of the divergence relations (7.41) and (7.40) with ∂_{n} instead of ∂_τ^α. We then estimate $\partial_{\mathrm{n}}\big(\partial_{\mathrm{n}}\partial_t^k(\vartheta\mathbf{E}(t))\big)_{\mathrm{n}}$ for $k \in \{0, 1\}$ as in inequalities (7.52) and (7.54). Here and in (7.36), (7.53) and (7.55), we take a small $\theta > 0$ to absorb the H^2-norms of $\partial_t^k\mathbf{E}$ on the right-hand side. Using also (7.50) for the magnetic field, for $k \in \{0, 1\}$ we derive the desired bound in H_x^2

$$\big\|\partial_t^k(\mathbf{E}(t),\mathbf{H}(t))\big\|_{H_x^2}^2 + \int_0^t \|\partial_t^k(\mathbf{E}(s),\mathbf{H}(s))\|_{H_x^2}^2\,\mathrm{d}s \tag{7.56}$$

$$\le c\big(z(0) + e(t) + z^2(t)\big) + c\int_0^t \big(e(s) + z^{3/2}(s)\big)\,\mathrm{d}s.$$

(5) *Estimate in* H^3_x. Since the reasoning is similar to step 4), we will omit some details here. Let $k = 0$.

(a) We again begin with the magnetic field $\mathbf{H}$. We first look at the tangential derivative $\partial_\tau^\alpha(\vartheta\mathbf{E})$ with $|\alpha| = 2$, where we proceed as in (7.47) using Proposition 7.8. For $\xi, \zeta \in \{\mathrm{n}, \tau^1, \tau^2\}$, differentiating the divergence relation (7.6) we obtain

$$\begin{aligned}\operatorname{div}\big(\mu(\mathbf{H})\partial_\xi\partial_\zeta(\vartheta\mathbf{H})\big) &= \partial_\xi\partial_\zeta([\operatorname{div},\vartheta]\mu(\mathbf{H})\mathbf{H}) - [\partial_\xi\partial_\zeta,\operatorname{div}](\mu(\mathbf{H})\vartheta\mathbf{H}) \\ &\quad - \operatorname{div}(\partial_\zeta\mu(\mathbf{H})\,\partial_\xi(\vartheta\mathbf{H})) - \operatorname{div}(\partial_\xi\mu(\mathbf{H})\,\partial_\zeta(\vartheta\mathbf{H})) \\ &\quad - \operatorname{div}(\partial_\xi\partial_\zeta\mu(\mathbf{H})\,\vartheta\mathbf{H}).\end{aligned}$$

Similarly, the magnetic boundary condition in (7.6) yields

$$\mathrm{tr_{no}}\big(\mu(\mathbf{H})\partial_\tau^\alpha(\vartheta\mathbf{H})\big) = [\mathrm{tr_{no}},\partial_\tau^\alpha](\mu(\mathbf{H})\vartheta\mathbf{H}) + \mathrm{tr_{no}}\sum_{0\le\beta<\alpha}\binom{\alpha}{\beta}\partial_\tau^{\alpha-\beta}\mu(\mathbf{H})\,\partial_\tau^\beta(\vartheta H).$$

Employing (7.16), we deduce the estimates

$$\begin{aligned}\|\operatorname{curl}\partial_\tau^\alpha(\vartheta\mathbf{H}(t))\|_{L^2_x} &\le c\big(\|\partial_t\mathbf{E}(t)\|_{H^2_x} + \|(\mathbf{E}(t),\mathbf{H}(t))\|_{H^2_x} + z(t)\big), \\ \|\operatorname{div}\big(\mu(\mathbf{H}(t))v\big)\|_{L^2_x} &\le c\|\mathbf{H}(t)\|_{H^2_x}, \\ \|\mathrm{tr_{no}}\big(\mu(\mathbf{H}(t))v\big)\|_{H^{1/2}(\Gamma)} &\le c\|\mathbf{H}(t)\|_{H^2_x}\end{aligned}$$

from (7.37) and the above formulas. The second-order bound (7.56) and Proposition 7.8 thus imply

$$\begin{aligned}&\big\|\partial_\tau^\alpha(\vartheta\mathbf{H}(t))\big\|^2_{H^1_x} + \int_0^t \|\partial_\tau^\alpha(\vartheta\mathbf{H}(s))\|^2_{H^1_x}\,\mathrm{d}s \\ &\qquad \le c\big(z(0) + e(t) + z^2(t)\big) + c\int_0^t \big(e(s) + z^{3/2}(s)\big)\,\mathrm{d}s.\end{aligned} \tag{7.57}$$

To include one normal derivative, we first use (7.37) with $|\alpha| = 1$ and the curl-formula (7.27). We can then bound the H^1_x-norm of $\partial_\mathrm{n}(\partial_\tau(\vartheta\mathbf{H}(t)))^\tau$ by

$$\|\partial_\tau^2(\vartheta\mathbf{H}(t))\|_{H^1_x} + \max_{j\le1}\|\partial_t^j(\mathbf{E}(t),\mathbf{H}(t))\big\|_{H^2_x} + z(t).$$

The normal component is treated as in (7.49), based on the divergence relation (7.34) with $|\alpha| = 1$, χ replaced by ϑ, and ∂_x^α by ∂_τ. By means of (7.28) and (7.16), the H^1_x-norm of the function $\partial_{\mathrm{n}}(\partial_\tau(\vartheta H(t)))_{\mathrm{n}}$ is thus controlled by that of $\partial_{\mathrm{n}}(\partial_\tau(\vartheta H(t)))^\tau$ and $\partial_\tau^2(\vartheta \mathbf{H}(t))$ plus lower order terms. Combining these inequalities with (7.56) and (7.57), we infer

$$\begin{aligned} &\big\|\partial_{\mathrm{n}}\partial_\tau(\vartheta \mathbf{H}(t))\big\|_{H_x^1}^2 + \int_0^t \|\partial_{\mathrm{n}}\partial_\tau(\vartheta \mathbf{H}(s))\|_{H_x^1}^2 \,\mathrm{d}s \\ &\qquad \le c\big(z(0)+e(t)+z^2(t)\big) + c\int_0^t \big(e(s)+z^{3/2}(s)\big)\,\mathrm{d}s. \end{aligned} \tag{7.58}$$

In this reasoning we can replace ∂_τ by ∂_{n}, arriving at

$$\begin{aligned} &\big\|\partial_{\mathrm{n}}^2(\vartheta \mathbf{H}(t))\big\|_{H_x^1}^2 + \int_0^t \|\partial_{\mathrm{n}}^2(\vartheta \mathbf{H}(s))\|_{H_x^1}^2 \,\mathrm{d}s \\ &\qquad \le c\big(z(0)+e(t)+z^2(t)\big) + c\int_0^t \big(e(s)+z^{3/2}(s)\big)\,\mathrm{d}s. \end{aligned} \tag{7.59}$$

Together with (7.35) and (7.56), the estimates (7.57), (7.58) and (7.59) lead to

$$\begin{aligned} &\|\mathbf{H}(t)\|_{H_x^3}^2 + \int_0^t \|\mathbf{H}(s)\|_{H_x^3}^2 \,\mathrm{d}s \\ &\qquad \le c\big(z(0)+e(t)+z^2(t)\big) + c\int_0^t \big(e(s)+z^{3/2}(s)\big)\,\mathrm{d}s. \end{aligned} \tag{7.60}$$

(b) We finally tackle $\mathbf{E}$ in H_x^3. The third-order tangential derivatives $\partial_\tau^\alpha(\vartheta \mathbf{E})$ were already treated in estimate (7.38) with $k=0$, where the lower order-terms on the right-hand side are now dominated by (7.56) and (7.60). Let $|\beta|=2$. The second equation in (7.37) with $|\alpha|=2$ and the curl-formula (7.27) allow us to bound $\partial_{\mathrm{n}}(\partial_\tau^\beta(\vartheta \mathbf{E}))^\tau$ in the same fashion. The normal component $\partial_{\mathrm{n}}(\partial_\tau^\beta(\vartheta E))_{\mathrm{n}}$ can also be estimated via equations (7.41) and (7.30) as in (7.42). We thus arrive at

$$\begin{aligned} &\big\|\partial_\tau^\beta(\vartheta \mathbf{E}(t))\big\|_{H_x^1}^2 + \int_0^t \big\|\partial_\tau^\beta(\vartheta \mathbf{E}(s))\big\|_{H_x^1}^2 \,\mathrm{d}s \\ &\quad \le c\big(z(0)+e(t)+z^2(t)\big) + \theta\int_0^t \big\|\mathbf{E}(s)\big\|_{H_x^3}^2\,\mathrm{d}s + c(\theta)\int_0^t \big(e(s)+z^{3/2}(s)\big)\,\mathrm{d}s. \end{aligned} \tag{7.61}$$

We replace the tangential derivative ∂_τ^α by $\partial_{\mathrm{n}}\partial_\tau$ in system (7.37). The second equation therein and formula (7.27) provide control of the tangential component $\partial_{\mathrm{n}}(\partial_{\mathrm{n}}\partial_\tau(\vartheta \mathbf{E}))^\tau$ in L_x^2 via inequalities (7.61) and (7.56). The related normal component can then be handled through the formula (7.30) and the divergence identity (7.41) with $\partial_{\mathrm{n}}\partial_\tau$ instead of ∂_τ^α. In this way we show the inequality

$$\begin{aligned}&\left\|\partial_\tau(\vartheta\mathbf{E}(t))\right\|_{H_x^2}^2+\int_0^t\left\|\partial_\tau(\vartheta\mathbf{E}(s))\right\|_{H_x^2}^2\,\mathrm{d}s\\&\quad\le c\big(z(0)+e(t)+z^2(t)\big)+\theta\int_0^t\|\mathbf{E}(s)\|_{H_x^3}^2\,\mathrm{d}s+c(\theta)\int_0^t\big(e(s)+z^{3/2}(s)\big)\,\mathrm{d}s.\end{aligned}$$

The remaining term $\partial_{\mathrm{n}}^3(\vartheta E)$ is managed analogously, resulting in

$$\begin{aligned}&\left\|\vartheta\mathbf{E}(t)\right\|_{H_x^3}^2+\int_0^t\left\|\vartheta\mathbf{E}(s)\right\|_{H_x^3}^2\,\mathrm{d}s\\&\quad\le c\big(z(0)+e(t)+z^2(t)\big)+\theta\int_0^t\|\mathbf{E}(s)\|_{H_x^3}^2\,\mathrm{d}s+c(\theta)\int_0^t\big(e(s)+z^{3/2}(s)\big)\,\mathrm{d}s.\end{aligned}$$

Fixing a sufficiently small number $\theta > 0$, the above inequalities and the interior estimate (7.36) combined with (7.44) and (7.56) lead to the bound

$$\|\mathbf{E}(t)\|_{H_x^3}^2+\int_0^t\|\mathbf{E}(s)\|_{H_x^3}^2\,\mathrm{d}s\le c\big(z(0)+e(t)+z^2(t)\big)+c\int_0^t\big(e(s)+z^{3/2}(s)\big)\,\mathrm{d}s.$$

The above equation and (7.60) now furnish our last result

$$\begin{aligned}&\left\|(\mathbf{E}(t),\mathbf{H}(t))\right\|_{H_x^3}^2+\int_0^t\left\|(\mathbf{E}(s),\mathbf{H}(s))\right\|_{H_x^3}^3\,\mathrm{d}s\\&\qquad\le c\big(z(0)+e(t)+z^2(t)\big)+c\int_0^t\big(e(s)+z^{3/2}(s)\big)\mathrm{d}s.\end{aligned}\tag{7.62}$$

Proposition 7.7 now follows from formulas (7.44), (7.56) and (7.62). □

Part III

Error Analysis of Second-Order Time Integration Methods for Discontinuous Galerkin Discretizations of Friedrichs' Systems

Marlis Hochbruck and Jonas Köhler

In these lecture notes, we study the full discretization in space and time of a general class of linear wave-type equations written in terms of Friedrichs' operators. For the space discretization, we consider a discontinuous Galerkin finite element method. As time integration schemes we restrict ourselves to second-order methods, namely the Crank–Nicolson, the leapfrog, the Peaceman–Rachford, and a locally implicit method. Our aim is to not only present rigorous error bounds measuring the quality of the fully discrete approximations, but also establish a systematic procedure to derive such bounds.

Introduction

8

Solving wave-type equations numerically requires their discretization either in space and time separately or in space-time. In these lecture notes, we follow a methods-of-lines approach, where we first discretize the problem in space and then in time. For the space discretization, we consider a discontinuous Galerkin finite element method. As time integration schemes, we restrict ourselves to four second-order methods, namely

- the Crank–Nicolson method, i.e., an implicit method,
- the leapfrog method, i.e., an explicit method,
- the Peaceman–Rachford method, i.e., a splitting method,
- a locally implicit method, i.e., a hybrid method, combining the Crank–Nicolson and the leapfrog method.

We study these methods for a quite general class of linear wave-type equations written in terms of Friedrichs' operators [59, 72]. Important applications fitting into this class are advection equations, various wave equations, and Maxwell equations for instance. Additionally, the Hodge-Dirac operator [154] is also a Friedrichs' operator. Some of these operators, like the Maxwell operator, exhibit a two-field structure which can be exploited in the construction of the methods as well as in the analysis.

In the space discretization, special emphasis is on the correct treatment of boundary conditions to make the analytical problem and its numerical approximation wellposed.

The aim of this work is to not only present rigorous error bounds measuring the quality of the fully discrete approximations, but also establish a systematic procedure to derive such bounds. This might also be useful for other methods. From the viewpoint of the analysis, the Crank–Nicolson method turns out to be the most simple one and this is why we start with it. The main idea is then to show that one can treat the other methods as

W. Dörfler et al., *Wave Phenomena*, Oberwolfach Seminars 49,
https://doi.org/10.1007/978-3-031-05793-9_8

perturbations of the Crank–Nicolson method. We will explain in detail the different types of perturbations and the techniques these differences require in the analysis.

The research we present here started with the PhD theses [113, 137, 161] and the publications [88–91]. The novelty in the current work is that we study all methods for a general class of Friedrichs' systems. In particular, locally implicit methods have not been considered yet for this class. Moreover, besides error bounds for the fully discrete solution, we also identify suitable approximations for the spatial and temporal derivatives of the exact solution and present appropriate error bounds for them. Furthermore, we were able to improve some of our previous results and proofs.

Note, however, that we do not consider the implementation and other practical issues of the methods in this work. For this we refer to our earlier work, in particular to [92] for the implicit midpoint rule (whose implementation is analogous to that of the Crank–Nicolson method) and exponential integrators, to [91, 161] for locally implicit methods, and to [87, 113] for the Peaceman–Rachford–ADI method.

These lecture notes are organized as follows. Each chapter starts with a short overview and references to the literature.

In Chap. 9, we introduce wave-type problems written in terms of Friedrichs' operators, present the analytical framework and review basic semigroup theory required for the following analysis. In addition, we present prototypical examples.

Chapter 10 is devoted to the spatial discretization of Friedrichs' operators by discontinuous Galerkin finite elements with central fluxes. We also collect important properties such as inverse inequalities and approximation results from the literature, in particular from [43]. The main results in this section include the stability analysis and error bounds for the spatially semidiscrete problem.

Our main results are presented in Chaps. 11 and 12, where we consider the time integration of the spatially semidiscrete problem, which then leads to a fully discrete method. All time integration schemes are first illustrated on a system of ordinary differential equations before we consider the more abstract formulations. Our analysis is based on properties of discrete semigroups and on the observation that all methods can be interpreted as perturbations of the Crank–Nicolson method. Chapter 11 contains the stability analysis, while in Chap. 12, we finally show the error bounds.

In Chap. 13, we comment on Friedrichs' operators exhibiting a two-field structure and provide more detailed bounds, which are postponed for the sake of presentation. We close this work by a glossary in Chap. 14, where we collect the symbols used together with short descriptions and the pages containing their definitions. Note that the electronic version redirects to this list whenever you click on such a symbol to facilitate studying the material.

Acknowledgment

We thank our colleagues Constantin Carle, Benjamin Dörich, Jan Leibold, and Konstantin Zerulla for their thorough proofreading and numerous suggestions that improved the quality of this work significantly. We further thank Constantin Carle for improving the bounds in Lemmas 11.14 and 11.28.

Funded by the Deutsche Forschungsgemeinschaft (DFG, German Research Foundation)—Project-ID 258734477—SFB 1173.

8.1 Notation

Before we start with the main part, we introduce the basic notation used throughout this work.

The spatial dimension is denoted by $d \in \mathbb{N}$, while we use $m \in \mathbb{N}$ for the number of components of the wave-type equation. Given an m-dimensional vector v, we denote its components by $v_1, \dots, v_m$. Further, $e_1, \dots, e_m$ are the canonical unit vectors and I is the identity matrix in $\mathbb{R}^m$. The boundary of a set S is denoted by ∂S and we write $\operatorname{ran}(\Phi)$ for the range of a map Φ and $\ker(\Phi)$ for its kernel.

We denote the distributional derivative in the ith coordinate direction of $\mathbb{R}^d$ by ∂_i, $i = 1, \dots, d$. Concatenations of such derivatives are denoted in the usual short-notation using a multi-index $\alpha \in \mathbb{N}_0^d$ as $\partial^\alpha v = \partial_1^{\alpha_1} \dots \partial_d^{\alpha_d} v$ with the convention $\partial^{(0,\dots,0)} v = v$. These derivatives are applied componentwise to vector- or matrix-valued functions and we denote the ℓ^1-norm of α by $|\alpha|$.

Let $\big(X, (\cdot \mid \cdot)_X\big)$ be a real Hilbert space. We denote the identity operator on X as $\mathcal{I}$, the dual space of X as X', and the canonical pairing between X and its dual by $\langle \cdot \mid \cdot \rangle \colon X' \times X \to \mathbb{R}$. Given a second Hilbert space $\big(Y, (\cdot \mid \cdot)_Y\big)$, we denote the set containing all bounded operators from X to Y by $\mathcal{B}(X, Y)$.

In the following, we will often restrict a Hilbert space operator $\mathcal{A}$ to its domain $D(\mathcal{A})$, which is usually chosen such that it incorporates the boundary values of the wave-type problem. Then, given two Hilbert space operators $\mathcal{A}$ and $\mathcal{B}$ with domains $D(\mathcal{A})$ and $D(\mathcal{B})$, we define the domain of the concatenation of $\mathcal{A}$ and $\mathcal{B}$ as $D(\mathcal{A}\mathcal{B}) := \big\{v \in D(\mathcal{B}) \mid \mathcal{B}v \in D(\mathcal{A})\big\}$. Given more than two such operators, this definition is recursively extended.

Throughout the rest of this part of the book, let $K \subset \mathbb{R}^d$ be open and $F \subset \partial K$. We denote the set of all polynomials of degree at most $j \in \mathbb{N}$ on K by $\mathbb{Q}_d^j(K)$.

Further, if we have two vector-valued functions $u, v \in L^2(K)^m$, the (standard) $L^2(K)$-inner product is denoted by

$$\big(u \mid v\big)_{L^2,K} = \int_K u \cdot v \, \mathrm{d}x,$$

and, in the case that the traces $u|_F$, $v|_F$ are well-defined elements of $L^2(F)^m$, the surface integral over F is denoted by

$$(u \mid v)_F = \int_F u|_F \cdot v|_F \, \mathrm{d}\sigma.$$

The induced norms are $\|\cdot\|_{L^2,K}$ and $\|\cdot\|_F$, respectively.

We also need L^2-Sobolev spaces on K, which we denote by $H^q(K)$, $q \in \mathbb{N}_0$. They are equipped with the norms

$$\|v\|_{L^2,q,K}^2 = \sum_{j=0}^{q} |v|_{L^2,j,K}^2, \qquad |v|_{L^2,j,K}^2 = \sum_{|\alpha|=j} \|\partial^\alpha v\|_{L^2,K}^2, \quad j = 0, \ldots, q,$$

where $\alpha \in \mathbb{N}_0^d$ is a multi-index.

The spectral norm of a matrix $A \in \mathbb{R}^{m\times m}$ is denoted by $\|A\|$ and the essential supremum of the spectral norm of a matrix-valued function $\Lambda \in L^\infty(K)^{m\times m}$ is denoted by

$$\|\Lambda\|_{\infty,K} = \operatorname*{ess\,sup}_{x\in K} \|\Lambda(x)\|.$$

Lastly, the L^∞-Sobolev spaces on K are denoted by $W^{q,\infty}(K)$, $q \in \mathbb{N}_0$. Given a matrix-valued function $\Lambda \in W^{q,\infty}(K)^{m\times m}$, we use the norm

$$\|\Lambda\|_{q,\infty,K} = \max_{|\alpha|\le q} \|\partial^\alpha \Lambda\|_{\infty,K}.$$

Lastly, we want to explain some guiding principles of our notation. Throughout these lecture notes, matrices and matrix-valued functions are denoted by capital roman letters. Operators that map from one function space to another are written in capital calligraphic letters. Discrete objects and operators mapping into discrete spaces are set in bold face, and we underline the discrete operators to better distinguish them from their continuous counterparts. Further, continuous objects that are projected into the discrete space are provided with the subscript π (and consequently written in bold face, since projected objects are discrete).

9 Linear Wave-Type Equations

In this chapter, we state and analyze the wave-type problem, which we consider within these lecture notes. As mentioned before, we state this problem in a rather general setting, namely in terms of Friedrichs' operators. We refer the interested reader to [72] for the original work introducing such operators and to [20,43,59–61,64,102] for a more detailed discussion of them.

Let the *spatial domain* $\Omega \subset \mathbb{R}^d$, $d = 1, 2, 3, \ldots$, be an open, bounded and connected Lipschitz domain with boundary $\Gamma = \partial\Omega$ and $\mathbb{R}_+$ be the *temporal domain*. Then, we seek the *solution* $u \colon \mathbb{R}_+ \times \Omega \to \mathbb{R}^m$ of the *wave-type problem*

$$\begin{cases} M\partial_t u = \widetilde{\mathcal{L}}u + \widetilde{f}, & \text{on } \mathbb{R}_+ \times \Omega\,, \quad (9.1\text{a}) \\ u(0) = u^0, & \text{on } \Omega\,, \quad (9.1\text{b}) \end{cases}$$

subject to an *initial value* $u^0 \colon \Omega \to \mathbb{R}^m$, an *inhomogeneity* or *source term* $\widetilde{f} \colon \mathbb{R}_+ \times \Omega \to \mathbb{R}^m$, and suitable *boundary conditions* (which we do not state explicitly, since they will be incorporated into the domain of the operator $\widetilde{\mathcal{L}}$). Further, the *material tensor* $M \in L^\infty(\Omega)^{m\times m}$ is symmetric and uniformly positive definite. The operator $\widetilde{\mathcal{L}}$, which governs the temporal evolution of the solution u, is a first-order differential operator (in space) of the form

$$\widetilde{\mathcal{L}}u = \sum_{i=1}^{d} L_i \partial_i u + L_0 u, \qquad L_i \in \mathbb{R}^{m\times m}, \quad i = 0, \ldots, d, \tag{9.2}$$

and henceforth called a *Friedrichs' operator* [72]. We point out that its *coefficients* $L_0, \ldots, L_d$ are chosen to be constant for the sake of presentation, but all results derived in this work hold true for more general coefficients fulfilling suitable assumptions. Details on

W. Dörfler et al., *Wave Phenomena*, Oberwolfach Seminars 49,
https://doi.org/10.1007/978-3-031-05793-9_9

the more general setting can be found in [102, 113]. If the symmetric part of L_0 is negative semidefinite, i.e., $x^\mathsf{T}(L_0 + L_0^\mathsf{T})x \le 0$ for all $x \in \mathbb{R}^m$, and the coefficients $L_1, \dots, L_d$ are symmetric, then (9.1) is wellposed [43,88] (if equipped with suitable boundary conditions) and thus this will be assumed in the following.

For the analysis of the wave-type problem (9.1) (and its discretizations), we employ the theory of (abstract) evolution equations and linear semigroups [138]. To do so, we reformulate (9.1) to fit this setting by eliminating the material tensor M in front of the time derivative and obtain the equivalent problem

$$\begin{cases} \partial_t u = \mathcal{L}u + f, & \text{on } \mathbb{R}_+ \times \Omega, \quad (9.3a) \\ u(0) = u^0, & \text{on } \Omega, \quad (9.3b) \end{cases}$$

where $f = M^{-1}\widetilde{f}$ and $\mathcal{L} = M^{-1}\widetilde{\mathcal{L}}$. Since $\mathcal{L}$ still exhibits the same structure as $\widetilde{\mathcal{L}}$ and is just a weighted version of it, we also refer to $\mathcal{L}$ as a Friedrichs' operator.

Due to the fact that we weight the Friedrichs' operator $\mathcal{L}$ by the inverse M^{-1} of the material tensor, the standard L^2-spaces are not suited for our analysis. In fact, we need to use an L^2-inner product on $K \subset \Omega$ that is weighted with the material tensor M. It is denoted by

$$\big(u \mid v\big)_K = \big(Mu \mid v\big)_{L^2,K}. \tag{9.4}$$

Since M is symmetric positive definite, the weighted and standard L^2-inner products are equivalent and thus, the space $\big(L^2(K)^m, (\cdot \mid \cdot)_K\big)$ is a Hilbert space. The norm induced by this weighted inner product is denoted as $\|\cdot\|_K$. Analogously, we define weighted Sobolev spaces and denote their norms and seminorms by

$$\|v\|_{q,K}^2 = \sum_{j=0}^{q} |v|_{j,K}^2, \qquad |v|_{j,K}^2 = \sum_{|\alpha|=j} \|\partial^\alpha v\|_K^2, \quad j = 0, \dots, q,$$

respectively, where $\alpha \in \mathbb{N}_0^d$ is a multi-index. By equivalence, we see that these norms still induce the standard Sobolev spaces.

We point out that most results in this section are taken from [43, 59–61]. However, the wellposedness theory found in these publications is performed either for stationary problems or in a space-time framework, where time-dependent problems are interpreted as elliptic problems posed on the space-time domain. As we are interested in the analysis of time-stepping methods, the framework of semigroups seems to be the more natural choice and hence, we adapt the results to this framework. A similar ansatz was also pursued in [20]. However, we follow the approach in [88, 113] which mostly differs in the treatment of boundary values.

Before we analyze the wave-type problem (9.3) itself, we present some of the theory we need to do so.

9.1 A Short Course on Semigroup Theory

In the following, let $\big(X, (\cdot \mid \cdot)_X\big)$ be a Hilbert space and denote by $\|\cdot\|_X$ the norm induced by its inner product. The aim of this section is to recap some wellposedness theory of *abstract Cauchy problems* given by

$$\begin{cases} \partial_t u(t) = \mathcal{A}u(t) + f(t), & t \in \mathbb{R}_+, \quad (9.5a) \\ u(0) = u^0, & \quad (9.5b) \end{cases}$$

with *solution* $u\colon \mathbb{R}_+ \to X$, given *initial value* $u^0 \in X$ and *inhomogeneity* or *source term* $f\colon \mathbb{R}_+ \to X$. Further, $\mathcal{A}\colon X \supset D(\mathcal{A}) \to X$ is a linear operator on the Hilbert space X with *domain* $D(\mathcal{A})$.

Most material in this section is taken from the textbooks [57, 101, 138] and our presentation of these topics closely follows the one in [161]. Before we study the abstract problem (9.5) we give an introductory example that aims to illustrate some core concepts of the theory.

Example 9.1 Given a matrix $A \in \mathbb{C}^{m\times m}$ and an initial value $u^0 \in \mathbb{C}^m$, we consider the following problem. Seek the solution $u\colon \mathbb{R}_+ \to \mathbb{C}^m$ of the system of linear, homogeneous, ordinary differential equations given by

$$\begin{cases} \partial_t u(t) = Au(t), & t \in \mathbb{R}_+, \\ u(0) = u^0. \end{cases} \tag{9.6}$$

As it is well known, the solution u of (9.6) is unique and given by

$$u(t) = \mathrm{e}^{tA}u^0, \tag{9.7}$$

with $\mathrm{e}^{tA} \in \mathbb{C}^{m\times m}$ being the matrix exponential of tA. Hence, in the case of a homogeneous initial value problem comprised of a system of ordinary differential equations, matrix-vector multiplication with the matrix exponential e^{tA} provides us with an operator that maps the initial value (i.e., the input-data) to the solution of the problem.

In fact, such an operator can also be derived for the inhomogeneous case

$$\begin{cases} \partial_t u(t) = Au(t) + f(t), & t \in \mathbb{R}_+, \\ u(0) = u^0, \end{cases} \tag{9.8}$$

with a sufficiently smooth inhomogeneity $f\colon \mathbb{R}_+ \to \mathbb{C}^m$, where the solution of (9.8) is given by the variation-of-constants formula (also called Duhamel's formula)

$$u(t) = \mathrm{e}^{tA}u^0 + \int_0^t \mathrm{e}^{(t-s)A} f(s)\,\mathrm{d}s.$$

Hence, again, we have a data-to-solution map comprising the matrix exponential, this time involving a convolution with the inhomogeneity.

Lastly, we observe that the matrix exponential satisfies some interesting properties that will become important later. In particular, we have the identities

$$\mathrm{e}^{tA}\big|_{t=0} = I, \qquad \text{and} \qquad \mathrm{e}^{(t+s)A} = \mathrm{e}^{tA}\,\mathrm{e}^{sA}, \tag{9.9}$$

for all $t, s \in \mathbb{R}_+$ and, using the first identity in (9.9), we have

$$\left(\partial_t\,\mathrm{e}^{tA}\right)\big|_{t=0} = \left(A\,\mathrm{e}^{tA}\right)\big|_{t=0} = A. \tag{9.10}$$

Hence, the matrix A can be recovered by differentiating and evaluating the matrix exponential at $t = 0$, and the derivative of the matrix exponential is given by application of the corresponding matrix.

If the field of values of A is contained in $\{z \in \mathbb{C} \mid \operatorname{Re} z \leq \omega\}$ for a real number ω, i.e., A satisfies

$$\operatorname{Re}\left(Au \mid u\right)_X \leq \omega\left(u \mid u\right)_X, \qquad \text{for all } u \in \mathbb{C}^m,$$

then we have

$$\|\,\mathrm{e}^{tA}\,\|_X \leq \mathrm{e}^{\omega t} \qquad \text{for all } t \geq 0. \tag{9.11}$$

This can be seen by considering $\varphi(t) = \mathrm{e}^{-2\omega t}\,\|u(t)\|_X^2$, where $u(t)$ denotes the solution (9.7) of (9.6). Differentiating yields

$$\begin{aligned}
\varphi'(t) &= \mathrm{e}^{-2\omega t}\left(2\operatorname{Re}\left(\partial_t u(t) \mid u(t)\right)_X - 2\omega\left(u(t) \mid u(t)\right)_X\right)\\
&= 2\,\mathrm{e}^{-2\omega t}\left(\operatorname{Re}\left(Au(t) \mid u(t)\right)_X - \omega\left(u(t) \mid u(t)\right)_X\right) \leq 0
\end{aligned}$$

and integrating this inequality over t shows that $\varphi(t) \leq \varphi(0)$. Hence we have

$$\|u(t)\|_X = \|\,\mathrm{e}^{tA}\,u^0\|_X \leq \mathrm{e}^{\omega t}\,\|u^0\|_X.$$

Since this is true for arbitrary vectors u^0, we showed (9.11). ◇

Now that we have seen how to solve problems given by systems of linear ordinary differential equations, the question comes to mind if one can apply similar ideas to obtain a formula for the solution of the abstract Cauchy problem (9.5). It turns out that this can be done by generalizing the concept of the matrix exponential to the abstract Hilbert space setting. This leads to the concept of strongly continuous semigroups, defined as follows.

Definition 9.2 A one-parameter family of bounded linear operators $(\mathcal{T}(t))_{t\geq 0}$ on X is called a **semigroup of bounded linear operators** if

(i) $\mathcal{T}(0) = \mathcal{I}$,
(ii) $\mathcal{T}(t+s) = \mathcal{T}(t)\mathcal{T}(s)$ for all $t, s \geq 0$.

A semigroup $(\mathcal{T}(t))_{t\geq 0}$ is called **strongly continuous** or a C_0**-semigroup** if

(iii) $\lim\limits_{t\to 0+} \|\mathcal{T}(t)x - x\|_X = 0$ for all $x \in X$.

Note that (i) and (ii) reflect the properties in (9.9), already establishing a first relation between the matrix exponential and semigroups. Further, the matrix exponential also fulfills (iii) in the spectral norm.

Next, we give some important properties of semigroups. The first one gives a bound on the operator norm of a semigroup, which can be seen as a stability bound and reflects the corresponding bound (9.11) on the matrix exponential. The second one shows the strong continuity already present in the nomenclature of Definition 9.2.

Lemma 9.3 *Let $(\mathcal{T}(t))_{t\geq 0}$ be a strongly continuous semigroup. Then the following holds.*

(i) *There exist constants $C_{\mathrm{sg}} \geq 1$ and $\omega \in \mathbb{R}$ such that for all $x \in X$ we have*

$$\|\mathcal{T}(t)x\|_X \leq C_{\mathrm{sg}}\, \mathrm{e}^{\omega t}\, \|x\|_X \qquad \textit{for all } t \geq 0. \tag{9.12}$$

(ii) *The mapping $t \mapsto \mathcal{T}(t)$ is strongly continuous on $\mathbb{R}_+$, i.e.,*

$$\lim_{s\to 0} \|\mathcal{T}(t+s)x - \mathcal{T}(t)x\|_X = 0 \qquad \textit{for all } t > 0.$$

So far, there is no visible connection between a semigroup and a Hilbert space operator $\mathcal{A}$. The next definition, however, introduces such a connection by mirroring relation (9.10), resulting in the generator of a semigroup.

Definition 9.4 Let $(\mathcal{T}(t))_{t\geq 0}$ be a strongly continuous semigroup and

$$\mathcal{D} = \left\{x \in X \mid \lim_{t\to 0+} \left(\tfrac{1}{t}(\mathcal{T}(t)x - x)\right) \in X\right\}.$$

The **infinitesimal generator** of $(\mathcal{T}(t))_{t\geq 0}$ is defined as the linear operator $\mathcal{A}\colon \mathcal{D} \to X$ given by

$$\mathcal{A}x = \lim_{t\to 0+} \frac{\mathcal{T}(t)x - x}{t} \qquad \text{for all } x \in \mathcal{D}.$$

The set $\mathcal{D}$ is called the **domain of** $\mathcal{A}$ and we denote it by $D(\mathcal{A})$.

We state some important properties of infinitesimal generators in the next lemma.

Lemma 9.5 *Let $(\mathcal{T}(t))_{t\geq 0}$ be a strongly continuous semigroup with infinitesimal generator $\mathcal{A}$. Then, for $x \in D(\mathcal{A})$ and $t \geq 0$, we have*

$$\mathcal{T}(t)x \in D(\mathcal{A}), \tag{9.13a}$$

$$\partial_t(\mathcal{T}(t)x) = \mathcal{A}\mathcal{T}(t)x = \mathcal{T}(t)\mathcal{A}x. \tag{9.13b}$$

Further, the domain $D(\mathcal{A})$ of $\mathcal{A}$ is dense in X and $\mathcal{A}$ is a closed operator.

Definition 9.4 defines a unique infinitesimal generator for each semigroup. The following corollary of Lemma 9.5 shows that each infinitesimal generator also generates a unique semigroup, establishing a one-to-one relation between a semigroup and its infinitesimal generator.

Corollary 9.6 *Let $(\mathcal{T}_1(t))_{t\geq 0}$ and $(\mathcal{T}_2(t))_{t\geq 0}$ be strongly continuous semigroups with infinitesimal generators $\mathcal{A}_1$ and $\mathcal{A}_2$, respectively. If $\mathcal{A}_1 = \mathcal{A}_2$, then we have $\mathcal{T}_1(t) = \mathcal{T}_2(t)$ for all $t \in \mathbb{R}_+$.*

Now, we have all the tools to establish a connection between abstract Cauchy problems involving the infinitesimal generator $\mathcal{A}$ of a semigroup $(\mathcal{T}(t))_{t\geq 0}$ and the semigroup itself. In fact, Lemma 9.5 together with Definition 9.2 (i) already show that the solution of the *homogeneous abstract Cauchy problem*

$$\begin{cases} \partial_t u(t) = \mathcal{A}u(t), \qquad t \in \mathbb{R}_+, & (9.14a) \\ u(0) = u^0, & (9.14b) \end{cases}$$

is given by $u(t) = \mathcal{T}(t)u^0$ if $u^0 \in D(\mathcal{A})$. Comparing this with our introductory example, we see that the semigroup is in fact a generalization of the matrix exponential. Also, the generator of this semigroup generalizes the corresponding matrix. Hence, in the following, we borrow the notation from the matrix exponential and denote a strongly continuous semigroup with generator $\mathcal{A}$ by $(\mathrm{e}^{t\mathcal{A}})_{t\geq 0}$ instead of $(\mathcal{T}(t))_{t\geq 0}$.

The next result yields existence and uniqueness of the solution of (9.14) for suitable initial values u^0.

Theorem 9.7 *Let $\mathcal{A}$ be the infinitesimal generator of the strongly continuous semigroup $\left(\mathrm{e}^{t\mathcal{A}}\right)_{t\geq 0}$ and let $u^0 \in D(\mathcal{A})$. Then there exists a unique solution $u \in C^1(\mathbb{R}_+; X) \cap C(\mathbb{R}_+; D(\mathcal{A}))$ of the homogeneous abstract Cauchy problem* (9.14). *It is given by*

$$u(t) = \mathrm{e}^{t\mathcal{A}} u^0. \tag{9.15}$$

Owing to the growth bound (9.12) for the semigroup $(\mathrm{e}^{t\mathcal{A}})_{t\geq 0}$, we also obtain stability of the solution, both in the L^2-norm as well as in a stronger norm involving the infinitesimal generator.

Theorem 9.8 *Let the assumptions of Theorem 9.7 be satisfied. Then, we have the following stability bounds for the solution u of* (9.14). *For all $t \in \mathbb{R}_+$ we have*

$$\|u(t)\|_X \leq C_{\text{sg}}\, \mathrm{e}^{\omega t}\, \|u^0\|_X, \qquad \text{and} \qquad \|\mathcal{A}u(t)\|_X \leq C_{\text{sg}}\, \mathrm{e}^{\omega t}\, \|\mathcal{A}u^0\|_X,$$

where $\omega \in \mathbb{R}$ and $C_{\text{sg}} > 1$ are the constants given by Lemma 9.3 (i).

The proof is an immediate consequence of the bound (9.12) and the fact that a semigroup and its generator commute, i.e., (9.13b). Note that since u is the solution of the Cauchy problem (9.14), the second stability bound also directly yields the same bound on $\|\partial_t u(t)\|_X$.

Similar to Example 9.1, this can also be transferred to the inhomogeneous case, where the simple solution formula (9.15) is replaced by a variation-of-constants formula. In particular, we consider the *inhomogeneous abstract Cauchy problem*

$$\begin{cases} \partial_t u(t) = \mathcal{A}u(t) + f(t), \qquad t \in \mathbb{R}_+, & (9.16a) \\ u(0) = u^0. & (9.16b) \end{cases}$$

Then, for suitable initial value u^0 and inhomogeneity f, we obtain the following result guaranteeing existence and uniqueness.

Theorem 9.9 *Let $\mathcal{A}$ be the infinitesimal generator of the strongly continuous semigroup $\left(\mathrm{e}^{t\mathcal{A}}\right)_{t\geq 0}$ and $u^0 \in D(\mathcal{A})$. Moreover, let $f \in C^1(\mathbb{R}_+; X) + C(\mathbb{R}_+; D(\mathcal{A}))$. Then there exists a unique solution $u \in C^1(\mathbb{R}_+; X) \cap C(\mathbb{R}_+; D(\mathcal{A}))$ of the inhomogeneous abstract Cauchy problem* (9.16) *given by*

$$u(t) = \mathrm{e}^{t\mathcal{A}} u^0 + \int_0^t \mathrm{e}^{(t-s)\mathcal{A}} f(s)\, \mathrm{d}s. \tag{9.17}$$

As in the homogeneous case, we also obtain stability bounds for the solution of (9.16). However, this time, the bound for the stronger norm is a bit more involved due to the convolution with the inhomogeneity f. Also, since the problem (9.16) does not yield equality of $\partial_t u$ and $\mathcal{A}u$ in this case, we give a bound on both terms.

Since the techniques employed to prove this result will be useful to illustrate concepts applied in the analysis of the discrete methods, we provide a proof here. It is taken from [111, Lem 2.4], where contractive semigroups (cf., Definition 9.13 below) were considered. We adapt it slightly to match the more general semigroups considered here and generalize it to inhomogeneities that (continuously) map into the domain of the operator $\mathcal{A}$.

Theorem 9.10 *Let the assumptions of Theorem 9.9 be satisfied. Then, the following stability bounds for the solution u of (9.16) hold. For all $t \in \mathbb{R}_+$ we have*

$$\|u(t)\|_X \leq C_{\mathrm{sg}}\, \mathrm{e}^{\omega t}\, \|u^0\|_X + C_{\mathrm{sg}} \int_0^t \mathrm{e}^{\omega(t-s)}\, \|f(s)\|_X \,\mathrm{d}s. \tag{9.18}$$

Further, writing $f = f_1 + f_2$ with $f_1 \in C^1(\mathbb{R}_+; X)$ and $f_2 \in C(\mathbb{R}_+; D(\mathcal{A}))$, we have

$$\begin{aligned}\|\partial_t u(t)\|_X \leq\ & C_{\mathrm{sg}}\, \mathrm{e}^{\omega t}\, \big(\|\mathcal{A}u^0\|_X + \|f_1(0)\|_X\big) + \|f_2(t)\|_X \\ & + C_{\mathrm{sg}} \int_0^t \mathrm{e}^{\omega(t-s)}\, \big(\|\partial_t f_1(s)\|_X + \|\mathcal{A} f_2(s)\|_X\big)\,\mathrm{d}s,\end{aligned} \tag{9.19}$$

and

$$\|\mathcal{A}u(t)\|_X \leq \|\partial_t u(t)\|_X + \|f(t)\|_X. \tag{9.20}$$

Here, $\omega \in \mathbb{R}$ and $C_{\mathrm{sg}} > 1$ are the constants given by Lemma 9.3 (i).

Proof The stability bounds (9.18) and (9.20) are immediate consequences of the variation-of-constants formula (9.17) and the differential equation (9.16a), respectively. Hence, it only remains to show (9.19).

To do so, we derive a representation of $\partial_t u$ by differentiating the variation-of-constants formula (9.17). Since we will treat the different parts f_1 and f_2 of the inhomogeneity differently, we first split the convolution in (9.17) and transform the integral containing f_1 so that the t-dependence lies in f_1. This yields

$$u(t) = \mathrm{e}^{t\mathcal{A}} u^0 + \int_0^t \mathrm{e}^{s\mathcal{A}}\, f_1(t-s)\,\mathrm{d}s + \int_0^t \mathrm{e}^{(t-s)\mathcal{A}}\, f_2(s)\,\mathrm{d}s.$$

Using the Leibniz integration rule to integrate the convolutions (and differentiating the semigroup via (9.13b)), we then obtain

$$\partial_t u(t) = \mathrm{e}^{t\mathcal{A}}\big(\mathcal{A}u^0 + f_1(0)\big) + f_2(t) + \int_0^t \mathrm{e}^{(t-s)\mathcal{A}}\big(\partial_t f_1(s) + \mathcal{A} f_2(s)\big)\,\mathrm{d}s,$$

where we have already transformed the integral containing f_1 back. From this, the claim follows by taking the norm and using the growth bound (9.12) on the semigroup. □

Remark 9.11 Solutions of (9.14) or (9.16), which are continuous with values in the domain $D(\mathcal{A})$, are called *strong solutions*. We point out that other solution concepts exist, which need less strict assumptions on the data, like *weak* or *mild solutions*. See e.g., the classical literature referenced in the beginning of this section. Here, we focus on strong solutions. ◇

Next, we investigate sufficient conditions under which a Hilbert space operator $\mathcal{A}$ is in fact the generator of a strongly continuous semigroup. In particular, the condition we will be working with is the following.

Definition 9.12 A linear operator $\mathcal{A}\colon D(\mathcal{A}) \to X$ is called **dissipative** if for every $x \in D(\mathcal{A})$ we have

$$\operatorname{Re}\big(\mathcal{A}x \mid x\big)_X \le 0. \tag{9.21}$$

If, in addition, we have $\operatorname{ran}(I - \lambda\mathcal{A}) = X$ for some $\lambda > 0$, the operator $\mathcal{A}$ is called **maximal dissipative**.

Maximal dissipative operators are strongly connected to the following class of semigroups.

Definition 9.13 A strongly continuous semigroup is called **contractive** or a **contraction semigroup** if (9.12) holds with $C_{\mathrm{sg}} = 1$ and $\omega = 0$.

The next result, the famous Lumer–Phillips theorem, details the aforementioned connection and states that maximal dissipative operators generate contraction semigroups and that all generators of contraction semigroups are maximal dissipative. It can be found, among others, in [101, Theorem 6.1.7] or [57, Theorem II.3.15 & Corollary II.3.20].

Theorem 9.14 (Lumer-Phillips) *Let $\mathcal{A}\colon D(\mathcal{A}) \to X$ be a linear operator. Then the following statements are equivalent.*

(i) *$\mathcal{A}$ is maximal dissipative.*
(ii) *$\mathcal{A}$ generates a contraction semigroup.*

Lastly, we state some important properties of dissipative operators that will be of use in the analysis of the considered numerical methods. They are proven in [57, Prop. II.3.14] and [141, Lem. 1.1.1], respectively.

Lemma 9.15 *Let $\mathcal{A}\colon D(\mathcal{A}) \to X$ be dissipative. Then the following holds for all $\lambda > 0$.*

(i) *The operator* $\mathcal{I} - \lambda\mathcal{A}$ *is injective, and for* $x \in \operatorname{ran}(\mathcal{I} - \lambda\mathcal{A})$ *we have*

$$\|(\mathcal{I} - \lambda\mathcal{A})^{-1}x\|_X \leq \|x\|_X.$$

(ii) *For* $x \in \operatorname{ran}(\mathcal{I} - \lambda\mathcal{A})$ *we have*

$$\|(\mathcal{I} + \lambda\mathcal{A})(\mathcal{I} - \lambda\mathcal{A})^{-1}x\|_X \leq \|x\|_X.$$

Lemma 9.16 *If* $\mathcal{A}\colon D(\mathcal{A}) \to X$ *is maximal dissipative, we have* $\operatorname{ran}(\mathcal{I} - \lambda\mathcal{A}) = X$ *for all* $\lambda > 0$.

Remark 9.17 We point out that the semigroups generated by operators of the form (9.2) considered in this work are in fact not only semigroups but groups. This means that the parameters t and s used in Definition 9.2 may also be negative. In practice, this corresponds to a reversal of time. This sets the problems considered here apart from, e.g., parabolic problems, in which such a time-reversal is not possible and the corresponding operators generate semigroups, not groups. ◇

9.2 Analytical Setting and Friedrichs' Operators

As we have seen in the last section, the Lumer–Phillips theorem provides sufficient conditions for wellposedness of problems like the wave-type problem (9.3). In particular, if the Friedrichs' operator $\mathcal{L}$ is a maximal dissipative Hilbert space operator and the initial data as well as the inhomogeneity are appropriate, we can apply the wellposedness results Theorem 9.9 and Theorem 9.10.

In this section, we present the analytical setting in which these conditions are fulfilled and the tools necessary to show this. The approach employed here is based on the work in [59] and [43, Chapter 7], where Friedrichs' systems [72] were analyzed in a weak setting suitable for our applications.

Once more we start with a small motivational example to demonstrate the concepts we employ in the general setting.

Example 9.18 We consider the *linear homogeneous advection problem* in $d = 2$ dimensions given by

$$\begin{cases} \partial_t u = \alpha \cdot \nabla u, & \text{on } \mathbb{R}_+ \times \Omega, \\ u(0) = u^0, & \text{on } \Omega. \end{cases} \tag{9.22}$$

Here, we seek the *solution* $u\colon \mathbb{R}_+ \times \Omega \to \mathbb{R}$ for given (constant) *advection velocity* $\alpha \in \mathbb{R}^2$ and *initial value* $u^0\colon \Omega \to \mathbb{R}$. First, observe that the advection equation is of the form (9.3) as we can write the spatial operator as

$$\alpha \cdot \nabla u = \alpha_1 \partial_1 u_1 + \alpha_2 \partial_2 u_2.$$

Moreover, (9.22) fits the structure of the wave-type problem (9.1a) with $M = I$ and thus we have $(\cdot \mid \cdot)_\Omega = (\cdot \mid \cdot)_{L^2,\Omega}$. Let us now investigate further properties of this problem or, more precisely, the spatial operator $\alpha \cdot \nabla$. For this, let $v, w\colon \Omega \to \mathbb{R}$ be sufficiently regular such that all expressions considered in the following make sense.

Firstly, denoting the outer unit normal vector to Ω as $\mathfrak{n}^\Omega$, we use the usual integration-by-parts formula to see

$$\big(\alpha \cdot \nabla v \mid w\big)_\Omega = -\big(v \mid \alpha \cdot \nabla w\big)_\Omega + \big((\alpha \cdot \mathfrak{n}^\Omega) v \mid w\big)_\Gamma. \tag{9.23}$$

In order to connect this with our general setting presented later, we introduce the following concepts and notation. We denote the spatial operator by

$$\mathcal{L} = \alpha \cdot \nabla,$$

and define its *formal adjoint* $\mathcal{L}^\circledast$ and the *boundary operator corresponding to* $\mathcal{L}$ by

$$\mathcal{L}^\circledast v = -\alpha \cdot \nabla v, \qquad \text{and} \qquad \mathcal{L}_\partial v = (\alpha \cdot \mathfrak{n}^\Omega) v,$$

respectively. Using this, we rewrite the integration-by-parts formula (9.23) as

$$\big(\mathcal{L} v \mid w\big)_\Omega = \big(v \mid \mathcal{L}^\circledast w\big)_\Omega + \big(\mathcal{L}_\partial v \mid w\big)_\Gamma. \tag{9.24}$$

To make the advection problem (9.22) wellposed, we need to prescribe suitable boundary conditions. One viable choice is to impose homogeneous *inflow boundary conditions.* They are given by

$$u = 0 \qquad \text{on } \mathbb{R}_+ \times \Gamma^+, \tag{9.25}$$

where the *inflow boundary* Γ^+ is defined by $\Gamma^+ = \{x \in \Gamma \mid \mathcal{L}_\partial(x) > 0\}$.

For our abstract setting we need to rewrite (9.25) as an equation posed on the whole boundary. This can be achieved by defining the boundary operator

$$\mathcal{L}_\Gamma v = -|\alpha \cdot \mathfrak{n}^\Omega| v,$$

and consequently reformulate (9.25) as

$$(\mathcal{L}_\partial - \mathcal{L}_\Gamma)u = 0 \qquad \text{on } \mathbb{R}_+ \times \Gamma. \tag{9.26}$$

As we see later, this is how we model boundary conditions in the general setting and it will be incorporated into the domain of the Hilbert space operator. ◇

After we have motivated some of the more abstract concepts with a concrete example, we now return to general Friedrichs' operators and systems like (9.3). We start by defining the *graph space* of a Friedrichs' operator $\mathcal{L}$ as

$$H(\mathcal{L}) = \{v \in L^2(\Omega)^m \mid \mathcal{L}v \in L^2(\Omega)^m\}. \tag{9.27}$$

Equipped with the (weighted) *graph norm*

$$\|\cdot\|_{\mathcal{L}} = \|\cdot\|_\Omega + \|\mathcal{L}\cdot\|_\Omega,$$

the graph space is a Hilbert space, cf., [43, Lem 7.2], and we immediately see that we have $\mathcal{L} \in \mathcal{B}(H(\mathcal{L}),\, L^2(\Omega)^m)$. We point out that we chose this notation for the graph space as a reference to the well-known spaces $H(\mathrm{div})$ and $H(\mathrm{curl})$, which are the graph spaces of the divergence and the curl, respectively.

Next, we want to introduce boundary conditions into this abstract setting in order to identify a subset of the graph space on which the Friedrichs' operator $\mathcal{L}$ is maximal dissipative (in accordance with semigroup theory, this subset will be called the domain of $\mathcal{L}$). In this weak setting, however, the traces of functions are not necessarily square-integrable, keeping us from setting boundary conditions in a standard way. Nevertheless, we have already seen in (9.24) that we can identify a boundary operator associated with a Friedrichs' operator $\mathcal{L}$ with the help of an integration-by-parts formula involving the formal adjoint of $\mathcal{L}$. This leads to the following definition.

Definition 9.19 We call $\mathcal{L}^{\circledast} \in \mathcal{B}(H(\mathcal{L}),\, L^2(\Omega)^m)$ defined via

$$M\mathcal{L}^{\circledast}u = -\sum_{i=1}^{d} L_i\partial_i u + L_0^{\mathsf{T}}u \tag{9.28}$$

the **formal adjoint** *of* $\mathcal{L}$.

Consequently, we now define the aforementioned boundary operator.

Definition 9.20 We call $\mathcal{L}_\partial \colon H(\mathcal{L}) \to H(\mathcal{L})'$ defined by

$$\big\langle \mathcal{L}_\partial\, u \,\big|\, v \big\rangle = \big(\mathcal{L} u \,\big|\, v\big)_\Omega - \big(u \,\big|\, \mathcal{L}^{\circledast} v\big)_\Omega \qquad \text{for all } u, v \in H(\mathcal{L}) \tag{9.29}$$

the **boundary operator associated with $\mathcal{L}$.**

By [59, Lem 2.2], the boundary operator $\mathcal{L}_\partial$ is both symmetric and bounded, i.e., $\mathcal{L}_\partial \in \mathcal{B}(H(\mathcal{L}), H(\mathcal{L})')$. Further, comparing (9.29) to (9.24) we realize that Definition 9.20 provides a generalization of the integration-by-parts formula to this weak setting.

We now identify and formulate suitable boundary conditions that will define the domain of the Friedrichs' operator. The way we do this is taken from [59, Sec. 2.1], which means that we make the following assumption.

Assumption 9.21 There exists a bounded operator $\mathcal{L}_\Gamma \in \mathcal{B}(H(\mathcal{L}), H(\mathcal{L})')$ with

$$\big\langle \mathcal{L}_\Gamma v \,\big|\, v \big\rangle \le 0 \qquad \text{for all } v \in H(\mathcal{L}), \tag{9.30a}$$

$$H(\mathcal{L}) = \ker(\mathcal{L}_\partial - \mathcal{L}_\Gamma) + \ker(\mathcal{L}_\partial + \mathcal{L}_\Gamma). \tag{9.30b}$$

◇

Note that both $\ker(\mathcal{L}_\partial - \mathcal{L}_\Gamma)$ as well as $\ker(\mathcal{L}_\partial + \mathcal{L}_\Gamma)$ are Hilbert spaces if endowed with the graph norm, since they are the kernels of bounded operators. With this, we define the *domain of* $\mathcal{L}$ as

$$D(\mathcal{L}) := \ker(\mathcal{L}_\partial - \mathcal{L}_\Gamma), \tag{9.31}$$

which incorporates a boundary condition of the form (9.26) in a weak sense. The next theorem shows that the properties of the boundary operator $\mathcal{L}_\Gamma$ are chosen in such a way that the Friedrichs' operator $\mathcal{L}$ is maximal dissipative on its domain $D(\mathcal{L})$.

Theorem 9.22 *The restriction of $\mathcal{L}$ to $D(\mathcal{L})$ is maximal dissipative.*

Proof For $v \in D(\mathcal{L})$, by the definition of the formal adjoint (9.28) and the boundary operator (9.29), we have

$$\begin{aligned} 2\big(\mathcal{L} v \,\big|\, v\big)_\Omega &= \big(\mathcal{L} v \,\big|\, v\big)_\Omega + \big(\mathcal{L}^{\circledast} v \,\big|\, v\big)_\Omega + \big(\mathcal{L} v \,\big|\, v\big)_\Omega - \big(\mathcal{L}^{\circledast} v \,\big|\, v\big)_\Omega \\ &= \big((L_0 + L_0^{\mathsf{T}}) v \,\big|\, v\big)_\Omega + \big\langle \mathcal{L}_\partial v \,\big|\, v \big\rangle. \end{aligned} \tag{9.32}$$

We now use $v \in D(\mathcal{L}) = \ker(\mathcal{L}_\partial - \mathcal{L}_\Gamma)$ (cf., (9.31)) to see that

$$\big\langle \mathcal{L}_\partial v \,\big|\, v \big\rangle = \big\langle (\mathcal{L}_\partial - \mathcal{L}_\Gamma) v \,\big|\, v \big\rangle + \big\langle \mathcal{L}_\Gamma v \,\big|\, v \big\rangle = \big\langle \mathcal{L}_\Gamma v \,\big|\, v \big\rangle$$

and thus, by the negative definiteness of L_0 and (9.30a), from (9.32) we obtain

$$\left(\mathcal{L}v \mid v\right)_\Omega = \tfrac{1}{2}\left(\left((L_0 + L_0^\top)v \mid v\right)_\Omega + \left\langle \mathcal{L}_\Gamma v \mid v\right\rangle\right) \le 0. \tag{9.33}$$

Hence, $\mathcal{L}$ is dissipative on $D(\mathcal{L})$. Maximality (i.e., the range condition of Definition 9.12) follows from [59, Thm 2.5] (note that the we use a different sign convention than the one in [59]). □

By Theorem 9.22, the restricted Friedrichs' operator $\mathcal{L}_{|D(\mathcal{L})}$ fulfills the conditions of the Lumer–Phillips Theorem 9.14 and thus generates a contraction semigroup, which we denote by $\left(e^{t\mathcal{L}}\right)_{t\ge 0}$. Therefore, provided $u^0 \in D(\mathcal{L})$ and that the inhomogeneity f is sufficiently smooth, we can deduce wellposedness (on $D(\mathcal{L})$) of the wave-type problem (9.3) immediately via Theorems 9.9 and 9.10. We still explicitly state the result because of its importance and since the growth bounds of the semigroup vanish due to the dissipative setting, which considerably simplifies the bounds.

Corollary 9.23 *Let the initial value u^0 and the inhomogeneity $f = f_1 + f_2$ be given such that $u^0 \in D(\mathcal{L})$, $f_1 \in C^1(\mathbb{R}_+; L^2(\Omega)^m)$, and $f_2 \in C(\mathbb{R}_+; D(\mathcal{L}))$. Then there exists a unique solution $u \in C^1(\mathbb{R}_+; L^2(\Omega)^m) \cap C(\mathbb{R}_+; D(\mathcal{L}))$ of (9.3) given by the variation-of-constants formula*

$$u(t) = e^{t\mathcal{L}} u^0 + \int_0^t e^{(t-s)\mathcal{L}} f(s)\, ds. \tag{9.34}$$

Further, we have the following stability bounds.

(a) The solution satisfies

$$\|u(t)\|_\Omega \le \|u^0\|_\Omega + \int_0^t \|f(s)\|_\Omega\, ds.$$

(b) The temporal derivative $\partial_t u$ satisfies

$$\|\partial_t u(t)\|_\Omega \le \|\mathcal{L}u^0\|_\Omega + \max_{s\in\mathbb{R}_+} \|f(s)\|_\Omega + \int_0^t \|\partial_t f_1(s)\|_\Omega + \|\mathcal{L} f_2(s)\|_\Omega\, ds.$$

(c) The spatial derivative $\mathcal{L}u$ satisfies

$$\|\mathcal{L}u(t)\|_\Omega \le \|\partial_t u(t)\|_\Omega + \|f(t)\|_\Omega.$$

Remark 9.24 If $\mathcal{L}$ is not only dissipative but satisfies the stronger bound

$$\big(\mathcal{L}v \,|\, v\big)_\Omega \le \omega\big(v \,|\, v\big)_\Omega \qquad \text{for all } v \in D(\mathcal{L})$$

for some $\omega < 0$ instead of (9.21), then we could slightly improve the above bounds (and all which follow, including those for the fully discrete methods). An example for this situation is a negative definite coefficient L_0, cf., (9.33).

By the same reasoning we used to show (9.11), we then have

$$\| \mathrm{e}^{t\mathcal{L}} \|_\Omega \le \mathrm{e}^{\omega t} \qquad \text{for all } t \ge 0.$$

Hence, the solution will even exhibit exponential damping over time. A similar behavior would then be encountered in the discrete settings from the following sections. However, to avoid unnecessarily cluttered bounds, we leave these more stringent results to the interested reader. ◇

An important special case in practice is the one, where the restriction of the operator $\mathcal{L}$ to its domain $D(\mathcal{L})$ becomes skew-adjoint. In particular, this leads to energy-preserving systems in the absence of the inhomogeneity f.

Corollary 9.25 *Let the boundary operator $\mathcal{L}_\Gamma$ fulfill*

$$\big\langle \mathcal{L}_\Gamma v \,|\, v \big\rangle = 0 \qquad \textit{for all } v \in H(\mathcal{L}),$$

and let L_0 be skew-symmetric, i.e., $L_0^\mathsf{T} = -L_0$. Then, the restriction of $\mathcal{L}$ to $D(\mathcal{L})$ is skew-adjoint.

If, additionally, the inhomogeneity f vanishes, the wave-type problem (9.3) *becomes norm-preserving, i.e., its solution u satisfies*

$$\|u(t)\|_\Omega = \|u^0\|_\Omega.$$

Proof By the assumptions on the boundary operator $\mathcal{L}_\Gamma$ and L_0, we see that (9.33) becomes an equality, implying the skew-symmetry of $\mathcal{L}_{|D(\mathcal{L})}$. The range condition is a consequence of [59, Thm. 2.5] and yields skew-adjointness. The remaining claim then follows by Stone's theorem [57, Thm. II.3.24], which states that skew-adjoint operators generate unitary semigroups. □

Note that Corollary 9.25 implies that (at least in the constant coefficient setting considered here) all damping exhibited by the homogeneous system is caused by the boundary condition and the coefficient L_0 (cf., also Remark 9.24). We further point out that the assumptions of Corollary 9.25 are fulfilled for the acoustic wave equation in

Example 9.3.2 below. Further, if the conductivity vanishes, they are also fulfilled for the (then undamped) Maxwell equations in Example 9.3.3.

Remark 9.26 We mention that the techniques presented in this work are not restricted to purely dissipative operators. In fact, they also hold for shift-dissipative operators, i.e., operators $\mathcal{L}$ for which $\mathcal{L} - \mu\mathcal{I}$ is dissipative for some positive constant $\mu > 0$. In this constant coefficient setting, this is the case if the coefficient L_0 fails to be negative semidefinite. Then, the corresponding semigroup is not contractive anymore, but fulfills (9.12) with $C_{\text{sg}} = 1$ and $\omega = \mu$. To avoid the technicalities associated with this generalization, we restrict ourselves to the purely dissipative case.

Lastly, note that if we restrict the formal adjoint $\mathcal{L}^{\circledast}$ of $\mathcal{L}$ to the Hilbert space $\ker(\mathcal{L}_\partial + \mathcal{L}_\Gamma)$, it is dissipative (or skew-adjoint or shift-dissipative in the settings of Corollary 9.25 or the last paragraph, respectively) as well. This can be easily seen by using the same techniques as for $\mathcal{L}$ itself. We refer to [59–61, 64] for more insight on the intrinsic structures associated with Friedrichs' systems. ◇

9.3 Examples

This section serves to present some concrete linear wave-type problems that fit the general framework considered in this work. We give three examples; the advection equation, which we already saw in Example 9.18, as well as the acoustic wave and Maxwell equations. We also point out that the elastic and other wave equations [27, 92] and equations governed by the Hodge–Dirac operator [154] are candidates for our framework if supplied with appropriate boundary conditions.

For the sake of presentation, we omit some technical details necessary to show that the presented examples do in fact fit the analytical framework (this mainly concerns the boundary conditions). Instead, we refer the interested reader to [113, Sec. 2.5], where results from [59] were used to work out these details.

9.3.1 Advection Equation

As a first example, we consider the d-dimensional *advection equation*. We have already seen in Example 9.18 that the two-dimensional problem exhibits the right structure. Now, we show that this is also the case for other space dimensions.

The d-dimensional advection problem is posed as

$$\begin{cases} \partial_t u = \alpha \cdot \nabla u + f & \text{on } \mathbb{R}_+ \times \Omega, \\ u(0) = u^0 & \text{on } \Omega, \end{cases} \tag{9.35}$$

where we seek the solution $u\colon \mathbb{R}_+ \times \Omega \to \mathbb{R}$ for given *advection velocity* $\alpha \in \mathbb{R}^d$ and inhomogeneity $f \in L^2(\Omega)$.

Since there is no material parameter involved in the advection problem (9.35) (i.e., $M = 1$), we show that the problem fits the wave-type problem (9.3) that is already freed of the material tensor. To do so, it suffices to show that the spatial operator $\alpha \cdot \nabla$ in (9.35) has the form (9.2). Using

$$\alpha \cdot \nabla = \sum_{i=1}^{d} \alpha_i \partial_i,$$

we readily see that this is indeed the case with coefficients given by

$$L_0 = 0, \qquad L_i = \alpha_i, \qquad i = 1, \dots, d.$$

An example for possible boundary conditions for problem (9.35) that fit our theory are (homogeneous) *inflow boundary conditions*. We refer to [113, Sec. 2.5.1], where these boundary conditions are given explicitly and where it was shown that, under the right conditions, they result in the applicability of the theory we presented here. Further, we point out that non-constant advection velocity can also be handled with some tweaks to the theory, cf., [113].

9.3.2 Acoustic Wave Equation

Next, we consider the linear *acoustic wave equation* in a d-dimensional space. Often, the acoustic wave equation is considered in a second-order (in time and space) formulation. However, since we analyze first-order problems it has to be rewritten to fit our theory. The derivation of this first-order formulation can be found in [92, Section 2.2].

Doing so leads to the acoustic wave problem in div-grad formulation stated as follows. We seek the *pressure* $p\colon \mathbb{R}_+ \times \Omega \to \mathbb{R}$ and the *velocity* $q\colon \mathbb{R}_+ \times \Omega \to \mathbb{R}^d$ determined by

$$\begin{cases} \rho \partial_t p = \nabla \cdot q + g & \text{on } \mathbb{R}_+ \times \Omega, \\ \partial_t q = \nabla p & \text{on } \mathbb{R}_+ \times \Omega, \\ p(0) = p^0, \quad q(0) = q^0 & \text{on } \Omega, \end{cases} \tag{9.36}$$

where $\rho \in L^\infty(\Omega)$ is the given, uniformly positive *density* and $g : \mathbb{R}_+ \times \Omega \to \mathbb{R}$.

To show that (9.36) indeed fits our theory, we write

$$u = \begin{pmatrix} p \\ q \end{pmatrix}, \qquad u^0 = \begin{pmatrix} p^0 \\ q^0 \end{pmatrix}, \qquad \tilde{f} = \begin{pmatrix} g \\ 0 \end{pmatrix}$$

and

$$\widetilde{\mathcal{L}} = \begin{pmatrix} 0 & \nabla\cdot \\ \nabla & 0 \end{pmatrix}, \qquad M = \begin{pmatrix} \rho & 0 \\ 0 & I \end{pmatrix}.$$

Using this in (9.36) immediately shows that the acoustic wave problem is of the form (9.1). Thus, by the properties of the density ρ, it can also be transformed into (9.3). Further, the spatial operator has the right structure, namely it fits the prototype (9.2) with coefficients given as

$$L_0 = 0, \qquad L_i = \begin{pmatrix} 0 & e_i^T \\ e_i & 0 \end{pmatrix}, \qquad i = 1, \dots, d.$$

Boundary conditions that are covered by our theory are, e.g., (homogeneous) *Dirichlet*, *Neumann* or *Robin boundary conditions*. We refer to [113, Sec. 2.5.2] for details on the Dirichlet case and [43, Section 7.1.2.2 & 7.2.5.2] for the necessary results to show the other two (as well as the Dirichlet case).

9.3.3 Maxwell Equations

As a last example, we consider the fundamental equations of electromagnetism, the *Maxwell equations*. In particular, we consider the linear version including external currents and damping.

The Maxwell problem is stated in a three-dimensional space as follows. We seek the unknown *electric* and *magnetic fields* $E\colon \mathbb{R}_+ \times \Omega \to \mathbb{R}^3$ and $H\colon \mathbb{R}_+ \times \Omega \to \mathbb{R}^3$, respectively, given by

$$\begin{cases} \varepsilon\partial_t E = \nabla\times H - \sigma E - J & \text{on } \mathbb{R}_+ \times \Omega, \\ \mu\partial_t H = -\nabla\times E & \text{on } \mathbb{R}_+ \times \Omega, \\ E(0) = E^0, \quad H(0) = H^0 & \text{on } \Omega. \end{cases} \tag{9.37}$$

Here, $J\colon \mathbb{R}_+ \times \Omega \to \mathbb{R}^3$ is the *current density*, $\sigma \in L^\infty(\Omega)^{3\times3}$ is the *conductivity*, and $\varepsilon, \mu \in L^\infty(\Omega)^{3\times3}$ are the *permittivity* and the *permeability*, respectively. In the following, we always assume that ε and μ are uniformly positive definite and that σ is uniformly positive semidefinite.

To bring the Maxwell problem (9.37) into the form of our wave-type problem (9.1), we define

$$u = \begin{pmatrix} E \\ H \end{pmatrix}, \qquad u^0 = \begin{pmatrix} E^0 \\ H^0 \end{pmatrix}, \qquad \tilde{f} = \begin{pmatrix} -J \\ 0 \end{pmatrix}$$

and

$$\widetilde{\mathcal{L}} = \begin{pmatrix} 0 & \nabla\times \\ -\nabla\times & 0 \end{pmatrix} - \begin{pmatrix} \sigma & 0 \\ 0 & 0 \end{pmatrix}, \qquad M = \begin{pmatrix} \varepsilon & 0 \\ 0 & \mu \end{pmatrix}. \tag{9.38}$$

Again, by the assumptions on the material parameter, we can also transform (9.37) into the material parameter-free system (9.3).

Moreover, the Maxwell operator $\widetilde{\mathcal{L}}$ defined in (9.38) also fits the structure of our Friedrichs' operator from (9.2), which can be seen by choosing the coefficients as

$$L_0 = \begin{pmatrix} -\sigma & 0 \\ 0 & 0 \end{pmatrix}, \qquad L_i = \begin{pmatrix} 0 & \ell_i^T \\ \ell_i & 0 \end{pmatrix}, \qquad i = 1, 2, 3,$$

with $\ell_1, \ell_2, \ell_3 \in \mathbb{R}^{3\times 3}$, $\ell_1 = e_2 e_3^T - e_3 e_2^T$, $\ell_2 = e_3 e_1^T - e_1 e_3^T$ and $\ell_3 = e_1 e_2^T - e_2 e_1^T$.

To complete the problem, we supply it with *perfectly conducting boundary conditions*, which are given as $\mathfrak{n}^\Omega \times E = 0$ on Γ if the solution is sufficiently regular to allow such traces. Once more, we refer to [113, Sec. 2.5.3] for proving that they are indeed contained in our abstract framework.

Spatial Discretization 10

In this chapter, we present the spatial discretization of the wave-type problem (9.3) and the corresponding Friedrichs' operator $\mathcal{L}$, respectively, via a central fluxes dG scheme [43, 86]. For the sake of presentation, we omit most of the technical details and proofs and only present the main results, which are necessary for the analysis of the full discretization. We refer the reader to [43, 59, 88, 89, 113] for more details.

Throughout the rest of these lecture notes, we give various results that do not require more regularity of the data than L^2-regularity for the initial value u^0 and the inhomogeneity f being continuous in time and L^2 in space (this mostly concerns wellposedness results of the discrete problems). Although we formulated our wellposedness theory in a strong setting, which requires stronger regularity (cf., the assumptions of Theorems 9.10 and 9.23 and Remark 9.11), in the following, we only assume this weaker regularity in the results whenever possible. Therefore, to avoid being unnecessarily repetitive, we assume that the data satisfy at least

$$u^0 \in L^2(\Omega) \qquad \text{and} \qquad f \in C(\mathbb{R}_+; L^2(\Omega)), \tag{10.1}$$

and only state regularity assumptions if more regularity is necessary for a result to hold.

10.1 The Discrete Setting

We start by setting up the general discrete framework, including the meshes and discrete spaces we investigate during the analysis. Firstly, for simplicity, we assume that the domain Ω is a polyhedron so that we do not need to approximate the domain.

To discretize the problem, we need to endow Ω with a *mesh* $\mathcal{T}$. Then, for each *mesh element* or *cell* $K \in \mathcal{T}$, we denote its diameter by h_K and the maximal and minimal

W. Dörfler et al., *Wave Phenomena*, Oberwolfach Seminars 49,
https://doi.org/10.1007/978-3-031-05793-9_10

diameter of all elements in $\mathcal{T}$ by $\hbar = \max_{K\in\mathcal{T}_\hbar} h_K$ and $\hbar_{\min} = \min_{K\in\mathcal{T}_\hbar} h_K$, respectively. The value $\hbar$ is also referred to as the *mesh size* of $\mathcal{T}$. We further define the piecewise constant function $h \in L^\infty(\Omega)$ by $h_{|K} \equiv h_K$, which is used throughout for a more convenient and concise notation of mesh-dependent norms.

Since we want to investigate the convergence of the dG method w.r.t. mesh refinement, we denote a *mesh with maximal element diameter* $\hbar$ by $\mathcal{T}_\hbar$. Using this, we consider a mesh sequence $\mathcal{T}_\mathcal{H} = (\mathcal{T}_\hbar)_{\hbar\in\mathcal{H}}$, where $\mathcal{H}$ is a countable collection of positive numbers $\hbar < 1$ with 0 as its only accumulation point. To ensure optimal convergence rates with constants which are robust w.r.t. mesh refinement we further assume that the mesh sequence $\mathcal{T}_\mathcal{H}$ is *admissible* [43, Def. 1.57]. This means that $\mathcal{T}_\mathcal{H}$ has *optimal polynomial approximation properties* and is *shape and contact regular* with *regularity parameter* ρ.

We collect the *faces* of a mesh $\mathcal{T}_\hbar$ in the set $\mathcal{F}_\hbar = \mathcal{F}_\hbar^{\text{int}} \cup \mathcal{F}_\hbar^{\text{bnd}}$, where $\mathcal{F}_\hbar^{\text{int}}$ contains *interior faces* or *interfaces* and $\mathcal{F}_\hbar^{\text{bnd}}$ contains the *boundary faces*. Given a mesh element $K \in \mathcal{T}_\hbar$ we further need the *set* $\mathcal{F}_\hbar^K$ *of all faces composing the boundary of* K. We denote the *maximum number of mesh faces per element* as $N_\partial = \max_{K\in\mathcal{T}_\hbar} |\mathcal{F}_\hbar^K|$. We point out that as a consequence of the regularity of our mesh sequence, N_∂ can be bounded independently of the mesh parameter $\hbar \in \mathcal{H}$, see [43, Lemma 1.41].

Next, we introduce some normal vectors, which are used for defining and handling the discrete operators. Firstly, the outward unit normal vector to an element $K \in \mathcal{T}_\hbar$ is denoted by $\mathfrak{n}^K$. For a more convenient notation of the discrete operators, we also need a fixed face normal vector $\mathfrak{n}^F$ to each face $F \in \mathcal{F}_\hbar$, defined as follows. For a boundary face $F \in \mathcal{F}_\hbar^{\text{bnd}}$, the face normal vector $\mathfrak{n}^F$ is simply defined as the outward unit normal vector to Γ. Given an interface $F \in \mathcal{F}_\hbar^{\text{int}}$, we denote the two elements sharing this face arbitrarily by K_1^F and K_2^F and fix this choice. With this, we define $\mathfrak{n}^F$ as the outward unit normal vector to K_1^F.

After having fixed the meshes and the corresponding notation, we now turn to the function spaces used in the dG method. The space we seek our approximations in is the *discrete approximation space* or *broken polynomial space*

$$\mathbb{V}_\hbar = \{\, \boldsymbol{v} \in L^2(\Omega) \mid \boldsymbol{v}_{|K} \in \mathbb{Q}_d^k(K) \text{ for all } K \in \mathcal{T}_\hbar \,\},$$

consisting of L^2-functions, which are polynomials of degree at most k in each variable if restricted to an element in the mesh. Note that for now we only consider the scalar case, but all results and concepts immediately carry over to vector-valued functions contained in $\mathbb{V}_\hbar^m$, which we predominantly use after this section.

At this point we once again mention the notation we use in the discrete setting. Namely, we always denote discrete objects in bold face. In particular, this applies to objects that are contained in or that are mapping into the discrete approximation space $\mathbb{V}_\hbar$. To better distinguish them from their continuous counterparts, we further underline the discrete operators. Because of this, as a slight deviation from the notation introduced in the beginning, we denote the identity on $\mathbb{V}_\hbar$ by $\underline{\boldsymbol{I}}$.

Remark 10.1 For the sake of presentation, we fix the same polynomial degree for all elements $K \in \mathcal{T}_\hbar$. However, we point out that the dG method is flexible enough to allow for different polynomial degrees on different elements of the mesh. Moreover, other polynomials, e.g., those of total degree at most k, may be chosen as the basis for the discrete approximation space (for more details we refer to [43, Sec. 1.2.4.3]).

In fact, other function spaces than polynomial ones could be employed on the elements, as long as the resulting approximation spaces fulfill the properties outlined in the following paragraphs (up until, including, Lemma 10.2). The following results also hold in these more general settings with only minor—if any—adjustments to the employed arguments. ◇

Due to the admissibility of the mesh sequence $\mathcal{T}_\mathcal{H}$, we can infer some important properties of the discrete approximation spaces corresponding to this sequence. Namely, by the shape and contact regularity, the inverse inequality and the discrete trace inequality

$$\|\nabla \boldsymbol{v}\|_K \leq C'_{\text{inv}} \|h^{-1}\boldsymbol{v}\|_K, \quad \text{and} \quad \|\boldsymbol{v}\|_F \leq C_{\text{tr}} \|h^{-1/2}\boldsymbol{v}\|_K, \qquad \boldsymbol{v} \in \mathbb{V}_\hbar \tag{10.2}$$

respectively, hold for all elements $K \in \mathcal{T}_\hbar$ and their corresponding faces $F \in \mathcal{F}_\hbar^K$. These results are proven in [43, Lem. 1.44 & 1.46] for the standard L^2-norm instead of the weighted one we employ here, which was defined in (9.4). However, by equivalence of norms, it is easy to see that they also hold in this setting.

From the inverse inequality we can further infer a similar result for the Friedrichs' operator $\mathcal{L}$ instead of the gradient. More precisely, we have

$$\|\mathcal{L}\boldsymbol{v}\|_K \leq C_\mathcal{L} C_{\text{inv}} \|h^{-1}\boldsymbol{v}\|_K, \qquad \boldsymbol{v} \in \mathbb{V}_\hbar \tag{10.3}$$

for all $K \in \mathcal{T}_\hbar$. The constants are given as $C_\mathcal{L} = \max_{i=0,\dots,d} \|M^{-1/2} L_i M^{-1/2}\|_{\infty,K}$ and $C_{\text{inv}} = d^{1/2} C'_{\text{inv}} + 1$.

Next, we investigate the quality of the approximations we can expect from functions in the discrete approximation space $\mathbb{V}_\hbar$. To this end, we define the (weighted) L^2-orthogonal projection $\pi_\hbar \colon L^2(\Omega) \to \mathbb{V}_\hbar$ w.r.t. the weighted inner product $\left(\cdot \mid \cdot \right)_\Omega$ such that for $v \in L^2(\Omega)$ we have

$$\left(v - \pi_\hbar v \mid \boldsymbol{\varphi} \right)_\Omega = 0 \qquad \text{for all } \boldsymbol{\varphi} \in \mathbb{V}_\hbar. \tag{10.4}$$

Using this, we define the *projection error*

$$e_\pi^v = v - \pi_\hbar v. \tag{10.5}$$

We point out that, due to the piecewise structure of the discrete approximation space $\mathbb{V}_\hbar$, this projection amounts to independent projections on each of the elements of the mesh.

Hence, with a slight abuse of notation, we use the same symbol for the (weighted) L^2-orthogonal projection of a function in $L^2(K)$ onto the polynomial space $\mathbb{Q}_d^k(K)$ of a mesh element $K \in \mathcal{T}_\hbar$.

The next result gives a bound on the projection error for sufficiently regular functions. Since the usual results in the literature (see, e.g., [43, Lem. 1.58 & 1.59]) are given for the L^2-projection without weights, we need to adapt them slightly to the (weighted) setting used here. We show the result for one element. However, due to the explanation in the preceding paragraph, this immediately gives the result for the whole mesh.

Lemma 10.2 *Let $\hbar \in \mathcal{H}$, $K \in \mathcal{T}_\hbar$, and $F \in \mathcal{F}_\hbar^K$. Then, for all $v \in H^{\ell+1}(K)$, $0 \le \ell \le k$, the projection error e_π^v satisfies*

$$\|e_\pi^v\|_K \le C_\pi |h^{\ell+1} v|_{\ell+1,K}, \qquad \|e_\pi^v\|_F \le C_{\pi,\partial} |h^{\ell+1/2} v|_{\ell+1,K},$$

where C_π and $C_{\pi,\partial}$ are independent of K and $\hbar$.

Proof We show the bound on the element K. The one on the face can be shown analogously.

Using the definition of the projection error and the triangle inequality, we obtain

$$\|e_\pi^v\|_K = \|v - \pi_\hbar v\|_K \le \|v - \pi_{L^2} v\|_K + \|\pi_\hbar v - \pi_{L^2} v\|_K,$$

where π_{L^2} is the L^2-projection onto $\mathbb{Q}_d^k(K)$ w.r.t. the standard L^2-inner product on K. For the second term, we use that $\pi_{L^2} v - \pi_\hbar v \in \mathbb{Q}_d^k(K)$ and hence

$$\begin{aligned}
\|\pi_\hbar v - \pi_{L^2} v\|_K &= \sup_{\substack{\|\varphi\|_K = 1 \\ \varphi \in \mathbb{Q}_d^k(K)}} \big(\pi_\hbar v - \pi_{L^2} v \,\big|\, \varphi\big)_K \\
&= \sup_{\substack{\|\varphi\|_K = 1 \\ \varphi \in \mathbb{Q}_d^k(K)}} \Big(\big(v - \pi_{L^2} v \,\big|\, \varphi\big)_K - \big(v - \pi_\hbar v \,\big|\, \varphi\big)_K\Big) \\
&= \sup_{\substack{\|\varphi\|_K = 1 \\ \varphi \in \mathbb{Q}_d^k(K)}} \big(v - \pi_{L^2} v \,\big|\, \varphi\big)_K \\
&\le \sup_{\substack{\|\varphi\|_K = 1 \\ \varphi \in L^2(K)}} \big(v - \pi_{L^2} v \,\big|\, \varphi\big)_K \\
&= \|v - \pi_{L^2} v\|_K,
\end{aligned}$$

where the third step is due to the definition (10.4) of the weighted L^2-projection. In summary, this yields

$$\|e_\pi^v\|_K \leq 2\|v - \pi_{L^2} v\|_K.$$

The claim follows after bounding the right-hand side by changing into the equivalent standard L^2-norm and using [43, Lem. 1.58]. For the bound on the faces, we use [43, Lem. 1.59] instead. □

Lemma 10.2 suggests that functions in $\mathbb{V}_h$ are suitable to approximate functions that are sufficiently regular on the mesh elements but need not to be regular across the faces of the mesh. Such functions are contained in the *broken Sobolev spaces* defined by

$$H^\ell(\mathcal{T}_h) = \{\, v \in L^2(\Omega) \mid v_{|K} \in H^\ell(K) \text{ for all } K \in \mathcal{T}_h \,\}, \qquad \ell \in \mathbb{N}_0.$$

For all $\ell \in \mathbb{N}_0$, the broken Sobolev spaces are Hilbert spaces with norms denoted by

$$\|v\|_{\ell,\mathcal{T}_h}^2 = \sum_{j=0}^{\ell} |v|_{j,\mathcal{T}_h}^2, \qquad |v|_{\ell,\mathcal{T}_h}^2 = \sum_{K \in \mathcal{T}_h} |v|_{\ell,K}^2.$$

By their respective definitions, functions contained in either the discrete approximation space $\mathbb{V}_h$ or the broken Sobolev spaces are not necessarily continuous across the faces of the mesh. Hence, the evaluation of such functions on an interface $F \in \mathcal{F}_h^{\text{int}}$ is not well-defined and the limits of the function approaching F from either K_1^F or K_2^F may differ. We denote these limits for a function $v \in H^1(\mathcal{T}_h)$ as $v_{|K_1^F}$ and $v_{|K_2^F}$, respectively, and with this define the *average* and the *jump of a function v across an interior face $F \in \mathcal{F}_h^{\text{int}}$ as*

$$\{\!\{v\}\!\}_F = \frac{v_{|K_1^F} + v_{|K_2^F}}{2}, \qquad \text{and} \qquad [\![v]\!]_F = v_{|K_1^F} - v_{|K_2^F}.$$

We point out that these values can be seen as a measure of the discontinuity of the function v. Further, they serve as a means to introduce the necessary coupling into the discrete problem.

Similar to the broken Sobolev spaces, we also need the following matrix-valued spaces, which serve to impose appropriate regularity assumptions on the material tensor. They are defined as

$$W^{\ell,\infty}(\mathcal{T}_h)^{m\times m} = \{\, A \in L^\infty(\Omega)^{m\times m} \mid A_{|K} \in W^{\ell,\infty}(K)^{m\times m} \text{ for all } K \in \mathcal{T}_h \,\},$$

$\ell \in \mathbb{N}_0$, and we equip them with the norms

$$\|A\|_{\ell,\infty,\mathcal{T}_h} = \max_{K \in \mathcal{T}_h} \|A\|_{\ell,\infty,K}.$$

10.2 Friedrichs' Operators in the Discrete Setting

As mentioned in the last section, in the discrete problem we use jumps and averages of functions to introduce coupling between the mesh elements. In order to do this, it is convenient to have a more concrete grasp of boundary values (and traces in general) than using the abstract boundary operator $\mathcal{L}_\partial$ from Definition 9.20.

Therefore, in the following, we restrict ourselves to the intersection of $H(\mathcal{L})$ (defined in (9.27)) with the broken Sobolev space $H^1(\mathcal{T}_h)^m$, whose elements, in particular, admit square-integrable traces on the faces of the mesh $\mathcal{T}_h$. This allows us to replace the dual pairing in the abstract integration-by-parts formula from Definition 9.20 by an L^2-inner product and the abstract boundary operator by a matrix-valued field comprised of a normal vector and the matrix-valued coefficients of $\mathcal{L}$. This boundary field, which we denote by L_∂^K, can be explicitly derived via the standard integration-by-parts formula (on each element $K \in \mathcal{T}_h$). More precisely, for $v, w \in H(\mathcal{L}) \cap H^1(\mathcal{T}_h)^m$, we obtain

$$\left(\mathcal{L}v \,|\, w\right)_K - \left(v \,|\, \mathcal{L}^\circledast w\right)_K = \left(L_\partial^K v \,|\, w\right)_{\partial K}, \qquad L_\partial^K = \sum_{i=1}^d \mathfrak{n}_i^K L_i,$$

by [88, Lem. 4.3] and therefore, by Definition 9.20, we have

$$\langle \mathcal{L}_\partial v \,|\, w \rangle = \sum_{K \in \mathcal{T}_h} \left(L_\partial^K v \,|\, w\right)_{\partial K} = \Big(\sum_{i=1}^d \mathfrak{n}^\Omega L_i v \,|\, w\Big)_\Gamma,$$

where $\mathfrak{n}^\Omega$ is the outward unit normal vector on Ω. Here, the last equality follows from [88, Lem. 4.7], and by this, we see that $\mathcal{L}_\partial$ does in fact act on the boundary of Ω. Note also that $L_\partial^K \in L^\infty(\partial K)^{m\times m}$ for all $K \in \mathcal{T}_h$.

In the definition and handling of the discrete operator, it is more convenient to have a similar representation of $\mathcal{L}_\partial$ w.r.t. the faces of the mesh instead of the elements. For each face $F \in \mathcal{F}_h$, we therefore define the boundary field $L_\partial^F \in L^\infty(F)^{m\times m}$ by

$$L_\partial^F = \sum_{i=1}^d \mathfrak{n}_i^F L_i.$$

To also write down boundary values in a more concrete way, we further want to replace the abstract boundary operator from Assumption 9.21 by a matrix-valued field. For this, we make the following assumption, which is fulfilled in many practical situations; see [59, Sec. 5] for some examples.

Assumption 10.3 The boundary operator $\mathcal{L}_\Gamma$ is associated with a matrix-valued boundary field $L_\Gamma \in L^\infty(\Gamma)^{m\times m}$ such that we have

$$\langle \mathcal{L}_\Gamma v \mid w \rangle = \big(L_\Gamma v \mid w\big)_\Gamma$$

for v, w sufficiently regular. ◇

To conclude this section, we provide a technical lemma, which we need to bound some terms arising in the error analysis of the fully discrete schemes. It basically formalizes the fact that the Friedrichs' operator $\mathcal{L}$ is a first-order spatial differential operator and thus its application can be bounded by the H^1-norm (albeit we give it in a slightly more general fashion).

Lemma 10.4 *Let $\ell \geq 0$, $v \in H(\mathcal{L}) \cap H^{\ell+1}(\mathcal{T}_h)^m$ and $M \in W^{\ell,\infty}(\mathcal{T}_h)^{m\times m}$. Then $\mathcal{L}v \in H^\ell(\mathcal{T}_h)^m$ and for all $j \in \mathbb{Z}$, we have*

$$\|h^j \mathcal{L} v\|_{\ell,\mathcal{T}_h} \leq C_{\mathcal{L},M,\ell} \|h^j v\|_{\ell+1,\mathcal{T}_h},$$

and, in particular,

$$\|\mathcal{L} v\|_{\ell,\mathcal{T}_h} \leq C_{\mathcal{L},M,\ell} \|v\|_{\ell+1,\mathcal{T}_h},$$

where $C_{\mathcal{L},M,\ell}$ only depends on ℓ, d, $C_{\mathcal{L}}$, $\|M\|_{\infty,\Omega}$, and $\|M^{-1}\|_{\ell,\infty,\mathcal{T}_h}$.

Proof By definition of the Sobolev norms we have

$$\|\mathcal{L} v\|_{\ell,K}^2 = \sum_{|\alpha|\leq \ell} \|\partial^\alpha(\mathcal{L} v)\|_K^2.$$

By $\mathcal{L} = M^{-1}\widetilde{\mathcal{L}}$ and by applying the product rule d times, for $|\alpha| \leq \ell$, we obtain

$$\partial^\alpha(\mathcal{L} v) = \sum_{i_1=0}^{\alpha_1} \cdots \sum_{i_d=0}^{\alpha_d} \binom{\alpha_1}{i_1} \cdots \binom{\alpha_d}{i_d} \partial_1^{\alpha_1-i_1} \cdots \partial_d^{\alpha_d-i_d}(M^{-1})\, \partial_1^{i_1} \cdots \partial_d^{i_d}(\widetilde{\mathcal{L}} v).$$

Hence, using the triangle inequality and

$$\begin{aligned}
&\|\partial_1^{\alpha_1-i_1}\dots\partial_d^{\alpha_d-i_d}(M^{-1})\,\partial_1^{i_1}\dots\partial_d^{i_d}(\widetilde{\mathcal{L}}v)\|_K\\
&\quad\le \|\partial_1^{\alpha_1-i_1}\dots\partial_d^{\alpha_d-i_d}(M^{-1})\|_{\infty,K}\|\partial_1^{i_1}\dots\partial_d^{i_d}(\widetilde{\mathcal{L}}v)\|_K\\
&\quad\le \|M^{-1}\|_{\ell,\infty,K}\,\|\partial_1^{i_1}\dots\partial_d^{i_d}(\widetilde{\mathcal{L}}v)\|_K
\end{aligned}$$

yields

$$\begin{aligned}
\|\mathcal{L}v\|_{\ell,K}^2 &\le \max_{0\le i\le\ell}\binom{\ell}{i}^{2d}\|M^{-1}\|_{\ell,\infty,K}^2\sum_{|\alpha|\le\ell}\Big(\sum_{i_1=0}^{\alpha_1}\dots\sum_{i_d=0}^{\alpha_d}\|\partial_1^{i_1}\dots\partial_d^{i_d}(\widetilde{\mathcal{L}}v)\|_K\Big)^2\\
&\le\binom{\ell}{\lceil\ell/2\rceil}^{2d}\Big(\sum_{|\alpha|\le\ell}\textstyle\prod_{i=1}^d(\alpha_i+1)\Big)\|M^{-1}\|_{\ell,\infty,K}^2\,\|\widetilde{\mathcal{L}}v\|_{\ell,K}^2,
\end{aligned}$$

where we have used the equivalence between the 1- and the 2-norm in the last inequality. Since $\|M^{-1}\|_{\ell,\infty,K}\le\|M^{-1}\|_{\ell,\infty,\Omega}$, it remains to give a bound on $\|\widetilde{\mathcal{L}}v\|_{\ell,K}$.

To do so, let $j\le\ell$. We use the triangle inequality and the fact that the coefficients of $\widetilde{\mathcal{L}}$ are constant to obtain

$$|\widetilde{\mathcal{L}}v|_{j,K}=|\sum_{i=1}^d L_i\partial_i v+L_0v|_{j,K}\le C_{\mathcal{L}}\big(\sum_{i=1}^d|\partial_i v|_{j,K}+|v|_{j,K}\big).$$

The right-hand side can be seen as the 1-norm of a vector with $d+1$ components. Using the equivalence (with constant $(d+1)^{1/2}$) of the 1- and the 2-norm therefore yields

$$\begin{aligned}
|\widetilde{\mathcal{L}}v|_{j,K}&\le C_{\mathcal{L}}(d+1)^{1/2}\big(\sum_{i=1}^d|\partial_i v|_{j,K}^2+|v|_{j,K}^2\big)^{1/2}\\
&\le C_{\mathcal{L}}(d+1)^{1/2}\big(|v|_{j+1,K}^2+|v|_{j,K}^2\big)^{1/2},
\end{aligned}$$

where the last inequality is due to the definition of the Sobolev seminorms. Using the definition of the full Sobolev norm (together with an index shift) yields the appropriate bound on $\|\widetilde{\mathcal{L}}v\|_{\ell,K}$ and thus the claim. □

10.3 Discrete Friedrichs' Operators

Now that we have investigated some properties of the Friedrichs' operator $\mathcal{L}$ in the discrete setting, we are ready to define its discrete counterpart. This *discrete Friedrichs' operator* is then used to define a spatially discrete version of the wave-type problem (9.3) on the discrete space $\mathbb{V}_{\hbar}^{m}$. However, instead of defining this discrete operator only on the space $\mathbb{V}_{\hbar}^{m}$, we extend the domain of definition to the *discrete operator domain* associated with $\mathcal{L}$ given by

$$V_{\hbar}^{\mathcal{L}} = \mathbb{V}_{\hbar}^{m} + (D(\mathcal{L}) \cap H^{1}(\mathcal{T}_{\hbar})^{m}).$$

This is a crucial step for the error analysis as we have to apply the discrete operator to the exact solution, which is, in general, not contained in the discrete space.

Definition 10.5 The **central fluxes dG discretization of** $\mathcal{L}$ is the operator $\underline{\mathcal{L}}\colon V_{\hbar}^{\mathcal{L}} \to \mathbb{V}_{\hbar}^{m}$ defined as

$$\begin{aligned}(\underline{\mathcal{L}}v \mid \varphi)_{\Omega} = \sum_{K\in\mathcal{T}_{\hbar}} (\mathcal{L}v \mid \varphi)_{K} - \sum_{F\in\mathcal{F}_{\hbar}^{\mathrm{int}}} \big(L_{\partial}^{F} [\![v]\!]_{F} \mid \{\!\{\varphi\}\!\}_{F}\big)_{F} \\ - \tfrac{1}{2} \sum_{F\in\mathcal{F}_{\hbar}^{\mathrm{bnd}}} \big((L_{\partial}^{F} - L_{\Gamma})v \mid \varphi\big)_{F} \qquad \text{for all } \varphi \in \mathbb{V}_{\hbar}^{m}.\end{aligned} \tag{10.6}$$

Note that the discrete operator $\underline{\mathcal{L}}$ is well-defined by the Riesz representation theorem, see [88, Sec. 4.3] for more details.

Let us give some insight into how this discrete operator is constructed. The first term in (10.6) is simply the original bilinear form corresponding to the operator $\mathcal{L}$. Due to the discontinuous nature of our discretization space $\mathbb{V}_{\hbar}^{m}$, however, taking only this bilinear form would completely decouple the elements of the mesh $\mathcal{T}_{\hbar}$. Hence, it is necessary to establish some coupling between the elements to ensure that the discrete operator approximates the original one. This is done (in a weak sense) by the sum over the interfaces in (10.6). Lastly, the sum over the boundary faces (again weakly) enforces the boundary condition. We refer to [43, 59–61] and [137] for more detailed derivations of dG discretizations.

In the remainder of this section, we give some important properties of the discrete operator. In fact, these are the key properties we need for the error analyses of both the spatially semidiscrete as well as the fully discrete problems. We skip the rather technical proofs for the sake of readability and refer the reader to Section 3 in both [88] and [89], where all proofs can be found.

The first property of the discrete operator is that its restriction to the discrete space $\mathbb{V}_{\hbar}^{m}$ is dissipative. Since $\mathbb{V}_{\hbar}^{m}$ is a finite-dimensional space, this immediately implies maximality (i.e., the range condition in Definition 9.12) and thus wellposedness of the spatially discrete problem defined later.

Proposition 10.6 (Maximal Dissipativity) *The restriction of the discrete Friedrichs' operator $\underline{\mathcal{L}}$ to $\mathbb{V}_{\hbar}^m$ is maximal dissipative, i.e., we have*

$$\left(\underline{\mathcal{L}}\boldsymbol{v} \mid \boldsymbol{v}\right)_{\Omega} \leq 0 \qquad \textit{for all } \boldsymbol{v} \in \mathbb{V}_{\hbar}^m,$$

and $\operatorname{ran}(\underline{\mathcal{I}} - \lambda\, \underline{\mathcal{L}}_{|\mathbb{V}_{\hbar}^m}) = \mathbb{V}_{\hbar}^m$ *for all* $\lambda > 0$.

Next, we give a result stating that the discrete operator is consistent in the sense that its application returns the weighted L^2-projection $\pi_{\hbar}$ of a function if said function is sufficiently regular. This leads to the consistency of the spatially discrete problem and is a key ingredient in obtaining spatial convergence.

Proposition 10.7 (Consistency) *The discrete Friedrichs' operator $\underline{\mathcal{L}}$ fulfills the consistency property*

$$\underline{\mathcal{L}}v = \pi_{\hbar}\mathcal{L}v \qquad \textit{for all } v \in D(\mathcal{L}) \cap H^1(\mathcal{T}_{\hbar})^m.$$

Similar to the continuous Friedrichs' operator, the discrete operator $\underline{\mathcal{L}}$ satisfies an inverse inequality. In fact, the original inverse inequality (10.3) is one of the main ingredients of the proof. However, due to the face terms occurring in the definition (10.6) of the discrete operator, one also needs the discrete trace inequality (10.2) to prove this.

Proposition 10.8 (Inverse Inequality) *Let $\boldsymbol{v} \in \mathbb{V}_{\hbar}^m$. Then, for all $j \in \mathbb{Z}$, the discrete Friedrichs' operator $\underline{\mathcal{L}}$ fulfills the inverse inequality*

$$\|h^j \underline{\mathcal{L}}\boldsymbol{v}\|_{\Omega} \leq C_{\mathrm{inv},\mathcal{L},j} \|h^{j-1}\boldsymbol{v}\|_{\Omega},$$

and, in particular,

$$\|\underline{\mathcal{L}}\boldsymbol{v}\|_{\Omega} \leq C_{\mathrm{inv},\mathcal{L}} \|h^{-1}\boldsymbol{v}\|_{\Omega}.$$

The constants are given by $C_{\mathrm{inv},\mathcal{L},j} = C_{\mathcal{L}} C_{\mathrm{inv}} + \frac{1}{2} C_{\mathrm{tr}}^2 \big(C_{\Gamma,\mathcal{L}} + N_{\partial} C_{\mathcal{L}} (1 + \rho^{j+1/2})\big)$ *with* $C_{\Gamma,\mathcal{L}} = \max_{F \in \mathcal{F}_{\hbar}^{\mathrm{bnd}}} \|L_{\partial}^F - L_{\Gamma}\|_{\infty,F}$, *and* $C_{\mathrm{inv},\mathcal{L}} = C_{\mathrm{inv},\mathcal{L},0}$.

The next result provides a bound on the application of the discrete operator $\underline{\mathcal{L}}$ to the projection error (10.5) of a sufficiently regular function. It can be seen as a measure of how well the discrete operator approximates its continuous counterpart, since we have $\underline{\mathcal{L}} e_{\pi}^v = \underline{\mathcal{L}}v - \underline{\mathcal{L}}\pi_{\hbar} v = (\pi_{\hbar}\mathcal{L} - \underline{\mathcal{L}}\pi_{\hbar})v$ for $v \in D(\mathcal{L}) \cap H^1(\mathcal{T}_{\hbar})^m$ by Proposition 10.7 (consistency).

Proposition 10.9 (Approximation Property) *Let $v \in D(\mathcal{L}) \cap H^{\ell+1}(\mathcal{T}_h)^m$ for $0 \leq \ell \leq k$, where k is the polynomial degree used in the dG discretization. Then, for all $j \in \mathbb{Z}$, we have*

$$\|h^j \underline{\mathcal{L}} e_\pi^v\|_\Omega \leq C_{\pi,\mathcal{L},j} |h^{j+\ell} v|_{\ell+1,\mathcal{T}_h},$$

and, in particular,

$$\|\underline{\mathcal{L}} e_\pi^v\|_\Omega \leq C_{\pi,\mathcal{L}} |h^\ell v|_{\ell+1,\mathcal{T}_h}.$$

The constants are given by $C_{\pi,\mathcal{L},j} = \frac{1}{2} N_\partial C_{\mathrm{tr}} C_{\pi,\partial}\big(C_{\Gamma,\mathcal{L}} + C_{\mathcal{L}}(1+\rho^{j+1/2})\big)$ and $C_{\pi,\mathcal{L}} = C_{\pi,\mathcal{L},0}$.

Lastly, we need a more general version of this statement involving more than one Friedrichs' operator.

Lemma 10.10 *Let $\mathcal{L}_1, \ldots, \mathcal{L}_r$ be Friedrichs' operators with corresponding domains $D(\mathcal{L}_1)$,..., $D(\mathcal{L}_r)$ and let $\underline{\mathcal{L}}_1, \ldots, \underline{\mathcal{L}}_r$ be their respective central fluxes dG discretizations. Further, let $M \in W^{\ell,\infty}(\mathcal{T}_h)$ and $v \in D(\mathcal{L}_r \ldots \mathcal{L}_1) \cap H^{\ell+1}(\mathcal{T}_h)^m$ for $r-1 \leq \ell \leq k$, where k is the polynomial degree used in the dG discretization. Then, for all $j \in \mathbb{Z}$ we have*

$$\|h^j(\underline{\mathcal{L}}_r \ldots \underline{\mathcal{L}}_2 \underline{\mathcal{L}}_1 \pi_h - \pi_h \mathcal{L}_r \ldots \mathcal{L}_1)v\|_\Omega \leq C\|h^{j+(\ell+1)-r} v\|_{\ell+1,\mathcal{T}_h},$$

and, in particular,

$$\|(\underline{\mathcal{L}}_r \ldots \underline{\mathcal{L}}_2 \underline{\mathcal{L}}_1 \pi_h - \pi_h \mathcal{L}_r \ldots \mathcal{L}_1)v\|_\Omega \leq C\|h^{(\ell+1)-r} v\|_{\ell+1,\mathcal{T}_h},$$

where the generic constants are independent of h.

We point out that we lose one power of h for each additional Friedrichs' operator we apply.

10.4 The Spatially Semidiscrete Problem

Now we have all ingredients to define the spatial discretization of the wave-type problem (9.3). We do this by projecting the data (i.e., the initial value u^0 and the inhomogeneity f) and replacing the continuous Friedrichs' operator $\mathcal{L}$ with its discretization $\underline{\mathcal{L}}$. This yields the *spatially discrete wave-type problem*, where we seek $\boldsymbol{u} \colon \mathbb{R}_+ \to \mathbb{V}_h^m$ fulfilling

$$\begin{cases} \partial_t \boldsymbol{u} = \underline{\mathcal{L}} \boldsymbol{u} + f_\pi, & \text{on } \mathbb{R}_+, \qquad (10.7a) \\ \boldsymbol{u}(0) = \boldsymbol{u}_\pi^0, & \qquad (10.7b) \end{cases}$$

with

$$\boldsymbol{u}_\pi^0 = \pi_{\hbar} u^0 \quad \text{and} \quad \boldsymbol{f}_\pi = \pi_{\hbar} f. \tag{10.7c}$$

By Proposition 10.6 (maximal dissipativity) and the Lumer–Phillips Theorem 9.14, the restriction of the discrete operator $\underline{\mathcal{L}}$ to $\mathbb{V}_{\hbar}^m$ generates a contractive semigroup. Using a slight abuse of notation, we denote this semigroup by $\left(e^{t\underline{\mathcal{L}}}\right)_{t\geq 0}$. Then, with the same reasoning as in the continuous case, i.e., Corollary 9.23, we obtain the following wellposedness result. Recall that we assumed some minimal regularity of the data in (10.1), which is necessary for the following result to hold.

Corollary 10.11 *There exists a unique solution* $\boldsymbol{u} \in C^1(\mathbb{R}_+; \mathbb{V}_{\hbar}^m)$ *of* (10.7) *given by the variation-of-constants formula*

$$\boldsymbol{u}(t) = e^{t\underline{\mathcal{L}}} \boldsymbol{u}_\pi^0 + \int_0^t e^{(t-s)\underline{\mathcal{L}}} \boldsymbol{f}_\pi(s)\, ds.$$

Further, we have the following stability bounds.

(a) The solution u satisfies

$$\|\boldsymbol{u}(t)\|_\Omega \leq \|\boldsymbol{u}_\pi^0\|_\Omega + \int_0^t \|\boldsymbol{f}_\pi(s)\|_\Omega\, ds \leq \|u^0\|_\Omega + \int_0^t \|f(s)\|_\Omega\, ds.$$

(b) If $f = f_1 + f_2$ *with* $f_1 \in C^1(\mathbb{R}_+; L^2(\Omega)^m)$ *and* $f_2 \in C(\mathbb{R}_+; L^2(\Omega)^m)$, *then the temporal derivative* $\partial_t \boldsymbol{u}$ *satisfies*

$$\begin{aligned} \|\partial_t \boldsymbol{u}(t)\|_\Omega \leq{}& \|\underline{\mathcal{L}} \boldsymbol{u}_\pi^0\|_\Omega + \max_{s\in\mathbb{R}_+} \|\boldsymbol{f}_\pi(s)\|_\Omega \\ &+ \int_0^t \|\partial_t \pi_{\hbar} f_1(s)\|_\Omega + \|\underline{\mathcal{L}} \pi_{\hbar} f_2(s)\|_\Omega\, ds. \end{aligned} \tag{10.8}$$

(c) The discrete spatial derivative $\underline{\mathcal{L}}\boldsymbol{u}$ *satisfies*

$$\|\underline{\mathcal{L}}\boldsymbol{u}(t)\|_\Omega \leq \|\partial_t \boldsymbol{u}(t)\|_\Omega + \|\boldsymbol{f}_\pi(t)\|_\Omega. \tag{10.9}$$

(d) If $u^0 \in D(\mathcal{L}) \cap H^1(\mathcal{T}_{\hbar})$ *and* $f = f_1 + f_2$ *with* $f_1 \in C^1(\mathbb{R}_+; L^2(\Omega)^m)$ *and* $f_2 \in C(\mathbb{R}_+; D(\mathcal{L}) \cap H^1(\mathcal{T}_{\hbar}))$, *we have*

$$\begin{aligned} \|\partial_t \boldsymbol{u}(t)\|_\Omega \leq{}& \|\mathcal{L}u^0\|_\Omega + C_{\pi,\mathcal{L}}|u^0|_{1,\mathcal{T}_{\hbar}} + \max_{s\in\mathbb{R}_+} \|f(s)\|_\Omega \\ &+ \int_0^t \|\partial_t f_1(s)\|_\Omega + \|\mathcal{L} f_2(s)\|_\Omega + C_{\pi,\mathcal{L}}|\mathcal{L} f_2(s)|_{1,\mathcal{T}_{\hbar}}\, ds. \end{aligned} \tag{10.10}$$

Before we prove this result, we point out that, due to the occurrence of the discrete operator $\underline{\mathcal{L}}$ in (10.8), this bound (and thus also (10.9)) is not necessarily independent of the spatial discretization parameter $\hbar$ and thus not robust under mesh refinement. This is why we give the bound (10.10), which shows that a robust stability bound on the discrete derivatives (in both space and time) can be achieved if we assume slightly more regularity on the data.

Proof Except for (10.10), all results are straightforward consequences of interpreting (10.7) as an abstract Cauchy problem on $\mathbb{V}_\hbar^m$ and using Theorem 9.10 (again, we obtain simpler bounds since the semigroup is contractive, i.e., $\omega = 0$ and $C_{\mathrm{sg}} = 1$). Hence, we only show (10.10).

To do so, we use that for all $v \in D(\underline{\mathcal{L}}) \cap H^1(\mathcal{T}_\hbar)$, we have

$$\begin{aligned} \|\underline{\mathcal{L}}\pi_\hbar v\|_\Omega &\leq \|\pi_\hbar \mathcal{L} v\|_\Omega + \|(\pi_\hbar \mathcal{L} - \underline{\mathcal{L}}\pi_\hbar)v\|_\Omega \\ &= \|\pi_\hbar \mathcal{L} v\|_\Omega + \|\underline{\mathcal{L}} e_\pi^v\|_\Omega \\ &\leq \|\mathcal{L} v\|_\Omega + C_{\pi,\mathcal{L}} |v|_{1,\mathcal{T}_\hbar}, \end{aligned} \tag{10.11}$$

where we have used Proposition 10.7 (consistency) together with the definition (10.5) of the projection error for the equality and Proposition 10.9 (approximation property) with $\ell = j = 0$ for the second inequality. With this, (10.10) is a direct consequence of (10.8) and the regularity of the data. □

We point out that the bound (10.11) (or analogous results, where ℓ and j are not necessarily zero in Proposition 10.9 (approximation property)) is a key ingredient of the analysis of the fully discrete method, as we will see later.

We now state a convergence result for the spatially semidiscrete problem (10.7). The proof of this result is the topic of [88] and we refer the interested reader to that paper for more details.

Theorem 10.12 *Assume that the exact solution u of the wave-type problem* (9.3) *satisfies $u \in C^1(\mathbb{R}_+; L^2(\Omega)^m) \cap C(\mathbb{R}_+; D(\mathcal{L}) \cap H^{k+1}(\mathcal{T}_\hbar)^m)$. Then, for $t \in \mathbb{R}_+$, the spatially semidiscrete error satisfies*

$$\begin{aligned} \|u(t) - \boldsymbol{u}(t)\|_\Omega &\leq C\big(|h^{k+1}u(t)|_{k+1,\mathcal{T}_\hbar} + \int_0^t |h^k u(s)|_{k+1,\mathcal{T}_\hbar}\, \mathrm{d}s\big) \\ &\leq C\hbar^k, \end{aligned}$$

where both generic constants are independent of the spatial discretization parameter $\hbar$ and the first constant is independent of the solution u.

We conclude the spatially semidiscrete analysis by mentioning that the stronger stability bounds (10.8)–(10.10) can be used to derive error bounds on the discrete derivatives $\partial_t \boldsymbol{u}$ and $\underline{\mathcal{L}}\boldsymbol{u}$ (compared to their continuous counterparts $\partial_t u$ and $\mathcal{L}u$). Since performing the analysis would simply recycle large parts of [88] and this work is focused on full discretizations, we abstain from doing so for the sake of brevity. They can be derived easily, however, by combining the techniques used in [88] with those of deriving the related stronger error bounds in Chap. 12 and are left to the interested reader.

Full Discretization 11

This chapter is devoted to the derivation and the stability analysis of fully discrete methods for approximating the solution to the wave-type problem (9.3) and thus, together with their error analysis in Chap. 12, constitutes the main part of this work. As mentioned before, we follow a method-of-lines ansatz, meaning that we now discretize the spatially semidiscrete problem (10.7) in time to obtain a fully discrete approximation to the solution of the original wave-type problem (9.3). To do so, we exploit that the semidiscrete problem is posed on a finite-dimensional space and can thus be interpreted as a first-order ordinary differential equation. This immediately enables us to apply standard methods that are designed for such problems, like Runge–Kutta or multistep methods. In this chapter, we present two such standard methods, the implicit Crank–Nicolson scheme (which is a Runge–Kutta method) and the explicit leapfrog scheme.

However, the semidiscrete problem preserves important properties and structures of the original one, which can be exploited to design more sophisticated time integration schemes. We present two such schemes, the Peaceman–Rachford method, a splitting scheme that can be used in *alternating direction implicit* (ADI) time integration [140], as well as a *locally implicit* integration scheme [163], which combines the explicit leapfrog and the implicit Crank–Nicolson method.

All time integration schemes we present in this chapter are classically of order two. This means that they are second-order accurate if applied to nonstiff systems of ordinary differential equations (ode), provided that the exact ode solution is sufficiently regular. However, simply splitting the full discretization error into an approximation error of the spatially discretized problem (cf., Theorem 10.12) and the time integration error applied to the system of ordinary equations (ode) falls short. The reason is that the ode error bounds require regularity of the spatially semidiscrete solution (i.e., the ode solution), which is not available in general. For a rigorous analysis, it is thus indispensable to intertwine the analysis of spatial and time discretization errors. The way to do so is to start with the

W. Dörfler et al., *Wave Phenomena*, Oberwolfach Seminars 49,
https://doi.org/10.1007/978-3-031-05793-9_11

exact solution, project it onto the finite dimensional dG space, and insert it into the time integration scheme. This finally yields error bounds where the constants only depend on the exact solution (and its (weak) derivatives) and do not deteriorate in the limit $\hbar \to 0$ (i.e., all constants are uniformly bounded in $\hbar$).

Throughout the remainder of this work, we use the following notation. The timestep size of the time integration schemes is denoted by $\tau > 0$ and discrete times by $t_n = n\tau$ for $n \in \mathbb{N}_0$. Further, for a more concise notation, we abbreviate the evaluation of the inhomogeneity f at the discrete times as well as its mean over two consecutive discrete times by

$$f^n = f(t_n) \qquad \text{and} \qquad \overline{f}^{n+1/2} = \frac{f^n + f^{n+1}}{2}, \qquad n \in \mathbb{N}_0, \tag{11.1}$$

respectively (and analogously for the projected inhomogeneity f_π). Since we will often also use these short notations at the midpoints between discrete times, we define $t_{n+1/2}$ and $f^{n+1/2}$ analogously.

11.1 Crank–Nicolson Scheme

The first scheme we consider is the well-known *Crank–Nicolson scheme*. As a matter of fact, we will see that this scheme serves as a basis for the other second-order time integration methods we present here, as all three schemes can be considered as a perturbation of this scheme. This is a key ingredient of the presented analysis.

The Crank–Nicolson scheme, also known as the implicit trapezoidal rule, is an implicit two-stage Runge–Kutta scheme with Butcher-Tableau given in Fig. 11.1. It has many favorable properties like unconditional stability and the preservation of the energy-structure of the original problem, cf., [80, 82].

Even though it can be seen and derived as a Runge–Kutta scheme, there are simpler ways to obtain it and to get some intuition on it because of the elementary structure of the method. For a motivation, we give a brief derivation of the scheme in the next example.

Example 11.1 We once again consider the inhomogeneous initial value problem (9.8) from Example 9.1 given by

$$\begin{cases} \partial_t u(t) = Au(t) + f(t), & t \in \mathbb{R}_+, \\ u(0) = u^0. \end{cases} \tag{11.2}$$

Fig. 11.1 Butcher-Tableau of the trapezoidal rule or Crank–Nicolson scheme

0	0	0
1	1/2	1/2
	1/2	1/2

We are interested in a time-stepping scheme that provides us with an approximation u_τ^{n+1} to the solution at time t_{n+1} if we already have an approximation $u_\tau^n \approx u(t_n)$. Starting at the given initial value u^0, we can then compute approximations step-by-step.

For the derivation of this scheme, we make use of the fundamental theorem of calculus to obtain

$$\begin{aligned} u(t_{n+1}) &= u(t_n) + \int_{t_n}^{t_{n+1}} \partial_t u(s)\,\mathrm{d}s \\ &= u(t_n) + \int_{t_n}^{t_{n+1}} Au(s) + f(s)\,\mathrm{d}s, \end{aligned}$$

where we have used that u solves (11.2). To obtain an approximation to $u(t_{n+1})$, we apply the trapezoidal rule to approximate the integral on the right-hand side, yielding

$$u(t_{n+1}) \approx u(t_n) + \tfrac{\tau}{2}\Big(\big(Au(t_n) + f^n\big) + \big(Au(t_{n+1}) + f^{n+1}\big)\Big).$$

After rearranging the terms and replacing the evaluations of the exact solution $u(t_{n+1})$ and $u(t_n)$ by their approximations u_τ^{n+1} and u_τ^n, we obtain the *Crank–Nicolson scheme*

$$\begin{cases} (I - \tfrac{\tau}{2}A)u_\tau^{n+1} = (I + \tfrac{\tau}{2}A)u_\tau^n + \tau \overline{f}^{n+1/2}, & n \in \mathbb{N}_0, \\ \qquad\qquad u_\tau^0 = u^0. \end{cases}$$

We point out that this scheme is implicit, since it requires the solution of a linear system with coefficient matrix $I - \frac{\tau}{2}A$ in each timestep. ◇

As mentioned before, the Crank–Nicolson method has many favorable properties. The main drawback of the method, however, is its implicitness. While it may be feasible to solve a linear system of equations in each timestep for moderately sized ordinary differential equations, it can severely restrict its usability for spatially discretized partial differential equations, which are often of large dimensions. This is due to the fact that the spatial resolution of the partial differential equation dictates the dimension of the linear systems needed to be solved and often one is forced to use a high number of degrees of freedom to resolve the spatial domain.

We apply the Crank–Nicolson method to the spatially discrete wave-type problem (10.7), yielding the *fully discrete dG-Crank–Nicolson (dG-CN) scheme* given by

$$\begin{cases} (\underline{I} - \tfrac{\tau}{2}\underline{\mathcal{L}})\boldsymbol{u}_{\mathrm{CN}}^{n+1} = (\underline{I} + \tfrac{\tau}{2}\underline{\mathcal{L}})\boldsymbol{u}_{\mathrm{CN}}^{n} + \tau \overline{\boldsymbol{f}}_\pi^{n+1/2}, & n \in \mathbb{N}_0, \qquad (11.3a) \\ \qquad\qquad \boldsymbol{u}_{\mathrm{CN}}^0 = \boldsymbol{u}_\pi^0. & \qquad (11.3b) \end{cases}$$

Note that the solution $\boldsymbol{u}_{\mathrm{CN}}^{n+1}$ of this scheme exists and is unique, since the resolvent $\left(\underline{\mathcal{I}} - \frac{\tau}{2}\underline{\mathcal{L}}\right)^{-1} : \mathbb{V}_{\hbar}^{m} \to \mathbb{V}_{\hbar}^{m}$ is a bijection by Proposition 10.6 (maximal dissipativity) and Lemma 9.15 (i). Also, recall the notation introduced for the projected data $\boldsymbol{u}_{\pi}^{0}$ and $\boldsymbol{f}_{\pi}$ in (10.7c) and for the mean $\overline{\boldsymbol{f}}_{\pi}^{n+1/2}$ in (11.1).

In order to obtain a concrete solution formula for the scheme, it is convenient to rewrite it in a more compact form. To this end, we define the discrete operators $\underline{\mathcal{R}}_{\mathrm{CN}}, \underline{\mathcal{S}}_{\mathrm{CN}} : \mathbb{V}_{\hbar}^{m} \to \mathbb{V}_{\hbar}^{m}$ by

$$\underline{\mathcal{R}}_{\mathrm{CN}} = \left(\underline{\mathcal{I}} - \tfrac{\tau}{2}\underline{\mathcal{L}}\right)^{-1}, \tag{11.4}$$

and

$$\underline{\mathcal{S}}_{\mathrm{CN}} = \left(\underline{\mathcal{I}} - \tfrac{\tau}{2}\underline{\mathcal{L}}\right)^{-1}\left(\underline{\mathcal{I}} + \tfrac{\tau}{2}\underline{\mathcal{L}}\right) = \left(\underline{\mathcal{I}} + \tfrac{\tau}{2}\underline{\mathcal{L}}\right)\left(\underline{\mathcal{I}} - \tfrac{\tau}{2}\underline{\mathcal{L}}\right)^{-1}. \tag{11.5}$$

Then, the Crank–Nicolson scheme (11.3) is equivalent to

$$\begin{cases} \boldsymbol{u}_{\mathrm{CN}}^{n+1} = \underline{\mathcal{S}}_{\mathrm{CN}}\,\boldsymbol{u}_{\mathrm{CN}}^{n} + \tau\underline{\mathcal{R}}_{\mathrm{CN}}\,\overline{\boldsymbol{f}}_{\pi}^{n+1/2}, \qquad n \in \mathbb{N}_0, & \text{(11.6a)} \\ \boldsymbol{u}_{\mathrm{CN}}^{0} = \boldsymbol{u}_{\pi}^{0}. & \text{(11.6b)} \end{cases}$$

Solving the recursion yields the following result.

Theorem 11.2 *For all $n \in \mathbb{N}_0$, all $\hbar \in \mathcal{H}$, and all $\tau > 0$, there exists a unique $\boldsymbol{u}_{\mathrm{CN}}^{n+1} \in \mathbb{V}_{\hbar}^{m}$ fulfilling the dG-CN scheme* (11.3). *It is given by the discrete variation-of-constants formula*

$$\boldsymbol{u}_{\mathrm{CN}}^{n+1} = \underline{\mathcal{S}}_{\mathrm{CN}}^{n+1}\boldsymbol{u}_{\pi}^{0} + \tau\sum_{j=0}^{n} \underline{\mathcal{S}}_{\mathrm{CN}}^{n-j}\,\underline{\mathcal{R}}_{\mathrm{CN}}\,\overline{\boldsymbol{f}}_{\pi}^{j+1/2}. \tag{11.7}$$

We point out that this is essentially a fully discrete counterpart to the existence and uniqueness results Theorem 9.9 and (the first part of) Corollary 10.11. Further, note the similarity between the continuous and the discrete variation-of-constants formulae (9.34) and (11.7). In the discrete variation-of-constants formula, the operator $\underline{\mathcal{S}}_{\mathrm{CN}}$ takes over the role of the semigroup $\mathrm{e}^{t\mathcal{L}}$ and thus, $\underline{\mathcal{S}}_{\mathrm{CN}}$ can be seen as a discrete semigroup. In fact, since $\dfrac{1+z/2}{1-z/2}$ is the (1,1)-Padé approximation to the exponential function, $\underline{\mathcal{S}}_{\mathrm{CN}}$ can be seen as a rational approximation of the semigroup. Lastly, the convolution in the continuous variation-of-constants formula is replaced by a discrete convolution of this discrete semigroup and the inhomogeneity.

As we have seen in Sect. 9.1, semigroups exhibit some important properties. Next, we show that the discrete semigroup $\underline{\mathcal{S}}_{\mathrm{CN}}$ retains some of these properties, although we need to adjust them to the discrete setting.

The first property we consider is the growth bound (9.12). Recall that the continuous semigroup is contractive, i.e., (9.12) is fulfilled with $C_{\mathrm{sg}} = 1$ and $\omega = 0$. We expect the discrete semigroup to fulfill a similar bound and as we see in the next result, it fulfills an equivalent one for the Crank–Nicolson method. Since (9.12) holds for all times $t \in \mathbb{R}_+$, we need such a bound for arbitrary powers of the discrete semigroup, which corresponds to propagating over arbitrarily many timesteps.

Proposition 11.3 *Let $\mathcal{h} \in \mathcal{H}$ and $\tau > 0$. Then, for all $\ell \in \mathbb{N}$ and all $\boldsymbol{v} \in \mathbb{V}_{\mathcal{h}}^{m}$, the discrete semigroup $\underline{\mathcal{S}}_{\mathrm{CN}}$ defined in* (11.5) *fulfills the power bound*

$$\|\underline{\mathcal{S}}_{\mathrm{CN}}^{\ell}\, \boldsymbol{v}\|_{\Omega} \leq \|\boldsymbol{v}\|_{\Omega}.$$

Proof Since $\mathcal{L}_{|\mathbb{V}_{\mathcal{h}}^{m}}$ is maximal dissipative by Proposition 10.6, the discrete semigroup $\underline{\mathcal{S}}_{\mathrm{CN}}$ is contractive by Lemma 9.15 (ii). Applying this ℓ times yields the claim. □

Another property that is carried over to the discrete setting is (9.13b), which states that the derivative of a semigroup is given by application of its generator (and that both commute). We have to adjust this to the discrete setting by replacing the derivative with a difference quotient as well as some other, minor adjustments.

Proposition 11.4 *Let $\mathcal{h} \in \mathcal{H}$ and $\tau > 0$. Then, for all $\ell \in \mathbb{N}_0$, the powers of the discrete semigroup $\underline{\mathcal{S}}_{\mathrm{CN}}$ defined in* (11.5) *satisfy*

$$\frac{\underline{\mathcal{S}}_{\mathrm{CN}}^{\ell+1} - \underline{\mathcal{S}}_{\mathrm{CN}}^{\ell}}{\tau} = (\underline{\mathcal{R}}_{\mathrm{CN}}\, \underline{\mathcal{L}})\, \underline{\mathcal{S}}_{\mathrm{CN}}^{\ell} = \underline{\mathcal{S}}_{\mathrm{CN}}^{\ell}\, (\underline{\mathcal{R}}_{\mathrm{CN}}\, \underline{\mathcal{L}}).$$

Proof Since all involved objects commute, the claim is easily seen after factoring out $\underline{\mathcal{S}}_{\mathrm{CN}}^{\ell}$ and multiplying with the inverse of the resolvent (11.4). □

Note that this result implies that in the fully discrete setting, the generator of the discrete semigroup $\underline{\mathcal{S}}_{\mathrm{CN}}$ is not given by the discrete operator $\underline{\mathcal{L}}$, but its concatenation with its resolvent. We will see that the generators of the other methods will take a similar form, each with a different operator replacing the resolvent $\underline{\mathcal{R}}_{\mathrm{CN}}$.

Next, we show that the Crank–Nicolson scheme is indeed unconditionally stable, meaning that we can bound the approximations given by (11.3) independently of the discretization parameters $\mathcal{h}$ and τ and the number of timesteps. Again, this is a discrete counterpart to results we have already seen in the continuous setting. Namely, the stability results in Theorem 9.10 and (the second part of) Corollary 10.11. As in the continuous case, the key ingredient is the growth bound of the (discrete) semigroup, i.e., Proposition 11.3. Further, we again obtain results in both the weighted L^2-norm as well as in a stronger norm, which involves both discrete spatial and temporal operators. For better readability, we split these results into two and start with the former.

Theorem 11.5 *Let $h \in \mathcal{H}$, $\tau > 0$. Then, for all $n \in \mathbb{N}_0$, the approximation $\boldsymbol{u}_{\mathrm{CN}}^{n+1}$ given by the dG-CN scheme* (11.3) *satisfies*

$$\|\boldsymbol{u}_{\mathrm{CN}}^{n+1}\|_\Omega \le \|u^0\|_\Omega + \tau \sum_{j=1}^{n} \|\overline{f}^{j+1/2}\|_\Omega .$$

Proof We take norms in the discrete variation-of-constants formula (11.7) and use the triangle inequality to obtain

$$\|\boldsymbol{u}_{\mathrm{CN}}^{n+1}\|_\Omega \le \|\underline{\boldsymbol{S}}_{\mathrm{CN}}^{n+1}\boldsymbol{u}_\pi^0\|_\Omega + \tau \sum_{j=1}^{n} \|\underline{\boldsymbol{S}}_{\mathrm{CN}}^{n-j}\underline{\mathcal{R}}_{\mathrm{CN}}\,\overline{f}_\pi^{j+1/2}\|_\Omega .$$

With this, the claim follows by using Proposition 11.3 for the powers of $\underline{\boldsymbol{S}}_{\mathrm{CN}}$ and Lemma 9.15 (i) for $\underline{\mathcal{R}}_{\mathrm{CN}}$ together with the contractivity of the L^2-orthogonal projection. □

Next, we show the stronger stability bound involving the spatial and temporal operators. Since the fully discrete approximation is no longer a function in time, but merely a sequence, we need to replace the time derivative in the (time-)continuous counterparts of this result (i.e., Theorem 9.10, Corollaries 9.23 and 10.11) by a discrete variant. It turns out that a difference quotient between consecutive approximations (which can be seen as a discrete time derivative) is the appropriate choice. Thus, we define the *discrete time derivative* and the *discrete space derivative* for the Crank–Nicolson approximations at time $t_{n+1/2}$ as

$$\underline{\partial_\tau \boldsymbol{u}}_{\mathrm{CN}}^{n+1/2} = \frac{\boldsymbol{u}_{\mathrm{CN}}^{n+1} - \boldsymbol{u}_{\mathrm{CN}}^{n}}{\tau}, \qquad \text{and} \qquad \underline{\mathcal{L}\boldsymbol{u}}_{\mathrm{CN}}^{n+1/2} = \underline{\mathcal{L}}\,\frac{\boldsymbol{u}_{\mathrm{CN}}^{n+1} + \boldsymbol{u}_{\mathrm{CN}}^{n}}{2}, \tag{11.8}$$

respectively. Note that the Crank–Nicolson iteration (11.3a) is thus equivalent to

$$\underline{\partial_\tau \boldsymbol{u}}_{\mathrm{CN}}^{n+1/2} = \underline{\mathcal{L}\boldsymbol{u}}_{\mathrm{CN}}^{n+1/2} + \overline{f}_\pi^{n+1/2}, \tag{11.9}$$

and note the similarity to the original differential equation (9.3a). With this, we show the desired stability bounds.

Theorem 11.6 *Let $\hbar \in \mathcal{H}$, $\tau > 0$, and $f = f_1 + f_2$ be a splitting of the inhomogeneity with $f_1, f_2 \in C(\mathbb{R}_+; L^2(\Omega))$. Then, the approximations $\{\underline{u}_{\mathrm{CN}}^n\}_{n\geq 0}$ given by the dG-CN scheme* (11.3) *satisfy*

$$\|\underline{\partial_\tau u}_{\mathrm{CN}}^{n+1/2}\|_\Omega \leq \|\underline{\mathcal{L}u}_\pi^0\|_\Omega + \max_{j=0,\dots,n+1} \|\underline{f}_\pi^j\|_\Omega \\ + \tau \sum_{j=1}^{n} \big(\|\pi_\hbar \frac{f_1^{j+1} - f_1^{j-1}}{2\tau}\|_\Omega + \|\underline{\mathcal{L}}\pi_\hbar \overline{f_2}^{j-1/2}\|_\Omega\big), \tag{11.10}$$

and

$$\|\underline{\mathcal{L}u}_{\mathrm{CN}}^{n+1/2}\|_\Omega \leq \|\underline{\partial_\tau u}_{\mathrm{CN}}^{n+1/2}\|_\Omega + \max_{j=0,\dots,n+1} \|\underline{f}_\pi^j\|_\Omega. \tag{11.11}$$

If further $u^0 \in D(\mathcal{L}) \cap H^1(\mathcal{T}_\hbar)^m$, $f_1 \in C^1(\mathbb{R}_+, L^2(\Omega)^m)$, and $f_2 \in C(\mathbb{R}_+, D(\mathcal{L}) \cap H^1(\mathcal{T}_\hbar)^m)$, then we have

$$\|\underline{\partial_\tau u}_{\mathrm{CN}}^{n+1/2}\|_\Omega \leq \|\mathcal{L}u^0\|_\Omega + C_{\pi,\mathcal{L}}|u^0|_{1,\mathcal{T}_\hbar} \\ + \max_{s\in\mathbb{R}_+} \|f(s)\|_\Omega + \int_0^{t_{n+1}} \|\partial_t f_1(s)\|_\Omega \, \mathrm{d}s \\ + \tau \sum_{j=1}^{n} \big(\|\mathcal{L}\overline{f_2}^{j-1/2}\|_\Omega + C_{\pi,\mathcal{L}}|\overline{f_2}^{j-1/2}|_{1,\mathcal{T}_\hbar}\big). \tag{11.12}$$

Proof We point out that this proof is the fully discrete analogue to the proof of the corresponding result in the continuous setting, i.e., Theorem 9.10. Of course, some adaptations have to be made in the discrete setting.

As in the continuous setting, we take the discrete time derivative (11.8) of the discrete variation-of-constants formula (11.7). This yields

$$\underline{\partial_\tau u}_{\mathrm{CN}}^{n+1/2} = \frac{\underline{\mathcal{S}}_{\mathrm{CN}}^{n+1} - \underline{\mathcal{S}}_{\mathrm{CN}}^{n}}{\tau} \underline{u}_\pi^0 + \sum_{j=0}^{n} \underline{\mathcal{S}}_{\mathrm{CN}}^{n-j} \underline{\mathcal{R}}_{\mathrm{CN}} \overline{\underline{f}_\pi}^{j+1/2} - \sum_{j=0}^{n-1} \underline{\mathcal{S}}_{\mathrm{CN}}^{n-1-j} \underline{\mathcal{R}}_{\mathrm{CN}} \overline{\underline{f}_\pi}^{j+1/2}.$$

We now take norms and use the triangle inequality to split the first term from the sums. The former contains the discrete derivative of the discrete semigroup and we thus use Proposition 11.4 together with Proposition 11.3 and Lemma 9.15 to bound it by

$$\|\frac{\underline{\mathcal{S}}_{\mathrm{CN}}^{n+1} - \underline{\mathcal{S}}_{\mathrm{CN}}^{n}}{\tau} \underline{u}_\pi^0\|_\Omega \leq \|\underline{\mathcal{L}u}_\pi^0\|_\Omega. \tag{11.13}$$

The remaining terms can be seen as the discrete derivative of a discrete convolution. We proceed analogously to the continuous case and use a discrete variant of the Leibniz integration rule to carry out the differentiation. As in the continuous case, we have two options (which were illustrated in the continuous case by transforming one of the integrals), which lead to either the inhomogeneity or the semigroup being differentiated.

In particular, this is done by merging the sums to obtain

$$\sum_{j=0}^{n} \underline{\mathcal{S}}_{\mathrm{CN}}^{n-j} \underline{\mathcal{R}}_{\mathrm{CN}} \overline{f}_{\pi}^{j+1/2} - \sum_{j=0}^{n-1} \underline{\mathcal{S}}_{\mathrm{CN}}^{(n-j)-1} \underline{\mathcal{R}}_{\mathrm{CN}} \overline{f}_{\pi}^{j+1/2}$$
$$= \begin{cases} \underline{\mathcal{R}}_{\mathrm{CN}} \overline{f}_{\pi}^{1/2} + \tau \sum_{j=1}^{n} \underline{\mathcal{S}}_{\mathrm{CN}}^{n-j} \underline{\mathcal{R}}_{\mathrm{CN}} \dfrac{f_{\pi}^{j+1} - f_{\pi}^{j-1}}{2\tau}, \\ \underline{\mathcal{R}}_{\mathrm{CN}} \overline{f}_{\pi}^{n+1/2} + \tau \sum_{j=0}^{n-1} \dfrac{\underline{\mathcal{S}}_{\mathrm{CN}}^{n-j} - \underline{\mathcal{S}}_{\mathrm{CN}}^{(n-j)-1}}{\tau} \underline{\mathcal{R}}_{\mathrm{CN}} \overline{f}_{\pi}^{j+1/2}, \end{cases} \tag{11.14}$$

where we have used an index shift and

$$\frac{\overline{f}_{\pi}^{j+1/2} - \overline{f}_{\pi}^{j-1/2}}{\tau} = \frac{f_{\pi}^{j+1} - f_{\pi}^{j-1}}{2\tau}, \qquad j \in \mathbb{N},$$

for the first case. We now split the inhomogeneity $f = f_1 + f_2$ and apply the first equality in (11.14) to f_1 and the second to f_2. Using Propositions 11.3 and 11.4 as well as Lemma 9.15 to bound the individual terms after using the triangle inequality yields the desired bound and thus (11.10) by (11.13).

The second bound (11.11) is derived by taking the norm in (11.9). Lastly, (11.12) follows from (11.10) by using (10.11) on both terms containing the discrete operator $\underline{\mathcal{L}}$, the contractivity of $\pi_{\tilde{h}}$ and an application of the fundamental theorem of calculus for f_1. □

We point out that we used the fact that the discrete operators $\underline{\mathcal{L}}$ and $\underline{\mathcal{R}}_{\mathrm{CN}}$ commute to obtain the bound for f_2 and that this does not work out for the other methods we investigate here. For the other methods, we therefore treat the whole inhomogeneity in the way we treated f_1 here.

11.2 Leapfrog Scheme

The second scheme we consider is the well-known explicit *leapfrog* or *Verlet scheme*, cf., among many others, [81, 82]. It is probably one of the most popular and most widely used time integration schemes for hyperbolic problems, mostly due to its favorable geometric properties, its hard-to-beat efficiency, as well as its rather easy implementation. Both last properties are the result of the explicit nature of this scheme—as is its main caveat. Namely, as all explicit schemes, the leapfrog scheme suffers from the fact that it is only stable under a timestep size restriction, the famous CFL condition (CFL stands for Courant,

Friedrichs, and Levy, who described this phenomenon in [33]). This can be especially detrimental if the method is applied to partial differential equations, as the stepsize is then governed by the smallest element diameter of the spatial mesh (as we will see later in Assumption 11.11). Hence, even one tiny mesh element may lead to prohibitively small timestep sizes that are governed solely by stability, not by the desired accuracy.

As for the Crank–Nicolson scheme, we start with a short derivation of the method to gain some intuition.

Example 11.7 As the most simple model problem, we consider a system of m_q undamped harmonic oscillators written in the following first-order formulation

$$\begin{cases} \partial_t p(t) = \Omega_q\, q(t), & t \in \mathbb{R}_+, \quad (11.15a) \\ \partial_t q(t) = \Omega_p\, p(t), & t \in \mathbb{R}_+, \quad (11.15b) \\ \quad p(0) = p^0, \quad q(0) = q^0, & \quad (11.15c) \end{cases}$$

where $\Omega_p \in \mathbb{R}^{m_q \times m_p}$ and $\Omega_q = -\Omega_p^T$. Here, we search for solutions $q\colon \mathbb{R}_+ \to \mathbb{R}^{m_q}$, $p\colon \mathbb{R}_+ \to \mathbb{R}^{m_p}$ for given initial values $q^0 \in \mathbb{R}^{m_q}$, $p^0 \in \mathbb{R}^{m_p}$. The equivalent second-order differential equation reads

$$\begin{cases} \partial_t^2 q(t) = \Omega_p \Omega_q\, q(t), & t \in \mathbb{R}_+, \quad (11.16a) \\ \quad q(0) = q^0, \quad \partial_t q(0) = \Omega_p p^0. & \quad (11.16b) \end{cases}$$

The leapfrog scheme is originally designed for second-order differential equations. Since we are interested in first-order problems, we only consider its one-step form. One of several possible ways to derive it is to interpret it as a Strang splitting method [82, Section II.5] applied to (11.15) with two half-steps for (11.15b) and a full step for (11.15a). This yields the *one-step leapfrog* or *Verlet scheme*

$$\begin{cases} q_\tau^{n+1/2} - q_\tau^n = \frac{\tau}{2}\, \Omega_p\, p_\tau^n, & (11.17a) \\ p_\tau^{n+1} - p_\tau^n = \tau\, \Omega_q\, q_\tau^{n+1/2}, & (11.17b) \\ q_\tau^{n+1} - q_\tau^{n+1/2} = \frac{\tau}{2}\, \Omega_p\, p_\tau^{n+1}, & (11.17c) \\ p_\tau^0 = p^0, \quad q_\tau^0 = q^0. & (11.17d) \end{cases}$$

Note that all approximations (i.e., discrete derivatives and averages) used in this derivation yield order two approximations to their respective continuous counterparts, as can be easily verified by Taylor expansion. Hence, we expect the resulting method to be of second order in time. Further, we see that the scheme is indeed explicit, since in each formula (11.17a)–(11.17c), the new unknown is given explicitly.

We point out that in the fully discrete case we will be looking at, we still need to solve systems, namely ones given by the mass matrix of the dG method. However, this mass matrix is block-diagonal with block sizes only depending upon the polynomial degree k used in the dG method and the spatial dimension d. Thus, the scheme will still be almost explicit (or quasi-explicit), since such systems are solved without much overhead. Alternatively, mass lumping techniques can be used to render the method essentially explicit with only a diagonal system left to be solved.

In this work, we are interested in more general problems that include inhomogeneities and damping. Hence, we now consider a more complex model problem, namely the system of m_q driven harmonic oscillators

$$\begin{cases} \partial_t p(t) = \Omega_q\, q(t) - \omega_0\, p(t) + g(t), & t \in \mathbb{R}_+, \quad (11.18a) \\ \partial_t q(t) = \Omega_p\, p(t), & t \in \mathbb{R}_+, \quad (11.18b) \\ p(0) = p^0, \quad q(0) = q^0. & \quad (11.18c) \end{cases}$$

which is a damped harmonic oscillator with external forcing $g\colon \mathbb{R}_+ \to \mathbb{R}^{m_p}$ and diagonal, positive semidefinite, damping matrix $\omega_0 \in \mathbb{R}^{m_p \times m_p}$.

Compared to (11.17), we only have to deal with the additional damping and forcing terms and we do so by using a Crank–Nicolson approximation (i.e., the trapezoidal rule). This gives the *one-step leapfrog* or *Verlet scheme* applied to the first-order system (11.18) as

$$\begin{cases} q_\tau^{n+1/2} - q_\tau^n = \frac{\tau}{2}\, \Omega_p\, p_\tau^n, & (11.19a) \\ p_\tau^{n+1} - p_\tau^n = \tau\, \Omega_q\, q_\tau^{n+1/2} - \frac{\tau}{2}\omega_0\, (p_\tau^{n+1} + p_\tau^n) + \tau \overline{g}^{n+1/2}, & (11.19b) \\ q_\tau^{n+1} - q_\tau^{n+1/2} = \frac{\tau}{2}\, \Omega_p\, p_\tau^{n+1}, & (11.19c) \\ p_\tau^0 = p^0, \quad q_\tau^0 = q^0. & (11.19d) \end{cases}$$

Here, for g, we used an analogous short notation to that introduced for f in (11.1). We stress that all approximations made to derive (11.19) are still of second order and we thus again expect the scheme to have second-order accuracy. However, because of the damping term in (11.19b), it now becomes implicit in the p-variable if damping is present. Despite that, in our application, it will still be quasi-explicit as the damping matrix ω_0 is diagonal. In practice, this leads to linear systems that are similarly efficient to solve as those with the mass matrix. Moreover, one can still eliminate these additional systems by using a mass lumping technique. ◇

As we have seen in Example 11.7, the leapfrog scheme requires a certain structure, since it is originally designed for second-order problems (in time). Thus, to apply the method, we assume that the wave-type problem (9.1) has a two-field structure. We point out that this is in fact a reasonable assumption, since many problems in practice exhibit this structure (e.g., Hamiltonian systems). See also the examples in Sect. 9.3 where both the acoustic wave equation as well as the Maxwell equations exhibit this structure. Further, the elastic wave equation and equations governed by the Hodge–Dirac operator are of such a two-field structure.

Assumption 11.8

(a) The Friedrichs' operator $\mathcal{L}$, the material tensor M and the inhomogeneity f are given as

$$\mathcal{L} = \begin{pmatrix} K & \mathcal{L}_q \\ \mathcal{L}_p & 0 \end{pmatrix}, \qquad M = \begin{pmatrix} M_p & 0 \\ 0 & M_q \end{pmatrix}, \qquad \text{and} \qquad f = \begin{pmatrix} g \\ 0 \end{pmatrix}, \tag{11.20}$$

where $g \in C(\mathbb{R}_+; L^2(\Omega)^{m_p})$, $M_p \in L^\infty(\Omega)^{m_p \times m_p}$, $M_q \in L^\infty(\Omega)^{m_q \times m_q}$, and $K = M_p^{-1}\widetilde{K}$ with $\widetilde{K} \in L^\infty(\Omega)^{m_p \times m_p}$ symmetric and negative semidefinite. Moreover, we have $m = m_p + m_q$. The operators $\mathcal{L}_p$ and $\mathcal{L}_q$ are given by

$$M_q \mathcal{L}_p p = \sum_{i=1}^{d} L_{i,p} \partial_i p, \qquad M_p \mathcal{L}_q q = \sum_{i=1}^{d} L_{i,p}^{\mathsf{T}} \partial_i q,$$

with $L_{i,p} \in \mathbb{R}^{m_q \times m_p}$ for $i = 0, \dots, d$ and their graph spaces are defined similarly to (9.27) as

$$H(\mathcal{L}_p) = \{p \in L^2(\Omega)^{m_p} \mid \mathcal{L}_p p \in L^2(\Omega)^{m_p}\}$$

and analogously for $\mathcal{L}_q$.

(b) The boundary condition does not introduce damping, i.e., the boundary operator $\mathcal{L}_\Gamma$ from Assumption 9.21 satisfies

$$\langle \mathcal{L}_\Gamma v \mid v \rangle = 0 \quad \text{for all } v \in H(\mathcal{L}). \tag{11.21}$$

Lastly, there exist $\mathcal{L}_{p,\Gamma} \in \mathcal{B}(H(\mathcal{L}_p), H(\mathcal{L}_p)')$ and $\mathcal{L}_{q,\Gamma} \in \mathcal{B}(H(\mathcal{L}_q), H(\mathcal{L}_q)')$ such that

$$\langle \mathcal{L}_\Gamma v \mid w \rangle = \langle \mathcal{L}_{p,\Gamma} p \mid \tilde{q} \rangle + \langle \mathcal{L}_{q,\Gamma} q \mid \tilde{p} \rangle \tag{11.22}$$

for all $v = \begin{pmatrix} p \\ q \end{pmatrix}, w = \begin{pmatrix} \tilde{p} \\ \tilde{q} \end{pmatrix} \in H(\mathcal{L}_p) \times H(\mathcal{L}_q)$. ◊

Note that Assumption 11.8 is consistent with the definition of the corresponding objects in Chap. 9. From now on, Assumption 11.8 is implicitly made each time we consider the leapfrog scheme. For the other methods we consider here, it does not have to be fulfilled if not stated otherwise (cf., Sect. 11.4).

The last two assumptions on the boundary operator $\mathcal{L}_\Gamma$ made in Assumption 11.8 might be expendable. However, one would have to adapt the design of the method (and probably its analysis) to the more general case.

Clearly, M_p and M_q inherit the properties of the full material tensor and, with a slight abuse of notation, we denote the L^2-inner product weighted by M_p as

$$\left(p \mid \tilde{p}\right)_K = \left(M_p p \mid \tilde{p}\right)_{L^2,K} \quad \text{for } p, \tilde{p} \in L^2(K)^{m_p},$$

and analogously for M_q. The induced norms are denoted accordingly. This does not introduce ambiguity since it is always clear, which weight is used by the arguments of the inner product and the norms, respectively.

We now state some important implications of this assumption. For the sake of presentation, we only provide the results that we need for the analysis of the leapfrog scheme and omit the technical details. However, for completeness, we give more detailed deductions as well as supplemental details in Sect. 13.1.

From Assumption 11.8, we obtain that the discrete operator $\underline{\mathcal{L}}$ inherits the structure of its continuous counterpart. More precisely, we have

$$\underline{\mathcal{L}} = \begin{pmatrix} K & \underline{\mathcal{L}}_q \\ \underline{\mathcal{L}}_p & 0 \end{pmatrix}. \tag{11.23}$$

We point out that $\underline{\mathcal{L}}_p$ and $\underline{\mathcal{L}}_q$ are the dG discretizations of $\mathcal{L}_p$ and $\mathcal{L}_q$, respectively. Moreover, as a consequence of Proposition 10.6 (maximal dissipativity), $\underline{\mathcal{L}}_p$ and $\underline{\mathcal{L}}_q$ share an adjointness property on the discrete space given by

$$\left(\underline{\mathcal{L}}_p \boldsymbol{p} \mid \boldsymbol{q}\right)_\Omega = -\left(\boldsymbol{p} \mid \underline{\mathcal{L}}_p \boldsymbol{q}\right)_\Omega \qquad \text{for all } \boldsymbol{p} \in \mathbb{V}_h^{m_p},\ \boldsymbol{q} \in \mathbb{V}_h^{m_q}, \tag{11.24}$$

where the inner products are L^2-inner products weighted by M_q and M_p, respectively, analogous to (9.4). In fact, this property is inherited from their continuous counterparts, which exhibit the same relation on their respective domains, cf., Sect. 13.1.

The discrete operators $\underline{\mathcal{L}}_p$ and $\underline{\mathcal{L}}_q$ also inherit the other properties of the full operator $\underline{\mathcal{L}}$ from Chap. 10. We only state them for $\underline{\mathcal{L}}_p$ as they are completely analogous for $\underline{\mathcal{L}}_q$.

Firstly, by Proposition 10.7 (consistency) we have

$$\underline{\mathcal{L}}_p p = \pi_{\hbar} \mathcal{L}_p p \quad \text{for all } p \in D(\mathcal{L}_p) \cap H^1(\mathcal{T}_{\hbar})^{m_p}. \tag{11.25}$$

Further, both Proposition 10.8 (inverse inequality) as well as Proposition 10.9 (approximation property) carry over, i.e., for $j \in \mathbb{Z}$ we have

$$\|h^j \underline{\mathcal{L}}_p \boldsymbol{p}\|_\Omega \leq C_{\mathrm{inv},\mathcal{L},j} \|h^{j-1} \boldsymbol{p}\|_\Omega \quad \text{for all } \boldsymbol{p} \in \mathbb{V}_{\hbar}^{m_p}, \tag{11.26}$$

and

$$\|h^j \underline{\mathcal{L}}_p e_\pi^p\|_\Omega \leq C_{\pi,\mathcal{L},j} |h^{j+\ell} p|_{\ell+1,\mathcal{T}_h} \quad \text{for all } p \in D(\mathcal{L}_p) \cap H^{\ell+1}(\mathcal{T}_h)^{m_p}. \tag{11.27}$$

By this, also an analogous result to Lemma 10.10 carries over to those operators, which we do not explicitly state here.

Now, by Assumption 11.8, the original problem (9.3) takes the form

$$\begin{cases} \partial_t p = \mathcal{L}_q q + Kp + g, & \text{on } \mathbb{R}_+, \quad (11.28a) \\ \partial_t q = \mathcal{L}_p p, & \text{on } \mathbb{R}_+, \quad (11.28b) \\ p(0) = p^0, \quad q(0) = q^0, & \quad (11.28c) \end{cases}$$

where $u^0 = \begin{pmatrix} p^0 \\ q^0 \end{pmatrix} \in L^2(\Omega)^{m_p} \times L^2(\Omega)^{m_q}$. Further, by (11.23), we can rewrite the spatially discrete wave-type problem (10.7) as

$$\begin{cases} \partial_t \boldsymbol{p} = \underline{\mathcal{L}}_q \boldsymbol{q} + K \boldsymbol{p} + \boldsymbol{g}_\pi, & \text{on } \mathbb{R}_+, \quad (11.29a) \\ \partial_t \boldsymbol{q} = \underline{\mathcal{L}}_p \boldsymbol{p}, & \text{on } \mathbb{R}_+, \quad (11.29b) \\ \boldsymbol{p}(0) = \boldsymbol{p}_\pi^0, \quad \boldsymbol{q}(0) = \boldsymbol{q}_\pi^0, & \quad (11.29c) \end{cases}$$

where

$$\boldsymbol{g}_\pi = \pi_h g, \qquad \boldsymbol{p}_\pi^0 = \pi_h p^0, \quad \text{and} \quad \boldsymbol{q}_\pi^0 = \pi_h q^0.$$

This enables us to apply the leapfrog scheme to the wave-type problem (9.3) in the following way. With this, we obtain the *fully discrete dG-leapfrog (dG-LF)* (or *Verlet*) *scheme* given by

$$\begin{cases} \boldsymbol{q}_{\mathrm{LF}}^{n+1/2} - \boldsymbol{q}_{\mathrm{LF}}^{n} = \frac{\tau}{2} \underline{\mathcal{L}}_p \boldsymbol{p}_{\mathrm{LF}}^{n}, & (11.30a) \\ \boldsymbol{p}_{\mathrm{LF}}^{n+1} - \boldsymbol{p}_{\mathrm{LF}}^{n} = \tau \underline{\mathcal{L}}_q \boldsymbol{q}_{\mathrm{LF}}^{n+1/2} + \frac{\tau}{2} K (\boldsymbol{p}_{\mathrm{LF}}^{n+1} + \boldsymbol{p}_{\mathrm{LF}}^{n}) + \tau \overline{\boldsymbol{g}}_\pi^{n+1/2}, & (11.30b) \\ \boldsymbol{q}_{\mathrm{LF}}^{n+1} - \boldsymbol{q}_{\mathrm{LF}}^{n+1/2} = \frac{\tau}{2} \underline{\mathcal{L}}_p \boldsymbol{p}_{\mathrm{LF}}^{n+1}, & (11.30c) \\ \boldsymbol{p}_{\mathrm{LF}}^0 = \boldsymbol{p}_\pi^0, \quad \boldsymbol{q}_{\mathrm{LF}}^0 = \boldsymbol{q}_\pi^0. & (11.30d) \end{cases}$$

Note that the approximations given by (11.30) exist and are unique. If no damping is present, i.e., $K = 0$, this can be seen directly as all unknowns are given explicitly in each step. With damping, the only value defined implicitly by (11.30) is the new approximation $\boldsymbol{p}_{\mathrm{LF}}^{n+1}$ of the p-field. To obtain it, we need to solve a linear system with $\underline{I} - \frac{\tau}{2} K$. Since

$\widetilde{K}$ is symmetric and negative semidefinite, the discrete operator $K_{|\mathbb{V}_h^{m_p}}$ is dissipative w.r.t. $(\cdot \mid \cdot)_\Omega$ (and maximal because of the finite dimensionality). Hence, by Lemma 9.15 (i), the resolvent $(\underline{\mathcal{I}} - \frac{\tau}{2}K)^{-1} : \mathbb{V}_h^{m_p} \to \mathbb{V}_h^{m_p}$ is an isomorphism and therefore, the system is uniquely solvable.

Next, we reformulate the leapfrog scheme (11.30) in a form similar to the Crank–Nicolson scheme (11.3). To do so, we first eliminate $\boldsymbol{q}_{\mathrm{LF}}^{n+1/2}$ by adding (11.30a) and (11.30c), yielding

$$\boldsymbol{q}_{\mathrm{LF}}^{n+1} - \boldsymbol{q}_{\mathrm{LF}}^{n} = \tfrac{\tau}{2}\underline{\mathcal{L}}_p(\boldsymbol{p}_{\mathrm{LF}}^{n+1} + \boldsymbol{p}_{\mathrm{LF}}^{n+1}), \tag{11.31}$$

and then subtract (11.30c) from (11.30a) to obtain

$$\boldsymbol{q}_{\mathrm{LF}}^{n+1/2} = \frac{\boldsymbol{q}_{\mathrm{LF}}^{n+1} + \boldsymbol{q}_{\mathrm{LF}}^{n}}{2} - \tfrac{\tau}{4}\underline{\mathcal{L}}_p(\boldsymbol{p}_{\mathrm{LF}}^{n+1} - \boldsymbol{p}_{\mathrm{LF}}^{n}). \tag{11.32}$$

Using (11.32) in (11.30b) yields, together with (11.31), an equivalent reformulation of the scheme. Rearranging the terms and using the vector notation $\boldsymbol{u}_{\mathrm{LF}}^n = \begin{pmatrix} \boldsymbol{p}_{\mathrm{LF}}^n \\ \boldsymbol{q}_{\mathrm{LF}}^n \end{pmatrix}$, $n \in \mathbb{N}_0$, we end up with

$$\begin{cases} (\underline{\mathcal{I}} - \frac{\tau}{2}\underline{\mathcal{L}} + \frac{\tau^2}{4}\underline{\mathfrak{D}})\boldsymbol{u}_{\mathrm{LF}}^{n+1} = (\underline{\mathcal{I}} + \frac{\tau}{2}\underline{\mathcal{L}} + \frac{\tau^2}{4}\underline{\mathfrak{D}})\boldsymbol{u}_{\mathrm{LF}}^{n} + \tau \boldsymbol{f}_\pi^{n+1/2}, & \text{(11.33a)} \\ \boldsymbol{u}_{\mathrm{LF}}^0 = \boldsymbol{u}_\pi^0, & \text{(11.33b)} \end{cases}$$

where the discrete perturbation $\underline{\mathfrak{D}}$ operator $\underline{\mathfrak{D}}$ of the leapfrog scheme is given by

$$\underline{\mathfrak{D}} = \begin{pmatrix} \underline{\mathcal{L}}_q\underline{\mathcal{L}}_p & 0 \\ 0 & 0 \end{pmatrix}. \tag{11.34}$$

Note that the reformulated leapfrog scheme (11.33) and the Crank–Nicolson scheme (11.3) only differ by a τ^2 perturbation on both sides.

Before we analyze the scheme, we give two important properties of the perturbation operator $\underline{\mathfrak{D}}$. They are both immediate consequences of the adjointness property (11.24), and the proof is therefore left to the reader.

Lemma 11.9 *The discrete perturbation operator $\underline{\mathfrak{D}}$ of the leapfrog method is self-adjoint and negative semidefinite, i.e., for $\boldsymbol{v}, \boldsymbol{w} \in \mathbb{V}_h^m = \mathbb{V}_h^{m_p} \times \mathbb{V}_h^{m_q}$, we have*

$$(\underline{\mathfrak{D}}\boldsymbol{v} \mid \boldsymbol{w})_\Omega = (\boldsymbol{v} \mid \underline{\mathfrak{D}}\boldsymbol{w})_\Omega,$$

and for $\boldsymbol{v} = \begin{pmatrix} \boldsymbol{p} \\ \boldsymbol{q} \end{pmatrix}$ *with* $\boldsymbol{p} \in \mathbb{V}_{\hbar}^{m_p}$ *and* $\boldsymbol{q} \in \mathbb{V}_{\hbar}^{m_q}$, *we have*

$$\left(\underline{\mathcal{D}}\boldsymbol{v} \mid \boldsymbol{v}\right)_{\Omega} = -\|\underline{\mathcal{L}}_p \boldsymbol{p}\|_{\Omega}^2 \leq 0.$$

Consequently, analogously to the Crank–Nicolson scheme, we rewrite the leapfrog scheme (11.33) in a more compact form by solving (11.33a) for $\boldsymbol{u}_{\mathrm{LF}}^{n+1}$. We point out that this is possible since (11.30) provides unique approximants and (11.33) is just an equivalent reformulation. In particular, this means that the discrete operators $\underline{\mathcal{R}}_{\mathrm{LF}}, \underline{\mathcal{S}}_{\mathrm{LF}} : \mathbb{V}_{\hbar}^{m} \to \mathbb{V}_{\hbar}^{m}$ defined by

$$\underline{\mathcal{R}}_{\mathrm{LF}} = (\underline{\mathcal{I}} - \tfrac{\tau}{2}\underline{\mathcal{L}} + \tfrac{\tau^2}{4}\underline{\mathcal{D}})^{-1}, \tag{11.35}$$

and

$$\underline{\mathcal{S}}_{\mathrm{LF}} = (\underline{\mathcal{I}} - \tfrac{\tau}{2}\underline{\mathcal{L}} + \tfrac{\tau^2}{4}\underline{\mathcal{D}})^{-1}(\underline{\mathcal{I}} + \tfrac{\tau}{2}\underline{\mathcal{L}} + \tfrac{\tau^2}{4}\underline{\mathcal{D}}) \tag{11.36}$$

exist. With this, we rewrite the leapfrog scheme (11.33) as

$$\begin{cases} \boldsymbol{u}_{\mathrm{LF}}^{n+1} = \underline{\mathcal{S}}_{\mathrm{LF}}\boldsymbol{u}_{\mathrm{LF}}^{n} + \tau\underline{\mathcal{R}}_{\mathrm{LF}}\overline{\boldsymbol{f}}_{\pi}^{n+1/2}, & n \in \mathbb{N}_0, \\ \boldsymbol{u}_{\mathrm{LF}}^{0} = \boldsymbol{u}_{\pi}^{0}. \end{cases}$$

Solving this recursion readily yields the following discrete variation-of-constants formula for the numerical solution.

Theorem 11.10 *For all* $n \in \mathbb{N}_0$, *all* $\hbar \in \mathcal{H}$, *and all* $\tau > 0$, *there exists a unique* $\boldsymbol{u}_{\mathrm{LF}}^{n+1} \in \mathbb{V}_{\hbar}^{m}$ *fulfilling the dG-LF scheme* (11.33). *It is given by the discrete variation-of-constants formula*

$$\boldsymbol{u}_{\mathrm{LF}}^{n+1} = \underline{\mathcal{S}}_{\mathrm{LF}}^{n+1}\boldsymbol{u}_{\pi}^{0} + \tau\sum_{j=0}^{n}\underline{\mathcal{S}}_{\mathrm{LF}}^{n-j}\underline{\mathcal{R}}_{\mathrm{LF}}\overline{\boldsymbol{f}}_{\pi}^{j+1/2}. \tag{11.37}$$

Note that this is structurally the same discrete variation-of-constants formula as the one for the Crank–Nicolson scheme (in Theorem 11.2) with the operators $\underline{\mathcal{S}}_{\mathrm{CN}}$ and $\underline{\mathcal{R}}_{\mathrm{CN}}$ replaced by their leapfrog counterparts. In particular, we can again interpret $\underline{\mathcal{S}}_{\mathrm{LF}}$ as a discrete semigroup.

However, a power bound for $\underline{\mathcal{S}}_{\mathrm{LF}}$ is not as easily derived as for the Crank–Nicolson method. In fact, this is where the conditional stability of the explicit leapfrog scheme comes into play, meaning that we need an additional condition on the discretization parameters, the famous CFL condition. We therefore make the following assumption.

Assumption 11.11 (CFL Condition for the Leapfrog Scheme) For an arbitrary but fixed parameter $0 < \theta < 1$, the timestep τ satisfies the CFL condition

$$\tau \leq \frac{2\theta}{C_{\mathrm{inv},\mathcal{L}}} h_{\min}, \tag{11.38}$$

where $h_{\min}$ is defined in the beginning of Sect. 10.1. ◇

Remark 11.12 We point out that, hidden in the constant $C_{\mathrm{inv},\mathcal{L}}$, we have a (reciprocal) dependency on the material parameters M_p and M_q. This can be seen since $C_{\mathrm{inv},\mathcal{L}}$ contains $C_{\mathcal{L}}$ defined in (10.3), where said dependency is evident. Therefore, the CFL condition is not only governed by the smallest diameter in the mesh, but also by the material parameters. ◇

Provided the CFL condition (11.38) holds, we can prove the following power bound on the discrete semigroup. We use an energy technique to show the result, but point out that a simpler proof based on the adjointness properties of $\underline{\mathcal{L}}_p$ and $\underline{\mathcal{L}}_q$ is possible in the absence of damping. This simpler proof can be found in [161, Lem. 4.15] for the undamped Maxwell equations.

Proposition 11.13 *Let $h \in \mathcal{H}$ and $\tau > 0$ fulfill Assumption 11.11 with $\theta \in (0, 1)$. Then, for all $j \in \mathbb{N}$ and all $\boldsymbol{v} \in \mathbb{V}_h^m$, the discrete semigroup $\underline{\boldsymbol{S}}_{\mathrm{LF}}$ defined in* (11.36) *fulfills the power bound*

$$\|\underline{\boldsymbol{S}}_{\mathrm{LF}}^j \boldsymbol{v}\|_\Omega \leq C_{\mathrm{stb}} \|\boldsymbol{v}\|_\Omega,$$

where $C_{\mathrm{stb}} = (1 - \theta^2)^{-1/2}$.

Proof We begin by decomposing

$$\underline{\mathcal{L}}_{|\mathbb{V}_h^m} = \underline{\mathcal{L}}_d + \underline{\mathcal{L}}_s, \qquad \underline{\mathcal{L}}_d = \begin{pmatrix} K & 0 \\ 0 & 0 \end{pmatrix} \quad \text{and} \quad \underline{\mathcal{L}}_s = \begin{pmatrix} 0 & \underline{\mathcal{L}}_q \\ \underline{\mathcal{L}}_p & 0 \end{pmatrix},$$

where $\underline{\mathcal{L}}_d$ is self-adjoint and dissipative w.r.t. $(\cdot \mid \cdot)_\Omega$ and $\underline{\mathcal{L}}_s$ is skew-adjoint by the properties of K and (11.24), respectively.

Next, we write $\boldsymbol{v}_j = \underline{\boldsymbol{S}}_{\mathrm{LF}}^j \boldsymbol{v}$, $j \in \mathbb{N}$, and thus have $\boldsymbol{v}_j = \underline{\boldsymbol{S}}_{\mathrm{LF}} \boldsymbol{v}_{j-1}$. Multiplying the latter by $\underline{\mathcal{R}}_{\mathrm{LF}}^{-1}$ and splitting $\underline{\mathcal{L}}$ as above, we end up with

$$\big(\underline{\mathcal{I}} - \tfrac{\tau}{2}\underline{\mathcal{L}}_s + \tfrac{\tau^2}{4}\underline{\mathcal{D}}\big)\boldsymbol{v}_j - \big(\underline{\mathcal{I}} + \tfrac{\tau}{2}\underline{\mathcal{L}}_s + \tfrac{\tau^2}{4}\underline{\mathcal{D}}\big)\boldsymbol{v}_{j-1} = \tfrac{\tau}{2}\underline{\mathcal{L}}_d\big(\boldsymbol{v}_j + \boldsymbol{v}_{j-1}\big).$$

We take the inner product with $\boldsymbol{v}_j + \boldsymbol{v}_{j-1}$ to obtain

$$\begin{aligned}\big(\boldsymbol{v}_j - \boldsymbol{v}_{j-1} \,\big|\, \boldsymbol{v}_j + \boldsymbol{v}_{j-1}\big)_\Omega + \tfrac{\tau^2}{4}\big(\underline{\mathcal{D}}(\boldsymbol{v}_j - \boldsymbol{v}_{j-1}) \,\big|\, \boldsymbol{v}_j + \boldsymbol{v}_{j-1}\big)_\Omega \\ = \tfrac{\tau}{2}\big(\underline{\mathcal{L}}_d\big(\boldsymbol{v}_j + \boldsymbol{v}_{j-1}\big) \,\big|\, \boldsymbol{v}_j + \boldsymbol{v}_{j-1}\big)_\Omega \le 0,\end{aligned} \tag{11.39}$$

where we have used the skew-adjointness of $\underline{\mathcal{L}}_s$ on the left-hand side and the dissipativity of $\underline{\mathcal{L}}_d$ for the inequality. We now write $\boldsymbol{v}_j = \begin{pmatrix} \boldsymbol{p}_j \\ \boldsymbol{q}_j \end{pmatrix}$ with $\boldsymbol{p}_j \in \mathbb{V}_{\hbar}^{m_p}$ and $\boldsymbol{q}_j \in \mathbb{V}_{\hbar}^{m_q}$ for all $j \in \mathbb{N}$. (and analogously for $\boldsymbol{v}$ below). By Lemma 11.9, this yields

$$\begin{aligned}\big(\underline{\mathcal{D}}(\boldsymbol{v}_j - \boldsymbol{v}_{j-1}) \,\big|\, \boldsymbol{v}_j + \boldsymbol{v}_{j-1}\big)_\Omega &= \big(\underline{\mathcal{D}}\boldsymbol{v}_j \,\big|\, \boldsymbol{v}_j\big)_\Omega - \big(\underline{\mathcal{D}}\boldsymbol{v}_{j-1} \,\big|\, \boldsymbol{v}_{j-1}\big)_\Omega \\ &= -\big(\|\underline{\mathcal{L}}_p \boldsymbol{p}_j\|_\Omega^2 - \|\underline{\mathcal{L}}_p \boldsymbol{p}_{j-1}\|_\Omega^2\big).\end{aligned}$$

Using $\big(\boldsymbol{v}_j - \boldsymbol{v}_{j-1} \,\big|\, \boldsymbol{v}_j + \boldsymbol{v}_{j-1}\big)_\Omega = \|\boldsymbol{v}_j\|_\Omega^2 - \|\boldsymbol{v}_{j-1}\|_\Omega^2$ in (11.39), we therefore obtain

$$\|\boldsymbol{v}_j\|_\Omega^2 - \tfrac{\tau^2}{4}\|\underline{\mathcal{L}}_p \boldsymbol{p}_j\|_\Omega^2 \le \|\boldsymbol{v}_{j-1}\|_\Omega^2 - \tfrac{\tau^2}{4}\|\underline{\mathcal{L}}_p \boldsymbol{p}_{j-1}\|_\Omega^2,$$

and thus, by induction,

$$\|\boldsymbol{v}_j\|_\Omega^2 - \tfrac{\tau^2}{4}\|\underline{\mathcal{L}}_p \boldsymbol{p}_j\|_\Omega^2 \le \|\boldsymbol{v}\|_\Omega^2 - \tfrac{\tau^2}{4}\|\underline{\mathcal{L}}_p \boldsymbol{p}\|_\Omega^2 \le \|\boldsymbol{v}\|_\Omega^2 \qquad \text{for all } j \in \mathbb{N}. \tag{11.40}$$

Lastly, using the inverse inequality (11.26) together with $h^{-1} \le \hbar_{\min}^{-1}$ a.e. on Ω and the CFL condition (11.38) yields

$$\tfrac{\tau^2}{4}\|\underline{\mathcal{L}}_p \boldsymbol{p}_j\|_\Omega^2 \le \Big(\frac{\tau C_{\mathrm{inv},\mathcal{L}}}{2\hbar_{\min}}\Big)^2 \|\boldsymbol{p}_j\|_\Omega^2 \le \theta^2 \|\boldsymbol{p}_j\|_\Omega^2 \le \theta^2 \|\boldsymbol{v}_j\|_\Omega^2. \tag{11.41}$$

With this, the assertion follows from (11.40), since $\boldsymbol{v}_j = \underline{\boldsymbol{S}}_{\mathrm{LF}}^j \boldsymbol{v}$. □

To show stability, we also need a bound on the discrete operator $\underline{\mathcal{R}}_{\mathrm{LF}}$ that is independent of the discretization parameters. This can again only be achieved under the CFL condition (11.38), once more reflecting the conditional stability of the leapfrog method. We give the appropriate bound in the following lemma.

Lemma 11.14 *Let $\hbar \in \mathcal{H}$ and $\tau > 0$ fulfill Assumption 11.11 with $\theta \in (0, 1)$. Then, for all $\boldsymbol{v} \in \mathbb{V}_{\hbar}^m$, the discrete operator $\underline{\mathcal{R}}_{\mathrm{LF}}$ defined in (11.35) fulfills*

$$\|\underline{\mathcal{R}}_{\mathrm{LF}} \boldsymbol{v}\|_\Omega \le \sqrt{3}\, \|\boldsymbol{v}\|_\Omega. \tag{11.42}$$

Proof Inverting $\underline{\mathcal{R}}_{\mathrm{LF}}^{-1} = \underline{I} - \frac{\tau}{2}\underline{\mathcal{L}} + \frac{\tau^2}{4}\underline{\mathfrak{D}}$ via a block Gauss procedure (exploiting the block structures (11.23) and (11.34) of $\underline{\mathcal{L}}$ and $\underline{\mathfrak{D}}$, respectively), we obtain

$$\underline{\mathcal{R}}_{\mathrm{LF}} = \begin{pmatrix} Y_{\mathrm{LF}} & \frac{\tau}{2} Y_{\mathrm{LF}} \underline{\mathcal{L}}_q \\ \frac{\tau}{2}\underline{\mathcal{L}}_p Y_{\mathrm{LF}} & \underline{I} + \frac{\tau^2}{4}\underline{\mathcal{L}}_p Y_{\mathrm{LF}} \underline{\mathcal{L}}_q \end{pmatrix}, \tag{11.43}$$

where $Y_{\mathrm{LF}} = (\underline{I} - \frac{\tau}{2}K)^{-1}$. Note that Y_{LF} is a contractive isomorphism on $\mathbb{V}_{\hbar}^{m_p}$ by Lemma 9.15 (i), since $K_{|\mathbb{V}_{\hbar}^{m_p}}$ is maximal dissipative (cf., the paragraph after (11.30)). Further, since $K_{|\mathbb{V}_{\hbar}^{m_p}}$ is self-adjoint, so is Y_{LF}. Therefore, its square root $Y_{\mathrm{LF}}^{1/2}$ exists and is a contractive, self-adjoint isomorphism as well.

To show (11.42), we write $\boldsymbol{v} = \begin{pmatrix} \boldsymbol{p} \\ \boldsymbol{q} \end{pmatrix}$ with $\boldsymbol{p} \in \mathbb{V}_{\hbar}^{m_p}$ and $\boldsymbol{q} \in \mathbb{V}_{\hbar}^{m_q}$. Then, by (11.43), we have

$$\begin{aligned} \|\underline{\mathcal{R}}_{\mathrm{LF}} \boldsymbol{v}\|_\Omega^2 &= \|Y_{\mathrm{LF}}\boldsymbol{p} + \tfrac{\tau}{2} Y_{\mathrm{LF}} \underline{\mathcal{L}}_q \boldsymbol{q}\|_\Omega^2 + \|\tfrac{\tau}{2}\underline{\mathcal{L}}_p Y_{\mathrm{LF}}\boldsymbol{p} + \boldsymbol{q} + \tfrac{\tau^2}{4}\underline{\mathcal{L}}_p Y_{\mathrm{LF}} \underline{\mathcal{L}}_q \boldsymbol{q}\|_\Omega^2 \\ &\le \|Y_{\mathrm{LF}}^{1/2}\boldsymbol{p} + \tfrac{\tau}{2} Y_{\mathrm{LF}}^{1/2} \underline{\mathcal{L}}_q \boldsymbol{q}\|_\Omega^2 + \|\tfrac{\tau}{2}\underline{\mathcal{L}}_p Y_{\mathrm{LF}}\boldsymbol{p} + \boldsymbol{q} + \tfrac{\tau^2}{4}\underline{\mathcal{L}}_p Y_{\mathrm{LF}} \underline{\mathcal{L}}_q \boldsymbol{q}\|_\Omega^2, \end{aligned}$$

where we have used the contractivity of $Y_{\mathrm{LF}}^{1/2}$. Using the binomial formula for inner products and the properties of $Y_{\mathrm{LF}}^{1/2}$, we obtain

$$\|Y_{\mathrm{LF}}^{1/2}\boldsymbol{p} + \tfrac{\tau}{2} Y_{\mathrm{LF}}^{1/2} \underline{\mathcal{L}}_q \boldsymbol{q}\|_\Omega^2 \le \|\boldsymbol{p}\|_\Omega^2 + 2\left(\boldsymbol{p} \,\middle|\, \tfrac{\tau}{2} Y_{\mathrm{LF}} \underline{\mathcal{L}}_q \boldsymbol{q}\right)_\Omega + \tfrac{\tau^2}{4}\|Y_{\mathrm{LF}}^{1/2}\underline{\mathcal{L}}_q \boldsymbol{q}\|_\Omega^2$$

and

$$\begin{aligned} &\|\tfrac{\tau}{2}\underline{\mathcal{L}}_p Y_{\mathrm{LF}}\boldsymbol{p} + \boldsymbol{q} + \tfrac{\tau^2}{4}\underline{\mathcal{L}}_p Y_{\mathrm{LF}} \underline{\mathcal{L}}_q \boldsymbol{q}\|_\Omega^2 \\ &\quad = \|\boldsymbol{q} + \tfrac{\tau^2}{4}\underline{\mathcal{L}}_p Y_{\mathrm{LF}} \underline{\mathcal{L}}_q \boldsymbol{q}\|_\Omega^2 + 2\left(\boldsymbol{q} + \tfrac{\tau^2}{4}\underline{\mathcal{L}}_p Y_{\mathrm{LF}} \underline{\mathcal{L}}_q \boldsymbol{q} \,\middle|\, \tfrac{\tau}{2}\underline{\mathcal{L}}_p Y_{\mathrm{LF}}\boldsymbol{p}\right)_\Omega + \tfrac{\tau^2}{4}\|\underline{\mathcal{L}}_p Y_{\mathrm{LF}}\boldsymbol{p}\|_\Omega^2. \end{aligned}$$

Further, additionally using the adjointness property (11.24), we have

$$\begin{aligned} \|\boldsymbol{q} + \tfrac{\tau^2}{4}\underline{\mathcal{L}}_p Y_{\mathrm{LF}} \underline{\mathcal{L}}_q \boldsymbol{q}\|_\Omega^2 &= \|\boldsymbol{q}\|_\Omega^2 + 2\left(\boldsymbol{q} \,\middle|\, \tfrac{\tau^2}{4}\underline{\mathcal{L}}_p Y_{\mathrm{LF}} \underline{\mathcal{L}}_q \boldsymbol{q}\right)_\Omega + \tfrac{\tau^4}{16}\|\underline{\mathcal{L}}_p Y_{\mathrm{LF}} \underline{\mathcal{L}}_q \boldsymbol{q}\|_\Omega^2 \\ &= \|\boldsymbol{q}\|_\Omega^2 - \tfrac{\tau^2}{2}\|Y_{\mathrm{LF}}^{1/2}\underline{\mathcal{L}}_q \boldsymbol{q}\|_\Omega^2 + \tfrac{\tau^4}{16}\|\underline{\mathcal{L}}_p Y_{\mathrm{LF}} \underline{\mathcal{L}}_q \boldsymbol{q}\|_\Omega^2 \end{aligned}$$

and

$$\begin{aligned} &2\left(\boldsymbol{q} + \tfrac{\tau^2}{4}\underline{\mathcal{L}}_p Y_{\mathrm{LF}} \underline{\mathcal{L}}_q \boldsymbol{q} \,\middle|\, \tfrac{\tau}{2}\underline{\mathcal{L}}_p Y_{\mathrm{LF}}\boldsymbol{p}\right)_\Omega \\ &\quad = -2\left(\boldsymbol{p} \,\middle|\, \tfrac{\tau}{2} Y_{\mathrm{LF}} \underline{\mathcal{L}}_q \boldsymbol{q}\right)_\Omega + \left(\tau Y_{\mathrm{LF}}^{1/2}\underline{\mathcal{L}}_q \boldsymbol{q} \,\middle|\, -\tfrac{\tau^2}{4} Y_{\mathrm{LF}}^{1/2}\underline{\mathcal{L}}_q \underline{\mathcal{L}}_p Y_{\mathrm{LF}}\boldsymbol{p}\right)_\Omega. \end{aligned}$$

Now, using the Cauchy–Schwarz and Young's inequality yields

$$\begin{aligned}\big(\tau Y_{\mathrm{LF}}^{1/2}\underline{\mathcal{L}}_q \boldsymbol{q} \mid -\tfrac{\tau^2}{4}Y_{\mathrm{LF}}^{1/2}\underline{\mathcal{L}}_q\underline{\mathcal{L}}_p Y_{\mathrm{LF}}\boldsymbol{p}\big)_\Omega &\leq \tau\|Y_{\mathrm{LF}}^{1/2}\underline{\mathcal{L}}_q\boldsymbol{q}\|_\Omega \tfrac{\tau^2}{4}\|Y_{\mathrm{LF}}^{1/2}\underline{\mathcal{L}}_q\underline{\mathcal{L}}_p Y_{\mathrm{LF}}\boldsymbol{p}\|_\Omega \\ &\leq \tfrac{\tau^2}{2}\|Y_{\mathrm{LF}}^{1/2}\underline{\mathcal{L}}_q\boldsymbol{q}\|_\Omega^2 + \tfrac{\tau^4}{32}\|Y_{\mathrm{LF}}^{1/2}\underline{\mathcal{L}}_q\underline{\mathcal{L}}_p Y_{\mathrm{LF}}\boldsymbol{p}\|_\Omega^2.\end{aligned}$$

Lastly, we piece everything back together, use the contractivity of $Y_{\mathrm{LF}}^{1/2}$ as well as (11.41) (and an analogous result for $\underline{\mathcal{L}}_q$) several times, and obtain

$$\|\underline{\mathcal{R}}_{\mathrm{LF}}\boldsymbol{v}\|_\Omega^2 \leq (1+\theta^2+\tfrac{1}{2}\theta^4)\|\boldsymbol{p}\|_\Omega^2 + (1+\theta^2+\theta^4)\|\boldsymbol{q}\|_\Omega^2 \leq (1+\theta^2+\theta^4)\|\boldsymbol{v}\|_\Omega^2.$$

The claim now follows by using $1+\theta^2+\theta^4 \leq 3$ for $\theta \in (0,1)$. □

As for the Crank–Nicolson scheme, we can also show a discrete version of (9.13b). Again, this means that the discrete derivative of the discrete semigroup $\underline{\mathcal{S}}_{\mathrm{LF}}$ yields its generator, which is given by $\underline{\mathcal{R}}_{\mathrm{LF}}\underline{\mathcal{L}}$. Note that this time, instead of the contractive resolvent $\underline{\mathcal{R}}_{\mathrm{CN}}$ of $\underline{\mathcal{L}}$ (cf., Proposition 11.15), the operator $\underline{\mathcal{R}}_{\mathrm{LF}}$ that can only be bounded uniformly under the CFL condition (11.38), plays a role.

Proposition 11.15 *Let $h \in \mathcal{H}$ and $\tau > 0$. Then, for all $\ell \in \mathbb{N}_0$, the powers of the discrete semigroup $\underline{\mathcal{S}}_{\mathrm{LF}}$ defined in* (11.36) *satisfy*

$$\frac{\underline{\mathcal{S}}_{\mathrm{LF}}^{\ell+1} - \underline{\mathcal{S}}_{\mathrm{LF}}^{\ell}}{\tau} = (\underline{\mathcal{R}}_{\mathrm{LF}}\underline{\mathcal{L}})\,\underline{\mathcal{S}}_{\mathrm{LF}}^{\ell} = \underline{\mathcal{S}}_{\mathrm{LF}}^{\ell}\,(\underline{\mathcal{R}}_{\mathrm{LF}}\underline{\mathcal{L}}). \tag{11.44}$$

Proof As in the proof of Proposition 11.4, we first factor out $\underline{\mathcal{S}}_{\mathrm{LF}}^{\ell}$ either to the left or to the right. Using

$$\frac{\underline{\mathcal{S}}_{\mathrm{LF}} - \underline{\mathcal{I}}}{\tau} = \underline{\mathcal{R}}_{\mathrm{LF}} \frac{\big(\underline{\mathcal{I}} + \tfrac{\tau}{2}\underline{\mathcal{L}} + \tfrac{\tau^2}{4}\underline{\mathcal{D}}\big) - \big(\underline{\mathcal{I}} - \tfrac{\tau}{2}\underline{\mathcal{L}} + \tfrac{\tau^2}{4}\underline{\mathcal{D}}\big)}{\tau} = \underline{\mathcal{R}}_{\mathrm{LF}}\underline{\mathcal{L}}$$

then yields the claim. □

Note that the CFL condition (11.38) is not required for this result. However, we need to bound the discrete derivative in (11.44) uniformly w.r.t. h later, and this can only be achieved if (11.38) is satisfied.

Now, we have all ingredients to show that the dG-LF scheme (11.33) is stable under the CFL condition (11.38). As for the Crank–Nicolson method, we obtain stability both in the $\|\cdot\|_\Omega$-norm as well as in a discrete graph norm, which will be defined below. Since the proofs are completely analogous to the Crank–Nicolson results, i.e., Theorems 11.5 and 11.6, we leave them to the interested reader. The only differences between the proofs are that we exchange the discrete variation-of-constants formula (11.7), Proposition 11.3, and Lemma 9.15 (i) by the corresponding leapfrog results, i.e., (11.37), Proposition 11.13, and Lemma 11.14.

Theorem 11.16 *Let $h \in \mathcal{H}$ and $\tau > 0$ fulfill Assumption 11.11. Then, for all $n \in \mathbb{N}_0$, the approximation $\boldsymbol{u}_{\mathrm{LF}}^{n+1}$ given by the dG-LF scheme* (11.33) *satisfies*

$$\|\boldsymbol{u}_{\mathrm{LF}}^{n+1}\|_\Omega \le C_{\mathrm{stb}}\|\boldsymbol{u}^0\|_\Omega + \sqrt{3}\, C_{\mathrm{stb}}\tau \sum_{j=1}^{n} \|\overline{\boldsymbol{f}}^{j+1/2}\|_\Omega.$$

Next, we state the result for the stronger norm, i.e., the one containing the discrete spatial and temporal derivatives. As before, we first need to define these discrete derivatives. The temporal derivative is again simply the difference quotient between two consecutive approximations. The spatial derivative, on the other hand, differs from the corresponding object used for the Crank–Nicolson method. This is due to the fact that we obtain the bound on the spatial derivative (11.11) through the corresponding scheme itself (see the proof of Theorem 11.6). Hence, we have to adjust this object for the different schemes. We define the *discrete time derivative* and the *discrete space derivative* for the leapfrog scheme at time $t_{n+1/2}$ as

$$\begin{aligned} \underline{\partial_\tau \boldsymbol{u}}_{\mathrm{LF}}^{n+1/2} &= \frac{\boldsymbol{u}_{\mathrm{LF}}^{n+1} - \boldsymbol{u}_{\mathrm{LF}}^{n}}{\tau}, \\ \underline{\mathcal{L}\boldsymbol{u}}_{\mathrm{LF}}^{n+1/2} &= \underline{\mathcal{L}}\, \frac{\boldsymbol{u}_{\mathrm{LF}}^{n+1} + \boldsymbol{u}_{\mathrm{LF}}^{n}}{2} + \tfrac{\tau^2}{4} \underline{\mathcal{D}}\, \underline{\partial_\tau \boldsymbol{u}}_{\mathrm{LF}}^{n+1/2}, \end{aligned} \tag{11.45}$$

respectively. Note that, with this, the leapfrog iteration (11.33a) is equivalent to

$$\underline{\partial_\tau \boldsymbol{u}}_{\mathrm{LF}}^{n+1/2} = \underline{\mathcal{L}\boldsymbol{u}}_{\mathrm{LF}}^{n+1/2} + \overline{\boldsymbol{f}}_{\pi}^{n+1/2}, \tag{11.46}$$

which replaces (11.9) for the bound on the discrete space derivative. Lastly, as already stated after Theorem 11.6, the operators $\underline{\mathcal{L}}$ and $\underline{\mathcal{R}}_{\mathrm{LF}}$ do not commute and hence, we are not able to get a uniform bound if f is not differentiable in time. Therefore, we do not split the inhomogeneity in this case.

Theorem 11.17 *Let $h \in \mathcal{H}$ and $\tau > 0$ such that Assumption 11.11 is fulfilled. Then, the approximations $\{\boldsymbol{u}_{\mathrm{LF}}^{n}\}_{n\ge 0}$ given by the dG-LF scheme* (11.33) *satisfy*

$$\begin{aligned} \|\underline{\partial_\tau \boldsymbol{u}}_{\mathrm{LF}}^{n+1/2}\|_\Omega \le{}& \sqrt{3}\, C_{\mathrm{stb}}\|\underline{\mathcal{L}\boldsymbol{u}}_{\pi}^{0}\|_\Omega + \sqrt{3} \max_{j=0,\dots,n+1} \|\boldsymbol{f}_{\pi}^{j}\|_\Omega \\ &+ \sqrt{3}\, C_{\mathrm{stb}}\tau \sum_{j=1}^{n} \Big\|\frac{\boldsymbol{f}_{\pi}^{j+1} - \boldsymbol{f}_{\pi}^{j-1}}{2\tau}\Big\|_\Omega, \end{aligned}$$

and

$$\|\underline{\mathcal{L}\boldsymbol{u}_{\mathrm{LF}}^{n+1/2}}\|_\Omega \leq \|\underline{\boldsymbol{\partial}_\tau \boldsymbol{u}_{\mathrm{LF}}^{n+1/2}}\|_\Omega + \max_{j=0,\dots,n+1} \|\boldsymbol{f}_\pi^j\|_\Omega.$$

If further $u^0 \in D(\mathcal{L}) \cap H^1(\mathcal{T}_h)^m$ *and* $f \in C^1(\mathbb{R}_+; L^2(\Omega)^m)$, *we have*

$$\|\underline{\boldsymbol{\partial}_\tau \boldsymbol{u}_{\mathrm{LF}}^{n+1/2}}\|_\Omega \leq \sqrt{3}\, C_{\mathrm{stb}}(\|\mathcal{L}u^0\|_\Omega + C_{\pi,\mathcal{L}}|u^0|_{1,\mathcal{T}_h}) + \sqrt{3} \max_{s\in\mathbb{R}_+} \|f(s)\|_\Omega$$

$$+ \sqrt{3}\, C_{\mathrm{stb}} \int_0^{t_{n+1}} \|\partial_t f(s)\|_\Omega \,\mathrm{d}s.$$

Now that we have shown how to analyze the leapfrog method by writing it as a perturbation of the Crank–Nicolson scheme, we will see in the next section that a similar approach with some modifications can also be used for the Peaceman–Rachford scheme.

11.3 Peaceman–Rachford Scheme

Next, we consider the *Peaceman–Rachford scheme*, proposed by Peaceman and Rachford in [140]. It is a splitting scheme, designed for problems that can be split into two distinct subproblems which enable a more efficient solution than the original problem. One famous application for the method are ADI schemes, which split the problems via the spatial dimension, effectively reducing them to lower-dimensional problems (most favorably into essentially one-dimensional problems). This is also the application we have in mind, but we point out that the Peaceman–Rachford scheme has also been used for other splittings, e.g., into a nonlinear and a linear part [84].

The main accomplishment of using the Peaceman–Rachford scheme in conjunction with an ADI splitting is the fact that it leads to unconditionally stable schemes that can be as efficient as an explicit method if carried out correctly. More precisely, this is possible if the original problem exhibits a certain structure, allowing it to be split into two essentially one-dimensional subproblems. We refer to [87] and [113, Chapter 6] for a discussion of this for dG space discretizations. Examples for such problems can be found in these publications as well as the original literature like [132, 140, 168], where finite differences are considered.

Again, we start by deriving the scheme in a rather simple way. This also gives us an intuition as to why the scheme is of classical order two.

Example 11.18 We consider the homogeneous initial value problem (9.6) from Example 9.1 given by

$$\begin{cases} \partial_t u(t) = Au(t), & t \in \mathbb{R}_+, \\ u(0) = u^0. \end{cases} \tag{11.47}$$

Since the Peaceman–Rachford scheme is a splitting scheme, we assume that we can split

$$A = A_1 + A_2 \tag{11.48}$$

in such a way that linear systems with coefficient matrix $I - \frac{\tau}{2}A_1$ or $I - \frac{\tau}{2}A_2$ are in some way easier to solve than those with the full coefficient matrix $I - \frac{\tau}{2}A$.

Similar to Example 11.1, we now use the fundamental theorem of calculus together with (11.47) and (11.48) to obtain

$$u(t_{n+1}) = u(t_n) + \int_{t_n}^{t_{n+1}} A_1 u(s)\,\mathrm{d}s + \int_{t_n}^{t_{n+1}} A_2 u(s)\,\mathrm{d}s.$$

This time, instead of only using the trapezoidal rule to approximate the integrals, we use the midpoint rule for the first and the trapezoidal rule for the second integral, yielding

$$u(t_{n+1}) \approx u(t_n) + \tau A_1 u(t_{n+1/2}) + \tfrac{\tau}{2} A_2\big(u(t_n) + u(t_{n+1})\big). \tag{11.49}$$

Since we are interested in a one-step method, we want to compute an approximation $u_\tau^{n+1} \approx u(t_{n+1})$ based solely on the former approximation $u_\tau^n \approx u(t_n)$. Hence, we need to approximate the intermediate value $u(t_{n+1/2})$ using only $u(t_n)$. To do so, we once again use the fundamental theorem to obtain

$$u(t_{n+1/2}) = u(t_n) + \int_{t_n}^{t_{n+1/2}} A_1 u(s)\,\mathrm{d}s + \int_{t_n}^{t_{n+1/2}} A_2 u(s)\,\mathrm{d}s.$$

Since $u(t_{n+1/2})$ is multiplied by τ in (11.49), we can use a rectangular rule to approximate the integrals, while still retaining (at least formally) second-order approximation overall. We use the right rectangular rule for the first and the left rectangular rule for the second integral and get

$$u(t_{n+1/2}) \approx u(t_n) + \tfrac{\tau}{2} A_1 u(t_{n+1/2}) + \tfrac{\tau}{2} A_2 u(t_n). \tag{11.50}$$

Replacing the function evaluations $u(t_n)$, $u(t_{n+1/2})$ and $u(t_{n+1})$ by their respective approximations u_τ^n, $u_\tau^{n+1/2}$ and u_τ^{n+1} in (11.49) and (11.50) yields

$$\begin{aligned} u_\tau^{n+1/2} &= u_\tau^n + \tfrac{\tau}{2} A_1 u_\tau^{n+1/2} + \tfrac{\tau}{2} A_2 u_\tau^n, \\ u_\tau^{n+1} &= u_\tau^n + \tau A_1 u_\tau^{n+1/2} + \tfrac{\tau}{2} A_2\big(u_\tau^n + u_\tau^{n+1}\big). \end{aligned}$$

Eliminating the terms involving u_τ^n in the second equation by subtracting the first and rearranging the terms, we end up with the Peaceman–Rachford scheme given by

$$\begin{cases} (I - \frac{\tau}{2}A_1)u_\tau^{n+1/2} = (I + \frac{\tau}{2}A_2)u_\tau^n, & \\ (I - \frac{\tau}{2}A_2)u_\tau^{n+1} = (I + \frac{\tau}{2}A_1)u_\tau^{n+1/2}, & n \in \mathbb{N}_0, \\ u_\tau^0 = u^0. & \end{cases}$$

Note that the Peaceman–Rachford scheme, like the Crank–Nicolson scheme, is still implicit and we even need to solve two linear systems in each step; one with coefficient matrix $I - \frac{\tau}{2}A_1$ and one with coefficient matrix $I - \frac{\tau}{2}A_2$. However, as we assumed that such systems are easier to solve than the full system with coefficient matrix $I - \frac{\tau}{2}A$, this may still be favorable over the Crank–Nicolson or other implicit schemes.

We are also interested in the inhomogeneous problem (9.8) given by

$$\begin{cases} \partial_t u(t) = Au(t) + f(t), & t \in \mathbb{R}_+, \\ u(0) = u^0. & \end{cases}$$

There are several ways to approximate the inhomogeneity in such a way that we get an overall approximation of second-order (in fact, this is also true for the other schemes). The *Peaceman–Rachford scheme* for inhomogeneous problems we consider here is given by

$$\begin{cases} (I - \frac{\tau}{2}A_1)u_\tau^{n+1/2} = (I + \frac{\tau}{2}A_2)u_\tau^n + \frac{\tau}{2}(I - \frac{\tau}{2}A_1)f^n, & \\ (I - \frac{\tau}{2}A_2)u_\tau^{n+1} = (I + \frac{\tau}{2}A_1)u_\tau^{n+1/2} + \frac{\tau}{2}(I + \frac{\tau}{2}A_1)f^{n+1}, & n \in \mathbb{N}_0, \\ u_\tau^0 = u^0. & \end{cases}$$

This particular scheme is taken from [51, 136]. However, we equivalently reformulated it in order to ensure that $u_\tau^{n+1/2} \approx u(t_{n+1/2})$ is also second-order accurate. Treating the inhomogeneity in this way has the advantage that no additional linear systems have to be solved. As mentioned before, other variants are possible and we refer to [113, Sec. 4.4] for an example. ◇

Before we can apply the Peaceman–Rachford scheme to the wave-type problem (9.3), or rather its spatially discrete counterpart (10.7), we need to split the (discrete) Friedrichs' operator. Although we could directly split $\underline{\mathcal{L}}$, we deem it to be more systematically to first split the continuous operator $\mathcal{L}$ and then discretize the continuous split operators (which we need anyway for the analysis later on).

In fact, we split the Friedrichs' operator $\mathcal{L} = \mathcal{A} + \mathcal{B}$ into two operators by splitting the coefficients of $\mathcal{L}$, i.e., $L_i = A_i + B_i$ for all $i = 0, \dots, d$, in such a way that $A_0, B_0 \in$

$\mathbb{R}^{m\times m}$ are negative semidefinite and $A_1, B_1, \ldots, A_d, B_d \in \mathbb{R}^{m\times m}$ are symmetric. From this, we obtain two operators $\mathcal{A}$ and $\mathcal{B}$ defined by

$$M\mathcal{A}u = \sum_{i=1}^{d} A_i\partial_i u + A_0 u, \qquad M\mathcal{B}u = \sum_{i=1}^{d} B_i\partial_i u + B_0 u.$$

Since, by construction, $\mathcal{A}$ and $\mathcal{B}$ exhibit exactly the same structure as $\mathcal{L}$, we extend all concepts from Chap. 9 to the split operators with corresponding notation.

To ensure that the theory derived in Chaps. 9 and 10 also holds for the split operators, we need to supply them with a suitable domain that contains appropriate boundary conditions. We thus assume that, similar to the Friedrichs' operator itself, we can split the boundary operator $\mathcal{L}_\Gamma = \mathcal{A}_\Gamma + \mathcal{B}_\Gamma$ such that $\mathcal{A}_\Gamma$ and $\mathcal{B}_\Gamma$ fulfill Assumption 9.21 (w.r.t. their respective operator) and Assumption 10.3 with the boundary fields denoted by A_Γ and B_Γ, respectively. Hence, by Chap. 9, we see that $\mathcal{A}$ and $\mathcal{B}$ are maximal dissipative on their respective domains $D(\mathcal{A}) := \ker(\mathcal{A}_\partial - \mathcal{A}_\Gamma)$ and $D(\mathcal{B}) := \ker(\mathcal{B}_\partial - \mathcal{B}_\Gamma)$.

Finally, we establish a splitting of the discrete operator $\underline{\mathfrak{L}}$ by discretizing the split operators $\mathcal{A}$ and $\mathcal{B}$ according to Definition 10.5. This yields the discrete split operators $\underline{\mathfrak{A}}$ and $\underline{\mathfrak{B}}$, respectively. By the bilinearity of the inner products and the way we designed the splitting, this immediately yields

$$\underline{\mathfrak{L}} = \underline{\mathfrak{A}} + \underline{\mathfrak{B}}, \tag{11.51}$$

i.e., $\underline{\mathfrak{A}}$ and $\underline{\mathfrak{B}}$ do in fact constitute a splitting of $\underline{\mathfrak{L}}$.

For the time discretization, we apply the Peaceman–Rachford scheme to the spatially discrete wave-type problem (10.7). This yields the *fully discrete dG-Peaceman–Rachford (dG-PR) scheme* given by

$$\begin{cases} (\underline{\mathcal{I}} - \frac{\tau}{2}\underline{\mathfrak{A}})\boldsymbol{u}_{\mathrm{PR}}^{n+1/2} = (\underline{\mathcal{I}} + \frac{\tau}{2}\underline{\mathfrak{B}})\boldsymbol{u}_{\mathrm{PR}}^{n} + \frac{\tau}{2}(\underline{\mathcal{I}} - \frac{\tau}{2}\underline{\mathfrak{A}})\boldsymbol{f}_\pi^n, & (11.52a)\\ (\underline{\mathcal{I}} - \frac{\tau}{2}\underline{\mathfrak{B}})\boldsymbol{u}_{\mathrm{PR}}^{n+1} = (\underline{\mathcal{I}} + \frac{\tau}{2}\underline{\mathfrak{A}})\boldsymbol{u}_{\mathrm{PR}}^{n+1/2} + \frac{\tau}{2}(\underline{\mathcal{I}} + \frac{\tau}{2}\underline{\mathfrak{A}})\boldsymbol{f}_\pi^{n+1}, & (11.52b)\\ \boldsymbol{u}_{\mathrm{PR}}^{0} = \boldsymbol{u}_\pi^0. & (11.52c) \end{cases}$$

As for the Crank–Nicolson method, we immediately see that both $\boldsymbol{u}_{\mathrm{PR}}^{n+1}$ and $\boldsymbol{u}_{\mathrm{PR}}^{n+1/2}$ exist and are uniquely defined for all $n \in \mathbb{N}$ and all initial values by Lemma 9.15 (i), since by construction both $\underline{\mathfrak{A}}$ and $\underline{\mathfrak{B}}$ are maximal dissipative on $\mathbb{V}_h^m$.

We now proceed similarly to the leapfrog method and first rewrite the dG-PR scheme in a form resembling the dG-CN scheme. To this purpose, we eliminate the intermediate approximation $\boldsymbol{u}_{\mathrm{PR}}^{n+1/2}$ by solving (11.52a) for $\boldsymbol{u}_{\mathrm{PR}}^{n+1/2}$ and inserting the result into (11.52b).

We subsequently multiply the resulting equation by $\underline{\mathcal{I}} - \frac{\tau}{2}\underline{\mathcal{A}}$ and, expanding all products, obtain the equivalent formulation of the dG-PR scheme given by

$$\begin{cases} (\underline{\mathcal{I}} - \frac{\tau}{2}\underline{\mathcal{L}} + \frac{\tau^2}{4}\underline{\mathcal{A}}\underline{\mathcal{B}})\boldsymbol{u}_{\text{PR}}^{n+1} = (\underline{\mathcal{I}} + \frac{\tau}{2}\underline{\mathcal{L}} + \frac{\tau^2}{4}\underline{\mathcal{A}}\underline{\mathcal{B}})\boldsymbol{u}_{\text{PR}}^{n} + (\underline{\mathcal{I}} - \frac{\tau^2}{4}\underline{\mathcal{A}}^2)\tau\overline{\boldsymbol{f}}_{\pi}^{n+1/2}, & (11.53a) \\ \boldsymbol{u}_{\text{PR}}^{0} = \boldsymbol{u}_{\pi}^{0}. & (11.53b) \end{cases}$$

Note that, similar to the dG-LF scheme, the dG-PR scheme coincides with the dG-CN scheme up to a τ^2 perturbation term and the treatment of the inhomogeneity.

Next, we cast the scheme into a more compact form, resembling the corresponding dG-CN formulation (11.6). To do so, we define the discrete operators $\underline{\mathcal{R}}_{\text{PR}}, \underline{\mathcal{S}}_{\text{PR}} : \mathbb{V}_{\hbar}^{m} \to \mathbb{V}_{\hbar}^{m}$ by

$$\underline{\mathcal{R}}_{\text{PR}} = (\underline{\mathcal{I}} - \tfrac{\tau}{2}\underline{\mathcal{L}} + \tfrac{\tau^2}{4}\underline{\mathcal{A}}\underline{\mathcal{B}})^{-1} = (\underline{\mathcal{I}} - \tfrac{\tau}{2}\underline{\mathcal{B}})^{-1}(\underline{\mathcal{I}} - \tfrac{\tau}{2}\underline{\mathcal{A}})^{-1},$$

and

$$\begin{aligned} \underline{\mathcal{S}}_{\text{PR}} &= (\underline{\mathcal{I}} - \tfrac{\tau}{2}\underline{\mathcal{L}} + \tfrac{\tau^2}{4}\underline{\mathcal{A}}\underline{\mathcal{B}})^{-1}(\underline{\mathcal{I}} + \tfrac{\tau}{2}\underline{\mathcal{L}} + \tfrac{\tau^2}{4}\underline{\mathcal{A}}\underline{\mathcal{B}}) \\ &= (\underline{\mathcal{I}} - \tfrac{\tau}{2}\underline{\mathcal{B}})^{-1}(\underline{\mathcal{I}} - \tfrac{\tau}{2}\underline{\mathcal{A}})^{-1}(\underline{\mathcal{I}} + \tfrac{\tau}{2}\underline{\mathcal{A}})(\underline{\mathcal{I}} + \tfrac{\tau}{2}\underline{\mathcal{B}}). \end{aligned} \quad (11.54)$$

Note that these are well-defined, since the resolvents of both $\underline{\mathcal{A}}$ and $\underline{\mathcal{B}}$ are isomorphisms by Lemma 9.15 (i). With this, the dG-PR scheme (11.53) is equivalent to

$$\begin{cases} \boldsymbol{u}_{\text{PR}}^{n+1} = \underline{\mathcal{S}}_{\text{PR}}\boldsymbol{u}_{\text{PR}}^{n} + (\underline{\mathcal{I}} - \frac{\tau}{2}\underline{\mathcal{B}})^{-1}(\underline{\mathcal{I}} + \frac{\tau}{2}\underline{\mathcal{A}})\tau\overline{\boldsymbol{f}}_{\pi}^{n+1/2}, & (11.55a) \\ \boldsymbol{u}_{\text{PR}}^{0} = \boldsymbol{u}_{\pi}^{0}. & (11.55b) \end{cases}$$

We solve this recursion to obtain the following solution formula.

Theorem 11.19 *For all $\hbar \in \mathcal{H}$, and all $\tau > 0$, there exists a unique $\boldsymbol{u}_{\text{PR}}^{n+1} \in \mathbb{V}_{\hbar}^{m}$ fulfilling the dG-PR scheme* (11.55). *It is given by the discrete variation-of-constants formula*

$$\boldsymbol{u}_{\text{PR}}^{n+1} = \underline{\mathcal{S}}_{\text{PR}}^{n+1}\boldsymbol{u}_{\pi}^{0} + \tau\sum_{j=0}^{n}\underline{\mathcal{S}}_{\text{PR}}^{n-j}(\underline{\mathcal{I}} - \tfrac{\tau}{2}\underline{\mathcal{B}})^{-1}(\underline{\mathcal{I}} + \tfrac{\tau}{2}\underline{\mathcal{A}})\overline{\boldsymbol{f}}_{\pi}^{j+1/2}. \quad (11.56)$$

Again, we point out the similarity between the continuous and discrete variation-of-constants formulae (9.34) and (11.56), respectively. Thus, $\underline{\boldsymbol{S}}_{\mathrm{PR}}$ can be seen as a discrete semigroup and we expect it to fulfill similar properties.

We start by giving a power bound. This result, which essentially provides unconditional stability in the $\|\cdot\|_\Omega$-norm, is well-known in the literature. There exist several proofs and we follow the strategy used in [69] and [93], where the Peaceman–Rachford scheme was considered in a finite difference context as well as an abstract Hilbert space setting (i.e., the spatially continuous case), respectively. We refer to the references in [93] for other strategies of proof.

Proposition 11.20 *Let $\hbar \in \mathcal{H}$ and $\tau > 0$. Then, for all $\ell \in \mathbb{N}$ and all $\boldsymbol{v} \in \mathbb{V}_\hbar^m$, the discrete semigroup $\underline{\boldsymbol{S}}_{\mathrm{PR}}$ defined in* (11.54) *fulfills the power bounds*

$$\|\underline{\boldsymbol{S}}_{\mathrm{PR}}^{\ell}\,\boldsymbol{v}\|_\Omega \leq \|\big(\underline{\mathcal{I}} + \tfrac{\tau}{2}\underline{\mathcal{B}}\big)\boldsymbol{v}\|_\Omega \tag{11.57a}$$

and

$$\|\underline{\boldsymbol{S}}_{\mathrm{PR}}^{\ell}\,\big(\underline{\mathcal{I}} - \tfrac{\tau}{2}\underline{\mathcal{B}}\big)^{-1}\boldsymbol{v}\|_\Omega \leq \|\boldsymbol{v}\|_\Omega. \tag{11.57b}$$

Proof For all $\ell \in \mathbb{N}$ and $\boldsymbol{v} \in \mathbb{V}_\hbar^m$, we have

$$\begin{aligned}\|\underline{\boldsymbol{S}}_{\mathrm{PR}}^{\ell}\,\boldsymbol{v}\|_\Omega &= \|\Big(\big(\underline{\mathcal{I}} - \tfrac{\tau}{2}\underline{\mathcal{B}}\big)^{-1}\big(\underline{\mathcal{I}} + \tfrac{\tau}{2}\underline{\mathcal{A}}\big)\big(\underline{\mathcal{I}} - \tfrac{\tau}{2}\underline{\mathcal{A}}\big)^{-1}\big(\underline{\mathcal{I}} + \tfrac{\tau}{2}\underline{\mathcal{B}}\big)\Big)^{\ell}\boldsymbol{v}\|_\Omega \\ &= \|\big(\underline{\mathcal{I}} - \tfrac{\tau}{2}\underline{\mathcal{B}}\big)^{-1}\boldsymbol{C}^{\ell-1}\big(\underline{\mathcal{I}} + \tfrac{\tau}{2}\underline{\mathcal{A}}\big)\big(\underline{\mathcal{I}} - \tfrac{\tau}{2}\underline{\mathcal{A}}\big)^{-1}\big(\underline{\mathcal{I}} + \tfrac{\tau}{2}\underline{\mathcal{B}}\big)\boldsymbol{v}\|_\Omega,\end{aligned}$$

where $\boldsymbol{C} = \big(\underline{\mathcal{I}} + \tfrac{\tau}{2}\underline{\mathcal{A}}\big)\big(\underline{\mathcal{I}} - \tfrac{\tau}{2}\underline{\mathcal{A}}\big)^{-1}\big(\underline{\mathcal{I}} + \tfrac{\tau}{2}\underline{\mathcal{B}}\big)\big(\underline{\mathcal{I}} - \tfrac{\tau}{2}\underline{\mathcal{B}}\big)^{-1}$. By Lemma 9.15, both the resolvents of $\underline{\mathcal{A}}$ and $\underline{\mathcal{B}}$ as well as the transforms $\big(\underline{\mathcal{I}} - \tfrac{\tau}{2}\underline{\mathcal{A}}\big)^{-1}\big(\underline{\mathcal{I}} + \tfrac{\tau}{2}\underline{\mathcal{A}}\big)$ and $\big(\underline{\mathcal{I}} - \tfrac{\tau}{2}\underline{\mathcal{B}}\big)^{-1}\big(\underline{\mathcal{I}} + \tfrac{\tau}{2}\underline{\mathcal{B}}\big)$ are contractive (since both operators are maximal dissipative), proving (11.57a). Using the same argument once more then shows (11.57b). □

Note that the bound given in (11.57a) is not independent of the discretization parameter $\hbar$ due to the occurrence of the discrete operator $\underline{\mathcal{B}}$, which only fulfills the inverse inequality in Proposition 10.8, but not a uniform bound. The second bound (11.57b), however, is independent of the discretization parameters and this is a crucial fact for the stability of the method. Note also that if $\boldsymbol{v}$ is the projection of a sufficiently regular function, we can bound the right hand side of (11.57a) uniformly in $\hbar$ by means of Proposition 10.9 (approximation property). This will be used frequently during the error analysis below.

Next, we show that the discrete semigroup $\underline{\boldsymbol{S}}_{\mathrm{PR}}$ again fulfills a discrete analogon of (9.13b), meaning that its (discrete) derivative is given by application of its generator.

Observe that this time, the generator involves the discrete operator $\underline{\mathcal{R}}_{\mathrm{PR}}$ instead of the corresponding Crank–Nicolson- or leapfrog-counterpart (cf., Propositions 11.4 and 11.15).

Proposition 11.21 *Let $\hbar \in \mathcal{H}$ and $\tau > 0$. Then, for all $\ell \in \mathbb{N}_0$, the powers of the discrete semigroup $\underline{\mathcal{S}}_{\mathrm{PR}}$ defined in* (11.54) *satisfy*

$$\frac{\underline{\mathcal{S}}_{\mathrm{PR}}^{\ell+1} - \underline{\mathcal{S}}_{\mathrm{PR}}^{\ell}}{\tau} = (\underline{\mathcal{R}}_{\mathrm{PR}}\,\underline{\mathcal{L}})\,\underline{\mathcal{S}}_{\mathrm{PR}}^{\ell} = \underline{\mathcal{S}}_{\mathrm{PR}}^{\ell}\,(\underline{\mathcal{R}}_{\mathrm{PR}}\,\underline{\mathcal{L}}).$$

Proof We factor out $\underline{\mathcal{S}}_{\mathrm{PR}}^{\ell}$ (either to the left or right) and calculate

$$\frac{\underline{\mathcal{S}}_{\mathrm{PR}} - \underline{\mathcal{I}}}{\tau} = \underline{\mathcal{R}}_{\mathrm{PR}}\,\frac{(\underline{\mathcal{I}} + \frac{\tau}{2}\underline{\mathcal{A}})(\underline{\mathcal{I}} + \frac{\tau}{2}\underline{\mathcal{B}}) - (\underline{\mathcal{I}} - \frac{\tau}{2}\underline{\mathcal{A}})(\underline{\mathcal{I}} - \frac{\tau}{2}\underline{\mathcal{B}})}{\tau} = \underline{\mathcal{R}}_{\mathrm{PR}}\,\underline{\mathcal{L}}.$$

This proves the claim. □

With these properties of the discrete semigroup, we can now show unconditional stability of the Peaceman–Rachford scheme. As before, we obtain two such results, one in the $\|\cdot\|_\Omega$-norm and one in a stronger norm involving discrete derivatives of the approximants in both space and time.

We start with the former, which, in principle, is a well-known result and a rather straightforward consequence of Proposition 11.20. However, due to the occurrence of the discrete operator $\underline{\mathcal{B}}$ in this result as well as the application of $\underline{\mathcal{A}}$ to the inhomogeneity in the scheme, we have to do some adjustments to obtain a robust statement with constants independent of the spatial discretization parameter.

Theorem 11.22 *For all $\hbar \in \mathcal{H}$ and all $\tau > 0$, the approximation $\boldsymbol{u}_{\mathrm{PR}}^{n+1}$ given by the dG-PR scheme* (11.55) *satisfies*

$$\|\boldsymbol{u}_{\mathrm{PR}}^{n+1}\|_\Omega \le \|\boldsymbol{u}_\pi^0\|_\Omega + \tfrac{\tau}{2}\|\underline{\mathcal{B}}\boldsymbol{u}_\pi^0\|_\Omega + \tau\sum_{j=0}^{n}\big(\|\overline{\boldsymbol{f}}_\pi^{j+1/2}\|_\Omega + \tfrac{\tau}{2}\|\underline{\mathcal{A}}\overline{\boldsymbol{f}}_\pi^{j+1/2}\|_\Omega\big).$$

If further $u^0 \in D(\mathcal{B}) \cap H^1(\mathcal{T}_\hbar)^m$ and $f \in C(\mathbb{R}_+; D(\mathcal{A}) \cap H^1(\mathcal{T}_\hbar)^m)$, we have

$$\|\boldsymbol{u}_{\mathrm{PR}}^{n+1}\|_\Omega \le \|u^0\|_\Omega + \tfrac{\tau}{2}\|\mathcal{B}u^0\|_\Omega + \tfrac{\tau}{2}C_{\pi,\mathcal{B}}|u^0|_{1,\mathcal{T}_\hbar}$$
$$+\,\tau\sum_{j=1}^{n}\Big(\|\overline{f}^{j+1/2}\|_\Omega + \tfrac{\tau}{2}\|\mathcal{A}\overline{f}^{j+1/2}\|_\Omega + C_{\pi,\mathcal{A}}|\overline{f}^{j+1/2}|_{1,\mathcal{T}_\hbar}\Big).$$

Proof We use the discrete variation-of-constants formula (11.56) and Proposition 11.20 together with the contractivity of $\big(\underline{\mathcal{I}}+\frac{\tau}{2}\underline{\mathcal{B}}\big)\big(\underline{\mathcal{I}}-\frac{\tau}{2}\underline{\mathcal{B}}\big)^{-1}$ to obtain

$$\begin{aligned}\|\boldsymbol{u}_{\mathrm{PR}}^{n+1}\|_\Omega &\le \|\underline{\mathcal{S}}_{\mathrm{PR}}^{n+1}\boldsymbol{u}_\pi^0\|_\Omega + \tau\sum_{j=0}^{n}\|\underline{\mathcal{S}}_{\mathrm{PR}}^{n-j}\big(\underline{\mathcal{I}}-\tfrac{\tau}{2}\underline{\mathcal{B}}\big)^{-1}\big(\underline{\mathcal{I}}+\tfrac{\tau}{2}\underline{\mathcal{A}}\big)\overline{\boldsymbol{f}}_\pi^{j+1/2}\|_\Omega\\ &\le \|\big(\underline{\mathcal{I}}+\tfrac{\tau}{2}\underline{\mathcal{B}}\big)\boldsymbol{u}_\pi^0\|_\Omega + \tau\sum_{j=0}^{n}\|\big(\underline{\mathcal{I}}+\tfrac{\tau}{2}\underline{\mathcal{A}}\big)\overline{\boldsymbol{f}}_\pi^{j+1/2}\|_\Omega.\end{aligned}$$

This shows the first bound after applying the triangle inequality. The second bound follows by treating both terms containing a discrete operator analogously to (10.11) and the contractivity of $\pi_{\hbar}$. □

Next, we show the result in the stronger norm, which consists of the discrete derivatives of the approximations given by the Peaceman–Rachford scheme (11.55). In principle, the proof consists of the same steps as the one for the dG-CN scheme. However, as for the leapfrog method, one has to adjust the discrete spatial derivative to the scheme. Doing so yields the *discrete time derivative* and the *discrete space derivative* for the Peaceman–Rachford scheme at time $t_{n+1/2}$ defined as

$$\underline{\partial_\tau \boldsymbol{u}}_{\mathrm{PR}}^{n+1/2} = \frac{\boldsymbol{u}_{\mathrm{PR}}^{n+1}-\boldsymbol{u}_{\mathrm{PR}}^{n}}{\tau}, \quad \text{and} \quad \underline{\mathcal{L}\boldsymbol{u}}_{\mathrm{PR}}^{n+1/2} = \underline{\mathcal{A}}\boldsymbol{u}_{\mathrm{PR}}^{n+1/2} + \underline{\mathcal{B}}\,\frac{\boldsymbol{u}_{\mathrm{PR}}^{n+1}+\boldsymbol{u}_{\mathrm{PR}}^{n}}{2}, \tag{11.58}$$

respectively. Note that we only apply $\underline{\mathcal{A}}$ to the half step approximations $\boldsymbol{u}_{\mathrm{PR}}^{n+1/2}$ and $\underline{\mathcal{B}}$ to the full step ones $\boldsymbol{u}_{\mathrm{PR}}^{n}$. This is no coincidence but based on the structure of the scheme (11.55). In fact, by solving (11.52a) and (11.52b) for the respective approximations, we see that $\underline{\mathcal{A}}\boldsymbol{u}_{\mathrm{PR}}^{n+1/2}$ and $\underline{\mathcal{B}}\boldsymbol{u}_{\mathrm{PR}}^{n}$ can be bounded independently of the space discretization parameter $\hbar$ by Lemma 9.15. This is not possible for the respective other or the full discrete operator $\underline{\mathcal{L}}$, indicating that such objects are not suitable to work with.

As it was the case for the Crank–Nicolson and leapfrog methods, we can rewrite the scheme with these discrete objects. More precisely, adding the two (half-step) iterations (11.52a) and (11.52b) yields

$$\underline{\partial_\tau \boldsymbol{u}}_{\mathrm{PR}}^{n+1/2} = \underline{\mathcal{L}\boldsymbol{u}}_{\mathrm{PR}}^{n+1/2} + \overline{\boldsymbol{f}}_\pi^{n+1/2} + \tfrac{\tau^2}{4}\underline{\mathcal{A}}\,\frac{\boldsymbol{f}_\pi^{n+1}-\boldsymbol{f}_\pi^{n}}{\tau}.$$

This slightly differs from the Crank–Nicolson and leapfrog versions (11.9) and (11.46), respectively. However, one can see that the additional term is only a τ^2 perturbation, which does not introduce significant problems if the inhomogeneity f is sufficiently regular. As for the leapfrog method, the discrete operators $\underline{\mathcal{L}}$ and $\underline{\mathcal{R}}_{\mathrm{PR}}$ do not commute and we therefore again abstain from splitting the inhomogeneity.

Theorem 11.23 *Let $\hbar \in \mathcal{H}$ and $\tau > 0$. Then, the approximations $\{\boldsymbol{u}_{\mathrm{PR}}^{n}\}_{n\geq 0}$ and $\{\boldsymbol{u}_{\mathrm{PR}}^{n+1/2}\}_{n\geq 0}$ given by the dG-PR scheme (11.55) satisfy*

$$\|\underline{\boldsymbol{\partial}_{\tau}\boldsymbol{u}}_{\mathrm{PR}}^{n+1/2}\|_{\Omega} \leq \|\underline{\boldsymbol{\mathcal{L}}\boldsymbol{u}}_{\pi}^{0}\|_{\Omega} + \max_{j=0,\dots,n+1}\big(\|\boldsymbol{f}_{\pi}^{j}\|_{\Omega} + \tfrac{\tau}{2}\|\underline{\boldsymbol{\mathcal{A}}}\boldsymbol{f}_{\pi}^{j}\|_{\Omega}\big)$$
$$+\,\tau\sum_{j=1}^{n}\big(\|\pi_{\hbar}\frac{f^{j+1}-f^{j-1}}{2\tau}\|_{\Omega} + \tfrac{\tau}{2}\|\underline{\boldsymbol{\mathcal{A}}}\pi_{\hbar}\frac{f^{j+1}-f^{j-1}}{2\tau}\|_{\Omega}\big),$$

and

$$\|\underline{\boldsymbol{\mathcal{L}}\boldsymbol{u}}_{\mathrm{PR}}^{n+1/2}\|_{\Omega} \leq \|\underline{\boldsymbol{\partial}_{\tau}\boldsymbol{u}}_{\mathrm{PR}}^{n+1/2}\|_{\Omega} + \max_{j=0,\dots,n+1}\big(\|\boldsymbol{f}_{\pi}^{j}\|_{\Omega} + \tfrac{\tau}{2}\|\underline{\boldsymbol{\mathcal{A}}}\boldsymbol{f}_{\pi}^{j}\|_{\Omega}\big).$$

If further $u^{0} \in D(\mathcal{L}) \cap H^{1}(\mathcal{T}_{\hbar})^{m}$ and $f \in C^{1}(\mathbb{R}_{+}; D(\mathcal{A}) \cap H^{1}(\mathcal{T}_{\hbar})^{m})$, we have

$$\|\underline{\boldsymbol{\partial}_{\tau}\boldsymbol{u}}_{\mathrm{PR}}^{n+1/2}\|_{\Omega} \leq \|\mathcal{L}u^{0}\|_{\Omega} + C_{\pi,\mathcal{L}}|u^{0}|_{1,\mathcal{T}_{\hbar}}$$
$$+\max_{s\in\mathbb{R}_{+}}\big(\|f(s)\|_{\Omega} + \tfrac{\tau}{2}\|\mathcal{A}f(s)\|_{\Omega} + \tfrac{\tau}{2}C_{\pi,\mathcal{A}}|f(s)|_{1,\mathcal{T}_{\hbar}}\big)$$
$$+\int_{0}^{t_{n+1}}\|\partial_{t}f(s)\|_{\Omega} + \tfrac{\tau}{2}\|\mathcal{A}\partial_{t}f(s)\|_{\Omega} + \tfrac{\tau}{2}C_{\pi,\mathcal{A}}|\partial_{t}f(s)|_{1,\mathcal{T}_{\hbar}}\,\mathrm{d}s,$$

and

$$\|\underline{\boldsymbol{\mathcal{L}}\boldsymbol{u}}_{\mathrm{PR}}^{n+1/2}\|_{\Omega} \leq \|\underline{\boldsymbol{\partial}_{\tau}\boldsymbol{u}}_{\mathrm{PR}}^{n+1/2}\|_{\Omega} + \max_{s\in\mathbb{R}_{+}}\big(\|f(s)\|_{\Omega} + \tfrac{\tau}{2}\|\mathcal{A}f(s)\|_{\Omega}$$
$$+\,\tfrac{\tau}{2}C_{\pi,\mathcal{A}}|f(s)|_{1,\mathcal{T}_{\hbar}}\big).$$

This is proven analogously to the proof of Theorem 11.6 with some minor adjustments. We hence leave the proof as an exercise for the reader. It can, however, also be found in [89].

11.4 Locally Implicit Scheme

The last method we investigate here is a locally implicit time integration method, which was proposed by Verwer in [163] for spatially semidiscrete wave-type equations of a particular structure (with special emphasis on the Maxwell equations). A similar scheme was proposed before in [144] but unfortunately did not retain second-order accuracy [39, 130]. An extension of these locally implicit schemes for Maxwell-Debye equations

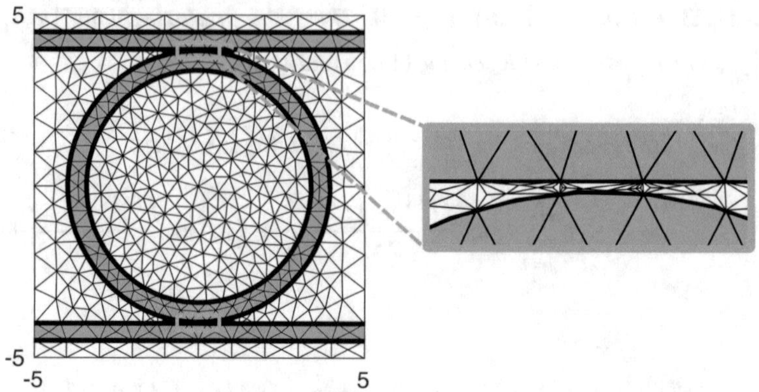

Fig. 11.2 Example with geometric constraints leading to a locally refined mesh (used for simulating a ring resonator between two waveguides). Picture taken from [161] based on an example in [45]

has been considered in [40, 41]. The error analysis in [41] is done on a fixed spatial mesh, where the fully discrete solution is compared to the spatially semidiscrete solution.

The rigorous analysis we present here is largely based on [91], where the method from [163] applied to the undamped Maxwell equations was considered. Here, we generalize it to Friedrichs' systems (including damping) and supplement the stronger stability results we already showed for the other methods.

As we have seen in Sect. 11.2, the leapfrog scheme (and explicit schemes in general) suffers from a timestep restriction in form of the CFL condition (11.38). This is particularly detrimental if only a few very small elements in the mesh dictate the maximal admissible timestep, e.g., due to geometric constraints as in Fig. 11.2. In this case of locally refined grids, the timestep is likely governed solely by stability, not by the desired accuracy, ruining the efficiency of the method. This could be remedied by using an implicit method, like the Crank–Nicolson method. However, as already mentioned in Sect. 11.1, this becomes unfeasible if too many degrees of freedom have to be used overall.

This motivated Verwer to construct a method that aims to combine the benefits of both methods, while simultaneously avoiding their main disadvantages. The idea is rather simple: small elements should be treated implicitly, while larger elements are treated explicitly. Of course, the realization of this idea is not as straightforward. Still, in [163], this goal was achieved by combining the leapfrog and Crank–Nicolson methods in a clever way.

We point out that other approaches to time integration on locally refined grids are possible, most notably e.g., local time-stepping methods. In these schemes, all elements are treated explicitly, but with a smaller timestep size for the small mesh elements than for the coarse ones, cf., e.g., [44, 78, 129, 144]. While some of them are fully explicit, there are also methods where Lagrange multipliers are introduced for the discretization of the interface conditions and for these methods, linear systems on this interface have to be solved, cf., e.g., [25, 29–31, 103]. However, most of these schemes are designed for wave equations in a second-order formulation and cannot be applied to Friedrichs' systems considered here.

Since the locally implicit method involves applying the leapfrog method on the fine part of the mesh, we again need the two-field structure of our wave-type problem already discussed in Sect. 11.2. In particular, this means that we require that Assumption 11.8 holds whenever we consider the locally implicit scheme.

In order to apply the locally implicit integrator, we first need to identify elements that should be treated explicitly and those that should be treated implicitly. If done correctly, this should lead to a CFL condition that is independent of the small elements of the mesh (or those with disadvantageous material properties).

Firstly, we assume that our mesh $\mathcal{T}_{\hbar}$ is divided into a part $\mathcal{T}_{\hbar,c}$ consisting of coarse elements and a part $\mathcal{T}_{\hbar,f}$ consisting of fine elements such that

$$\mathcal{T}_{\hbar} = \mathcal{T}_{\hbar,c} \,\dot{\cup}\, \mathcal{T}_{\hbar,f}.$$

Note that, although we call them fine and coarse elements, the material parameters on each element should also be taken into account for this partitioning of the mesh, since they too dictate the CFL condition (11.38), cf., Remark 11.12. More precisely, a large element on which the material parameters are small should also be classified as a fine element.

We further point out that locally implicit schemes are especially attractive if the number of fine elements is much smaller than the number of coarse ones. This is due to the fact that in this case, the linear systems, which have to be solved in each timestep, are small and efficient to solve. Otherwise it may be better to resort to other approaches such as explicit local time-stepping methods, see the references given above.

We also need mesh-dependent norms corresponding to either the fine or the coarse part of the mesh in a similar way to those corresponding to the full mesh. Hence, for a consistent notation, we also define the piecewise constant mesh-diameter functions $h_c, h_f \in L^\infty(\Omega)$ via

$$h_c \equiv \begin{cases} h_K, & \text{if } K \in \mathcal{T}_{\hbar,c}, \\ 0, & \text{else,} \end{cases} \qquad \text{and} \qquad h_f \equiv \begin{cases} h_K, & \text{if } K \in \mathcal{T}_{\hbar,f}, \\ 0, & \text{else.} \end{cases}$$

Further, $\hbar_{\text{min,c}} = \min_{K \in \mathcal{T}_{\hbar,c}} h_K$ is the minimal element diameter of the coarse mesh, which will enter the CFL condition of the locally implicit method.

It turns out that it is not sufficient to simply treat only the fine part of the mesh implicitly to obtain a CFL condition independent of the fine elements. Namely, since the discrete operator introduces coupling between the elements over faces (cf., Definition 10.5), we also need to treat the direct neighbors (which share a joint face) of the fine mesh implicitly. Hence, we further need the decomposition of $\mathcal{T}_{\hbar}$ into implicitly treated and explicitly treated elements, which does not coincide with the one into fine and coarse elements. The implicitly and explicitly treated parts of the mesh are then given as

$$\mathcal{T}_{h,i} = \{ K \in \mathcal{T}_h \mid \exists K_f \in \mathcal{T}_{h,f} \text{ s.t. } \mathrm{vol}_{d-1}(\partial K \cap \partial K_f) \neq 0 \}, \quad \mathcal{T}_{h,e} = \mathcal{T}_h \setminus \mathcal{T}_{h,i},$$

respectively.

To properly define the objects involved in the locally implicit scheme, we further define cut-off operators that restrict the support of a given function to either the explicitly or implicitly treated part of the mesh by

$$\chi_i v \equiv \begin{cases} v, & \text{on } K \in \mathcal{T}_{h,i}, \\ 0, & \text{on } K \in \mathcal{T}_{h,e}, \end{cases} \quad \text{and} \quad \chi_e v \equiv \begin{cases} 0, & \text{on } K \in \mathcal{T}_{h,i}, \\ v, & \text{on } K \in \mathcal{T}_{h,e}, \end{cases}$$

respectively. Note, in particular, that the application of these cut-off operators to discrete functions in the ansatz space $\mathbb{V}_h$ again yields a discrete function in $\mathbb{V}_h$, i.e., we have

$$\chi_i \boldsymbol{v}, \chi_e \boldsymbol{v} \in \mathbb{V}_h, \quad \text{for all } \boldsymbol{v} \in \mathbb{V}_h.$$

This can be seen immediately since the action of the cut-off operators is matched to the elements of $\mathcal{T}_h$.

Using the cut-off functions χ_e and χ_i, we split the discrete Friedrichs' operator $\underline{\mathcal{L}}$ into a part that is treated by the explicit leapfrog scheme and one that is treated by the implicit Crank–Nicolson scheme. We generalize the splitting proposed in [91] for the Maxwell equations to our more general setting. This boils down to splitting the two discrete operators $\underline{\mathcal{L}}_p$ and $\underline{\mathcal{L}}_q$ as follows. The explicit operators $\underline{\mathcal{L}}_p^e$ and $\underline{\mathcal{L}}_q^e$ are defined as

$$\underline{\mathcal{L}}_p^e = \chi_e \circ \underline{\mathcal{L}}_p, \quad \text{and} \quad \underline{\mathcal{L}}_q^e = \underline{\mathcal{L}}_q \circ \chi_e,$$

and the implicit operators $\underline{\mathcal{L}}_p^i$ and $\underline{\mathcal{L}}_q^i$ as

$$\underline{\mathcal{L}}_p^i = \chi_i \circ \underline{\mathcal{L}}_p, \quad \text{and} \quad \underline{\mathcal{L}}_q^i = \underline{\mathcal{L}}_q \circ \chi_i.$$

Clearly, since we have $\chi_e + \chi_i = \underline{I}$ (on $\mathbb{V}_h^{m_p}$ or $\mathbb{V}_h^{m_q}$, respectively), this is indeed a splitting, i.e., we have

$$\underline{\mathcal{L}}_p = \underline{\mathcal{L}}_p^e + \underline{\mathcal{L}}_p^i, \quad \text{and} \quad \underline{\mathcal{L}}_q = \underline{\mathcal{L}}_q^e + \underline{\mathcal{L}}_q^i.$$

With this, we split $\underline{\mathcal{L}}$ into the explicit and implicit part

$$\underline{\mathcal{L}}^e = \begin{pmatrix} 0 & \underline{\mathcal{L}}_q^e \\ \underline{\mathcal{L}}_p^e & 0 \end{pmatrix}, \qquad \text{and} \qquad \underline{\mathcal{L}}^i = \begin{pmatrix} K & \underline{\mathcal{L}}_q^i \\ \underline{\mathcal{L}}_p^i & 0 \end{pmatrix},$$

respectively. Note that we included the damping K in the implicit operator, since we treat it implicitly (as we already did for the leapfrog scheme).

These split operators inherit some properties of the original ones. In fact, they are constructed such that they still exhibit an adjointness property like (11.24), which is a key ingredient in the analysis of the method. More precisely, for all $\boldsymbol{p} \in \mathbb{V}_{\hbar}^{m_p}$, $\boldsymbol{q} \in \mathbb{V}_{\hbar}^{m_q}$, we have

$$(\underline{\mathcal{L}}_p^e \boldsymbol{p} \mid \boldsymbol{q})_\Omega = -(\boldsymbol{p} \mid \underline{\mathcal{L}}_q^e \boldsymbol{q})_\Omega, \quad \text{and} \quad (\underline{\mathcal{L}}_p^i \boldsymbol{p} \mid \boldsymbol{q})_\Omega = -(\boldsymbol{p} \mid \underline{\mathcal{L}}_q^i \boldsymbol{q})_\Omega, \tag{11.59}$$

which is easily verified using the original adjointness property (11.24) and the symmetry of χ_e and χ_i. In particular, this immediately yields the dissipativity of both $\underline{\mathcal{L}}^e$ and $\underline{\mathcal{L}}^i$ on $\mathbb{V}_{\hbar}^m$, the former even being skew-adjoint.

The split operators also fulfill inverse inequalities. In contrast to the full operators, only parts of the element diameters enter the bound instead of those belonging to the whole mesh. Due to the cut-off operators one might expect that only either the implicit or explicit parts of the mesh play a role. However, since the discrete operators couple elements over faces, the elements neighboring the explicitly or implicitly treated mesh have to be taken into account as well. This is the reason why the implicit part of the mesh also contains coarse neighbors of fine elements.

We state the inequalities for the explicit operators only, since the implicit counterparts do not matter for our theory. Also, we omit the rather technical proofs as they can be performed similarly to [88, Prop. 4.13] with some minor tweaks. Additionally, the proof for the corresponding objects in the Maxwell case can be found in [161, Thm. 5.6]. The inequalities are given such that, for all $\boldsymbol{p} \in \mathbb{V}_{\hbar}^{m_p}$ and $\boldsymbol{q} \in \mathbb{V}_{\hbar}^{m_q}$, we have

$$\|h^j \underline{\mathcal{L}}_p^e \boldsymbol{p}\|_\Omega \leq C_{\mathrm{inv},\mathcal{L},c,j} \|h_c^{j-1} \boldsymbol{p}\|_\Omega, \quad \|h^j \underline{\mathcal{L}}_q^e \boldsymbol{q}\|_\Omega \leq C_{\mathrm{inv},\mathcal{L},c,j} \|h_c^{j-1} \boldsymbol{q}\|_\Omega, \tag{11.60}$$

where the constant $C_{\mathrm{inv},\mathcal{L},c,j}$ does not depend on the fine mesh. Moreover, note that only the element diameters of the coarse mesh enter, since the norms on the right-hand side involve h_c, which vanishes on the fine elements.

We point out that this bound could be slightly refined. Namely, there might be coarse elements completely surrounded by fine elements or even patches of coarse elements which all have a fine neighbor. These can be neglected here as they and their neighbors are completely contained in the implicitly treated mesh. See the bound in [161, Thm. 5.6], where only the explicitly treated elements and the buffer layer of coarse neighbors are involved. However, since these inequalities have the same qualitative behavior, we abstain from doing so in favor of a more concise notation.

Unfortunately, not all properties of the full operators are carried over. In particular, while the p-operators are still consistent in some sense, i.e., for $p \in D(\mathcal{L}_p) \cap H^1(\mathcal{T}_h)^{m_p}$ we have

$$\underline{\mathcal{L}}^e_p p = \chi_e(\pi_h \mathcal{L}_p p), \quad \text{and} \quad \underline{\mathcal{L}}^i_p p = \chi_i(\pi_h \mathcal{L}_p p),$$

the q-operators do not fulfill a similar property.

Now, we have all ingredients necessary to derive (and analyze) the locally implicit integrator. We start by rewriting the dG-CN scheme (11.3) in a similar form as the dG-LF scheme (11.30) so we can afterwards combine them. Firstly, note that, due to Assumption 11.8, the Crank–Nicolson iteration (11.3a) can be written as

$$\begin{aligned} \boldsymbol{p}^{n+1}_{\text{CN}} - \boldsymbol{p}^n_{\text{CN}} &= \tau \underline{\mathcal{L}}_q(\boldsymbol{q}^{n+1}_{\text{CN}} + \boldsymbol{q}^n_{\text{CN}}) + \tfrac{\tau}{2} K(\boldsymbol{p}^{n+1}_{\text{CN}} + \boldsymbol{p}^n_{\text{CN}}) + \tau \overline{\boldsymbol{g}}^{n+1/2}_\pi, \\ \boldsymbol{q}^{n+1}_{\text{CN}} - \boldsymbol{q}^n_{\text{CN}} &= \tfrac{\tau}{2} \underline{\mathcal{L}}_p(\boldsymbol{p}^{n+1}_{\text{CN}} + \boldsymbol{p}^n_{\text{CN}}), \end{aligned}$$

with $\boldsymbol{u}^n_{\text{CN}} = \begin{pmatrix} \boldsymbol{p}^n_{\text{CN}} \\ \boldsymbol{q}^n_{\text{CN}} \end{pmatrix}$. By introducing the intermediate variable $\boldsymbol{q}^{n+1/2}_{\text{CN}}$ via

$$\boldsymbol{q}^{n+1/2}_{\text{CN}} = \boldsymbol{q}^n_{\text{CN}} + \tfrac{\tau}{2} \underline{\mathcal{L}}_p \boldsymbol{p}^n_{\text{CN}},$$

we can therefore rewrite the dG-CN scheme (11.3) as

$$\left\{ \begin{aligned} &\boldsymbol{q}^{n+1/2}_{\text{CN}} - \boldsymbol{q}^n_{\text{CN}} = \tfrac{\tau}{2} \underline{\mathcal{L}}_p \boldsymbol{p}^n_{\text{CN}}, && \text{(11.61a)} \\ &\boldsymbol{p}^{n+1}_{\text{CN}} - \boldsymbol{p}^n_{\text{CN}} = \tau \underline{\mathcal{L}}_q(\boldsymbol{q}^{n+1}_{\text{CN}} + \boldsymbol{q}^n_{\text{CN}}) + \tfrac{\tau}{2} K(\boldsymbol{p}^{n+1}_{\text{CN}} + \boldsymbol{p}^n_{\text{CN}}) + \tau \overline{\boldsymbol{g}}^{n+1/2}_\pi, && \text{(11.61b)} \\ &\boldsymbol{q}^{n+1}_{\text{CN}} - \boldsymbol{q}^{n+1/2}_{\text{CN}} = \tfrac{\tau}{2} \underline{\mathcal{L}}_p \boldsymbol{p}^{n+1}_{\text{CN}}, && \text{(11.61c)} \\ &\boldsymbol{p}^0_{\text{CN}} = \boldsymbol{p}^0_\pi, \quad \boldsymbol{q}^0_{\text{CN}} = \boldsymbol{q}^0_\pi. && \text{(11.61d)} \end{aligned} \right.$$

Comparing this with the dG-LF scheme (11.30), we see that the only difference between both schemes lies in the treatment of the $\underline{\mathcal{L}}_q$-term in (11.61b) and (11.30b), respectively.

Hence, we combine both schemes by using the explicit operator $\underline{\mathcal{L}}^e_q$ as in the dG-LF scheme (11.30) and the implicit one, $\underline{\mathcal{L}}^i_q$, as in the dG-CN scheme (11.61). We apply this idea to the spatially discrete wave-type problem (10.7) to obtain the *fully discrete dG-locally implicit (dG-LI) scheme* given by

$$\begin{cases} \boldsymbol{q}_{\mathrm{LI}}^{n+1/2} - \boldsymbol{q}_{\mathrm{LI}}^{n} = \frac{\tau}{2}\underline{\mathcal{L}}_p \boldsymbol{p}_{\mathrm{LI}}^{n}, \\ \boldsymbol{p}_{\mathrm{LI}}^{n+1} - \boldsymbol{p}_{\mathrm{LI}}^{n} = \tau\,\underline{\mathcal{L}}_q^e \boldsymbol{q}_{\mathrm{LI}}^{n+1/2} + \frac{\tau}{2}\underline{\mathcal{L}}_q^i(\boldsymbol{q}_{\mathrm{LI}}^{n+1} + \boldsymbol{q}_{\mathrm{LI}}^{n}) \\ \qquad + \frac{\tau}{2}K(\boldsymbol{p}_{\mathrm{LI}}^{n+1} + \boldsymbol{p}_{\mathrm{LI}}^{n}) + \tau\overline{\boldsymbol{g}}_{\pi}^{n+1/2}, \\ \boldsymbol{q}_{\mathrm{LI}}^{n+1} - \boldsymbol{q}_{\mathrm{LI}}^{n+1/2} = \frac{\tau}{2}\underline{\mathcal{L}}_p \boldsymbol{p}_{\mathrm{LI}}^{n+1}, \\ \boldsymbol{p}_{\mathrm{LI}}^{0} = \boldsymbol{p}_{\pi}^{0}, \quad \boldsymbol{q}_{\mathrm{LI}}^{0} = \boldsymbol{q}_{\pi}^{0}. \end{cases}$$

As for the other methods, we can directly deduce that the scheme delivers unique approximations at each timestep. In particular, we see that to obtain the new approximations $\boldsymbol{p}_{\mathrm{LI}}^{n+1}$ and $\boldsymbol{q}_{\mathrm{LI}}^{n+1}$, a linear system with $\underline{\mathcal{I}} - \frac{\tau}{2}\underline{\mathcal{L}}^i$ has to be solved. Since $\underline{\mathcal{L}}^i$ is dissipative on $\mathbb{V}_h^m$, this system is uniquely solvable by Lemma 9.15 (i). We point out that in practice, it is possible to further reduce the size of the linear system by using a Schur complement formulation, cf., [161, Chap. 6].

As for the leapfrog scheme, we next formulate the locally implicit scheme as a perturbation of the Crank–Nicolson scheme on the full mesh. We proceed completely analogously and obtain

$$\begin{cases} (\underline{\mathcal{I}} - \frac{\tau}{2}\underline{\mathcal{L}} + \frac{\tau^2}{4}\underline{\mathfrak{D}}^e)\boldsymbol{u}_{\mathrm{LI}}^{n+1} = (\underline{\mathcal{I}} + \frac{\tau}{2}\underline{\mathcal{L}} + \frac{\tau^2}{4}\underline{\mathfrak{D}}^e)\boldsymbol{u}_{\mathrm{LI}}^{n} + \tau\overline{\boldsymbol{f}}_{\pi}^{n+1/2}, & \text{(11.62a)} \\ \boldsymbol{u}_{\mathrm{LI}}^{0} = \boldsymbol{u}_{\pi}^{0}, & \text{(11.62b)} \end{cases}$$

$\underline{\mathfrak{D}}^e$ with $\boldsymbol{u}_{\mathrm{LI}}^{n} = \begin{pmatrix} \boldsymbol{p}_{\mathrm{LI}}^{n} \\ \boldsymbol{q}_{\mathrm{LI}}^{n} \end{pmatrix}$, $n \in \mathbb{N}_0$, and where

$$\underline{\mathfrak{D}}^e = \begin{pmatrix} \underline{\mathcal{L}}_q^e \underline{\mathcal{L}}_p & 0 \\ 0 & 0 \end{pmatrix} = \begin{pmatrix} \underline{\mathcal{L}}_q^e \underline{\mathcal{L}}_p^e & 0 \\ 0 & 0 \end{pmatrix} \tag{11.63}$$

is the discrete perturbation operator of the locally implicit scheme. The next lemma, similarly to Lemma 11.9, shows two important properties of the discrete perturbation operator. It can be proven in the same way and the proof is therefore again left to the reader.

Lemma 11.24 *The discrete perturbation operator $\underline{\mathfrak{D}}^e$ of the locally implicit method is self-adjoint and negative semidefinite, i.e., for $\boldsymbol{v}, \boldsymbol{w} \in \mathbb{V}_h^m$, we have*

$$(\underline{\mathfrak{D}}^e \boldsymbol{v} \mid \boldsymbol{w})_\Omega = (\boldsymbol{v} \mid \underline{\mathfrak{D}}^e \boldsymbol{w})_\Omega,$$

and for $\boldsymbol{v} = \begin{pmatrix} \boldsymbol{p} \\ \boldsymbol{q} \end{pmatrix} \in \mathbb{V}_h^m$ *with* $\boldsymbol{p} \in \mathbb{V}_h^{m_p}$ *and* $\boldsymbol{q} \in \mathbb{V}_h^{m_q}$*, we have*

$$\left(\underline{\mathcal{D}}^e \boldsymbol{v} \mid \boldsymbol{v}\right)_\Omega = -\|\underline{\mathcal{L}}_p^e \boldsymbol{p}\|_\Omega^2 \leq 0. \tag{11.64}$$

Note, in particular, that only the explicit operator $\underline{\mathcal{L}}_p^e$ occurs in (11.64). This is key to avoiding a CFL condition on the whole mesh, since $\underline{\mathcal{L}}_p^e$ fulfills the inverse inequality (11.60), which only depends on the coarse mesh.

Following the same reasoning as for the dG-LF scheme, the discrete operator $\underline{\mathcal{I}} - \frac{\tau}{2}\underline{\mathcal{L}} + \frac{\tau^2}{4}\underline{\mathcal{D}}^e$ is invertible (on $\mathbb{V}_h^m$) and we thus define $\underline{\mathcal{R}}_{\mathrm{LI}}, \underline{\mathcal{S}}_{\mathrm{LI}} : \mathbb{V}_h^m \to \mathbb{V}_h^m$ by

$$\underline{\mathcal{R}}_{\mathrm{LI}} = (\underline{\mathcal{I}} - \tfrac{\tau}{2}\underline{\mathcal{L}} + \tfrac{\tau^2}{4}\underline{\mathcal{D}}^e)^{-1} \tag{11.65}$$

and

$$\underline{\mathcal{S}}_{\mathrm{LI}} = (\underline{\mathcal{I}} - \tfrac{\tau}{2}\underline{\mathcal{L}} + \tfrac{\tau^2}{4}\underline{\mathcal{D}}^e)^{-1}(\underline{\mathcal{I}} + \tfrac{\tau}{2}\underline{\mathcal{L}} + \tfrac{\tau^2}{4}\underline{\mathcal{D}}^e). \tag{11.66}$$

Using these operators, the locally implicit scheme (11.62) can be rewritten as

$$\begin{cases} \boldsymbol{u}_{\mathrm{LI}}^{n+1} = \underline{\mathcal{S}}_{\mathrm{LI}} \boldsymbol{u}_{\mathrm{LI}}^n + \tau \underline{\mathcal{R}}_{\mathrm{LI}} \overline{\boldsymbol{f}}_\pi^{n+1/2}, & n \in \mathbb{N}_0, \\ \boldsymbol{u}_{\mathrm{LI}}^0 = \boldsymbol{u}_\pi^0. \end{cases}$$

Once again, this is (in principle) of the same form as all former methods and thus, we can solve the recursion to obtain the following result.

Theorem 11.25 *For all* $n \in \mathbb{N}_0$*, all* $h \in \mathcal{H}$*, and all* $\tau > 0$*, there exists a unique* $\boldsymbol{u}_{\mathrm{LI}}^{n+1} \in \mathbb{V}_h^m$ *fulfilling the dG-LI scheme* (11.62)*. It is given by the discrete variation-of-constants formula*

$$\boldsymbol{u}_{\mathrm{LI}}^{n+1} = \underline{\mathcal{S}}_{\mathrm{LI}}^{n+1} \boldsymbol{u}_\pi^0 + \tau \sum_{j=0}^{n} \underline{\mathcal{S}}_{\mathrm{LI}}^{n-j} \underline{\mathcal{R}}_{\mathrm{LI}} \overline{\boldsymbol{f}}_\pi^{j+1/2}.$$

As in the other methods, $\underline{\mathcal{S}}_{\mathrm{LI}}$ can be seen as a discrete semigroup.

Since the scheme is explicit on parts of the mesh, it is only stable if a CFL condition is fulfilled. In the following, we show that this CFL condition is independent of the fine mesh elements.

Assumption 11.26 (CFL Condition for the Locally Implicit Scheme) For an arbitrary but fixed parameter $0 < \theta_c < 1$, the timestep τ satisfies the CFL condition

$$\tau \leq \frac{2\theta_c}{C_{\mathrm{inv},\mathcal{L},c}} h_{\mathrm{min,c}}, \tag{11.67}$$

where $C_{\mathrm{inv},\mathcal{L},c} = C_{\mathrm{inv},\mathcal{L},c,0}$ (cf., (11.60)). ◇

We point out that both the minimal mesh diameter $h_{\mathrm{min,c}}$ as well as the constant $C_{\mathrm{inv},\mathcal{L},c}$ only depend on the coarse mesh elements, in contrast to the leapfrog method, i.e., Assumption 11.11, where the whole mesh had to be considered.

Now, we provide the desired properties of the discrete semigroup as well as the two stability bounds resulting from them. Note that the structure of the locally implicit and the leapfrog scheme are almost identical with the exception of the perturbation operators $\underline{\mathcal{D}}^e$ and $\underline{\mathcal{D}}$, respectively. Since both operators share analogous properties, the proofs of the following results are still identical to their leapfrog counterparts (i.e., Propositions 11.13 and 11.15, Lemma 11.14 and Theorems 11.16 and 11.17). Thus, we omit them here. We start with the power bound on $\underline{\mathcal{S}}_{\mathrm{LI}}$.

Proposition 11.27 *Let $h \in \mathcal{H}$ and $\tau > 0$ fulfill Assumption 11.26 with $\theta_c \in (0, 1)$. Then, for all $\ell \in \mathbb{N}$ and all $\boldsymbol{v} \in \mathbb{V}_h^m$, the discrete semigroup $\underline{\mathcal{S}}_{\mathrm{LI}}$ defined in (11.66) fulfills the power bound*

$$\|\underline{\mathcal{S}}_{\mathrm{LI}}^{\ell} \boldsymbol{v}\|_\Omega \leq C_{\mathrm{stb,c}} \|\boldsymbol{v}\|_\Omega,$$

where $C_{\mathrm{stb,c}} = (1 - \theta_c^2)^{-1/2}$.

Again, we need a bound on the discrete operator $\underline{\mathcal{R}}_{\mathrm{LI}}$, which is independent of the mesh width. This can be achieved under the CFL condition (11.67), and we state it in the next lemma.

Lemma 11.28 *Let $h \in \mathcal{H}$ and $\tau > 0$ such that Assumption 11.26 is fulfilled with $\theta_c \in (0, 1)$. Then, for all $\boldsymbol{v} \in \mathbb{V}_h^m$, the discrete operator $\underline{\mathcal{R}}_{\mathrm{LI}}$ defined in (11.65) fulfills*

$$\|\underline{\mathcal{R}}_{\mathrm{LI}} \boldsymbol{v}\|_\Omega \leq \sqrt{3}\, \|\boldsymbol{v}\|_\Omega.$$

Proof Observe that (11.43) still holds true if we replace $\underline{\mathcal{R}}_{\mathrm{LF}}$ by $\underline{\mathcal{R}}_{\mathrm{LI}}$ and Y_{LF} by $Y_{\mathrm{LI}} = (\underline{\mathcal{I}} - \frac{\tau}{2} K - \frac{\tau^2}{4} \underline{\mathcal{L}}_q^i \underline{\mathcal{L}}_p^i)^{-1}$. Since $\underline{\mathcal{L}}_q^i \underline{\mathcal{L}}_p^i$ is self-adjoint by the adjointness property (11.59) and maximal dissipative, the sum $K + \frac{\tau}{2} \underline{\mathcal{L}}_q^i \underline{\mathcal{L}}_p^i$ is self-adjoint and maximal dissipative as well. Thus, Y_{LI} has the same properties as Y_{LF} and the rest of the proof relies on the proof of Lemma 11.14, but the treatment of the split operators increases the technical effort considerably. □

The last property we state is again a discrete version of the fact that the time derivative of a semigroup is given by the application of its generator, i.e., (9.13b).

Proposition 11.29 *Let* $\mathfrak{h} \in \mathcal{H}$ *and* $\tau > 0$. *Then, for all* $\ell \in \mathbb{N}_0$, *the powers of the discrete semigroup* $\underline{\mathcal{S}}_{\mathrm{LI}}$ *defined in* (11.66) *satisfy*

$$\frac{\underline{\mathcal{S}}_{\mathrm{LI}}^{\ell+1} - \underline{\mathcal{S}}_{\mathrm{LI}}^{\ell}}{\tau} = (\underline{\mathcal{R}}_{\mathrm{LI}} \underline{\mathcal{L}})\, \underline{\mathcal{S}}_{\mathrm{LI}}^{\ell} = \underline{\mathcal{S}}_{\mathrm{LI}}^{\ell}\, (\underline{\mathcal{R}}_{\mathrm{LI}} \underline{\mathcal{L}}).$$

We point out that, as it was the case for the leapfrog method, this result does not require the CFL condition (11.67).

With this, we are able to show the two stability results. Again, the proofs are performed completely analogous to the Crank–Nicolson method (with exception of the second bound on the inhomogeneity, which is therefore omitted again). Thus, we state the results without proof.

Theorem 11.30 *Let* $\mathfrak{h} \in \mathcal{H}$ *and* $\tau > 0$ *such that Assumption 11.26 is fulfilled. Then, for all* $n \in \mathbb{N}_0$, *the approximation* $\boldsymbol{u}_{\mathrm{LI}}^{n+1}$ *given by the dG-LI scheme* (11.62) *satisfies*

$$\|\boldsymbol{u}_{\mathrm{LI}}^{n+1}\|_{\Omega} \le C_{\mathrm{stb,c}} \|u^0\|_{\Omega} + \sqrt{3}\, C_{\mathrm{stb,c}} \tau \sum_{j=1}^{n} \|\overline{f}^{j+1/2}\|_{\Omega}.$$

Lastly, we define the *discrete time derivative* and the *discrete space derivative* for the locally implicit scheme at time $t_{n+1/2}$ as

$$\begin{aligned} \underline{\partial_\tau \boldsymbol{u}}_{\mathrm{LI}}^{n+1/2} &= \frac{\boldsymbol{u}_{\mathrm{LI}}^{n+1} - \boldsymbol{u}_{\mathrm{LI}}^{n}}{\tau}, \\ \underline{\mathcal{L}\boldsymbol{u}}_{\mathrm{LI}}^{n+1/2} &= \underline{\mathcal{L}}\, \frac{\boldsymbol{u}_{\mathrm{LI}}^{n+1} + \boldsymbol{u}_{\mathrm{LI}}^{n}}{2} + \frac{\tau^2}{4} \underline{\mathcal{D}}^e\, \underline{\partial_\tau \boldsymbol{u}}_{\mathrm{LI}}^{n+1/2}, \end{aligned} \tag{11.68}$$

respectively. This yields the equivalent form of the locally implicit iteration (11.62a) given by

$$\underline{\partial_\tau \boldsymbol{u}}_{\mathrm{LI}}^{n+1/2} = \underline{\mathcal{L}\boldsymbol{u}}_{\mathrm{LI}}^{n+1/2} + \overline{\boldsymbol{f}}_{\pi}^{n+1/2}.$$

We state bounds on these discrete derivatives in the next theorem.

Theorem 11.31 *Let $h \in \mathcal{H}$ and $\tau > 0$ such that Assumption 11.26 is fulfilled. Then, the approximations $\{\boldsymbol{u}_{\mathrm{LI}}^n\}_{n\geq 0}$ given by the dG-LI scheme* (11.62) *satisfy*

$$\|\underline{\boldsymbol{\partial}_\tau \boldsymbol{u}}_{\mathrm{LI}}^{n+1/2}\|_\Omega \leq \sqrt{3}\, C_{\mathrm{stb,c}} \|\underline{\mathcal{L}\boldsymbol{u}}_\pi^0\|_\Omega + \sqrt{3} \max_{j=0,\dots,n+1} \|\boldsymbol{f}_\pi^j\|_\Omega + \sqrt{3}\, C_{\mathrm{stb,c}} \tau \sum_{j=1}^{n} \|\frac{\boldsymbol{f}_\pi^{j+1} - \boldsymbol{f}_\pi^{j-1}}{2\tau}\|_\Omega,$$

and

$$\|\underline{\mathcal{L}\boldsymbol{u}}_{\mathrm{LI}}^{n+1/2}\|_\Omega \leq \|\underline{\boldsymbol{\partial}_\tau \boldsymbol{u}}_{\mathrm{LI}}^{n+1/2}\|_\Omega + \max_{j=0,\dots,n+1} \|\boldsymbol{f}_\pi^j\|_\Omega.$$

If further $u^0 \in D(\mathcal{L}) \cap H^1(\mathcal{T}_h)^m$ and $f \in C^1(\mathbb{R}_+; L^2(\Omega)^m)$, we have

$$\|\underline{\boldsymbol{\partial}_\tau \boldsymbol{u}}_{\mathrm{LI}}^{n+1/2}\|_\Omega \leq \sqrt{3}\, C_{\mathrm{stb,c}} (\|\mathcal{L}u^0\|_\Omega + C_{\pi,\mathcal{L}} |u^0|_{1,\mathcal{T}_h}) + \sqrt{3} \max_{s\in\mathbb{R}_+} \|f(s)\|_\Omega + \sqrt{3}\, C_{\mathrm{stb,c}} \int_0^{t_{n+1}} \|\partial_t f(s)\|_\Omega \,\mathrm{d}s.$$

11.5 Addendum

As we have seen, all considered methods exhibit the same basic structure and similar properties, albeit with some adaptations from scheme to scheme. In fact, as we will see in this section, there is even more similarity between the Peaceman–Rachford, the leapfrog, and the locally implicit method. This new viewpoint, in particular, provides us with alternative formulae for the discrete derivatives $\underline{\mathcal{L}\boldsymbol{u}}_{\mathrm{LF}}^{n+1/2}$, $\underline{\mathcal{L}\boldsymbol{u}}_{\mathrm{PR}}^{n+1/2}$, and $\underline{\mathcal{L}\boldsymbol{u}}_{\mathrm{LI}}^{n+1/2}$ defined in (11.45), (11.58), and (11.58), respectively.

More precisely, we introduce a splitting of the discrete operator $\underline{\mathcal{L}} = \underline{\mathcal{A}}_{\mathrm{LF}} + \underline{\mathcal{B}}_{\mathrm{LF}}$ with

$$\underline{\mathcal{A}}_{\mathrm{LF}} = \begin{pmatrix} 0 & \underline{\mathcal{L}}_q \\ 0 & 0 \end{pmatrix} \quad \text{and} \quad \underline{\mathcal{B}}_{\mathrm{LF}} = \begin{pmatrix} K & 0 \\ \underline{\mathcal{L}}_p & 0 \end{pmatrix}. \tag{11.69}$$

This allows us to rewrite the dG-LF scheme (11.30) as

$$\begin{cases} (\underline{\mathcal{I}} - \frac{\tau}{2}\underline{\mathcal{A}}_{\mathrm{LF}})\boldsymbol{u}_{\mathrm{LF}}^{n+1/2} = (\underline{\mathcal{I}} + \frac{\tau}{2}\underline{\mathcal{B}}_{\mathrm{LF}})\boldsymbol{u}_{\mathrm{LF}}^n + \frac{\tau}{2}\boldsymbol{f}_\pi^n, & \text{(11.70a)} \\ (\underline{\mathcal{I}} - \frac{\tau}{2}\underline{\mathcal{B}}_{\mathrm{LF}})\boldsymbol{u}_{\mathrm{LF}}^{n+1} = (\underline{\mathcal{I}} + \frac{\tau}{2}\underline{\mathcal{A}}_{\mathrm{LF}})\boldsymbol{u}_{\mathrm{LF}}^{n+1/2} + \frac{\tau}{2}\boldsymbol{f}_\pi^{n+1}, & \text{(11.70b)} \\ \boldsymbol{u}_{\mathrm{LF}}^0 = \boldsymbol{u}_\pi^0, & \text{(11.70c)} \end{cases}$$

where the intermediate step is given by $\boldsymbol{u}_{\mathrm{LF}}^{n+1/2} = \begin{pmatrix} \boldsymbol{p}_{\mathrm{LF}}^{n+1/2} \\ \boldsymbol{q}_{\mathrm{LF}}^{n+1/2} \end{pmatrix}$, $n \in \mathbb{N}_0$. The yet to be determined value $\boldsymbol{p}_{\mathrm{LF}}^{n+1/2}$ can actually be chosen arbitrarily, since it does not change the other values due to the structure of $\underline{\mathcal{A}}_{\mathrm{LF}}$ and can be completely eliminated by adding (11.70a) and (11.70b). However, a suitable and natural choice would be

$$\boldsymbol{p}_{\mathrm{LF}}^{n+1/2} = \frac{\boldsymbol{p}_{\mathrm{LF}}^{n+1} + \boldsymbol{p}_{\mathrm{LF}}^{n}}{2},$$

which also provides an approximation to $p(t_{n+1/2})$.

Comparing (11.70) with the dG-PR scheme (11.55) we see that the dG-leapfrog scheme can in fact be interpreted as the Peaceman–Rachford scheme with splitting given by (11.69) (and a slightly different treatment of the inhomogeneity). In particular, from this observation, we immediately obtain that the discrete space derivative $\underline{\mathcal{L}\boldsymbol{u}}_{\mathrm{LF}}^{n+1/2}$ of the leapfrog scheme (defined in (11.45)) is also given by

$$\underline{\mathcal{L}\boldsymbol{u}}_{\mathrm{LF}}^{n+1/2} = \underline{\mathcal{A}}_{\mathrm{LF}}\boldsymbol{u}_{\mathrm{LF}}^{n+1/2} + \underline{\mathcal{B}}_{\mathrm{LF}}\,\frac{\boldsymbol{u}_{\mathrm{LF}}^{n+1} + \boldsymbol{u}_{\mathrm{LF}}^{n}}{2},$$

cf., (11.58). Similarly, by acknowledging $\underline{\mathcal{D}} = \underline{\mathcal{A}}_{\mathrm{LF}}\underline{\mathcal{B}}_{\mathrm{LF}}$ in (11.45), we see that we can also calculate the discrete space derivative $\underline{\mathcal{L}\boldsymbol{u}}_{\mathrm{PR}}^{n+1/2}$ of the dG-Peaceman–Rachford scheme (defined in (11.58)) by

$$\underline{\mathcal{L}\boldsymbol{u}}_{\mathrm{PR}}^{n+1/2} = \underline{\mathcal{L}}\,\frac{\boldsymbol{u}_{\mathrm{PR}}^{n+1} + \boldsymbol{u}_{\mathrm{PR}}^{n}}{2} + \tfrac{\tau^2}{4}\underline{\mathcal{A}}\underline{\mathcal{B}}\,\partial_\tau \boldsymbol{u}_{\mathrm{PR}}^{n+1/2}. \tag{11.71}$$

This can of course also be derived from the method directly, which we leave as an exercise.

Moreover, the locally implicit scheme can also be interpreted as a Peaceman–Rachford scheme by introducing the splitting $\underline{\mathcal{L}} = \underline{\mathcal{A}}_{\mathrm{LI}} + \underline{\mathcal{B}}_{\mathrm{LI}}$ with

$$\underline{\mathcal{A}}_{\mathrm{LI}} = \begin{pmatrix} 0 & \underline{\mathcal{L}}_q^e \\ 0 & 0 \end{pmatrix} \quad \text{and} \quad \underline{\mathcal{B}}_{\mathrm{LI}} = \begin{pmatrix} K & \underline{\mathcal{L}}_q^i \\ \underline{\mathcal{L}}_p & 0 \end{pmatrix}.$$

Using this, the dG-LI scheme is equivalent to

$$\begin{cases} (\underline{\mathcal{I}} - \frac{\tau}{2}\underline{\mathcal{A}}_{\mathrm{LI}})\boldsymbol{u}_{\mathrm{LI}}^{n+1/2} = (\underline{\mathcal{I}} + \frac{\tau}{2}\underline{\mathcal{B}}_{\mathrm{LI}})\boldsymbol{u}_{\mathrm{LI}}^{n} + \frac{\tau}{2}\boldsymbol{f}_\pi^{n}, \\ (\underline{\mathcal{I}} - \frac{\tau}{2}\underline{\mathcal{B}}_{\mathrm{LI}})\boldsymbol{u}_{\mathrm{LI}}^{n+1} = (\underline{\mathcal{I}} + \frac{\tau}{2}\underline{\mathcal{A}}_{\mathrm{LI}})\boldsymbol{u}_{\mathrm{LI}}^{n+1/2} + \frac{\tau}{2}\boldsymbol{f}_\pi^{n+1}, \\ \boldsymbol{u}_{\mathrm{LI}}^{0} = \boldsymbol{u}_\pi^{0}, \end{cases}$$

where $\boldsymbol{u}_{\mathrm{LI}}^{n+1/2} = \begin{pmatrix} \boldsymbol{p}_{\mathrm{LF}}^{n+1/2} \\ \boldsymbol{q}_{\mathrm{LI}}^{n+1/2} \end{pmatrix}$, $n \in \mathbb{N}_0$. Again, the intermediate value $\boldsymbol{p}_{\mathrm{LI}}^{n+1/2}$ can be chosen arbitrarily. A suitable choice is given by the average of $\boldsymbol{p}_{\mathrm{LI}}^{n}$ and $\boldsymbol{p}_{\mathrm{LI}}^{n+1}$, cf., (11.71). Further, analogously to the leapfrog method, this means that the discrete space derivative $\underline{\mathcal{L}}\boldsymbol{u}_{\mathrm{LI}}^{n+1/2}$ of the dG-locally implicit scheme (defined in (11.68)) is equivalently given by

$$\underline{\mathcal{L}}\boldsymbol{u}_{\mathrm{LI}}^{n+1/2} = \underline{\mathcal{A}}_{\mathrm{LI}}\boldsymbol{u}_{\mathrm{LI}}^{n+1/2} + \underline{\mathcal{B}}_{\mathrm{LI}} \frac{\boldsymbol{u}_{\mathrm{LI}}^{n+1} + \boldsymbol{u}_{\mathrm{LI}}^{n}}{2}.$$

Lastly, we point out that even the dG-CN scheme can be interpreted as a Peaceman–Rachford scheme by choosing the (somewhat artificial) splitting $\underline{\mathcal{L}} = \frac{1}{2}\underline{\mathcal{L}} + \frac{1}{2}\underline{\mathcal{L}}$ or $\underline{\mathcal{L}} = \underline{\mathcal{L}} + 0$. The fact that the split operators commute in this case simplifies things significantly.

12 Error Analysis

In this chapter, we analyze the full discretization error caused by the four schemes presented in Chap. 11. As stated above, we show that, on a finite time interval $[0, T]$, the approximations given by all four methods converge to the exact solution of the wave-type problem (9.3) with order two in time and order k in space. This is both the classical order of the individual time integration method as well as the expected (and in this general setting optimal [120]) order of the central flux dG space discretization.

The key idea is to exploit that—as we have seen in the previous section—the leapfrog, Peaceman–Rachford, and locally implicit schemes can be written as perturbations of the Crank–Nicolson scheme. Therefore, we first consider the latter. Subsequently, we adjust our analysis to the perturbed schemes, which each need their individual treatments.

The basic procedure for all methods is to first insert the (projected) exact solution into the numerical scheme in order to derive an error recursion. This recursion is of the same form as the corresponding scheme, but with the inhomogeneity replaced by a defect caused by the numerical approximation. Once the defect is bounded in the right order in space and time, we derive the desired error bounds by using the stability properties of the method we showed in Chap. 11.

Unfortunately, for the dG-LI scheme, a direct bound on its defect can only be obtained with a suboptimal order in either time or space, see [161, Lem. 5.12]. Hence, we follow [91, 161], where summation-by-parts is used to get rid of the problematic discrete operator causing this loss of order. To present the ideas behind this approach in a more accessible way, we also apply it to the leapfrog method, as is done in [161, Sec. 4.2.2]. Alternatively, the dG-LF scheme can also be analyzed in the same way as the dG-PR scheme (although with some rather mild additional assumptions).

W. Dörfler et al., *Wave Phenomena*, Oberwolfach Seminars 49,
https://doi.org/10.1007/978-3-031-05793-9_12

To distinguish quantities appearing in all four methods considered here, we use the subscript M, which stands for either CN, LF, PR, or LI. First, we define the *full discretization error* e_{M}^n at time t_n as

$$e_{\mathrm{M}}^n = u(t_n) - \boldsymbol{u}_{\mathrm{M}}^n = e_\pi^n + \boldsymbol{e}_{\mathrm{M}}^n,$$

which is split into the *projection error*

$$e_\pi^n = u(t_n) - \pi_h u(t_n)$$

and the *discretization error*

$$\boldsymbol{e}_{\mathrm{M}}^n = \pi_h u(t_n) - \boldsymbol{u}_{\mathrm{M}}^n$$

at time t_n (recall definition (10.4) of the projection π_h). Note that the projection error e_π^n is already bounded by Lemma 10.2 with order $k+1$ in h if the exact solution is sufficiently regular and thus, it suffices to derive a bound on the discretization error $\boldsymbol{e}_{\mathrm{M}}^n$.

We also derive error bounds for the discrete derivatives given by the methods (except for the locally implicit method, where our analysis does not deliver such bounds) and show that they converge to their continuous counterparts with the same order as e_{M}^n (given some additional regularity of the exact solution). Recall the definitions of these discrete derivatives in (11.8), (11.45), (11.58), and (11.68), respectively. The corresponding errors are defined as the *time derivative error*

$$\boldsymbol{e}_{\tau,\mathrm{M}}^{n+1/2} = \pi_h \partial_t u(t_{n+1/2}) - \underline{\boldsymbol{\partial}_\tau \boldsymbol{u}}_{\mathrm{M}}^{n+1/2},$$

and the *space derivative error*

$$\boldsymbol{e}_{\mathcal{L},\mathrm{M}}^{n+1/2} = \pi_h \mathcal{L} u(t_{n+1/2}) - \underline{\boldsymbol{\mathcal{L}} \boldsymbol{u}}_{\mathrm{M}}^{n+1/2}.$$

Lastly, during the analysis we frequently encounter discrete derivatives (i.e., differential quotients) of sequences representing (time-)discrete functions. To keep notation concise and intuitive and to emphasize this interpretation, we denote the first- and second-order discrete derivative of a sequence $\{v^n\}$ by

$$\underline{\boldsymbol{\partial}_\tau} v^{n+1/2} = \frac{v^{n+1} - v^n}{\tau} \quad \text{and} \quad \underline{\boldsymbol{\partial}_\tau^2} v^n = \frac{v^{n+1} - 2v^n + v^{n-1}}{\tau^2},$$

respectively. The notation is motivated as follows. If $v^n = v(t_n)$ for a sufficiently regular function v, then $\underline{\boldsymbol{\partial}_\tau} v^{n+1/2} \approx \partial_t v(t_{n+1/2})$ and $\underline{\boldsymbol{\partial}_\tau^2} v^n \approx \partial_t^2 v(t_n)$, and both approximations are of second-order accuracy.

12.1 Crank–Nicolson Scheme

We start with the Crank–Nicolson scheme since it serves as a basis for all other schemes we consider and also because it is the most simple to analyze.

12.1.1 Error Recursion

The first step consists of deriving an error recursion for the discretization error $\boldsymbol{e}_{\mathrm{CN}}^{n}$. In fact, this recursion coincides with the dG-CN scheme (11.3) itself if the inhomogeneity is replaced by a defect. This defect is determined by inserting the projected exact solution into the scheme and is closely related to the local error of the scheme, i.e., the error made by performing a single step starting on the exact solution.

Lemma 12.1 *Let $\mathfrak{h} \in \mathcal{H}$ and $\tau > 0$. Then, for all $n \in \mathbb{N}_0$, the discretization error $\boldsymbol{e}_{\mathrm{CN}}^{n+1} = \pi_{\mathfrak{h}} u(t_{n+1}) - \boldsymbol{u}_{\mathrm{CN}}^{n+1}$ of the dG-Crank–Nicolson scheme satisfies*

$$\begin{cases} \boldsymbol{e}_{\mathrm{CN}}^{n+1} = \underline{\mathcal{S}}_{\mathrm{CN}}\, \boldsymbol{e}_{\mathrm{CN}}^{n} + \underline{\mathcal{R}}_{\mathrm{CN}}\, \boldsymbol{d}_{\mathrm{CN}}^{n}, & n \in \mathbb{N}_0, \quad (12.1\mathrm{a}) \\ \boldsymbol{e}_{\mathrm{CN}}^{0} = 0, & \quad (12.1\mathrm{b}) \end{cases}$$

where the ***dG-Crank–Nicolson defect*** $\boldsymbol{d}_{\mathrm{CN}}^{n}$ *is given by*

$$\boldsymbol{d}_{\mathrm{CN}}^{n} = \pi_{\mathfrak{h}}\big(u(t_{n+1}) - u(t_n)\big) - \tau\big(\underline{\mathcal{L}}\pi_{\mathfrak{h}}\frac{u(t_{n+1}) + u(t_n)}{2} + \overline{\boldsymbol{f}}_{\pi}^{n+1/2}\big). \tag{12.2}$$

Recall the notations $\boldsymbol{f}_{\pi}$ and $\overline{f}^{n+1/2}$, which were defined in (10.7c) and (11.1), respectively.

Proof The initial value (12.1b) vanishes since the initial value of the numerical scheme (11.3b) and the projected exact initial value coincide. To derive the error recursion, we insert the projected exact solution $\pi_{\mathfrak{h}} u$ into the dG-CN iteration (11.3a) to obtain

$$\big(\underline{\mathcal{I}} - \tfrac{\tau}{2}\underline{\mathcal{L}}\big)\pi_{\mathfrak{h}} u(t_{n+1}) = \big(\underline{\mathcal{I}} + \tfrac{\tau}{2}\underline{\mathcal{L}}\big)\pi_{\mathfrak{h}} u(t_n) + \tau\, \overline{\boldsymbol{f}}_{\pi}^{n+1/2} + \boldsymbol{d}_{\mathrm{CN}}^{n}. \tag{12.3}$$

Here, the defect $\boldsymbol{d}_{\mathrm{CN}}^{n}$ is caused by the fact that $\pi_{\mathfrak{h}} u$ does not fulfill the numerical scheme exactly. Solving for $\pi_{\mathfrak{h}} u(t_{n+1})$ then yields

$$\pi_{\mathfrak{h}} u(t_{n+1}) = \underline{\mathcal{S}}_{\mathrm{CN}}\, \pi_{\mathfrak{h}} u(t_n) + \tau \underline{\mathcal{R}}_{\mathrm{CN}}\, \overline{\boldsymbol{f}}_{\pi}^{n+1/2} + \underline{\mathcal{R}}_{\mathrm{CN}}\, \boldsymbol{d}_{\mathrm{CN}}^{n},$$

and, by subtracting the original dG-CN iteration (11.6a), we obtain the error recursion (12.1a). The formula (12.2) of the defect $\boldsymbol{d}^n_{\mathrm{CN}}$ is verified by solving (12.3) for $\boldsymbol{d}^n_{\mathrm{CN}}$. □

As stated before, the error recursion (12.1) is essentially the dG-CN scheme (11.6) with the term involving the inhomogeneity replaced by the defect $\boldsymbol{d}^n_{\mathrm{CN}}$. In particular, this means that by Theorem 11.2, or more precisely the discrete variation-of-constants formula (11.7), the error is given by

$$\boldsymbol{e}^{n+1}_{\mathrm{CN}} = \sum_{j=0}^{n} \underline{\boldsymbol{S}}_{\mathrm{CN}}^{n-j} \underline{\mathcal{R}}_{\mathrm{CN}} \boldsymbol{d}^j_{\mathrm{CN}}. \tag{12.4}$$

Thus, by the power bound on $\underline{\boldsymbol{S}}_{\mathrm{CN}}$ from Proposition 11.3, we obtain an error bound if the defect $\boldsymbol{d}^n_{\mathrm{CN}}$ is bounded appropriately. Moreover, by proceeding as for the stability bounds in Theorem 11.6, we can also derive error bounds for the discrete derivatives.

12.1.2 Bounds on the Defect

Before providing the bounds on the defect, we point out a crucial observation, namely, that the defect $\boldsymbol{d}^n_{\mathrm{CN}}$ resembles the quadrature error δ^n_{CN} of the trapezoidal rule applied to $\partial_t u$. In particular, if u solves the original differential equation (9.3a), this quadrature error is given by

$$\begin{aligned}\delta^n_{\mathrm{CN}} &= \int_{t_n}^{t_{n+1}} \partial_t u(s)\,\mathrm{d}s - \tau \frac{\partial_t u(t_{n+1}) + \partial_t u(t_n)}{2} \\ &= \big(u(t_{n+1}) - u(t_n)\big) - \tau\Big(\mathcal{L}\frac{u(t_{n+1}) + u(t_n)}{2} + \overline{f}^{n+1/2}\Big),\end{aligned} \tag{12.5}$$

and thus, the difference between the projection of δ^n_{CN} and $\boldsymbol{d}^n_{\mathrm{CN}}$ takes the form

$$\boldsymbol{d}^n_{\mathrm{CN}} - \pi_h \delta^n_{\mathrm{CN}} = \tfrac{\tau}{2}\big(\pi_h \mathcal{L} - \underline{\mathcal{L}}\pi_h\big)\big(u(t_{n+1}) + u(t_n)\big). \tag{12.6}$$

This term gives rise to the spatial error of the scheme and we have already seen how to deal with it, see (10.11).

In the next result, we use this observation to bound the defect $\boldsymbol{d}^n_{\mathrm{CN}}$ and its discrete time derivative (or difference quotient). These bounds then imply a bound on the discretization error $\boldsymbol{e}^n_{\mathrm{CN}}$ and the errors in the discrete derivatives $\boldsymbol{e}^{n+1/2}_{\tau,\mathrm{CN}}$ and $\boldsymbol{e}^{n+1/2}_{\mathcal{L},\mathrm{CN}}$, respectively.

Lemma 12.2 *Let $h \in \mathcal{H}$ and $\tau > 0$. Further, assume that*

$$u \in C^3(\mathbb{R}_+; L^2(\Omega)^m) \cap C(\mathbb{R}_+; D(\mathcal{L}) \cap H^{k+1}(\mathcal{T}_h)^m).$$

Then we have

$$\|\boldsymbol{d}_{\mathrm{CN}}^{n}\|_{\Omega} \leq \frac{\tau^2}{8} \int_{t_n}^{t_{n+1}} \|\partial_t^3 u(s)\|_{\Omega}\, \mathrm{d}s + C_{\pi,\mathcal{L}} \tfrac{\tau}{2} |h^k \big(u(t_{n+1}) + u(t_n)\big)|_{k+1,\mathcal{T}_h}. \tag{12.7}$$

Further, if

$$u \in C^4(\mathbb{R}_+; L^2(\Omega)^m) \cap C^1(\mathbb{R}_+; D(\mathcal{L}) \cap H^{k+1}(\mathcal{T}_h)^m),$$

the discrete derivative satisfies

$$\begin{aligned} \|\underline{\boldsymbol{\partial}_\tau} \boldsymbol{d}_{\mathrm{CN}}^{n-1/2}\|_{\Omega} &\leq \frac{\tau^2}{8} \int_{t_{n-1}}^{t_n} \int_0^1 \|\partial_t^4 u(s + \varsigma\tau)\|_{\Omega}\, \mathrm{d}\varsigma\, \mathrm{d}s \\ &\quad + \frac{1}{2} C_{\pi,\mathcal{L}} \int_{t_{n-1}}^{t_{n+1}} |h^k \partial_t u(s)|_{k+1,\mathcal{T}_h}\, \mathrm{d}s. \end{aligned} \tag{12.8}$$

Proof As stated in the beginning of this section, the defect $\boldsymbol{d}_{\mathrm{CN}}^{n}$ is already almost in the form of the quadrature error of the trapezoidal rule. We exploit this observation and add and subtract the missing part with the operator $\pi_h \mathcal{L}$. More precisely, by (12.6) and Proposition 10.7 (consistency), we have

$$\boldsymbol{d}_{\mathrm{CN}}^{n} = \pi_h \delta_{\mathrm{CN}}^{n} - \tfrac{\tau}{2} \underline{\mathcal{L}} \big(e_\pi^{n+1} + e_\pi^{n}\big).$$

A standard Peano kernel representation of the quadrature error δ_{CN}^{n} together with the contractivity of π_h now yields the first term in (12.7). The remaining terms are bounded by Proposition 10.9 (approximation property). This shows (12.7).

To show the second bound (12.8) we take the discrete time derivative of the defect $\boldsymbol{d}_{\mathrm{CN}}^{n}$ and obtain

$$\underline{\boldsymbol{\partial}_\tau} \boldsymbol{d}_{\mathrm{CN}}^{n-1/2} = \pi_h \underline{\boldsymbol{\partial}_\tau} \delta_{\mathrm{CN}}^{n-1/2} - \tfrac{1}{2} \underline{\mathcal{L}} \big(e_\pi^{n+1} - e_\pi^{n-1}\big).$$

Again, we bound the term involving the quadrature error δ_{CN}^{n} and the one involving the spatial operator $\underline{\mathcal{L}}$ separately.

For the former, we use the definition (12.5) of δ_{CN}^n and the fundamental theorem of calculus to obtain

$$\begin{aligned}
\delta_{\mathrm{CN}}^n - \delta_{\mathrm{CN}}^{n-1} &= \int_{t_n}^{t_{n+1}} \partial_t u(s)\,\mathrm{d}s - \int_{t_{n-1}}^{t_n} \partial_t u(s)\,\mathrm{d}s \\
&\quad - \tau\Big(\frac{\partial_t u(t_{n+1}) + \partial_t u(t_n)}{2} - \frac{\partial_t u(t_n) + \partial_t u(t_{n-1})}{2}\Big) \\
&= \int_{t_{n-1}}^{t_n} \Big(\partial_t u(s+\tau) - \partial_t u(s) - \tfrac{\tau}{2}\big(\partial_t^2 u(s+\tau) + \partial_t^2 u(s)\big)\Big)\,\mathrm{d}s \\
&= \int_{t_{n-1}}^{t_n} \Big(\int_s^{s+\tau} \partial_t^2 u(\varsigma)\,\mathrm{d}\varsigma - \tfrac{\tau}{2}\big(\partial_t^2 u(s+\tau) + \partial_t^2 u(s)\big)\Big)\,\mathrm{d}s.
\end{aligned} \tag{12.9}$$

We recognize that the integrand w.r.t. s is again the quadrature error of the trapezoidal rule, this time applied to $\partial_t^2 u$. Hence, we bound it as before and obtain (after further transformation of the integrals) the first term in (12.8).

To bound the term involving the spatial operator $\underline{\mathcal{L}}$, we write

$$\begin{aligned}
\underline{\mathcal{L}}(e_\pi^{n+1} - e_\pi^{n-1}) &= \underline{\mathcal{L}}\Big(u(t_{n+1}) - u(t_{n-1}) - \pi_{\hbar}\big(u(t_{n+1}) - u(t_{n-1})\big)\Big) \\
&= \int_{t_{n-1}}^{t_{n+1}} \underline{\mathcal{L}}\big(\partial_t u(s) - \pi_{\hbar}\partial_t u(s)\big)\,\mathrm{d}s \\
&= \int_{t_{n-1}}^{t_{n+1}} \underline{\mathcal{L}} e_\pi^{\partial_t u}(s)\,\mathrm{d}s.
\end{aligned}$$

Thus, application of Proposition 10.9 (approximation property) yields the desired bound. □

12.1.3 Error Bounds for the dG-Crank–Nicolson Scheme

Having established the error formula (12.4) as well as appropriate bounds on the defects, we now have all ingredients to derive the desired error bounds for the dG-CN scheme. We state three separate results. The first one gives the bound on the full discretization error e_{CN}^n and shows that, under suitable regularity assumptions on the exact solution, the approximations given by the dG-CN scheme converge to the exact solution of the wave-type problem (9.3) with order two in time and k in space.

Theorem 12.3 *Let $\hbar \in \mathcal{H}$ and $\tau > 0$. Further, assume that the exact solution of the wave-type problem* (9.3) *satisfies*

$$u \in C^3(\mathbb{R}_+; L^2(\Omega)^m) \cap C(\mathbb{R}_+; D(\mathcal{L}) \cap H^{k+1}(\mathcal{T}_\hbar)^m).$$

Then, for all $n \in \mathbb{N}_0$, the error of the dG-Crank–Nicolson method satisfies

$$\begin{aligned}
\|u(t_{n+1}) - \boldsymbol{u}_{\mathrm{CN}}^{n+1}\|_\Omega &\le C_\pi |h^{k+1} u(t_{n+1})|_{k+1,\mathcal{T}_\hbar} \\
&\quad + C_{\pi,\mathcal{L}} \tfrac{\tau}{2} \sum_{j=0}^{n} |h^k \big(u(t_{j+1}) + u(t_j)\big)|_{k+1,\mathcal{T}_\hbar} \\
&\quad + \frac{\tau^2}{8} \int_0^{t_{n+1}} \|\partial_t^3 u(s)\|_\Omega \,\mathrm{d}s \\
&\le C\big(\hbar^k + \tau^2\big),
\end{aligned}$$

where the constant C only depends on t_{n+1}, C_π, $C_{\pi,\mathcal{L}}$, $\|\partial_t^3 u(\cdot)\|_\Omega$, and $|u(\cdot)|_{k+1,\mathcal{T}_\hbar}$.

Proof We first split the full error into a projection part and a discretization part as described at the beginning of Chap. 12. The projection error is bounded by Lemma 10.2.

For the remaining bound on the discretization error $\boldsymbol{e}_{\mathrm{CN}}^{n+1}$, we take norms in the error formula (12.4) to obtain

$$\|\boldsymbol{e}_{\mathrm{CN}}^{n+1}\|_\Omega \le \sum_{j=0}^{n} \|\boldsymbol{d}_{\mathrm{CN}}^{j}\|_\Omega,$$

where we have also used the power bound in Proposition 11.3 and the contractivity of $\underline{\mathcal{R}}_{\mathrm{CN}}$ by Lemma 9.15 (i). The claim now follows by the bound on the defect given in Lemma 12.2. □

The next two results show that, under stricter regularity assumptions, both the discrete time as well as the space derivative errors converge with the same rate. In other words, this shows that the discrete derivatives are indeed approximations of their continuous counterparts of the same quality as the approximation of the exact solution. For the sake of presentation, we only state the qualitative bounds with a generic constant. However, for completeness, the full bounds can be found in Sect. 13.2.

The main part of the proof basically utilizes Theorem 11.6, but because of its brevity, we still explicitly carry it out for better understanding. However, to do so, we first have to work a bit to write the discrete derivative errors in terms of the discretization error $\boldsymbol{e}_{\mathrm{CN}}^{n+1}$. This then enables us to make use of the error representation (12.4).

Theorem 12.4 *Let $\hbar \in \mathcal{H}$, $\tau > 0$, and $T > 0$. Further, assume that the exact solution of the wave-type problem* (9.3) *satisfies*

$$u \in C^4(\mathbb{R}_+; L^2(\Omega)^m) \cap C^1(\mathbb{R}_+; D(\mathcal{L}) \cap H^{k+1}(\mathcal{T}_\hbar)^m).$$

Then, for all $n \in \mathbb{N}_0$, the full error of the discrete time derivative of the dG-Crank–Nicolson method satisfies

$$\|\partial_t u(t_{n+1/2}) - \underline{\boldsymbol{\partial_\tau u}}_{\mathrm{CN}}^{n+1/2}\|_\Omega \le C\big(\hbar^k + \tau^2\big),$$

where the constant C is independent of $\hbar$, τ and n if $t_{n+1} \le T$.

Proof We already discussed the projection part of the error (i.e., $\partial_t u - \pi_\hbar \partial_t u$) in the introductory part of this section. Hence, it suffices to bound $\boldsymbol{e}_{\tau,\mathrm{CN}}^{n+1/2}$.

To do so, note that the derivative error $\boldsymbol{e}_{\tau,\mathrm{CN}}^{n+1/2}$ satisfies

$$\begin{aligned}
\|\boldsymbol{e}_{\tau,\mathrm{CN}}^{n+1/2}\|_\Omega &= \|\pi_\hbar \partial_t u(t_{n+1/2}) - \underline{\boldsymbol{\partial_\tau u}}_{\mathrm{CN}}^{n+1/2}\|_\Omega \\
&\le \|\partial_t u(t_{n+1/2}) - \frac{u(t_{n+1}) - u(t_n)}{\tau}\|_\Omega + \|\underline{\boldsymbol{\partial_\tau e}}_{\mathrm{CN}}^{n+1/2}\|_\Omega \\
&\le \frac{\tau}{8}\int_{t_n}^{t_{n+1}} \|\partial_t^3 u(s)\|_\Omega \,\mathrm{d}s + \|\underline{\boldsymbol{\partial_\tau e}}_{\mathrm{CN}}^{n+1/2}\|_\Omega,
\end{aligned}$$

where we have used Taylor expansion to obtain the last inequality. The first term is already of the right order (since the integral is of length τ with a bounded integrand) and therefore, it only remains to derive a bound on $\|\underline{\boldsymbol{\partial_\tau e}}_{\mathrm{CN}}^{n+1/2}\|_\Omega$.

To do so, we proceed analogous to the proof of Theorem 11.6 and use the error formula (12.4) to obtain

$$\begin{aligned}
\underline{\boldsymbol{\partial_\tau e}}_{\mathrm{CN}}^{n+1/2} &= \frac{1}{\tau}\Big(\sum_{j=0}^{n} \underline{\mathcal{S}}_{\mathrm{CN}}^{n-j} \underline{\mathcal{R}}_{\mathrm{CN}} \boldsymbol{d}_{\mathrm{CN}}^{j} - \sum_{j=0}^{n-1} \underline{\mathcal{S}}_{\mathrm{CN}}^{n-1-j} \underline{\mathcal{R}}_{\mathrm{CN}} \boldsymbol{d}_{\mathrm{CN}}^{j}\Big) \\
&= \frac{1}{\tau}\underline{\mathcal{S}}_{\mathrm{CN}}^{n} \underline{\mathcal{R}}_{\mathrm{CN}} \boldsymbol{d}_{\mathrm{CN}}^{0} + \sum_{j=1}^{n} \underline{\mathcal{S}}_{\mathrm{CN}}^{n-j} \underline{\mathcal{R}}_{\mathrm{CN}} \underline{\boldsymbol{\partial_\tau d}}_{\mathrm{CN}}^{j-1/2}.
\end{aligned}$$

Taking norms together with Proposition 11.3, Lemma 9.15 (i) (boundedness of the discrete operators) and Lemma 12.2 (bound on defects) now yields the claim. □

Theorem 12.5 *Let $\hbar \in \mathcal{H}$, $\tau > 0$, and $T > 0$. Further, assume that the exact solution of the wave-type problem (9.3) satisfies*

$$u \in C^4(\mathbb{R}_+; L^2(\Omega)^m) \cap C^1(\mathbb{R}_+; D(\mathcal{L}) \cap H^{k+1}(\mathcal{T}_\hbar)^m)$$

and that the inhomogeneity fulfills

$$f \in C(\mathbb{R}_+; H^{k+1}(\mathcal{T}_\hbar)^m) \cap C^2(\mathbb{R}_+; L^2(\Omega)^m).$$

Then, for all $n \in \mathbb{N}_0$, the full error of the discrete space derivative of the dG-Crank–Nicolson method satisfies

$$\|\mathcal{L}u(t_{n+1/2}) - \underline{\mathcal{L}}\boldsymbol{u}_{\mathrm{CN}}^{n+1/2}\|_\Omega \leq C\big(h^k + \tau^2\big),$$

where the constant C is independent of h, τ and n if $t_{n+1} \leq T$.

Proof Once more, we first bound the projection error $e_\pi^{\mathcal{L}u(t_{n+1/2})}$ (recall the notation defined in (10.5)). By our regularity assumptions, we have $\partial_t u, f \in C(\mathbb{R}_+; H^{k+1}(\mathcal{T}_h)^m)$ and thus, via the original differential equation (9.3a), also $\mathcal{L}u \in C(\mathbb{R}_+; H^{k+1}(\mathcal{T}_h)^m)$. This yields an appropriate bound by Lemma 10.2.

It again remains to bound the discretization part, i.e., $\boldsymbol{e}_{\mathcal{L},\mathrm{CN}}^{n+1/2}$. Subtracting the numerical scheme (11.9) from the projected original differential equation (9.3a) yields

$$\boldsymbol{e}_{\mathcal{L},\mathrm{CN}}^{n+1/2} = \boldsymbol{e}_{\tau,\mathrm{CN}}^{n+1/2} - \pi_h\big(f(t_{n+1/2}) - \overline{f}^{n+1/2}\big).$$

The claim now follows by Theorem 12.4 for the first and Taylor expansion for the second term. □

12.2 Leapfrog Scheme

Next, we perform the error analysis of the leapfrog scheme. As stated before, the overarching strategy is the same as the one used for the Crank–Nicolson method and the analysis can, in principle, be performed as for the Peaceman–Rachford method (by using Lemma 10.10 to tackle the defect caused by the perturbation), which we will see in Sect. 12.3 below.

However, the leapfrog scheme admits a special structure which we are able to exploit following the approach in [91, 161, 163]. This enables us to get rid of some of the (milder) conditions, which are necessary for the Peaceman–Rachford method. But, more importantly, this strategy is still viable for the locally implicit method, where Lemma 10.10 is of no use due to the loss of consistency in the split operators.

To put it simply, the idea behind this approach is to split the defect into a "nice" and a "problematic" part. The latter is problematic, since it contains a discrete operator that cannot be bounded uniformly in the spatial discretization parameter h due to Proposition 10.8 (inverse inequality).

Our idea is to exploit that the discrete operator $\underline{\mathcal{L}}$ satisfies (11.44) and we can therefore replace it by a discrete derivative of the discrete semigroup. Using summation-by-parts to transfer the discrete derivative from the semigroup to the remaining defect leaves us with objects that can easily be bounded in the right order by means of Taylor expansion. A more formal explanation of this procedure can be found after Lemma 12.7.

12.2.1 Error Recursion

As for the Crank–Nicolson method, we start by deriving an error recursion. We proceed completely analogously, but due to the perturbation present in the dG-LF scheme, cf., (11.33), an additional perturbation defect emerges.

Lemma 12.6 *Let $\mathfrak{h} \in \mathcal{H}$ and $\tau > 0$. Then, for all $n \in \mathbb{N}_0$, the discretization error $\boldsymbol{e}_{\mathrm{LF}}^{n+1} = \pi_{\mathfrak{h}} u(t_{n+1}) - \boldsymbol{u}_{\mathrm{LF}}^{n+1}$ of the dG-leapfrog scheme satisfies*

$$\begin{cases} \boldsymbol{e}_{\mathrm{LF}}^{n+1} = \underline{\mathcal{S}}_{\mathrm{LF}} \boldsymbol{e}_{\mathrm{LF}}^{n} + \underline{\mathcal{R}}_{\mathrm{LF}} \big(\boldsymbol{d}_{\mathrm{CN}}^{n} + \boldsymbol{d}_{\mathrm{LF}}^{n} \big), & n \in \mathbb{N}_0, \quad (12.10\mathrm{a}) \\ \boldsymbol{e}_{\mathrm{LF}}^{0} = 0, & \quad (12.10\mathrm{b}) \end{cases}$$

where the ***(dG-leapfrog) perturbation defect*** $\boldsymbol{d}_{\mathrm{LF}}^{n}$ *is given by*

$$\boldsymbol{d}_{\mathrm{LF}}^{n} = \frac{\tau^2}{4} \underline{\mathcal{D}} \pi_{\mathfrak{h}} \big(u(t_{n+1}) - u(t_n) \big). \tag{12.11}$$

The proof is left as an exercise to the reader. It is performed completely analogously to the proof of Lemma 12.1 (using (11.33a) instead of (11.3a)).

Again, like at the end of Sect. 12.1.1 we derive a closed expression for the error by using the discrete variation-of-constants formula (11.37) applied to the error recursion (12.10a). This yields

$$\boldsymbol{e}_{\mathrm{LF}}^{n+1} = \sum_{j=0}^{n} \underline{\mathcal{S}}_{\mathrm{LF}}^{n-j} \underline{\mathcal{R}}_{\mathrm{LF}} \big(\boldsymbol{d}_{\mathrm{CN}}^{j} + \boldsymbol{d}_{\mathrm{LF}}^{j} \big). \tag{12.12}$$

Note that the perturbation defect $\boldsymbol{d}_{\mathrm{LF}}^{n}$ in (12.11) is, in principle, already of third (and thus sufficient) order in τ after using the fundamental theorem of calculus. However, the perturbation operator $\underline{\mathcal{D}}$ comprises the discrete operators $\underline{\mathcal{L}}_p$ and $\underline{\mathcal{L}}_q$ and is therefore not bounded independently of $\mathfrak{h}$. This could be remedied by assuming some additional regularity of the exact solution and applying Lemma 10.10. We will see how to do that in Sect. 12.3, where a similar situation occurs. However, as stated before, we choose a different strategy in the present case.

12.2.2 Bounds on and Splitting of the Defect

In the next lemma, we perform the splitting of the defect into two parts, a projection and a discretization part. Note that we leave the discrete operator $\underline{\mathcal{L}}$ out of the discretization defect. This will turn out to be crucial later, where we show how to compensate this operator in a clever way.

Lemma 12.7 *The perturbation defect* $\boldsymbol{d}_{\mathrm{LF}}^{n}$ *can be rewritten as*

$$\boldsymbol{d}_{\mathrm{LF}}^{n} = \boldsymbol{d}_{\mathrm{LF},\pi}^{n} + \underline{\boldsymbol{\mathcal{L}}}\,\boldsymbol{d}_{\mathrm{LF},\hbar}^{n} \tag{12.13}$$

with the projection and discretization part given by

$$\boldsymbol{d}_{\mathrm{LF},\pi}^{n} = -\frac{\tau^3}{4}\underline{\boldsymbol{\mathcal{D}}}\,\underline{\boldsymbol{\partial}_\tau} e_\pi^{n+1/2}, \quad \text{and} \quad \boldsymbol{d}_{\mathrm{LF},\hbar}^{n} = \frac{\tau^2}{4}\begin{pmatrix} 0 \\ \pi_\hbar\big(\partial_t q(t_{n+1}) - \partial_t q(t_n)\big)\end{pmatrix},$$

respectively.

Proof Adding and subtracting the (unprojected) exact solution in the definition (12.11) of $\boldsymbol{d}_{\mathrm{LF}}^{n}$ yields

$$\boldsymbol{d}_{\mathrm{LF}}^{n} = \tfrac{\tau^2}{4}\underline{\boldsymbol{\mathcal{D}}}\big(u(t_{n+1}) - u(t_n)\big) - \tfrac{\tau^3}{4}\underline{\boldsymbol{\mathcal{D}}}\,\underline{\boldsymbol{\partial}_\tau} e_\pi^{n+1/2}$$

The second term is already the projection part $\boldsymbol{d}_{\mathrm{LF},\pi}^{n}$ and it remains to rewrite the first term. Recalling the definition (11.34) of $\underline{\boldsymbol{\mathcal{D}}}$, we see that the q-component vanishes. Moreover, the p-component of $\underline{\boldsymbol{\mathcal{D}}}\big(u(t_{n+1}) - u(t_n)\big)$ can be rewritten as

$$\begin{aligned}\underline{\mathcal{L}}_q\underline{\mathcal{L}}_p\big(p(t_{n+1}) - p(t_n)\big) &= \underline{\mathcal{L}}_q\pi_\hbar\mathcal{L}_p\big(p(t_{n+1}) - p(t_n)\big)\\ &= \underline{\mathcal{L}}_q\pi_\hbar\big(\partial_t q(t_{n+1}) - \partial_t q(t_n)\big),\end{aligned}$$

where we have used the consistency property (11.25) of the discrete operator $\underline{\mathcal{L}}_p$ for the first and the fact that the exact solution satisfies the differential equation (11.28b) for the second equality. Rewriting this in terms of the full operator $\underline{\boldsymbol{\mathcal{L}}}$ concludes the proof. □

With this, we are able to present the main idea behind our strategy for the error analysis, namely, using the splitting (12.13) in the error formula (12.12). This yields

$$\boldsymbol{e}_{\mathrm{LF}}^{n+1} = \sum_{j=0}^{n} \underline{\boldsymbol{\mathcal{S}}}_{\mathrm{LF}}^{n-j}\underline{\boldsymbol{\mathcal{R}}}_{\mathrm{LF}}\big(\boldsymbol{d}_{\mathrm{CN}}^{j} + \boldsymbol{d}_{\mathrm{LF},\pi}^{j}\big) + \sum_{j=0}^{n} \underline{\boldsymbol{\mathcal{S}}}_{\mathrm{LF}}^{n-j}\underline{\boldsymbol{\mathcal{R}}}_{\mathrm{LF}}\underline{\boldsymbol{\mathcal{L}}}\,\boldsymbol{d}_{\mathrm{LF},\hbar}^{j}. \tag{12.14}$$

Using Proposition 11.15 yields

$$\begin{aligned}\sum_{j=0}^{n} \underline{\boldsymbol{\mathcal{S}}}_{\mathrm{LF}}^{n-j}\underline{\boldsymbol{\mathcal{R}}}_{\mathrm{LF}}\underline{\boldsymbol{\mathcal{L}}}\,\boldsymbol{d}_{\mathrm{LF},\hbar}^{j} &= \sum_{j=0}^{n} \frac{\underline{\boldsymbol{\mathcal{S}}}_{\mathrm{LF}}^{n-j+1} - \underline{\boldsymbol{\mathcal{S}}}_{\mathrm{LF}}^{n-j}}{\tau}\,\boldsymbol{d}_{\mathrm{LF},\hbar}^{j}\\ &= \frac{1}{\tau}\big(\underline{\boldsymbol{\mathcal{S}}}_{\mathrm{LF}}^{n+1}\boldsymbol{d}_{\mathrm{LF},\hbar}^{0} - \boldsymbol{d}_{\mathrm{LF},\hbar}^{n}\big) + \sum_{j=1}^{n} \underline{\boldsymbol{\mathcal{S}}}_{\mathrm{LF}}^{n-j+1}\,\underline{\boldsymbol{\partial}_\tau}\boldsymbol{d}_{\mathrm{LF},\hbar}^{j-1/2},\end{aligned} \tag{12.15}$$

where we have used summation-by-parts (or, in other words, an index shift after splitting the sum) for the second equality.

Having eliminated the problematic discrete operator $\underline{\mathcal{L}}$, we see that it suffices to give a bound of order 3 in τ for the defect $\boldsymbol{d}^n_{\mathrm{LF},\hbar}$ as well as its discrete derivative. Recall that we need a second-order bound in τ overall and lose one τ to compensate the denominator and the sum in (12.15), respectively. Further, to bound the discrete derivative errors $\boldsymbol{e}^n_{\tau,\mathrm{LF}}$ and $\boldsymbol{e}^n_{\mathcal{L},\mathrm{LF}}$, we also need an appropriate bound on the second-order discrete derivative. We give these bounds in the next lemma. The proof is a rather straightforward consequence of Taylor expansion and is therefore left to the reader.

Lemma 12.8 *Let $\hbar \in \mathcal{H}$ and $\tau > 0$. Then, if $q \in C^2(\mathbb{R}_+; L^2(\Omega)^{m_q})$, the discretization part $\boldsymbol{d}^n_{\mathrm{LF},\hbar}$ fulfills*

$$\|\boldsymbol{d}^n_{\mathrm{LF},\hbar}\|_\Omega \leq \frac{\tau^2}{4} \int_{t_n}^{t_{n+1}} \|\partial_t^2 q(s)\|_\Omega \,\mathrm{d}s,$$

if $q \in C^3(\mathbb{R}_+; L^2(\Omega)^{m_q})$, its first-order discrete derivative fulfills

$$\|\underline{\boldsymbol{\partial}_\tau} \boldsymbol{d}^{n-1/2}_{\mathrm{LF},\hbar}\|_\Omega \leq \frac{\tau^2}{4} \int_{t_{n-1}}^{t_{n+1}} \|\partial_t^3 q(s)\|_\Omega \,\mathrm{d}s,$$

and, if $q \in C^4(\mathbb{R}_+; L^2(\Omega)^{m_q})$, its second-order discrete derivative fulfills

$$\|\underline{\boldsymbol{\partial}_\tau^2} \boldsymbol{d}^{n-1}_{\mathrm{LF},\hbar}\|_\Omega \leq \frac{\tau^2}{8} \int_{t_{n-2}}^{t_{n+1}} \|\partial_t^4 q(s)\|_\Omega \,\mathrm{d}s.$$

Moreover, we also need appropriate bounds on the projection part $\boldsymbol{d}^n_{\mathrm{LF},\pi}$ of the perturbation defect. Since we treat this part analogously to the dG-CN defect, we give the same kind of bounds as in this case, i.e., one on the defect itself as well as one on its discrete derivative. We point out that in order to obtain the bounds on this defect, we need the CFL condition (11.38) to hold, which we did not need for the discretization part.

Lemma 12.9 *Let $\hbar \in \mathcal{H}$ and $\tau > 0$ fulfill Assumption 11.11. Then, if $p \in C(\mathbb{R}_+; D(\mathcal{L}_p) \cap H^{k+1}(\mathcal{T}_\hbar)^{m_p})$, the projection part $\boldsymbol{d}^n_{\mathrm{LF},\pi}$ fulfills*

$$\|\boldsymbol{d}^n_{\mathrm{LF},\pi}\|_\Omega \leq \tfrac{\tau}{2} C_{\pi,\mathcal{L}} |h^k\big(p(t_{n+1}) - p(t_n)\big)|_{k+1,\mathcal{T}_\hbar}, \tag{12.16}$$

and, if $p \in C^1(\mathbb{R}_+; D(\mathcal{L}_p) \cap H^{k+1}(\mathcal{T}_\hbar)^{m_p})$, its discrete derivative fulfills

$$\|\underline{\boldsymbol{\partial}_\tau} \boldsymbol{d}^{n-1/2}_{\mathrm{LF},\pi}\|_\Omega \leq \tfrac{1}{2} C_{\pi,\mathcal{L}} \int_{t_{n-1}}^{t_{n+1}} |h^k \partial_t p(s)|_{k+1,\mathcal{T}_\hbar} \,\mathrm{d}s. \tag{12.17}$$

Proof We begin with the proof of (12.16). As already observed in the proof of Lemma 12.7, due to the structure of $\underline{\mathfrak{D}}$, the q-component of the defect is 0. Hence, we only need to consider the p-component

$$\tfrac{\tau^2}{4}\underline{\mathcal{L}}_q\underline{\mathcal{L}}_p\big(e_{\pi,p}^{n+1} - e_{\pi,p}^{n}\big) = \tfrac{\tau^3}{4}\underline{\mathcal{L}}_q\underline{\mathcal{L}}_p\underline{\boldsymbol{\partial}_\tau} e_{\pi,p}^{n+1/2},$$

where $e_{\pi,p}^n$, $n \in \mathbb{N}_0$, is the p-component of the projection error e_π^n. Note that, since the projection acts componentwise, $e_{\pi,p}^n$ is in fact the projection error of the p-component of the solution.

We take norms and proceed analogously to (11.41), i.e., we use the inverse inequality (11.26) (for $\underline{\mathcal{L}}_q$) together with the CFL condition (11.38), to obtain

$$\tfrac{\tau^3}{4}\|\underline{\mathcal{L}}_q\underline{\mathcal{L}}_p\underline{\boldsymbol{\partial}_\tau} e_{\pi,p}^{n+1/2}\|_\Omega \le \tfrac{\tau^2}{2}\theta\|\underline{\mathcal{L}}_p\underline{\boldsymbol{\partial}_\tau} e_{\pi,p}^{n+1/2}\|_\Omega.$$

Using the approximation property (11.27) of $\underline{\mathcal{L}}_p$ and $\theta < 1$ shows (12.16).

The second bound (12.17) is easily shown by using the same ideas as above and those in the last part of the proof of Lemma 12.2. □

Note that all regularity assumptions we assume for these bounds are already present in the regularity assumptions of the corresponding Crank–Nicolson result Lemma 12.2. Since the bounds given here are already sufficient to show the bound on the discretization error $\boldsymbol{e}_{\mathrm{LF}}^n$, we do not need to assume more regularity than for the Crank–Nicolson method, i.e., in Theorem 12.3.

Lastly, for the bounds on the discrete derivatives $\boldsymbol{e}_{\tau,\mathrm{LF}}^n$ and $\boldsymbol{e}_{\mathcal{L},\mathrm{LF}}^n$, we need an additional bound on the discretization part of the defect. We point out that this result requires the consistency (11.25) (and the approximation property (11.27)) of the discrete operator $\underline{\mathcal{L}}_q$. Further, we need to assume additional regularity of the exact solution for the result to hold. Hence, we impose the same assumptions for the error bounds on $\boldsymbol{e}_{\tau,\mathrm{LF}}^n$ and $\boldsymbol{e}_{\mathcal{L},\mathrm{LF}}^n$.

Lemma 12.10 *Let $\hbar \in \mathcal{H}$ and $\tau > 0$. Then, if $q \in C^2(\mathbb{R}_+; D(\underline{\mathcal{L}}_q) \cap H^1(\mathcal{T}_\hbar)^{m_p})$, the discretization part $\boldsymbol{d}_{\mathrm{LF},\hbar}^n$ fulfills*

$$\|\underline{\mathcal{L}}\boldsymbol{d}_{\mathrm{LF},\hbar}^n\|_\Omega \le \frac{\tau^2}{4}\int_{t_n}^{t_{n+1}} \|\mathcal{L}_q\partial_t^2 q(s)\|_\Omega + C_{\pi,\mathcal{L}}|\partial_t^2 q(s)|_{1,\mathcal{T}_\hbar}\,\mathrm{d}s.$$

Proof Taking norms and using the fundamental theorem of calculus yields

$$\|\underline{\mathcal{L}}\boldsymbol{d}_{\mathrm{LF},\hbar}^n\|_\Omega \le \frac{\tau^2}{4}\int_{t_n}^{t_{n+1}} \|\underline{\mathcal{L}}_q\pi_\hbar\partial_t^2 q(s)\|_\Omega\,\mathrm{d}s.$$

We now proceed similar to the proof of Lemma 12.2, meaning we use the consistency property (11.25) of $\underline{\mathcal{L}}_q$, to obtain

$$\|\underline{\mathcal{L}}_q \pi_h \partial_t^2 q(s)\|_\Omega \leq \|\mathcal{L}_q \partial_t^2 q(s)\|_\Omega + \|\underline{\mathcal{L}}_q e_\pi^{\partial_t^2 q(s)}\|_\Omega.$$

Applying the approximation property (11.27) (with $\ell = j = 0$) proves the claim. □

12.2.3 Error Bounds for the dG-Leapfrog Scheme

Now that we have shown all necessary preliminary results, we are able to give the full error bounds for the dG-leapfrog scheme. As before, we provide three separate results, one on the full discretization error and one for both the (full) spatial and temporal discrete derivative errors.

We start with the former, i.e., the full discretization error. As for the Crank–Nicolson method, we show that the error converges to 0 (and thus, the dG-LF approximation to the exact solution) with order two in time and k in space.

Theorem 12.11 *Let $h \in \mathcal{H}$ and $\tau > 0$ fulfill Assumption 11.11 (with $\theta \in (0, 1)$). Further, assume that the exact solution of the (two-field) wave-type problem* (11.28) *satisfies*

$$u = \begin{pmatrix} p \\ q \end{pmatrix} \in C^3(\mathbb{R}_+; L^2(\Omega)^m) \cap C(\mathbb{R}_+; D(\mathcal{L}) \cap H^{k+1}(\mathcal{T}_h)^m).$$

Then, for all $n \in \mathbb{N}_0$, the error of the dG-leapfrog method satisfies

$$\begin{aligned}
\|u(t_{n+1}) - \boldsymbol{u}_{\mathrm{LF}}^{n+1}\|_\Omega &\leq C_\pi |h^{k+1} u(t_{n+1})|_{k+1,\mathcal{T}_h} \\
&+ \sqrt{3} C_{\mathrm{stb}} C_{\pi,\mathcal{L}} \tfrac{\tau}{2} \sum_{j=0}^{n} \big(|h^k \big(u(t_{j+1}) + u(t_j)\big)|_{k+1,\mathcal{T}_h} + |h^k \big(p(t_{j+1}) - p(t_j)\big)|_{k+1,\mathcal{T}_h} \big) \\
&+ \frac{\tau^2}{4} \int_0^1 C_{\mathrm{stb}} \|\partial_t^2 q(\tau s)\|_\Omega + \|\partial_t^2 q(t_n + \tau s)\|_\Omega \,\mathrm{d}s \\
&+ C_{\mathrm{stb}} \frac{\tau^2}{8} \int_0^{t_{n+1}} \sqrt{3}\, \|\partial_t^3 u(s)\|_\Omega + 4 \|\partial_t^3 q(s)\|_\Omega \,\mathrm{d}s \\
&\leq C\big(h^k + \tau^2\big),
\end{aligned}$$

where the constant C only depends on t_{n+1}, C_π, θ, $C_{\pi,\mathcal{L}}$, $\|\partial_t^3 u(\cdot)\|_\Omega$, $\|\partial_t^2 q(\cdot)\|_\Omega$, and $|u(\cdot)|_{k+1,\mathcal{T}_h}$.

Proof Again, the projection error is bounded by Lemma 10.2 and it remains to bound the discretization error $\boldsymbol{e}_{\mathrm{LF}}^{n+1}$. To do so, we take norms in the error representation (12.14) to obtain

$$\|\boldsymbol{e}_{\mathrm{LF}}^{n+1}\|_\Omega \le \sum_{j=0}^{n} \sqrt{3}\, C_{\mathrm{stb}}\big(\|\boldsymbol{d}_{\mathrm{CN}}^{j}\|_\Omega + \|\boldsymbol{d}_{\mathrm{LF},\pi}^{j}\|_\Omega\big) + \|\sum_{j=0}^{n} \underline{\boldsymbol{S}}_{\mathrm{LF}}^{n-j} \underline{\boldsymbol{\mathcal{R}}}_{\mathrm{LF}} \underline{\boldsymbol{\mathcal{L}}} \boldsymbol{d}_{\mathrm{LF},\hbar}^{j}\|_\Omega,$$

where we have already used the stability results Proposition 11.13 and Lemma 11.14 for the first term. Using the bounds in Lemmas 12.2 and 12.9 already provides the appropriate bound for this term.

For the second term, we use (12.15) and the stability result Proposition 11.13 to obtain

$$\|\sum_{j=0}^{n} \underline{\boldsymbol{S}}_{\mathrm{LF}}^{n-j} \underline{\boldsymbol{\mathcal{R}}}_{\mathrm{LF}} \underline{\boldsymbol{\mathcal{L}}} \boldsymbol{d}_{\mathrm{LF},\hbar}^{j}\|_\Omega \le \tfrac{1}{\tau}\big(C_{\mathrm{stb}}\|\boldsymbol{d}_{\mathrm{LF},\hbar}^{0}\|_\Omega + \|\boldsymbol{d}_{\mathrm{LF},\hbar}^{n}\|_\Omega\big) + C_{\mathrm{stb}} \sum_{j=1}^{n} \|\underline{\boldsymbol{\partial}_\tau} \boldsymbol{d}_{\mathrm{LF},\hbar}^{j-1/2}\|_\Omega.$$

Now, using the appropriate bounds in Lemma 12.8 together with

$$\tfrac{\tau^2}{4}\sum_{j=1}^{n} \int_{t_{j-1}}^{t_{j+1}} \|\partial_t^3 q(s)\|_\Omega \,\mathrm{d}s \le \tfrac{\tau^2}{2} \int_0^{t_{n+1}} \|\partial_t^3 q(s)\|_\Omega \,\mathrm{d}s$$

and transformation of the remaining integrals yields the claim. □

Next, we give the bounds on the time derivative error. As pointed out before, we need slightly more regularity than in the Crank–Nicolson counterpart in Theorem 12.4. Again, we do not state the full bounds here for the sake of presentation. They can, however, be found in Appendix 13.2.

Theorem 12.12 *Let $\hbar \in \mathcal{H}$ and $\tau > 0$ fulfill Assumption 11.11 and let $T > 0$. Further, assume that the exact solution of the (two-field) wave-type problem* (11.28) *satisfies*

$$u = \begin{pmatrix} p \\ q \end{pmatrix} \in C^4(\mathbb{R}_+; L^2(\Omega)^m) \cap C^1(\mathbb{R}_+; D(\mathcal{L}) \cap H^{k+1}(\mathcal{T}_\hbar)^m)$$

and

$$q \in C^2([0, t_1]; D(\mathcal{L}_q) \cap H^1(\mathcal{T}_\hbar)^{m_p}).$$

Then, for all $n \in \mathbb{N}_0$, the full error of the discrete time derivative of the dG-leapfrog method satisfies

$$\|\partial_t u(t_{n+1/2}) - \underline{\boldsymbol{\partial}_\tau} \boldsymbol{u}_{\mathrm{LF}}^{n+1/2}\|_\Omega \le C\big(\hbar^k + \tau^2\big),$$

where the constant C is independent of $\hbar$, τ and n if $t_{n+1} \le T$.

Proof Using the same reasoning as in the beginning of the proof of Theorem 12.4, we only need to derive a bound on $\|\underline{\partial_\tau \boldsymbol{e}}_{\mathrm{LF}}^{n+1/2}\|_\Omega$. We proceed similarly and use the error formula (12.14) together with (12.15) to obtain

$$\begin{aligned}\underline{\partial_\tau \boldsymbol{e}}_{\mathrm{LF}}^{n+1/2} = \frac{1}{\tau}\Big(&\sum_{j=0}^{n} \underline{\boldsymbol{S}}_{\mathrm{LF}}^{n-j} \underline{\boldsymbol{R}}_{\mathrm{LF}} (\boldsymbol{d}_{\mathrm{CN}}^{j} + \boldsymbol{d}_{\mathrm{LF},\pi}^{j}) - \sum_{j=0}^{n-1} \underline{\boldsymbol{S}}_{\mathrm{LF}}^{n-j-1} \underline{\boldsymbol{R}}_{\mathrm{LF}} (\boldsymbol{d}_{\mathrm{CN}}^{j} + \boldsymbol{d}_{\mathrm{LF},\pi}^{j})\\ &+ \frac{\underline{\boldsymbol{S}}_{\mathrm{LF}}^{n+1} - \underline{\boldsymbol{S}}_{\mathrm{LF}}^{n}}{\tau} \boldsymbol{d}_{\mathrm{LF},\hbar}^{0} - \underline{\partial_\tau \boldsymbol{d}}_{\mathrm{LF},\hbar}^{n-1/2}\\ &+ \sum_{j=1}^{n} \underline{\boldsymbol{S}}_{\mathrm{LF}}^{n-j+1} \underline{\partial_\tau \boldsymbol{d}}_{\mathrm{LF},\hbar}^{j-1/2} - \sum_{j=1}^{n-1} \underline{\boldsymbol{S}}_{\mathrm{LF}}^{n-j} \underline{\partial_\tau \boldsymbol{d}}_{\mathrm{LF},\hbar}^{j-1/2}\Big).\end{aligned}$$

Applying the discrete version of the Leibniz integral rule as in the proof of Theorem 11.6 in the first and third line as well as Proposition 11.15 in the second therefore gives

$$\begin{aligned}\underline{\partial_\tau \boldsymbol{e}}_{\mathrm{LF}}^{n+1/2} =& \frac{1}{\tau}\underline{\boldsymbol{S}}_{\mathrm{LF}}^{n} \underline{\boldsymbol{R}}_{\mathrm{LF}} (\boldsymbol{d}_{\mathrm{CN}}^{0} + \boldsymbol{d}_{\mathrm{LF},\pi}^{0}) + \sum_{j=1}^{n} \underline{\boldsymbol{S}}_{\mathrm{LF}}^{n-j} \underline{\boldsymbol{R}}_{\mathrm{LF}} \Big(\underline{\partial_\tau \boldsymbol{d}}_{\mathrm{CN}}^{j-1/2} + \underline{\partial_\tau \boldsymbol{d}}_{\mathrm{LF},\pi}^{j-1/2}\Big)\\ &+ \frac{1}{\tau}\Big(\underline{\boldsymbol{S}}_{\mathrm{LF}}^{n} \underline{\boldsymbol{R}}_{\mathrm{LF}} \underline{\boldsymbol{\mathcal{L}}}\, \boldsymbol{d}_{\mathrm{LF},\hbar}^{0} + \underline{\boldsymbol{S}}_{\mathrm{LF}}^{n} \underline{\partial_\tau \boldsymbol{d}}_{\mathrm{LF},\hbar}^{1/2} - \underline{\partial_\tau \boldsymbol{d}}_{\mathrm{LF},\hbar}^{n-1/2}\Big) + \sum_{j=1}^{n-1} \underline{\boldsymbol{S}}_{\mathrm{LF}}^{n-j} \underline{\partial_\tau^2 \boldsymbol{d}}_{\mathrm{LF},\hbar}^{j}.\end{aligned}$$

Taking norms and using the stability results Proposition 11.13 and Lemma 11.14 together with the bounds provided by Lemmas 12.8–12.10 now yields the claim. □

The counterpart of Theorem 12.5 for the space derivative error reads as follows.

Theorem 12.13 *Let* $\hbar \in \mathcal{H}$ *and* $\tau > 0$ *fulfill Assumption 11.11 and let* $T > 0$. *Further, assume that the exact solution of the (two-field) wave-type problem* (11.28) *satisfies*

$$u = \begin{pmatrix} p \\ q \end{pmatrix} \in C^4(\mathbb{R}_+; L^2(\Omega)^m) \cap C^1(\mathbb{R}_+; D(\mathcal{L}) \cap H^{k+1}(\mathcal{T}_\hbar)^m)$$

and

$$q \in C^2([0, t_1]; D(\mathcal{L}_q) \cap H^1(\mathcal{T}_\hbar)^{m_p}),$$

and that the inhomogeneity fulfills $f \in C(\mathbb{R}_+; H^{k+1}(\mathcal{T}_\hbar)^m) \cap C^2(\mathbb{R}_+; L^2(\Omega)^m)$. *Then, for all* $n \in \mathbb{N}_0$, *the full error of the discrete space derivative of the dG-leapfrog method satisfies*

$$\|\mathcal{L}u(t_{n+1/2}) - \underline{\mathcal{L}\boldsymbol{u}}_{\mathrm{LF}}^{n+1/2}\|_\Omega \le C\big(\hbar^k + \tau^2\big),$$

where the constant C *is independent of* $\hbar$, τ *and* n *if* $t_{n+1} \le T$.

Proof The proof is performed completely analogously to the corresponding Crank–Nicolson result in Theorem 12.5. We just replace (11.9) by (11.46) and apply Theorem 12.12 instead of Theorem 12.4. □

12.3 Peaceman–Rachford Scheme

In this section, we analyze the error of the dG-PR scheme (11.52). Once more, the overarching strategy is the same as in all other schemes we consider. However, similar to the leapfrog method, an additional defect caused by the perturbation present in the scheme occurs. This time, however, we are not able to exploit an additional structure of the scheme, since the perturbation only involves the split operators $\underline{\mathcal{A}}$ and $\underline{\mathcal{B}}$, not the full operator $\underline{\mathcal{L}}$.

Instead, as already indicated, we use Lemma 10.10 to control this perturbation defect. This requires some additional regularity of the solution but turns out to be a more versatile approach than the one used for the leapfrog method. In fact, the strategy presented in this section is able to handle general second-order (in time) perturbations of the dG-CN scheme that involve (consistent) discrete Friedrichs' operators.

12.3.1 Error Recursion

Once more, we start by deriving an error recursion. Similarly to the leapfrog method, an additional defect arises, which is caused by the perturbation in the dG-PR scheme (11.53). However, since we also treat the inhomogeneity differently, this perturbation defect also contains a part involving the inhomogeneity.

Lemma 12.14 *Let $\mathcal{h} \in \mathcal{H}$ and $\tau > 0$. Then, for all $n \in \mathbb{N}_0$, the discretization error $\boldsymbol{e}_{\mathrm{PR}}^{n+1} = \pi_{\mathcal{h}} u(t_{n+1}) - \boldsymbol{u}_{\mathrm{PR}}^{n+1}$ of the dG-Peaceman–Rachford scheme satisfies*

$$\begin{cases} \boldsymbol{e}_{\mathrm{PR}}^{n+1} = \underline{\mathcal{S}}_{\mathrm{PR}}\, \boldsymbol{e}_{\mathrm{PR}}^{n} + \underline{\mathcal{R}}_{\mathrm{PR}}\, (\boldsymbol{d}_{\mathrm{CN}}^{n} + \boldsymbol{d}_{\mathrm{PR}}^{n}), & n \in \mathbb{N}_0, \qquad (12.18a) \\ \boldsymbol{e}_{\mathrm{PR}}^{0} = 0, & \qquad (12.18b) \end{cases}$$

*where the **(dG-Peaceman–Rachford) perturbation defect** $\boldsymbol{d}_{\mathrm{PR}}^{n}$ is given by*

$$\boldsymbol{d}_{\mathrm{PR}}^{n} = \frac{\tau^2}{4} \underline{\mathcal{A}\mathcal{B}} \pi_{\mathcal{h}} \big(u(t_{n+1}) - u(t_n)\big) + \frac{\tau^3}{4} \underline{\mathcal{A}}^2 \pi_{\mathcal{h}} \overline{f}^{n+1/2}.$$

Proof Proceeding analogously to the proof of the corresponding Crank–Nicolson result Lemma 12.1 yields

$$\pi_h u(t_{n+1}) = \underline{S}_{\mathrm{PR}}\, \pi_h u(t_n) + \left(\underline{\mathcal{I}} - \tfrac{\tau}{2}\underline{\mathcal{B}}\right)^{-1}\left(\underline{\mathcal{I}} + \tfrac{\tau}{2}\underline{\mathcal{A}}\right)\tau \overline{f}_\pi^{n+1/2} + \underline{\mathcal{R}}_{\mathrm{PR}}\left(\boldsymbol{d}_{\mathrm{CN}}^n + \boldsymbol{d}_{\mathrm{PR}}^n\right)$$

instead of (12.3). Comparing this to the dG-PR iteration (11.55a) we readily obtain the error recursion (12.18a). Further, solving for the defects $\boldsymbol{d}_{\mathrm{CN}}^n + \boldsymbol{d}_{\mathrm{PR}}^n$ yields

$$\begin{aligned}\boldsymbol{d}_{\mathrm{CN}}^n + \boldsymbol{d}_{\mathrm{PR}}^n &= \left(\underline{\mathcal{I}} - \tfrac{\tau}{2}\underline{\mathcal{A}}\right)\left(\underline{\mathcal{I}} - \tfrac{\tau}{2}\underline{\mathcal{B}}\right)\pi_h u(t_{n+1}) - \left(\underline{\mathcal{I}} + \tfrac{\tau}{2}\underline{\mathcal{A}}\right)\left(\underline{\mathcal{I}} + \tfrac{\tau}{2}\underline{\mathcal{B}}\right)\pi_h u(t_n)\\ &\quad - \tau\left(\underline{\mathcal{I}} - \tfrac{\tau}{2}\underline{\mathcal{A}}\right)\left(\underline{\mathcal{I}} + \tfrac{\tau}{2}\underline{\mathcal{A}}\right)\overline{f}_\pi^{n+1/2}.\end{aligned}$$

Expanding the products and using the splitting property (11.51) proves the claim. □

This again provides us with an explicit representation of the discretization error $\boldsymbol{e}_{\mathrm{PR}}^n$ in terms of the defects and the system operators, namely

$$\boldsymbol{e}_{\mathrm{PR}}^{n+1} = \sum_{j=0}^{n} \underline{S}_{\mathrm{PR}}^{n-j}\, \underline{\mathcal{R}}_{\mathrm{PR}}\left(\boldsymbol{d}_{\mathrm{CN}}^j + \boldsymbol{d}_{\mathrm{PR}}^j\right). \tag{12.19}$$

As for the leapfrog method, we observe that the perturbation defect $\boldsymbol{d}_{\mathrm{PR}}^n$ of the dG-PR scheme is essentially already of the right order in τ, the only problem being the involved discrete operators. This time, this is the key observation for the analysis and we will use Lemma 10.10 to handle these discrete operators.

Note also that the strategy used for the leapfrog method is not viable in this setting. More precisely, we have used the original differential equation to get rid of one discrete operator and Proposition 11.15 for the other. Both approaches fail here, since the operators occurring in the perturbation defect are not the full operator $\underline{\mathcal{L}}$ but merely the split ones, $\underline{\mathcal{A}}$ and $\underline{\mathcal{B}}$. Both neither occur in the original differential equation nor in the corresponding result Proposition 11.21.

12.3.2 Bounds on the Defect

Since we already have the necessary bounds on the dG-CN defect $\boldsymbol{d}_{\mathrm{CN}}^n$, it remains to bound the perturbation defect $\boldsymbol{d}_{\mathrm{PR}}^n$. We provide them in the following lemma.

Lemma 12.15 *Let $h \in \mathcal{H}$, $\tau > 0$ and $k \geq 1$. Further, assume that we have $M \in W^{1,\infty}(\mathcal{T}_h)^{m\times m}$, $u \in C^1(\mathbb{R}_+; D(\mathcal{A}\mathcal{B}) \cap H^2(\mathcal{T}_h)^m)$, and $f \in C(\mathbb{R}_+; D(\mathcal{A}^2) \cap H^2(\mathcal{T}_h)^m)$. Then we have*

$$\begin{aligned}\|\boldsymbol{d}_{\mathrm{PR}}^n\|_\Omega &\leq \frac{\tau^2}{4}\int_{t_n}^{t_{n+1}} \|\mathcal{A}\mathcal{B}\partial_t u(s)\|_\Omega + C_{\mathrm{PR},u}\|\partial_t u(s)\|_{2,\mathcal{T}_h}\,\mathrm{d}s\\ &\quad + \frac{\tau^3}{4}\left(\|\mathcal{A}^2\overline{f}^{n+1/2}\|_\Omega + C_{\mathrm{PR},f}\|\overline{f}^{n+1/2}\|_{2,\mathcal{T}_h}\right).\end{aligned}$$

Further, if $u \in C^2(\mathbb{R}_+; D(\mathcal{AB}) \cap H^2(\mathcal{T}_h)^m)$ *and* $f \in C^1(\mathbb{R}_+; D(\mathcal{A}^2) \cap H^2(\mathcal{T}_h)^m)$*, for the discrete derivative we have*

$$\|\underline{\partial_\tau} \boldsymbol{d}_{\mathrm{PR}}^{n-1/2}\|_\Omega \le \frac{\tau^2}{4} \int_{t_{n-1}}^{t_n} \int_0^1 \|\mathcal{AB}\partial_t^2 u(s+\varsigma\tau)\|_\Omega + C_{\mathrm{PR},u} \|\partial_t^2 u(s+\varsigma\tau)\|_{2,\mathcal{T}_h} \,\mathrm{d}\varsigma \,\mathrm{d}s$$
$$+ \frac{\tau^2}{4} \int_{t_{n-1}}^{t_{n+1}} \|\mathcal{A}^2 \partial_t f(s)\|_\Omega + C_{\mathrm{PR},f} \|\partial_t f(s)\|_{2,\mathcal{T}_h} \,\mathrm{d}s,$$

where the constants are given by $C_{\mathrm{PR},u} = C_{\mathrm{inv},\mathcal{A}} C_{\pi,\mathcal{B},-1} + C_{\pi,\mathcal{A}} C_{\mathcal{B},M,1}$ *and* $C_{\mathrm{PR},f} = C_{\mathrm{inv},\mathcal{A}} C_{\pi,\mathcal{A},-1} + C_{\pi,\mathcal{A}} C_{\mathcal{A},M,1}$.

Proof The fundamental theorem of calculus together with taking the norm yields

$$\|\boldsymbol{d}_{\mathrm{PR}}^n\|_\Omega \le \frac{\tau^2}{4} \int_{t_n}^{t_{n+1}} \|\underline{\mathcal{AB}} \pi_h \partial_t u(s)\|_\Omega \,\mathrm{d}s + \frac{\tau^3}{4} \|\underline{\mathcal{A}}^2 \pi_h \overline{f}^{n+1/2}\|_\Omega.$$

Further, by treating the double integral analogously to (12.9), we obtain

$$\|\underline{\partial_\tau} \boldsymbol{d}_{\mathrm{PR}}^{n-1/2}\|_\Omega \le \frac{\tau^2}{4} \Big(\int_{t_{n-1}}^{t_n} \int_0^1 \|\underline{\mathcal{AB}} \pi_h \partial_t^2 u(s+\tau\varsigma)\|_\Omega \,\mathrm{d}\varsigma \,\mathrm{d}s$$
$$+ \int_{t_{n-1}}^{t_{n+1}} \|\underline{\mathcal{A}}^2 \pi_h \partial_t f(s)\|_\Omega \,\mathrm{d}s \Big).$$

Hence, it remains to bound four terms, each involving two discrete operators (either $\underline{\mathcal{AB}}$ or $\underline{\mathcal{A}}^2$). They can all be bounded by the same procedure, which we show exemplary for $\|\underline{\mathcal{AB}} \pi_h \partial_t u(s)\|_\Omega$. We add and subtract $\pi_h \mathcal{AB} \partial_t u(s)$ and use the contractivity of the L^2-projection to obtain

$$\|\underline{\mathcal{AB}} \pi_h \partial_t u(s)\|_\Omega \le \|\mathcal{AB} \partial_t u(s)\|_\Omega + \|(\underline{\mathcal{AB}} \pi_h - \pi_h \mathcal{AB}) \partial_t u(s)\|_\Omega.$$

The proof is concluded by using Lemma 10.10 (with $r = 2$, $j = 0$ and $\ell = 1$) to bound the second term. The exact constants for this bound are obtained by explicitly performing the first step of the induction in the proof of Lemma 10.10 (which can be found in [89, Lem 3.9]). □

12.3.3 Error Bounds for the dG-Peaceman–Rachford Scheme

After having derived the explicit error representation (12.19) from the error recursion (12.18) and all necessary bounds on the occurring defects, we are now able to state the main results for the dG-PR scheme. Since the regularity assumptions for this case are a bit lengthy, we gather them in the following assumption.

Assumption 12.16 For $\ell \in \mathbb{N}$, the exact solution of the wave-type problem (9.3) satisfies

$$u \in C^{\ell+2}(\mathbb{R}_+; L^2(\Omega)^m) \cap C^{\ell}(\mathbb{R}_+; D(\mathcal{A}\mathcal{B}) \cap H^2(\mathcal{T}_h)^m) \\ \cap C^{\ell-1}(\mathbb{R}_+; D(\mathcal{A}) \cap H^{k+1}(\mathcal{T}_h)^m),$$

and the inhomogeneity satisfies $f \in C^{\ell-1}(\mathbb{R}_+; D(\mathcal{A}^2) \cap H^2(\mathcal{T}_h)^m)$. Further, the material tensor fulfills $M \in W^{1,\infty}(\mathcal{T}_h)^{m\times m}$. ◇

As for the previous two schemes, we give both a bound for the discretization error as well as the error in the discrete derivatives. The proofs of these results are performed completely analogously to those for the Crank–Nicolson method (with the corresponding auxiliary results switched by those for the Peaceman–Rachford method), which is why we leave them as an exercise for the reader.

Theorem 12.17 *Let $h \in \mathcal{H}$, $\tau > 0$, and $k \geq 1$ be the polynomial degree of the dG method, and let Assumption 12.16 be fulfilled with $\ell = 1$. Then, for all $n \in \mathbb{N}_0$, the dG-Peaceman–Rachford error satisfies*

$$\begin{aligned}\|u(t_{n+1}) - \boldsymbol{u}_{\mathrm{PR}}^{n+1}\|_\Omega &\leq C_\pi |h^{k+1} u(t_{n+1})|_{k+1,\mathcal{T}_h} + C_{\pi,\mathcal{L}} \tfrac{\tau}{2} \sum_{j=0}^{n} |h^k \big(u(t_{j+1}) + u(t_j)\big)|_{k+1,\mathcal{T}_h} \\ &\quad + \frac{\tau^2}{4}\Big(\int_0^{t_{n+1}} \tfrac{1}{2}\|\partial_t^3 u(s)\|_\Omega + \|\mathcal{A}\mathcal{B}\partial_t u(s)\|_\Omega + C_{\mathrm{PR},u}\|\partial_t u(s)\|_{2,\mathcal{T}_h}\, \mathrm{d}s \\ &\quad + \tau \sum_{j=0}^{n} \big(\|\mathcal{A}^2 \overline{f}^{j+1/2}\|_\Omega + C_{\mathrm{PR},f}\|\overline{f}^{j+1/2}\|_{2,\mathcal{T}_h}\big)\Big) \\ &\leq C\big(h^k + \tau^2\big),\end{aligned}$$

where the constant C only depends on t_{n+1}, C_π, $C_{\pi,\mathcal{L}}$, $C_{\mathrm{PR},u}$, $C_{\mathrm{PR},f}$, $\|\partial_t^3 u(\cdot)\|_\Omega$, $\|\mathcal{A}\mathcal{B}\partial_t u(\cdot)\|_\Omega$, $|u(\cdot)|_{k+1,\mathcal{T}_h}$, $\|\partial_t u(\cdot)\|_{2,\mathcal{T}_h}$, $\|\mathcal{A}^2 f(\cdot)\|_\Omega$ and $\|f(\cdot)\|_{2,\mathcal{T}_h}$.

Once more, for better readability, we leave out the full bounds for the derivative errors, but they can be found in Appendix 13.2. We start with the discrete time derivative.

Theorem 12.18 *Let $h \in \mathcal{H}$, $\tau > 0$, $T > 0$, let $k \geq 1$ be the polynomial degree of the dG method, and let Assumption 12.16 be fulfilled with $\ell = 2$. Then, for all $n \in \mathbb{N}_0$, the full error of the discrete time derivative of the dG-Peaceman–Rachford method satisfies*

$$\|\partial_t u(t_{n+1/2}) - \underline{\boldsymbol{\partial}_\tau \boldsymbol{u}}_{\mathrm{PR}}^{n+1/2}\|_\Omega \leq C\big(h^k + \tau^2\big),$$

where the constant C is independent of h, τ and n if $t_{n+1} \leq T$.

For the spatial derivative the following bounds hold.

Theorem 12.19 *Let the assumptions of Theorem 12.18 be satisfied and, in addition, assume that* $f \in C(\mathbb{R}_+; H^{k+1}(\mathcal{T}_h)^m) \cap C^2(\mathbb{R}_+; L^2(\Omega)^m)$. *Then, for all* $n \in \mathbb{N}_0$, *the full error of the discrete space derivative of the dG-Peaceman–Rachford method satisfies*

$$\|\mathcal{L}u(t_{n+1/2}) - \underline{\mathcal{L}}\boldsymbol{u}_{\text{PR}}^{n+1/2}\|_\Omega \leq C\big(h^k + \tau^2\big),$$

where the constant C *is independent of* h, τ *and* n *if* $t_{n+1} \leq T$.

12.4 Locally Implicit Scheme

Lastly, we consider the dG-locally implicit scheme. The error analysis is performed completely analogously to the leapfrog method in Sect. 12.2 and therefore, we omit all proofs in this section.

In contrast to the other three methods, we do not give bounds on the error for the discrete derivatives, i.e., $\boldsymbol{e}_{\tau,\text{LI}}^n$ and $\boldsymbol{e}_{\mathcal{L},\text{LI}}^n$. This is due to the fact that in order to prove Theorems 12.12 and 12.13, we needed the consistency (and approximation property) of the operator $\underline{\mathcal{L}}_q$ to obtain optimal convergence rates. Unfortunately, in the analysis of the dG-LI scheme, the explicit operator $\underline{\mathcal{L}}_q^e$, which suffers from a loss of consistency, takes over the role of $\underline{\mathcal{L}}_q$. In contrast to the other methods, this prevents us from obtaining the same convergence rates for the discrete derivatives as for the discrete solution itself.

We point out, however, that one can exploit the CFL condition (11.67) to obtain bounds on $\boldsymbol{e}_{\tau,\text{LI}}^n$ and $\boldsymbol{e}_{\mathcal{L},\text{LI}}^n$ with a suboptimal order in time. We refer to [161, Lems. 5.11 & 5.12] for the general idea how to do this. However, it is not clear if this loss of order is factual or merely an artefact of the error analysis.

12.4.1 Error Recursion

As for all other schemes, we start with deriving an error recursion, which coincides with the dG-LI scheme with the inhomogeneity replaced by the corresponding defect.

Lemma 12.20 *Let* $h \in \mathcal{H}$ *and* $\tau > 0$. *Then, for all* $n \in \mathbb{N}_0$, *the discretization error* $\boldsymbol{e}_{\text{LI}}^{n+1} = \pi_h u(t_{n+1}) - \boldsymbol{u}_{\text{LI}}^{n+1}$ *of the dG-locally implicit scheme satisfies*

$$\begin{cases} \boldsymbol{e}_{\text{LI}}^{n+1} = \underline{\boldsymbol{S}}_{\text{LI}}\boldsymbol{e}_{\text{LI}}^n + \underline{\boldsymbol{\mathcal{R}}}_{\text{LI}}\big(\boldsymbol{d}_{\text{CN}}^n + \boldsymbol{d}_{\text{LI}}^n\big), & n \in \mathbb{N}_0, \\ \boldsymbol{e}_{\text{LI}}^0 = 0, \end{cases}$$

where the **(dG-locally implicit) perturbation defect** $\boldsymbol{d}_{\text{LI}}^n$ *is given by*

$$\boldsymbol{d}_{\text{LI}}^n = \frac{\tau^2}{4}\underline{\mathcal{D}}^e\pi_h\big(u(t_{n+1}) - u(t_n)\big).$$

Recall the definition of $\underline{\mathcal{D}}^e$ in (11.63). Note that the only difference to the leapfrog method is that the perturbation operator $\underline{\mathcal{D}}^e$ occurs instead of $\underline{\mathcal{D}}$, i.e., the explicit operators are replaced by the full ones.

12.4.2 Bounds on and Splitting of the Defect

The fact that the explicit operator $\underline{\mathcal{L}}_q^e$ suffers from a loss of consistency prevents us from using Lemma 10.10 and thus the approach used in Sect. 12.3 for the dG-PR scheme. However, we can apply the approach used for the leapfrog method.

Hence, we again start by splitting the defect into a "nice" and a "problematic" part.

Lemma 12.21 *The perturbation defect $\boldsymbol{d}_{\mathrm{LI}}^n$ can be rewritten as*

$$\boldsymbol{d}_{\mathrm{LI}}^n = \boldsymbol{d}_{\mathrm{LI},\pi}^n + \underline{\mathcal{L}}\, \boldsymbol{d}_{\mathrm{LI},\hbar}^n$$

with the projection and discretization part given by

$$\boldsymbol{d}_{\mathrm{LI},\pi}^n = -\frac{\tau^3}{4}\underline{\mathcal{D}}^e \underline{\partial_\tau} e_\pi^{n+1/2}, \quad \text{and} \quad \boldsymbol{d}_{\mathrm{LI},\hbar}^n = \frac{\tau^2}{4}\begin{pmatrix} 0 \\ \chi_e \pi_\hbar\big(\partial_t q(t_{n+1}) - \partial_t q(t_n)\big)\end{pmatrix},$$

respectively.

Note that the discretization part of the defect now contains the cut-off operator χ_e, so that we can still take out the full operator $\underline{\mathcal{L}}$.

Again, this splitting enables us to give an explicit formula for the discretization error $\boldsymbol{e}_{\mathrm{LI}}^n$. More precisely, proceeding analogously to the leapfrog method and directly combining (the locally implicit counterparts) of (12.14) and (12.15), we obtain

$$\begin{aligned}\boldsymbol{e}_{\mathrm{LI}}^{n+1} &= \frac{1}{\tau}\big(\underline{\mathcal{S}}_{\mathrm{LI}}^{n+1}\boldsymbol{d}_{\mathrm{LI},\hbar}^0 - \boldsymbol{d}_{\mathrm{LI},\hbar}^n\big) + \sum_{j=1}^{n} \underline{\mathcal{S}}_{\mathrm{LI}}^{n-j+1}\, \underline{\partial_\tau} \boldsymbol{d}_{\mathrm{LI},\hbar}^{j-1/2} \\ &\quad + \sum_{j=0}^{n} \underline{\mathcal{S}}_{\mathrm{LI}}^{n-j}\underline{\mathcal{R}}_{\mathrm{LI}}\big(\boldsymbol{d}_{\mathrm{CN}}^j + \boldsymbol{d}_{\mathrm{LI},\pi}^j\big).\end{aligned}$$

It remains to give bounds of appropriate order for the defects $\boldsymbol{d}_{\mathrm{LI},\pi}^n$ and $\boldsymbol{d}_{\mathrm{LI},\hbar}^n$. We gather them in the next lemma.

Lemma 12.22 *Let $\hbar \in \mathcal{H}$ and $\tau > 0$. Then, if $q \in C^2(\mathbb{R}_+; L^2(\Omega)^{m_q})$, the discretization part $\boldsymbol{d}_{\mathrm{LI},\hbar}^n$ fulfills*

$$\|\boldsymbol{d}_{\mathrm{LI},\hbar}^n\|_\Omega \le \frac{\tau^2}{4}\int_{t_n}^{t_{n+1}} \|\chi_e \partial_t^2 q(s)\|_\Omega \,\mathrm{d}s,$$

and if $q \in C^3(\mathbb{R}_+; L^2(\Omega)^{m_q})$, its discrete derivative fulfills

$$\|\underline{\partial_\tau} \boldsymbol{d}_{\mathrm{LI},\hbar}^{n-1/2}\|_\Omega \le \frac{\tau^2}{4} \int_{t_{n-1}}^{t_{n+1}} \|\chi_e \partial_t^3 q(s)\|_\Omega \, \mathrm{d}s.$$

Further, if Assumption 11.26 is fulfilled and we have $p \in C(\mathbb{R}_+; D(\mathcal{L}_p) \cap H^{k+1}(\mathcal{T}_\hbar)^{m_p})$, the projection part $\boldsymbol{d}_{\mathrm{LI},\pi}^n$ fulfills

$$\|\boldsymbol{d}_{\mathrm{LI},\pi}^n\|_\Omega \le \frac{\tau}{2} C_{\pi,\mathcal{L}} |h^k \big(p(t_{n+1}) - p(t_n)\big)|_{k+1,\mathcal{T}_\hbar}.$$

Note that, as for the leapfrog method, the regularity assumptions we need for these bounds are already included in those for the Crank–Nicolson result Lemma 12.2.

We point out that all other bounds given for the leapfrog method are still valid except for Lemma 12.10. In fact, the proof of the latter breaks down due to the loss of consistency in $\underline{\mathcal{L}}_q^e$. Since this result was necessary for the error bounds of the discrete derivatives, we are not able to show them in the locally implicit setting.

12.4.3 Error Bounds for the dG-Locally Implicit Scheme

Finally, we give the bound on the full discretization error of the dG-locally implicit scheme.

Theorem 12.23 *Let $\hbar \in \mathcal{H}$ and $\tau > 0$ fulfill Assumption 11.26 (with $\theta_c \in (0, 1)$). Further, assume that the exact solution of the (two-field) wave-type problem* (11.28) *satisfies*

$$u = \begin{pmatrix} p \\ q \end{pmatrix} \in C^3(\mathbb{R}_+; L^2(\Omega)^m) \cap C(\mathbb{R}_+; D(\mathcal{L}) \cap H^{k+1}(\mathcal{T}_\hbar)^m).$$

Then, for all $n \in \mathbb{N}_0$, the error of the dG-locally implicit method satisfies

$$\begin{aligned}
\|u(t_{n+1}) - \boldsymbol{u}_{\mathrm{LI}}^{n+1}\|_\Omega &\le C_\pi |h^{k+1} u(t_{n+1})|_{k+1,\mathcal{T}_\hbar} \\
&+ \sqrt{3}\, C_{\mathrm{stb}} C_{\pi,\mathcal{L}} \tfrac{\tau}{2} \sum_{j=0}^{n} \big(|h^k \big(u(t_{j+1}) + u(t_j)\big)|_{k+1,\mathcal{T}_\hbar} + |h^k \big(p(t_{j+1}) - p(t_j)\big)|_{k+1,\mathcal{T}_\hbar} \big) \\
&+ \frac{\tau^2}{4} \int_0^1 C_{\mathrm{stb}} \|\chi_e \partial_t^2 q(\tau s)\|_\Omega + \|\chi_e \partial_t^2 q(t_n + \tau s)\|_\Omega \, \mathrm{d}s \\
&+ C_{\mathrm{stb}} \frac{\tau^2}{8} \int_0^{t_{n+1}} \sqrt{3}\, \|\partial_t^3 u(s)\|_\Omega + 4 \|\chi_e \partial_t^3 q(s)\|_\Omega \, \mathrm{d}s \\
&\le C\big(\hbar^k + \tau^2\big),
\end{aligned}$$

where the constant C only depends on t_{n+1}, C_π, θ_c, $C_{\pi,\mathcal{L}}$, $\|\partial_t^3 u(\cdot)\|_\Omega$, $\|\chi_e \partial_t^2 q(\cdot)\|_\Omega$, and $|u(\cdot)|_{k+1,\mathcal{T}_{\hslash}}$.

12.5   Concluding Remarks

In this section, we remark on some straightforward extensions of the presented theory.

12.5.1 Less Regular Solutions

Throughout, we have only shown the optimal bounds under optimal regularity assumptions. However, it is a rather straightforward task to augment the employed techniques to show suboptimal bounds if the exact solution is not sufficiently regular to obtain the full order, but still regular enough for a lower convergence rate.

Let us illustrate this with examples. Throughout, we assume some evaluation of the exact solution to lie in the broken Sobolev space $H^{k+1}(\mathcal{T}_{\hslash})^m$, where k is the polynomial degree used in the dG method. However, if we only assume $H^{\ell+1}(\mathcal{T}_{\hslash})^m$ for some $\ell < k$ we can still easily show convergence of order ℓ in space.

Similarly, we often assume the solution to be three times continuously differentiable in time. If we instead only assume the solution to be twice continuously differentiable in time, we can still show first-order convergence in time by using lower order quadrature formulae in the analysis.

12.5.2 Approximations of Initial Values

All presented methods use the projected exact initial values as their starting value. However, it is also possible to use other approximations to the initial value that are of the right order (namely the order we want to achieve globally) in $\hslash$ and τ and still achieve the full global convergence orders.

Similarly, as we also see in the Peaceman–Rachford scheme, other treatments of the inhomogeneity are possible without losing the desired order (possibly requiring more regularity of the inhomogeneity). The only requirement these treatments have to fulfill is that they are approximations of the right order to the inhomogeneity at the intermediate timestep $t_{n+1/2}$.

12.5.3 Approximations at Half Time Steps

Lastly, we point out that the various intermediate variables ($\boldsymbol{u}_{\mathrm{PR}}^{n+1/2}$, $\boldsymbol{q}_{\mathrm{LF}}^{n+1/2}$, and $\boldsymbol{q}_{\mathrm{LI}}^{n+1/2}$) provided by the different schemes do in fact provide approximations to the solution at the intermediate timestep $t_{n+1/2}$ of the same order as those at the full steps. This can also be shown by rather straightforward augmentations of the presented theory. As a consequence, by using similar formulae to compute them, such intermediate approximations may also be provided for the methods that do not inherently provide them (like the dG-CN scheme or $\boldsymbol{p}_{\mathrm{M}}^{n+1/2}$ for the dG-LF or dG-LI scheme).

Appendix 13

In this appendix, we provide several results, which were postponed in the other chapters for the sake of presentation.

13.1 Friedrichs' Operators Exhibiting a Two-Field Structure

This section serves to discuss the consequences of Assumption 11.8 on the two-field structure of the Friedrichs' operator $\mathcal{L}$ and to derive the resulting properties used in the corresponding parts of Chaps. 11 and 12. Thus, throughout this section, we make the aforementioned assumption.

First, we point out that the operators $\mathcal{L}_p$ and $\mathcal{L}_q$ do not exhibit the exact same structure as the Friedrichs' operator $\mathcal{L}$, since their coefficients are not necessarily symmetric. Since this is the only difference to the setting in Sect. 9.2, we are still able to define the formal adjoint $\mathcal{L}_p^{\circledast}$ and the associated boundary operator $\mathcal{L}_{p,\partial}$ for $\mathcal{L}_p$. The corresponding objects for $\mathcal{L}_q$ are defined and denoted analogously. Note, however, that, in general, Theorem 9.22 does not hold for $\mathcal{L}_p$ and $\mathcal{L}_q$. Therefore, these operators do not necessarily generate semigroups.

Comparing the coefficients of $\mathcal{L}_p$ and $\mathcal{L}_q$, one immediately sees that the formal adjoint $\mathcal{L}_p$ is given by $\mathcal{L}_p^{\circledast} = -\mathcal{L}_q$ and vice versa for $\mathcal{L}_q$. Consequently, the formal adjoint of $\mathcal{L}$ takes the form

$$\mathcal{L}^{\circledast} = \begin{pmatrix} K & -\mathcal{L}_q \\ -\mathcal{L}_p & 0 \end{pmatrix} = \begin{pmatrix} K & \mathcal{L}_p^{\circledast} \\ \mathcal{L}_q^{\circledast} & 0 \end{pmatrix},$$

where the first equality follows from the definition of the formal adjoint and the properties of K, $\mathcal{L}_p$ and $\mathcal{L}_q$.

W. Dörfler et al., *Wave Phenomena*, Oberwolfach Seminars 49,
https://doi.org/10.1007/978-3-031-05793-9_13

Further, by the assumed structure (11.20) of $\mathcal{L}$, it is apparent that its graph space admits a product structure, i.e., we have $H(\mathcal{L}) = H(\mathcal{L}_p) \times H(\mathcal{L}_q)$ and the boundary operator $\mathcal{L}_\partial$ also exhibits such a structure. More precisely, for $u = (p, q)^\top, v = (\tilde{p}, \tilde{q})^\top \in H(\mathcal{L}_p) \times H(\mathcal{L}_q)$, we have

$$\langle \mathcal{L}_\partial u \mid v \rangle = \langle \mathcal{L}_{p,\partial} p \mid \tilde{q} \rangle + \langle \mathcal{L}_{q,\partial} q \mid \tilde{p} \rangle, \tag{13.1}$$

and, by the relation between the coefficients of $\mathcal{L}_p$ and $\mathcal{L}_q$ (or the symmetry of $\mathcal{L}_\partial$), we deduce

$$\langle \mathcal{L}_{p,\partial} p \mid \tilde{q} \rangle = \langle \mathcal{L}_{q,\partial} \tilde{q} \mid p \rangle, \tag{13.2}$$

for $p \in H(\mathcal{L}_p)$ and $q \in H(\mathcal{L}_q)$.

Similarly, since the boundary operator $\mathcal{L}_\Gamma$ fulfills (11.22) and the anti-symmetry property (11.21) we obtain

$$\langle \mathcal{L}_{p,\Gamma} p \mid q \rangle = -\langle \mathcal{L}_{q,\Gamma} q \mid p \rangle. \tag{13.3}$$

for $p \in H(\mathcal{L}_p)$ and $q \in H(\mathcal{L}_q)$. Moreover, by (13.1) and (11.22), we deduce that, analogously to the graph space, the domain $D(\mathcal{L})$ exhibits a product structure. This motivates to define the domains of $\mathcal{L}_p$ and $\mathcal{L}_q$ as

$$D(\mathcal{L}_p) := \ker(\mathcal{L}_{p,\partial} - \mathcal{L}_{p,\Gamma}) \quad \text{and} \quad D(\mathcal{L}_q) := \ker(\mathcal{L}_{q,\partial} - \mathcal{L}_{q,\Gamma}),$$

respectively, and thus $D(\mathcal{L}) = D(\mathcal{L}_p) \times D(\mathcal{L}_q)$.

With this, we are able to show that $\mathcal{L}_p$ and $\mathcal{L}_q$ in fact share an adjointness property if restricted to their respective domains.

Lemma 13.1 *For all $p \in D(\mathcal{L}_p)$ and $q \in D(\mathcal{L}_q)$, we have*

$$(\mathcal{L}_p p \mid q)_\Omega = -(\mathcal{L}_q q \mid p)_\Omega.$$

Proof By definition of the boundary operator $\mathcal{L}_{p,\partial}$ and $\mathcal{L}_p^{\circledast} = -\mathcal{L}_q$, we obtain

$$(\mathcal{L}_p p \mid q)_\Omega = -(\mathcal{L}_q q \mid p)_\Omega + \langle \mathcal{L}_{p,\partial} p \mid q \rangle.$$

We now show that $\langle \mathcal{L}_{p,\partial} p \mid q \rangle = 0$ for all $p \in D(\mathcal{L}_p)$ and $q \in D(\mathcal{L}_q)$. To do so, we use (13.2) to obtain

$$\begin{aligned} 2\langle \mathcal{L}_{p,\partial} p \mid q \rangle &= \langle \mathcal{L}_{p,\partial} p \mid q \rangle + \langle \mathcal{L}_{q,\partial} q \mid p \rangle \\ &= \langle (\mathcal{L}_{p,\partial} - \mathcal{L}_{p,\Gamma}) p \mid q \rangle + \langle (\mathcal{L}_{q,\partial} - \mathcal{L}_{q,\Gamma}) q \mid p \rangle \\ &\quad + \langle \mathcal{L}_{p,\Gamma} p \mid q \rangle + \langle \mathcal{L}_{q,\Gamma} q \mid p \rangle \\ &= 0, \end{aligned}$$

where we have used the definitions of the domains $D(\mathcal{L}_p)$ and $D(\mathcal{L}_q)$ and (13.3) in the last step. This proves the claim.

With this groundwork laid, all necessary ingredients to show the desired structure (11.23) of the discrete operator $\underline{\mathbf{L}}$ are already available. In fact, the only remaining steps to be taken are to derive the more concrete representation of the boundary operators. For this, we use (13.1) for $\mathcal{L}_\partial$, and Assumption 10.3 together with (11.22) for $\mathcal{L}_\Gamma$ to deduce

$$L_\partial^F = \begin{pmatrix} 0 & L_{q,\partial}^F \\ L_{p,\partial}^F & 0 \end{pmatrix}, \qquad L_\Gamma = \begin{pmatrix} 0 & L_{q,\Gamma} \\ L_{p,\Gamma} & 0 \end{pmatrix}.$$

Here, the nonzero entries are the associated boundary fields to the corresponding boundary operators, cf., Sect. 10.2. We point out that due to (13.2) and (13.3), we have $(L_{p,\partial}^F)^\mathsf{T} = L_{q,\partial}^F$ and $(L_{p,\Gamma})^\mathsf{T} = -L_{q,\Gamma}$, respectively.

Now, we can define the discrete operators $\underline{\mathbf{L}}_p$ and $\underline{\mathbf{L}}_q$ completely analogously to the full operator $\underline{\mathbf{L}}$. Since, by the preceding paragraphs, all objects used in the definition of these discrete operators decouple completely in the full operator, this then immediately yields the desired structure (11.23). The adjointness property (11.24), as well as all remaining properties, are an easy consequence and therefore left as an exercise to the reader.

13.2 Full Bounds for the Discrete Derivative Errors

In this section, we give full bounds defining the constants for the discrete derivative errors from Chap. 12.

Crank–Nicolson Method If the assumptions of Theorem 12.4 are satisfied, then we have

$$\begin{aligned}\|\partial_t u(t_{n+1/2}) - \underline{\boldsymbol{\partial}_\tau \boldsymbol{u}}_{\text{CN}}^{n+1/2}\|_\Omega &\le C_\pi |h^{k+1}\partial_t u(t_{n+1/2})|_{k+1,\mathcal{T}_h}\\ &+ C_{\pi,\mathcal{L}}\Big(\tfrac{1}{2}|h^k\big(u(\tau)+u^0\big)|_{k+1,\mathcal{T}_h} + \int_0^{t_{n+1}} |h^k\partial_t u(s)|_{k+1,\mathcal{T}_h}\,\mathrm{d}s\Big)\\ &+ \frac{\tau^2}{8}\Big(\int_0^1 \|\partial_t^3 u(t_n+\tau s)\|_\Omega + \|\partial_t^3 u(\tau s)\|_\Omega\,\mathrm{d}s\\ &\quad + \int_0^1\int_{\tau s}^{t_n+\tau s} \|\partial_t^4 u(\varsigma)\|_\Omega\,\mathrm{d}\varsigma\,\mathrm{d}s\Big),\end{aligned}$$

and if the assumptions of Theorem 12.5 are satisfied, we have

$$\begin{aligned}\|\mathcal{L}u(t_{n+1/2}) - \underline{\mathcal{L}\boldsymbol{u}}_{\text{CN}}^{n+1/2}\|_\Omega &\le C_\pi |h^{k+1}\mathcal{L}u(t_{n+1/2})|_{k+1,\mathcal{T}_h}\\ &+ C_{\pi,\mathcal{L}}\Big(\tfrac{1}{2}|h^k\big(u(\tau)+u^0\big)|_{k+1,\mathcal{T}_h} + \int_0^{t_{n+1}} |h^k\partial_t u(s)|_{k+1,\mathcal{T}_h}\,\mathrm{d}s\Big)\\ &+ \frac{\tau^2}{8}\Big(\int_0^1 \|\partial_t^2 f(t_n+\tau s)\|_\Omega + \|\partial_t^3 u(t_n+\tau s)\|_\Omega + \|\partial_t^3 u(\tau s)\|_\Omega\,\mathrm{d}s\\ &\quad + \int_0^1\int_{\tau s}^{t_n+\tau s} \|\partial_t^4 u(\varsigma)\|_\Omega\,\mathrm{d}\varsigma\,\mathrm{d}s\Big).\end{aligned}$$

Leapfrog Method If the assumptions of Theorem 12.12 are satisfied, then we have

$$\begin{aligned}\|\partial_t u(t_{n+1/2}) - \underline{\boldsymbol{\partial}_\tau \boldsymbol{u}}_{\text{LF}}^{n+1/2}\|_\Omega &\le C_\pi |h^{k+1}\partial_t u(t_{n+1/2})|_{k+1,\mathcal{T}_h}\\ &+ \sqrt{3}\,C_{\text{stb}}C_{\pi,\mathcal{L}}\Big(\tfrac{1}{2}\big(|h^k\big(u(\tau)+u^0\big)|_{k+1,\mathcal{T}_h} + |h^k\big(p(\tau)-p^0\big)|_{k+1,\mathcal{T}_h}\big)\\ &\qquad + \int_0^{t_{n+1}} |h^k\partial_t u(s)|_{k+1,\mathcal{T}_h} + |h^k\partial_t p(s)|_{k+1,\mathcal{T}_h}\,\mathrm{d}s\Big)\\ &+ \frac{\tau^2}{8}\Big(\sqrt{3}\,C_{\text{stb}}\int_0^1 \|\partial_t^3 u(t_n+\tau s)\|_\Omega + \|\partial_t^3 u(\tau s)\|_\Omega\,\mathrm{d}s\\ &\quad + \sqrt{3}\,C_{\text{stb}}\int_0^1\int_{\tau s}^{t_n+\tau s} \|\partial_t^4 u(\varsigma)\|_\Omega\,\mathrm{d}\varsigma\,\mathrm{d}s\\ &\quad + 2\sqrt{3}\,C_{\text{stb}}\int_0^1 \|\mathcal{L}_q\partial_t^2 q(\tau s)\|_\Omega + C_{\pi,\mathcal{L}}|\partial_t^2 q(\tau s)|_{1,\mathcal{T}_h}\,\mathrm{d}s\\ &\quad + 2\int_0^1 C_{\text{stb}}\|\partial_t^3 q(\tau s)\|_\Omega + \|\partial_t^3 q(t_n+\tau s)\|_\Omega\,\mathrm{d}s\\ &\quad + 3C_{\text{stb}}\int_0^{t_{n+1}} \|\partial_t^4 q(s)\|_\Omega\,\mathrm{d}s\Big),\end{aligned}$$

and if the assumptions of Theorem 12.13 are satisfied, we have

$$
\begin{aligned}
\|\mathcal{L}u(t_{n+1/2}) - \underline{\boldsymbol{\mathcal{L}u}}_{\mathrm{LF}}^{n+1/2}\|_\Omega &\le C_\pi |h^{k+1}\mathcal{L}u(t_{n+1/2})|_{k+1,\mathcal{T}_h} \\
&+ \sqrt{3}\, C_{\mathrm{stb}} C_{\pi,\mathcal{L}} \Big(\tfrac{1}{2}\big(|h^k\big(u(\tau)+u^0\big)|_{k+1,\mathcal{T}_h} + |h^k\big(p(\tau)-p^0\big)|_{k+1,\mathcal{T}_h}\big) \\
&\qquad + \int_0^{t_{n+1}} |h^k \partial_t u(s)|_{k+1,\mathcal{T}_h} + |h^k \partial_t p(s)|_{k+1,\mathcal{T}_h}\, \mathrm{d}s \Big) \\
&+ \frac{\tau^2}{8}\Big(\|\partial_t^2 f(t_n+\tau s)\|_\Omega + \sqrt{3}\, C_{\mathrm{stb}} \int_0^1 \|\partial_t^3 u(t_n+\tau s)\|_\Omega + \|\partial_t^3 u(\tau s)\|_\Omega \,\mathrm{d}s \\
&\qquad + \sqrt{3}\, C_{\mathrm{stb}} \int_0^1 \int_{\tau s}^{t_n+\tau s} \|\partial_t^4 u(\varsigma)\|_\Omega \,\mathrm{d}\varsigma\, \mathrm{d}s \\
&\qquad + 2\sqrt{3}\, C_{\mathrm{stb}} \int_0^1 \|\mathcal{L}_q \partial_t^2 q(\tau s)\|_\Omega + C_{\pi,\mathcal{L}} |\partial_t^2 q(\tau s)|_{1,\mathcal{T}_h}\, \mathrm{d}s \\
&\qquad + 2 \int_0^1 C_{\mathrm{stb}} \|\partial_t^3 q(\tau s)\|_\Omega + \|\partial_t^3 q(t_n+\tau s)\|_\Omega \,\mathrm{d}s \\
&\qquad + 3 C_{\mathrm{stb}} \int_0^{t_{n+1}} \|\partial_t^4 q(s)\|_\Omega \,\mathrm{d}s \Big).
\end{aligned}
$$

Peaceman–Rachford Method If the assumptions of Theorem 12.18 are satisfied, then we have

$$
\begin{aligned}
\|\partial_t u(t_{n+1/2}) - \underline{\boldsymbol{\partial_\tau u}}_{\mathrm{PR}}^{n+1/2}\|_\Omega &\le C_\pi |h^{k+1}\partial_t u(t_{n+1/2})|_{k+1,\mathcal{T}_h} \\
&\quad + C_{\pi,\mathcal{L}} \Big(\tfrac{1}{2} |h^k\big(u(\tau)+u^0\big)|_{k+1,\mathcal{T}_h} + \int_0^{t_{n+1}} |h^k \partial_t u(s)|_{k+1,\mathcal{T}_h}\, \mathrm{d}s \Big) \\
&\quad + \frac{\tau^2}{8}\Big(\int_0^1 \|\partial_t^3 u(t_n+\tau s)\|_\Omega + \|\partial_t^3 u(\tau s)\|_\Omega \,\mathrm{d}s \\
&\qquad + 2 \int_0^1 \|\mathcal{A}\mathcal{B}\partial_t u(\tau s)\|_\Omega + C_{\mathrm{PR},u} \|\partial_t u(\tau s)\|_{2,\mathcal{T}_h}\, \mathrm{d}s \\
&\qquad + 2\left(\|\mathcal{A}^2 \overline{f}^{1/2}\|_\Omega + C_{\mathrm{PR},f} \|\overline{f}^{1/2}\|_{2,\mathcal{T}_h} \right) \\
&\qquad + \int_0^1 \int_{\tau s}^{t_n+\tau s} \|\partial_t^4 u(\varsigma)\|_\Omega \,\mathrm{d}\varsigma\, \mathrm{d}s \\
&\qquad + 2 \int_0^1 \int_{\tau s}^{t_n+\tau s} \|\mathcal{A}\mathcal{B}\partial_t^2 u(\varsigma)\|_\Omega + C_{\mathrm{PR},u} \|\partial_t^2 u(s)\|_{2,\mathcal{T}_h}\, \mathrm{d}\varsigma\, \mathrm{d}s \\
&\qquad + 4 \int_0^{t_{n+1}} \|\mathcal{A}^2 \partial_t f(s)\|_\Omega + C_{\mathrm{PR},f} \|\partial_t f(s)\|_{2,\mathcal{T}_h}\, \mathrm{d}s \Big),
\end{aligned}
$$

and if the assumptions of Theorem 12.19 are satisfied, we have

$$
\begin{aligned}
\|\mathcal{L}u(t_{n+1/2}) - \underline{\mathcal{L}\boldsymbol{u}}_{\mathrm{PR}}^{n+1/2}\|_\Omega \le C_\pi |h^{k+1}\mathcal{L}u(t_{n+1/2})|_{k+1,\mathcal{T}_h} \\
+ C_{\pi,\mathcal{L}}\Big(\tfrac{1}{2}|h^k\big(u(\tau)+u^0\big)|_{k+1,\mathcal{T}_h} + \int_0^{t_{n+1}} |h^k \partial_t u(s)|_{k+1,\mathcal{T}_h}\,\mathrm{d}s\Big) \\
+ \frac{\tau^2}{8}\Big(\int_0^1 \|\partial_t^2 f(t_n+\tau s)\|_\Omega + \|\partial_t^3 u(t_n+\tau s)\|_\Omega + \|\partial_t^3 u(\tau s)\|_\Omega\,\mathrm{d}s \\
+ 2\int_0^1 \big(\|\mathcal{A}\mathcal{B}\partial_t u(\tau s)\|_\Omega + C_{\mathrm{PR},u}\|\partial_t u(\tau s)\|_{2,\mathcal{T}_h}\big)\,\mathrm{d}s \\
+ 2\int_0^1 \big(\|\mathcal{A}\partial_t f(t_n+\tau s)\|_\Omega + C_{\pi,\mathcal{A}}|\partial_t f(t_n+\tau s)|_{1,\mathcal{T}_h}\big)\,\mathrm{d}s \\
+ 2\big(\|\mathcal{A}^2\overline{f}^{1/2}\|_\Omega + C_{\mathrm{PR},f}\|\overline{f}^{1/2}\|_{2,\mathcal{T}_h}\big) \\
+ \int_0^1\int_{\tau s}^{t_n+\tau s} \|\partial_t^4 u(\varsigma)\|_\Omega\,\mathrm{d}\varsigma\,\mathrm{d}s \\
+ 2\int_0^1\int_{\tau s}^{t_n+\tau s} \|\mathcal{A}\mathcal{B}\partial_t^2 u(\varsigma)\|_\Omega + C_{\mathrm{PR},u}\|\partial_t^2 u(s)\|_{2,\mathcal{T}_h}\,\mathrm{d}\varsigma\,\mathrm{d}s \\
+ 4\int_0^{t_{n+1}} \|\mathcal{A}^2\partial_t f(s)\|_\Omega + C_{\mathrm{PR},f}\|\partial_t f(s)\|_{2,\mathcal{T}_h}\,\mathrm{d}s\Big).
\end{aligned}
$$

List of Definitions 14

For better readability, in this section, we present a list of defined symbols together with a short description and the page the symbol is defined on.

Symbol	Description	Page
$\mathbb{Q}_d^k(K)$	Space consisting of d-dimensional polynomials on K of degree at most k in each variable	165
Ω	Physical and computational domain	167
Γ	Boundary of Ω, i.e., $\Gamma = \partial\Omega$	167
M	Material tensor	167
$\widetilde{\mathcal{L}}, \widetilde{\mathcal{A}}, \widetilde{\mathcal{B}}$	Friedrichs' operator; $\widetilde{\mathcal{L}} = \sum_{i=1}^d L_i \partial_i + L_0$	167
$\mathcal{L}, \mathcal{A}, \mathcal{B}$	Friedrichs' operator weighted by M^{-1}	168
$H(\cdot)$	Graph space of a Friedrichs' operator	178
$\mathcal{L}^{\circledast}, \mathcal{A}^{\circledast}, \mathcal{B}^{\circledast}$	Formal adjoint of $\mathcal{L}, \mathcal{A}, \mathcal{B}$	178
$\mathcal{L}_\partial, \mathcal{A}_\partial, \mathcal{B}_\partial$	Boundary operator corresponding to $\mathcal{L}, \mathcal{A}, \mathcal{B}$	179
$\mathcal{L}_\Gamma, \mathcal{A}_\Gamma, \mathcal{B}_\Gamma$	Boundary operator modeling boundary conditions	179
$D(\cdot)$	Domain of a Friedrichs' operator (also used for general operators in Chap. 9)	179
$\mathcal{T}, \mathcal{T}_\hbar$	Mesh of Ω	187
h_K	Diameter of a mesh element $K \in \mathcal{T}_\hbar$	187
$\hbar$	Maximal diameter of all elements in a mesh $\mathcal{T}_\hbar$	188
$\hbar_{\min}$	Minimal diameter of all elements in a mesh $\mathcal{T}_\hbar$	188
h	Piecewise constant function that returns element diameters	188
$\mathcal{T}_\mathcal{H}$	Admissible mesh sequence	188
$\mathcal{H}$	Set containing all mesh diameters of a mesh sequence	188

W. Dörfler et al., *Wave Phenomena*, Oberwolfach Seminars 49,
https://doi.org/10.1007/978-3-031-05793-9_14

Symbol	Description	Page
ρ	Shape- and contact regularity parameter of the mesh sequence $\mathcal{T}_{\mathcal{H}}$	188
$\mathcal{F}_{\hbar}$	Faces of $\mathcal{T}_{\hbar}$	188
$\mathcal{F}_{\hbar}^{\text{int}}$	Interior faces of $\mathcal{T}_{\hbar}$	188
$\mathcal{F}_{\hbar}^{\text{bnd}}$	Boundary faces of $\mathcal{T}_{\hbar}$	188
$\mathcal{F}_{\hbar}^{K}$	Faces composing the boundary ∂K of a mesh element $K \in \mathcal{T}_{\hbar}$	188
N_∂	Maximal number of faces composing the boundary ∂K of a mesh element $K \in \mathcal{T}_{\hbar}$	188
$\mathfrak{n}^K$	Outer unit normal vector on the boundary ∂K of a mesh element $K \in \mathcal{T}_{\hbar}$	188
$\mathfrak{n}^F$	Unit normal vector on a face of the mesh (fixed, but arbitrary orientation)	188
$\mathbb{V}_{\hbar}$	Broken polynomial space; approximation space for the dG method	188
k	Polynomial degree used in the dG method	188
$\pi_{\hbar}$	Weighted L^2-orthogonal projection onto the discrete space $\mathbb{V}_{\hbar}$	189
π_{L^2}	Standard L^2-orthogonal projection onto the discrete space $\mathbb{V}_{\hbar}$	190
$L_\partial^K, A_\partial^K, B_\partial^K$	Boundary field associated with $\mathcal{L}_\partial, \mathcal{A}_\partial, \mathcal{B}_\partial$ w.r.t. elements	192
$L_\partial^F, A_\partial^F, B_\partial^F$	Boundary field associated with $\mathcal{L}_\partial, \mathcal{A}_\partial, \mathcal{B}_\partial$ w.r.t. faces	192
$L_\Gamma, A_\Gamma, B_\Gamma$	Boundary field associated with $\mathcal{L}_\Gamma, \mathcal{A}_\Gamma, \mathcal{B}_\Gamma$	193
$\underline{\mathfrak{L}}, \underline{\mathfrak{A}}, \underline{\mathfrak{B}}$	Central fluxes dG discretization of the Friedrichs' operators $\mathcal{L}, \mathcal{A}, \mathcal{B}$	195
τ	Timestep used in the time integration schemes	202
$\mathcal{L}_q$	Off-diagonal operator of the two-field operator applied to the q-field	211
$\mathcal{L}_p$	Off-diagonal operator of the two-field operator applied to the p-field	211
M_p	Material tensor corresponding to the p-field	211
M_q	Material tensor corresponding to the q-field	211
$\mathcal{L}_{p,\Gamma}$	Boundary operator modelling the boundary condition for the p-field	211
$\underline{\mathfrak{L}}_q$	Central fluxes dG discretization of $\mathcal{L}_q$	212
$\underline{\mathfrak{L}}_p$	Central fluxes dG discretization of $\mathcal{L}_p$	212
$\underline{\mathfrak{D}}$	Perturbation operator of the dG-LF scheme	214

Symbol	Description	Page
θ	Coefficient measuring the strictness of the CFL condition (11.38) in the dG-LF scheme	216
$\mathcal{T}_{\hbar,c}$	Coarse part of the mesh $\mathcal{T}_\hbar$	231
$\mathcal{T}_{\hbar,f}$	Fine part of the mesh $\mathcal{T}_\hbar$	231
h_c	Piecewise constant function that returns element diameters on the coarse part of the mesh	231
h_f	Piecewise constant function that returns element diameters on the fine part of the mesh	231
$\hbar_{\mathrm{min,c}}$	Minimal diameter of all elements in the coarse part of the mesh $\mathcal{T}_{\hbar,c}$	231
$\mathcal{T}_{\hbar,i}$	Implicitly treated part of the mesh $\mathcal{T}_\hbar$	232
$\mathcal{T}_{\hbar,e}$	Explicitly treated part of the mesh $\mathcal{T}_\hbar$	232
χ_i	Cut-off function onto the implicitly treated mesh	232
χ_e	Cut-off function onto the explicitly treated mesh	232
$\underline{\mathfrak{L}}_p^e$	Explicitly treated part of $\underline{\mathfrak{L}}_p$	232
$\underline{\mathfrak{L}}_q^e$	Explicitly treated part of $\underline{\mathfrak{L}}_p$	232
$\underline{\mathfrak{L}}_p^i$	Implicitly treated part of $\underline{\mathfrak{L}}_p$	232
$\underline{\mathfrak{L}}_q^i$	Implicitly treated part of $\underline{\mathfrak{L}}_p$	232
$\underline{\mathfrak{D}}^e$	Perturbation operator of the dG-LI scheme	235
θ_c	Coefficient measuring the strictness of the CFL condition (11.67) in the dG-LI scheme	237
$\mathcal{L}_{p,\partial}$	Boundary operator corresponding to $\mathcal{L}_p$	269
$\mathcal{L}_{q,\partial}$	Boundary operator corresponding to $\mathcal{L}_q$	270

Part IV

An Abstract Framework for Inverse Wave Problems with Applications

Andreas Rieder

The purpose of this chapter is twofold. In the first part we consider the theory of nonlinear inverse problems. Here, we introduce the concept of local ill-posedness in Banach spaces and identify criteria being sufficient for local ill-posedness. The usability of theses criteria is demonstrated by two examples, one from medical and one from seismic imaging. Further, we analyze an iterative scheme of inexact Newton type for the stable solution of ill-posed problems in Hilbert spaces. The second part is devoted to inverse problems arising from wave propagation. Here, we rigorously provide all ingredients to set up a Newton type solver for full waveform inversion (FWI) of seismic imaging in the visco-elastic regime. FWI is the inverse problem of finding material parameters like wave speed and attenuation from recorded wave fields. We study the nonlinear operator which maps these parameters to the wave field. In particular, we derive explicit analytic expressions for its Fréchet derivative and the corresponding adjoint operator. Since we provide an abstract framework our results apply to an inverse electromagnetic scattering problem as well.

What Is an Inverse and Ill-Posed Problem? 15

In an inverse problem we draw conclusions about the cause from its observed (measured) effect. This kind of task is typically ill-posed, that is, small changes (noisy measurements) in the effects lead to dramatic changes in the corresponding causes. Yet to obtain meaningful results the inversion or reconstruction process needs stabilization which is commonly referred to as regularization. All these basic concepts will be introduced first in some detail before we study advanced inverse wave problems.

We start by giving two examples of inverse problems which are also ill-posed as we will learn later.

15.1 Electric Impedance Tomography: The Continuum Model

In electric impedance tomography (EIT) one applies currents through the boundary of an object and measures the resulting voltages at the boundary as well. From this information one seeks the conductivity inside the object.

Let $\Omega \subset \mathbb{R}^d$ be a bounded Lipschitz domain and let $\sigma : \Omega \to [0, \infty[$ be the isotropic conductivity coefficient. Further, $f : \partial\Omega \to \mathbb{R}$ and $u : \Omega \to \mathbb{R}$ denote the boundary current and the electric potential in the domain, respectively. All these quantities are connected via the potential equation

$$\begin{aligned} -\operatorname{div}(\sigma \nabla u) &= 0 \quad \text{in } \Omega, \\ \sigma \partial_{\mathbf{n}} u &= f \quad \text{on } \partial\Omega. \end{aligned}$$

Let $f \in L^2_\diamond(\partial\Omega) := \{g \in L^2(\partial\Omega) : \int_{\partial\Omega} g \, \mathrm{d}S = 0\}$ and

$$\sigma \in L^\infty_+(\Omega) := \{\gamma \in L^\infty(\Omega) : \gamma \geq c_0 \text{ a.e.}\} \tag{15.1}$$

W. Dörfler et al., *Wave Phenomena*, Oberwolfach Seminars 49,
https://doi.org/10.1007/978-3-031-05793-9_15

where $c_0 > 0$ is a constant. Then, the governing equation in weak formulation is

$$\int_\Omega \sigma \nabla u \cdot \nabla v \,\mathrm{d}x = \int_{\partial\Omega} f v \,\mathrm{d}S \quad \text{for all } v \in H^1_\diamond(\Omega) \tag{15.2}$$

and it has a unique solution $u \in H^1_\diamond(\Omega) := \{v \in H^1(\Omega) : \int_{\partial\Omega} v \,\mathrm{d}S = 0\}$, see, e.g., [67].

The inverse EIT problem in the continuum model can now be phrased as: given the *Neumann-to-Dirichlet operator*

$$\Lambda_\sigma : f \mapsto u|_{\partial\Omega}$$

find the conductivity σ. By classical results from the theory of partial differential equations, Λ_σ is known to be a bounded linear operator between $H_\diamond^{-1/2}(\partial B)$ and $H_\diamond^{1/2}(\partial B)$ (where the former space is the dual of the latter).

Mathematically, we have to solve an equation with the nonlinear operator F describing the forward problem, that is, we need to solve $F(\sigma) = \Lambda_\sigma$ where

$$F\colon \mathsf{D}(F) \subset L^\infty(\Omega) \to \mathcal{L}(H_\diamond^{-1/2}(\partial\Omega), H_\diamond^{1/2}(\partial\Omega)), \quad \gamma \mapsto \Lambda_\gamma, \tag{15.3}$$

with $\mathsf{D}(F) := L^\infty_+(\Omega)$. Note that $F(\gamma) = \Lambda$ is uniquely solvable in case $d = 2$ [5].

In practice one does not have full knowledge of Λ_σ. But one applies, say, m currents $f_1, \dots, f_m$ and measures $g_j = \Lambda_\sigma f_j$, $j = 1, \dots, m$. Thus, one needs to solve

$$F(\sigma) = \begin{pmatrix} g_1 \\ \vdots \\ g_m \end{pmatrix}$$

where now

$$F\colon L^\infty_+(\Omega) \subset L^\infty(\Omega) \to H_\diamond^{1/2}(\partial\Omega)^m, \quad \gamma \mapsto \begin{pmatrix} \Lambda_\gamma f_1 \\ \vdots \\ \Lambda_\gamma f_m \end{pmatrix}. \tag{15.4}$$

15.2 Seismic Tomography

Seismic tomography entails the reconstruction of subsurface material parameters (e.g. mass density, Lamé constants) from reflected waves measured on a part of the propagation medium (typically an area on the earth's surface or in the ocean). From a mathematical point of view we have to solve a nonlinear parameter identification problem for the elastic wave equation (with damping).

If we assume that the medium only supports normal stress we are in the regime of acoustic waves which are solutions u of the acoustic wave equation

$$c\,\partial_t^2 u - \nabla_{\mathbf{x}} \cdot \big(r \nabla_{\mathbf{x}} u\big) = f(\mathbf{x}, t) \tag{15.5}$$

with

$$c = \frac{1}{\varrho\, \nu^2} \quad \text{and} \quad r = \frac{1}{\varrho}$$

where ν and ϱ are the wave speed and the mass density of the medium, respectively. Both coefficients are real valued and spatially varying: $\nu = \nu(\mathbf{x})$, $\varrho = \varrho(\mathbf{x})$. The forcing term f is the energy source which excites wave propagation.

We consider wave propagation in $\Omega \subset \mathbb{R}^d$, an open and bounded set with a Lipschitz boundary. Moreover, we require a Dirichlet boundary condition,

$$u|_{\partial\Omega} = 0, \tag{15.6}$$

as well as initial data

$$u(\cdot, 0) = u_0; \qquad \partial_t u(\cdot, 0) = u_1. \tag{15.7}$$

Let $V = H_0^1(\Omega)$, $H = L^2(\Omega)$, and $T > 0$. The weak formulation of (15.5) with (15.6) and (15.7) reads:

> Given $c, r \in L^\infty(\Omega)$ and $u_0 \in V$ and $u_1 \in H$ and $f \in L^2\big([0, T] \times \Omega\big) = L^2\big((0, T), H\big)$ find $u \in L^2\big((0, T), V\big)$ with $u(0) = u_0$ and $\dot{u}(0) = u_1$ such that
>
> $$\int_0^T \Big(a_r\big(u(t), v(t)\big) - \langle c\dot{u}(t), \dot{v}(t)\rangle_H\Big)\mathrm{d}t = \int_0^T \langle f(t), v(t)\rangle_H \,\mathrm{d}t \tag{15.8}$$
>
> for all $v \in C_0^\infty([0, T], V)$.

The dot on top of a time-dependent function indicates the time derivative, $\langle\cdot,\cdot\rangle_H$ denotes the inner product in H, and

$$a_r \colon V \times V \to \mathbb{R}, \quad a_r(\psi, \varphi) = \int_\Omega r \nabla_{\mathbf{x}} \psi \cdot \nabla_{\mathbf{x}} \varphi \,\mathrm{d}\mathbf{x}.$$

If $c, r \in L_+^\infty(\Omega)$ then this weak formulation admits a unique solution, see [110]. Hence, the forward map

$$F \colon L_+^\infty(\Omega)^2 \subset L^\infty(\Omega)^2 \to L^2([0, T] \times \Omega), \quad (c, r) \mapsto u, \tag{15.9}$$

is well defined. Let $\Psi\colon L^2([0,T]\times\Omega)\to\mathbb{R}^N$ be the observation/measurement operator. Then, the inverse problems reads: given $w\in\mathbb{R}^N$ find $(c,r)\in L^\infty_+(\Omega)^2$ such that

$$\Psi F(c,r) = w.$$

Local Ill-Posedness 16

In this section we give a mathematical definition and characterizations of (local) ill-posedness of operator equations. Roughly speaking, ill-posedness means that the preimages do not depend continuously on the images of the underlying operator. Ill-posedness is the negation of well-posedness which is easier to comprehend and which we, for that reason, define first. We refer to [94, 95] for more details.

Let $F\colon \mathsf{D}(F) \subset X \to Y$ be a nonlinear mapping between the Banach spaces X and Y.

Definition 16.1 The equation $F(\cdot) = y$ is called **locally well-posed** at $x^+ \in \mathsf{D}(F)$ satisfying $F(x^+) = y$ if there is a neighborhood U of x^+ such that for any sequence $\{x_k\} \subset U \cap \mathsf{D}(F)$ with $\lim_{k\to\infty} \|F(x_k) - y\|_Y = 0$ we have that $\lim_{k\to\infty} \|x_k - x^+\|_X = 0$.

If $F(\cdot) = y$ is not locally well-posed at $x^+ \in \mathsf{D}(F)$ we call the equation **locally ill-posed**, that is, in *any* neighborhood U of x^+ there exist a sequence $\{\xi_k\} \subset U \cap \mathsf{D}(F)$ with $\lim_{k\to\infty} \|F(\xi_k) - y\|_Y = 0$ but $\{\xi_k\}$ *does not* converge to x^+.

There is a neat equivalence characterization of local ill-posedness for linear operator equations.

Lemma 16.2 *Let $A \in \mathcal{L}(X, Y)$ where X and Y are Banach spaces. The problem $A(\cdot) = y$ is locally ill-posed at x^+ (satisfying $Ax^+ = y$) if and only if*

$$\mathsf{R}(A) \textit{ is non-closed in } Y \textit{ or } \mathsf{N}(A) \textit{ is nontrivial.}$$

Proof First we show the 'if'-direction. If $\mathsf{N}(A) \neq \{0\}$ we take a $v \in \mathsf{N}(A)\backslash\{0\}$ and set $x_k := x^+ + \varrho_k v$ where $\varrho_k \searrow \varrho > 0$. Then, $x_k \not\to x^+$ but $Ax_k = Ax^+ = y$.

We consider now the case $\mathsf{R}(A) \neq \overline{\mathsf{R}(A)}$ and $\mathsf{N}(A) = \{0\}$. Then, $A^{-1}\colon \mathsf{R}(A) \subset Y \to X$ is well defined and unbounded (otherwise $\mathsf{R}(A)$ would be closed). Hence,

W. Dörfler et al., *Wave Phenomena*, Oberwolfach Seminars 49,
https://doi.org/10.1007/978-3-031-05793-9_16

there is $\{w_k\} \subset \mathsf{R}(A)$ with $\|w_k\|_Y = 1$ but $\|A^{-1}w_k\|_X \to \infty$. Define $\xi_k := x^+ + rA^{-1}w_k/\|A^{-1}w_k\|_X$ for $r > 0$. Then, $\|\xi_k - x^+\|_X = r$ and

$$\|A\xi_k - Ax^+\|_Y = \frac{r}{\|A^{-1}w_k\|_X} \xrightarrow{k\to\infty} 0.$$

The 'only if'-direction follows immediately from the bounded inverse theorem. □

The above lemma shows two sources for local ill-posedness of linear problems which are of a different nature:

- If $\mathsf{N}(A) \neq \{0\}$ then an algebraic property, the non-uniqueness of a solution, gives rise to the local ill-posedness. In Hilbert spaces, for instance, this source can be eliminated by introducing the concept of minimum norm solutions. In a general Banach space the algebraic deficiency may be resolved by exchanging X with the quotient space $X/\mathsf{N}(A)$ and by defining the restriction $\widehat{A}\colon X/\mathsf{N}(A) \to Y$ in the natural way. Then, $\mathsf{N}(\widehat{A}) = \{0\}$ and $\mathsf{R}(\widehat{A})$ is non-closed if and only if $\mathsf{R}(A)$ is non-closed. So the algebraic ill-posedness is not a real threat.
- The topological property $\mathsf{R}(A) \neq \overline{\mathsf{R}(A)}$ is a cause for instability in the reconstruction process which is not straightforwardly resolved.

For nonlinear problems we provide only sufficient conditions for local ill-posedness. Note that there is also an algebraic source of ill-posedness for nonlinear problems: $F(\cdot) = y$ is locally ill-posed at any x^+ which is not an isolated solution of the equation.

Lemma 16.3 *The nonlinear problem $F(\cdot) = y$ in infinite dimensional Banach spaces is locally ill-posed at $x^+ \in \operatorname{int}(\mathsf{D}(F))$ if X is reflexive, F is weak-to-weak sequentially continuous*[1] *and compact.*[2]

Proof According to the reflexivity of X there is a sequence $\{e_k\} \subset X$ with $\|e_k\|_X = 1$ and $e_k \rightharpoonup 0$ (this is consequence of the Josefson-Nissenzweig theorem, see, e.g., [46, Chap. XII]). As x^+ is an interior point of $\mathsf{D}(F)$ there exists an $r > 0$ such that $B_r(x^+) \subset \mathsf{D}(F)$. Then, $x_k := x^+ + \varrho e_k \in B_r(x^+)$ for any $\varrho \in]0, r[$. Obviously, $\|x_k - x^+\|_X = \varrho$ but $x_k \rightharpoonup x^+$. Hence, $F(x_k) \rightharpoonup F(x^+)$. Since $\{\|x_k\|_X\}$ is bounded, $\{F(x_k)\}$ admits a strongly convergent subsequence, say, $F(x_{k_j}) \to \eta$. Taking into account that $F(x_{k_j}) \rightharpoonup F(x^+)$ we must have that $F(x^+) = \eta$. As any subsequence of $\{F(x_k)\}$ contains a subsequence which converges strongly to $F(x^+)$, the whole sequence must converge: $\lim_{k\to\infty} \|F(x_k) - F(x^+)\|_Y = 0$. □

[1] $x_k \rightharpoonup x$ implies that $F(x_k) \rightharpoonup F(x)$ (The arrow $\rightharpoonup$ symbolizes weak convergence).

[2] F maps bounded sets to relatively compact ones.

The above lemma can neither be applied to EIT nor to seismic tomography as $L^\infty(\Omega)$ fails to be reflexive. However, we can weaken the assumptions of the former lemma. Suppose, X can be equipped with the weak-$\star$ topology, that is, X is the dual of a Banach space Z: $X = Z'$. A sequence $\{x_k\} \subset X$ converges weakly-$\star$ to x (notation: $x_k \overset{\star}{\rightharpoonup} x$) if

$$\langle x_k, z\rangle_{Z'\times Z} \xrightarrow{k\to\infty} \langle x, z\rangle_{Z'\times Z} \quad \text{for all } z \in Z.$$

In case of $X = L^\infty(\Omega)$ we have $Z = L^1(\Omega)$, that is,

$$w_k \overset{\star}{\rightharpoonup} w \quad :\Longleftrightarrow \quad \int_\Omega w_k(x)v(x)\,\mathrm{d}x \longrightarrow \int_\Omega w(x)v(x)\,\mathrm{d}x \quad \text{for all } v \in L^1(\Omega).$$

Lemma 16.4 *Let $X = Z'$ for a Banach space Z. The nonlinear problem $F(\cdot) = y$ in infinite dimensional Banach spaces is locally ill-posed at $x^+ \in \operatorname{int}\big(\mathsf{D}(F)\big)$ if F is weak-$\star$-to-weak sequentially continuous[3] and compact.*

Proof Again, the Josefson-Nissenzweig theorem guarantees the existence of a sequence $\{e_k\} \subset X$ with $\|e_k\|_X = 1$ and $e_k \overset{\star}{\rightharpoonup} 0$. Now the proof of Lemma 16.3 carries over. □

For the application to EIT and seismic tomography we still need a further generalization. We say that $\{x_k\} \subset X$ converges weakly-$\star$ to x with respect to a cone $K \subset X$ if $\{x_k - x\} \subset K$ and $x_k \overset{\star}{\rightharpoonup} x$.

For $X = L^\infty(\Omega)$ and the cone $K = \{v \in L^\infty(\Omega) : v \geq 0 \text{ a.e.}\}$ of non-negative functions, a sequence $\{w_k\} \subset L^\infty(\Omega)$ converges to $w \in L^\infty(\Omega)$ weakly-$\star$ with respect to K if

$$w_k \geq w \text{ a.e. and } \int_\Omega w_k(\mathbf{x})v(\mathbf{x})\mathrm{d}\mathbf{x} \xrightarrow{k\to\infty} \int_\Omega w(\mathbf{x})v(\mathbf{x})\mathrm{d}\mathbf{x} \quad \text{for any } v \in L^1(\Omega). \tag{16.1}$$

Especially, if Ω is bounded, $\{w_k\}$ entails a subsequence which converges point-wise a.e. to w. Indeed, as Ω is bounded we can set $v = 1$ and obtain $\|w_k - w\|_{L^1(\Omega)} = \int_\Omega (w_k - w)\mathrm{d}\mathbf{x} \to 0$ as $k \to \infty$. Finally, L^1-convergence forces the existence of a point-wise convergent subsequence, see, e.g. [149, Th. 3.12].

Lemma 16.5 *Let $X = Z'$ for a Banach space Z. The nonlinear problem $F(\cdot) = y$ in infinite dimensional Banach spaces is locally ill-posed at $x^+ \in \mathsf{D}(F)$ if the following hold true:*

[3] $x_k \overset{\star}{\rightharpoonup} x$ implies that $F(x_k) \rightharpoonup F(x)$.

- *there exists a sequence* $\{e_k\}_{k\in\mathbb{N}} \subset X$, *which is normalized*, $\|e_k\|_X = 1$, *and which converges weakly-*$\star$ *to* 0 *with respect to a cone* K *such that* $\{x^+ + re_k\} \subset \mathsf{D}(F)$ *for any* $r \in]0, 1]$,
- F *is compact and weak-*$\star$*-to-weak sequentially continuous with respect to* K, *i.e*, $x_k \overset{\star}{\rightharpoonup} x$ *with respect to* K *implies that* $F(x_k) \rightharpoonup F(x)$.

16.1 Examples for Local Ill-Posedness

16.1.1 Electric Impedance Tomography

With the help of Lemma 16.5 we demonstrate that the inverse EIT problem $F(\cdot) = (g_1, \ldots, g_m)^\top$ with

$$F \colon L^\infty_+(\Omega) \subset L^\infty(\Omega) \to L^2(\partial\Omega)^m \tag{16.2}$$

from (15.4) is locally ill-posed at any conductivity $\sigma_0 \in L^\infty_+(\Omega)$ satisfying $F(\sigma_0) = (g_1, \ldots, g_m)^\top$.

Without loss of generality we may consider $m = 1$. Using the linear and bounded trace operator $\operatorname{tr} \colon H^1(\Omega) \to H^{1/2}(\partial\Omega)$ we get $F(\sigma) = \operatorname{tr}(u)$ where u uniquely solves (15.2).

Proposition 16.6 *The mapping* F *as defined in* (16.2) *is weak-*$\star$*-to-weak sequentially continuous with respect to the cone of positive functions.*

Proof Let $\{\sigma_k\}_{k\in\mathbb{N}} \subset L^\infty_+(\Omega)$ converge weakly-$\star$ to $\sigma \in L^\infty_+(\Omega)$ in $L^\infty(\Omega)$ with $\sigma_k \geq \sigma$ a.e. The sequence $\{\sigma_k\}_{k\in\mathbb{N}}$ is bounded in $L^\infty(\Omega)$ by the uniform boundedness principle.

Let u and u_k be the solutions of (15.2) belonging to σ and σ_k, respectively. Recall that $u, u_k \in H^1_\diamond(\Omega)$. Further,

$$\|u_k\|_{H^1(\Omega)} \lesssim \|f\|_{L^2(\partial\Omega)} \quad \text{for all } k \in \mathbb{N}. \tag{16.3}$$

The constant in the above equation does not depend on $\|\sigma_k\|_\infty$. Thus, there exists a subsequence $\{u_{k_l}\}_l$ which converges weakly in H^1, say, to η. We have that

$$\begin{aligned}\langle f, v\rangle_{L^2(\partial\Omega)} - a_\sigma(\eta, v) &= a_{\sigma_{k_l}}(u_{k_l}, v) - a_\sigma(\eta, v)\\ &= a_{\sigma_{k_l}-\sigma}(u_{k_l}, v) + a_\sigma(u_{k_l} - \eta, v)\end{aligned} \tag{16.4}$$

where $a_\sigma(w, v) = \int_\Omega \sigma \nabla w \cdot \nabla v \,\mathrm{d}x$ for $v, w \in H^1(\Omega)$. Since $a_\sigma(\cdot, v)$ is a bounded linear functional on $H^1(\Omega)$ for any $v \in H^1(\Omega)$ we get

$$\lim_{l\to\infty} a_\sigma(u_{k_l} - \eta, v) = 0 \quad \text{for any } v \in H^1(\Omega).$$

Moreover,

$$|a_{\sigma_{k_l}-\sigma}(u_{k_l}, v)| = \left| \int_\Omega (\sigma_{k_l} - \sigma)\nabla u_{k_l} \cdot \nabla v \, \mathrm{d}x \right| \le \int_\Omega (\sigma_{k_l} - \sigma)|\nabla u_{k_l}|\,|\nabla v| \, \mathrm{d}x$$

$$\lesssim \|f\|_{L^2(\partial\Omega)} \Big(\int_\Omega (\sigma_{k_l} - \sigma)|\nabla v|^2 \, \mathrm{d}x \Big)^{1/2}.$$

As explained in the lines following (16.1), $\{\sigma_{k_l}\}_l$ admits a subsequence which converges point-wise a.e. to σ. Without loss of generality we denote this subsequence again by $\{\sigma_{k_l}\}_l$. Now, by applying the dominated convergence theorem,

$$\lim_{l\to\infty} a_{\sigma_{k_l}-\sigma}(u_{k_l}, v) = 0 \quad \text{for any } v \in H^1(\Omega).$$

In view of (16.4),

$$a_\sigma(\eta, v) = \langle f, v\rangle_{L^2(\partial\Omega)} \quad \text{for any } v \in H^1(\Omega).$$

Since $\operatorname{tr}(u_{k_l}) \rightharpoonup \operatorname{tr}(\eta)$ as $l \to \infty$ we obtain first $\int_{\partial\Omega} \operatorname{tr}(\eta)\mathrm{d}S = 0$ and then $\eta = u$.

Thus, we have shown that every subsequence of $\{F(\sigma_k)\}_{k\in\mathbb{N}}$ has a subsequence which converges weakly to $F(\sigma)$. Hence the whole sequence converges weakly to $F(\sigma)$. □

Proposition 16.7 *The mapping F as defined in* (16.2) *is compact.*

Proof Let $Q \subset L^\infty_+(\Omega)$ be bounded. Then, $F(Q) \subset H^{1/2}(\partial\Omega)$ is bounded as well which follows from (16.3) and the boundedness of the trace operator. Since the embedding $H^{1/2}(\partial\Omega) \hookrightarrow L^2(\partial\Omega)$ is compact, $F(Q)$ is relatively compact in $L^2(\partial\Omega)$ and we are done. □

Theorem 16.8 *Let the mapping F be defined in* (16.2). *Then, the inverse problem of EIT $F(\cdot) = (g_1, \ldots, g_m)^\top$ is locally ill-posed at any conductivity $\sigma_0 \in L^\infty_+(\Omega)$ satisfying $F(\sigma_0) = (g_1, \ldots, g_m)^\top$.*

Proof The assertion follows readily from Lemma 16.5 as soon as we have found a sequence $\{e_n\} \subset L^\infty(\Omega)$ with $e_n \ge 0$ a.e., $\|e_n\|_\infty = 1$, and $e_n \overset{\star}{\rightharpoonup} 0$.

Take a point $\xi \in \Omega$. Then there is a ball $B_\rho(\xi)$ with radius $\rho > 0$ completely contained in Ω. Let $\{\rho_n\}$ be a monotone zero sequence with $\rho_1 \le \rho$. Define e_n to be the indicator function of $B_{\rho_n}(\xi)$. Obviously, $e_n \ge 0$ and $\|e_n\|_\infty = 1$. It remains to show that $\{e_n\}$ converges weakly-$\star$ to 0. Let $v \in L^1(\Omega)$. The convergence of $\int_\Omega e_n v \, \mathbf{d}x$ to 0 as $n \to \infty$ is a straightforward consequence of the dominated convergence theorem. □

16.1.2 Seismic Tomography

Recall the weak formulation (15.8) of the acoustic wave equation (15.5). Its unique solution u has the following properties:

- We have $u \in X := C^0([0, T], V) \cap C^1([0, T], H)$ and the energy estimate holds

$$\|u(t)\|_V^2 + \|\dot{u}(t)\|_H^2 \lesssim \|u_0\|_V^2 + \|u_1\|_H^2 + \int_0^T \|f(\tau)\|_H^2 \,\mathrm{d}\tau \tag{16.5}$$

  where the involved constant only depends on c_0 from (15.1), $\|c\|_\infty$, and $\|r\|_\infty$.
- For almost all $s \in]0, T[$,

$$a_r(u(s), w) + \langle c\ddot{u}(s), w\rangle_{V' \times V} = \langle f(s), w\rangle_H \quad \text{for all } w \in V. \tag{16.6}$$

- We have $c\ddot{u} \in L^2\big([0, T], V'\big)$ and $\ddot{u} \in L^2\big([0, T], V'\big)$ provided that $c \in W^{1,\infty}(\Omega)$.

We are going to show that F from (15.9) satisfies the requirements of Lemma 16.5.

Proposition 16.9 *The mapping F as defined in* (15.9) *is weak-$\star$-to-weak sequentially continuous with respect to the cone of positive functions.*

Proof Let $\{(c_m, r_m)\}_{m\in\mathbb{N}} \subset \mathsf{D}(F)$ converge weakly-$\star$ to $(c, r) \in \mathsf{D}(F)$ in $L^\infty(\Omega)^2$ with $c_m \geq c$ and $r_m \geq r$ a.e. Recall that weakly-$\star$ and weakly convergent sequences are bounded which is a consequence of the uniform boundedness principle.

Denote by $u_m \in X$ and $u \in X$ the solutions of (15.8) with parameters (c_m, r_m) and (c, r), respectively. By (16.5) both sequences $\{u_m\}$ and $\{\dot{u}_m\}$ are bounded in $L^2([0, T], V)$ and $L^2([0, T], H)$, respectively. As both spaces are reflexive, each sequence admits weakly convergent subsequences $\{u_{m_l}\}_{l\in\mathbb{N}}$ and $\{\dot{u}_{m_l}\}_{l\in\mathbb{N}}$ with limits η and ξ, respectively. Since the derivative is defined weakly, we even have that $\dot{\eta} = \xi$.

We will show now that η solves (15.8). To this end let $v \in C_0^\infty([0, T], V)$ and consider

$$\int_0^T \Big(a_{r_{m_l}}\big(u_{m_l}(t), v(t)\big) - a_r\big(\eta(t), v(t)\big)\Big)\mathrm{d}t =$$
$$\int_0^T a_{r_{m_l}-r}\big(u_{m_l}(t), v(t)\big)\mathrm{d}t + \int_0^T a_r\big(u_{m_l}(t) - \eta(t), v(t)\big)\mathrm{d}t.$$

The second term on the right hand side converges to zero because $u_{m_l} \rightharpoonup \eta$ in $L^2([0,T],V)$. For the first term we get

$$\left|\int_0^T a_{r_{m_l}-r}\big(u_{m_l}(t), v(t)\big)\mathrm{d}t\right| \le \|(r_{m_l}-r)\nabla_{\mathbf{x}} v\|_{L^2([0,T],H^d)}\|u_{m_l}\|_{L^2([0,T],V)}$$
$$\lesssim \|(r_{m_l}-r)\nabla_{\mathbf{x}} v\|_{L^2([0,T],H^d)}$$

where the latter estimate holds true due to the boundedness of $\{u_{m_l}\}$. As $\{r_{m_l}\}$ converges weakly-$\star$ to r and $r_{m_l} \ge r$ a.e., it, in particular, admits a subsequence which converges to r point-wise a.e. and which we denote again by $\{r_{m_l}\}$ for simplicity. Further, $|(r_{m_l} - r)\nabla_{\mathbf{x}} v|^2 \lesssim |\nabla_{\mathbf{x}} v|^2$ a.e. in $\Omega \times [0,T]$. Applying the dominated convergence theorem yields that

$$\int_0^T \Big(a_{r_{m_l}}\big(u_{m_l}(t), v(t)\big) - a_r\big(\eta(t), v(t)\big)\Big)\mathrm{d}t \xrightarrow{l\to\infty} 0.$$

Analogously,

$$\int_0^T \Big(\big\langle c_{m_l}\dot{u}_{m_l}(t), \dot{v}(t)\big\rangle_H - \big\langle c\dot{\eta}(t), \dot{v}(t)\big\rangle_H\Big)\mathrm{d}t \xrightarrow{l\to\infty} 0.$$

Hence, η satisfies the wave equation in the weak form (15.8). Moreover, η attains the initial values

$$\eta(0) = u_0 \text{ and } \dot{\eta}(0) = u_1 \tag{16.7}$$

which we validate by a standard argument: We emphasize that (16.6) holds for η. It also holds for u_{m_l} when we replace c and r by c_{m_l} and r_{m_l}, respectively. Both resulting equations we test with $w = w(s) = \phi(s)v$, $v \in V$, $\phi \in C^2(0,T)$, $\phi(T) = \dot{\phi}(T) = 0$, integrate then with respect to s over $[0,T]$, and, finally, integrate in parts twice to obtain

$$\int_0^T \Big(a_r(\eta(s), w(s)) + \langle c\eta(s), \ddot{w}(s)\rangle_H\Big)\mathrm{d}s = \int_0^T \langle f(s), w(s)\rangle_H \mathrm{d}s$$
$$+\, \phi(0)\,\langle c\dot{\eta}(0), v\rangle_H - \dot{\phi}(0)\,\langle c\eta(0), v\rangle_H$$

as well as

$$\int_0^T \Big(a_{r_{m_l}}(u_{m_l}(s), w(s)) + \langle c_{m_l}u_{m_l}(s), \ddot{w}(s)\rangle_H\Big)\mathrm{d}s = \int_0^T \langle f(s), w(s)\rangle_H \mathrm{d}s$$
$$+\, \phi(0)\,\langle c_{m_l}u_1, v\rangle_H - \dot{\phi}(0)\,\langle c_{m_l}u_0, v\rangle_H.$$

In the limit we conclude (16.7) since v and ϕ are arbitrary.

Thus, $u = \eta$ and the whole sequence $\{u_m\}$ converges weakly to u because all convergent subsequences of $\{u_m\}$ have the limit u. □

Remark 16.10 Note that we have actually validated the weak-⋆-to-weak continuity of F as a mapping from $L^\infty(\Omega)^2$ to $L^2([0,T],V)$. However, a sequence, which converges weakly in $L^2([0,T],V)$, converges, in particular, weakly in $L^2([0,T],H)$. This can be seen from the fact that $\langle v, w\rangle_{V'\times V} = \langle v, w\rangle_H$ in case $v \in H$.[4]

Proposition 16.11 *The map F as defined in* (15.9) *is compact.*

Proof Let $Q \subset \mathsf{D}(F)$ be bounded, that is, there exist $\mathrm{k}_- > 0$ with $\mathrm{k}_- \le c(\mathbf{x}) \le \mathrm{k}_+$ and $\mathrm{k}_- \le r(\mathbf{x}) \le \mathrm{k}_+$ for almost all $\mathbf{x} \in \Omega$. Let $F(c,r) = u \in X$ be the corresponding solution. We show that $F(Q)$ is relatively compact in $C([0,T],H)$ by the (general) theorem of Arzela-Ascoli. Indeed, for any $t \in [0,T]$ we have, in view of (16.5), that $\{u(t) : u \in F(Q)\} \subset \{v \in V : \|\nabla_{\mathbf{x}} v\|_{L^2(\Omega)^d} \le \hat{c}\}$ for some $\hat{c}$, and the latter set is relatively compact in $H = L^2(\Omega)$. Furthermore, $F(Q)$ is equi-continuous because, for $u \in F(Q)$,

$$\begin{aligned}\|u(t_2) - u(t_1)\|_H &= \sup_{\|\psi\|_H=1} \big\langle u(t_2) - u(t_1), \psi\big\rangle_H = \sup_{\|\psi\|_H=1} \int_{t_1}^{t_2} \frac{\mathrm{d}}{\mathrm{d}s}\big\langle u(s), \psi\big\rangle_H \,\mathrm{d}s \\ &= \sup_{\|\psi\|_H=1} \int_{t_1}^{t_2} \big\langle \dot{u}(s), \psi\big\rangle_H \,\mathrm{d}s \le |t_2 - t_1| \, \|\dot{u}\|_{C([0,T],H)}\end{aligned}$$

and $\|\dot{u}\|_{C([0,T],H)}$ is uniformly bounded for $(c,r) \in Q$ due to (16.5).

The continuous embedding $C([0,T],H) \hookrightarrow L^2([0,T],H)$ finishes the proof. □

Theorem 16.12 *Let the mapping F be defined in* (15.9). *Then, the inverse problem of seismic imaging $F(\cdot,\cdot) = u$ is locally ill-posed at any point $(c_0, r_0) \in \mathsf{D}(F)$ satisfying $F(c_0, r_0) = u$.*

Proof See the proof of Theorem 16.8. □

16.2 Linearization and Ill-Posedness

In order to linearize nonlinear problems we introduce the Fréchet-derivative.

[4] The three Hilbert spaces V, H, and V' constitute a Gelfand triple: the spaces are continuously embedded, $V \hookrightarrow H \hookrightarrow V'$, and the duality pairing $\langle\cdot,\cdot\rangle_{V'\times V}$ is an extension of the inner product in H.

Definition 16.13 Let X and Y be Banach spaces. A mapping $\Phi\colon \mathsf{D}(\Phi) \subset X \to Y$ is called **Fréchet-differentiable** at an inner point $x \in \mathsf{D}(\Phi)$, if there are a ball $B_r(x) \subset \mathsf{D}(\Phi)$ and a linear operator $A \in \mathcal{L}(X, Y)$ such that

$$\lim_{B_r(0)\ni h\to 0} \frac{\|\Phi(x+h) - \Phi(x) - Ah\|_Y}{\|h\|_X} = 0.$$

The operator $\Phi'(x) := A$ is called the **Fréchet-derivative** or the **linearization** of Φ at x. The mapping Φ is **Fréchet-differentiable** in the open set $U \subset \mathsf{D}(\Phi)$, if it has a Fréchet-derivative at all points of U. If the mapping $\Phi'\colon U \to \mathcal{L}(X, Y)$ is continuous, then Φ is **continuously Fréchet-differentiable** in U.

We present three examples. As a starter we consider a mapping between finite dimensional spaces.

Example 16.14 (Fréchet-Derivative of the Matrix Inverse) Let $\mathrm{GL}(n)$ be the general linear group, that is, the open subset of $\mathbb{R}^{n\times n}$ containing the regular matrices. Define $\Phi\colon \mathrm{GL}(n) \subset \mathbb{R}^{n\times n} \to \mathbb{R}^{n\times n}$ by $A \mapsto A^{-1}$ and equip $\mathbb{R}^{n\times n}$ with the spectral norm $\|\cdot\|$. The Fréchet-derivative $\Phi'\colon \mathrm{GL}(n) \to \mathcal{L}(\mathbb{R}^{n\times n})$ is given by $\Phi'(A)H = -A^{-1}HA^{-1}$. Indeed, let $H \in \mathbb{R}^{n\times n}$ with $\|H\| < \|A^{-1}\|^{-1}$. Then, $A + H \in \mathrm{GL}(n)$ and

$$(A+H)^{-1} - A^{-1} + A^{-1}HA^{-1} = (A^{-1}H)^2(A+H)^{-1}$$

from which the assertion follows.

Example 16.15 (Fréchet-Derivative of the EIT-Operator for the Continuum Model) Recall the EIT forward mapping $F\colon \sigma \mapsto \Lambda_\sigma$ from (15.3) where $\Lambda_\sigma\colon f \mapsto u|_{\partial\Omega}$ and $u \in H^1_\diamond(\Omega)$ solves

$$a_\sigma(u, v) = \int_{\partial\Omega} f v \,\mathrm{d}S \quad \forall v \in H^1_\diamond(\Omega). \tag{16.8}$$

Here, $a_\sigma(w, v) = \int_\Omega \sigma \nabla w \nabla v \,\mathrm{d}x$. The derivative $F'(\sigma) \in \mathcal{L}\big(L^\infty(\Omega), \mathcal{L}(L^2_\diamond(\partial\Omega))\big)$ is given by

$$F'(\sigma)[h]f = u'|_{\partial\Omega}$$

where $u' \in H^1_\diamond(\Omega)$ uniquely solves

$$a_\sigma(u', v) = -a_h(u, v) \quad \forall v \in H^1_\diamond(\Omega)$$

with u being the solution of (16.8).

Example 16.16 (Fréchet-Derivative of the Seismic Tomography Operator) Let $u_0 = u_1 = 0$, $f \in C^2([0,T], H)$ with $f(0) = 0$. Recall the seismic tomography operator $F: (c, r) \mapsto u$ from (15.9) where $u \in V$ solves the wave equation

$$\mathcal{A}_{c,r}(u, v) = \int_0^T \langle f(t), v(t)\rangle_H \, \mathrm{d}t \quad \forall v \in C_0^\infty([0,T], V).$$

Here, $\mathcal{A}_{c,r}(w, v) := \int_0^T \big(a_r(w(t), v(t)) - \langle c\dot{w}(t), \dot{v}(t)\rangle_H\big) \, \mathrm{d}t$. Let $h_1, h_2 \in L^\infty(\Omega)$. Then,

$$F'(c, r)\begin{pmatrix} h_1 \\ h_2 \end{pmatrix} = u'$$

where $u' \in L^2([0,T], V)$ uniquely solves

$$\mathcal{A}_{c,r}(u', v) = -\mathcal{A}_{h_1,h_2}(F(c,r), v) \quad \forall v \in C_0^\infty([0,T], V)$$

with $u'(0) = \dot{u}'(0) = 0$.

In the sequel we need a mean value theorem for Fréchet-derivatives. To this end we introduce the concept of the Riemann integral for functions $v: [a, b] \to Y$ with values in a Banach space. The integral

$$\int_a^b v(t) \, \mathrm{d}t \in Y$$

is defined via Riemann sums in the usual way and it is well defined for continuous v. Moreover,

$$\left\| \int_a^b v(t) \, \mathrm{d}t \right\|_Y \le \int_a^b \|v(t)\|_Y \, \mathrm{d}t, \tag{16.9}$$

see, for instance, [166, Chap. 3].

Theorem 16.17 (Mean-Value Theorem) *Let $\Phi: \mathsf{D}(\Phi) \subset X \to Y$ be continuously Fréchet-differentiable in $\mathsf{D}(\Phi)$. Let f, $f + h$, and their connecting line segment be in $\mathsf{D}(\Phi)$. Then,*

$$\Phi(f + h) - \Phi(f) = \int_0^1 \Phi'(f + th)h \, \mathrm{d}t.$$

For a proof see again [166].

In this section we consider the following question: Can the local ill-posedness of

$$\Phi(x) = y \tag{16.10}$$

at x^+ be seen from the ill-posedness of

$$\Phi'(x^+)x = y \tag{16.11}$$

and vice versa?

Theorem 16.18 *Let X and Y be Banach spaces. Let $\Phi\colon \mathsf{D}(\Phi) \subset X \to Y$ be Fréchet-differentiable at $f^+ \in \operatorname{int}(\mathsf{D}(\Phi))$ with a local Hölder-continuous Fréchet-derivative, that is, there are positive constants α, L, r with*

$$\|\Phi'(f) - \Phi'(f^+)\|_{\mathcal{L}(X,Y)} \leq L\,\|f - f^+\|_X^\alpha \quad \textit{for all } f \in B_r(f^+) \subset \mathsf{D}(\Phi). \tag{16.12}$$

If (16.10) is locally ill-posed at f^+ then (16.11) is locally ill-posed anywhere, i.e., $\mathsf{R}(\Phi'(f^+)) \neq \overline{\mathsf{R}(\Phi'(f^+))}$ or $\mathsf{N}(\Phi'(f^+)) \neq \{0\}$.

Proof We employ a contradiction argument. To this end assume (16.10) to be locally ill-posed at f^+ but (16.11) to be locally well-posed, that is, $\mathsf{R}(A) = \overline{\mathsf{R}(A)}$ and $\mathsf{N}(A) = \{0\}$ with $A := \Phi'(f^+)$. The bounded inverse theorem yields that $A^{-1}\colon \mathsf{R}(A) \to X$ is well-defined and bounded. For any $f \in X$ we may write

$$f - f^+ = A^{-1}w \quad \text{with } w = A(f - f^+).$$

Choose $\mu \in]0, 1[$ such that

$$r^* := \frac{1}{\|A\|}\left(\frac{\mu\,(1+\alpha)}{L\,\|A^{-1}\|^{1+\alpha}}\right)^{1/\alpha} \leq r.$$

Then, $B_{r^*}(f^+) \subset \mathsf{D}(\Phi)$ and

$$\|w\|_Y^\alpha \leq \frac{\mu\,(1+\alpha)}{L\,\|A^{-1}\|^{1+\alpha}} \quad \text{for all } f \in B_{r^*}(f^+).$$

In the sequel we prove that

$$\|\Phi'(f^+)(f - f^+)\|_Y \leq \frac{1}{1-\mu}\,\|\Phi(f) - \Phi(f^+)\|_Y \quad \text{for all } f \in B_{r^*}(f^+) \tag{16.13}$$

which contradicts the assumed well-posedness of (16.11).

By the mean-value theorem,

$$E(f, f^+) := \Phi(f) - \Phi(f^+) - \Phi'(f^+)(f - f^+)$$
$$= \int_0^1 \left(\Phi'\left(f^+ + t(f - f^+)\right) - \Phi'(f^+)\right)(f - f^+)\,\mathrm{d}t \quad \text{for all } f \in B_{r^*}(f^+).$$

In view of (16.12) we estimate

$$\|E(f, f^+)\|_Y \le \frac{L}{1+\alpha}\|f - f^+\|_X^{1+\alpha} = \frac{L}{1+\alpha}\|A^{-1}w\|_X^{1+\alpha}$$
$$\le \frac{L}{1+\alpha}\|A^{-1}\|^{1+\alpha}\,\|w\|_Y^{\alpha}\,\|A(f - f^+)\|_Y$$
$$\le \mu\,\|A(f - f^+)\|_Y.$$

Using the triangle inequality yields

$$\|A(f - f^+)\|_Y \le \|E(f, f^+)\|_Y + \|\Phi(f) - \Phi(f^+)\|_Y$$
$$\le \mu\,\|A(f - f^+)\|_Y + \|\Phi(f) - \Phi(f^+)\|_Y$$

implying (16.13). □

The converse of Theorem 16.18 does not hold in general as the following example indicates.

Example 16.19 Let $\mathsf{D} = \{\xi \in \ell^2(\mathbb{N}) : \xi_i \ge 0,\ i \in \mathbb{N}\}$ and define

$$\Phi\colon \mathsf{D} \subset \ell^2(\mathbb{N}) \to \ell^1(\mathbb{N}), \quad \Phi(\xi)_i := \xi_i^2,\ i \in \mathbb{N}.$$

The mapping Φ is continuous and continuously invertible on its range $\mathsf{R}(\Phi) = \{\eta \in \ell^1 : \eta_i \ge 0,\ i \in \mathbb{N}\}$ by $\Phi^{-1}(\eta)_i = \sqrt{\eta_i}$:

$$\|\Phi(\xi) - \Phi(\eta)\|_{\ell^1} \le \|\xi + \eta\|_{\ell^2}\,\|\xi - \eta\|_{\ell^2} \quad \text{for all } \xi, \eta \in \mathsf{D}$$

and

$$\|\Phi^{-1}(\xi) - \Phi^{-1}(\eta)\|_{\ell^2} \le \|\xi - \eta\|_{\ell^1} \quad \text{for all } \xi, \eta \in \mathsf{R}(\Phi).$$

Thus, the nonlinear problem $\Phi(\xi) = v$ is locally well-posed in D. The operator $\Phi'(\xi) \in \mathcal{L}(\ell^2, \ell^1)$,

$$\big(\Phi'(\xi)\eta\big)_i = 2\,\xi_i\,\eta_i,$$

is the (right-sided[5]) Fréchet-derivative of Φ in $\xi \in \mathsf{D}$.

Let $\{e_k\}_{k\in\mathbb{N}}$ be the canonical basis, $(e_k)_i = \delta_{k,i}$:

$$\|e_k\|_{\ell^2} = 1 \quad \text{as well as} \quad \lim_{k\to\infty} \|\Phi'(\xi)e_k\|_{\ell^1} = \lim_{k\to\infty} 2\,\xi_k = 0.$$

Thus, the linear problem $\Phi'(\xi)x = y$ is locally ill-posed for any $\xi \in \mathsf{D}$.

Definition 16.20 The mapping $\Phi\colon \mathsf{D}(\Phi) \subset X \to Y$ satisfies the **tangential cone condition**[6] (TCC) locally at $x^+ \in \mathsf{D}(\Phi)$ if there are positive constants ϱ and ω with $\omega < 1$ such that

$$\|\Phi(v) - \Phi(u) - \Phi'(u)(v-u)\|_Y \le \omega\|\Phi(v) - \Phi(u)\|_Y \quad \text{for all } v, u \in B_\varrho(x^+) \cap \mathsf{D}(\Phi).$$

If ω of Definition 16.20 is smaller than $1/2$ then the TCC is essentially equivalent to

$$\|\Phi(v) - \Phi(u) - \Phi'(u)(v-u)\|_Y \le L\|\Phi'(u)(v-u)\|_Y \quad \text{for all } v, u \in B_\varrho(x^+) \cap \mathsf{D}(\Phi)$$

where $L < 1$. A schematic illustration of this condition is displayed in Fig. 16.1. As can be guessed from the sketch, the TCC holds on finite dimensional spaces at x^+ if $\Phi'(x^+)$ is one-to-one and Φ' is (Hölder-)continuous at x^+, see [55, Lem. C.1].

The proof of the following theorem is left as an exercise.

Theorem 16.21 *Let the TCC hold for $\Phi\colon \mathsf{D}(\Phi) \subset X \to Y$ at $x^+ \in \mathsf{D}(\Phi)$. Then, the following statements are equivalent:*

- *$\Phi(x) = y$ is locally ill-posed at x^+,*
- *$\Phi'(x^+)x = y$ is locally ill-posed anywhere.*

[5] D has no interior points: $\operatorname{int}(\mathsf{D}) = \emptyset$.

[6] Sometimes called weak Scherzer condition since it appeared for the first time in [150].

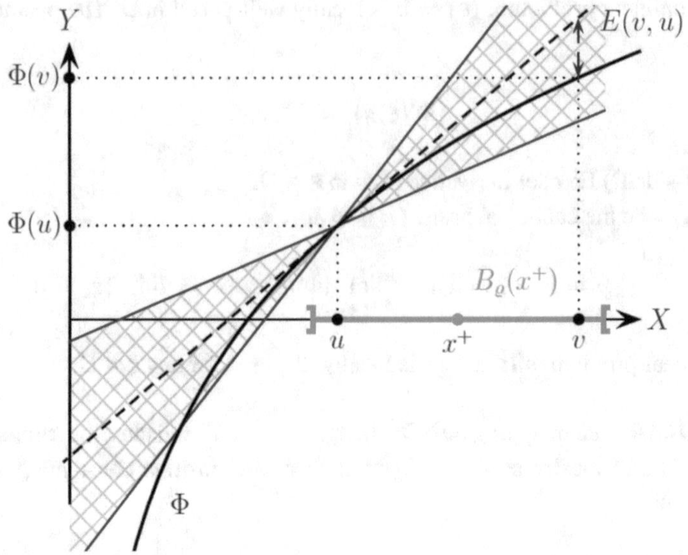

Fig. 16.1 Schematic illustration of the TCC about $x^+ \in \mathrm{D}(\Phi) \subset X$. The red-hatched area is the tangential cone at $u \in B_\varrho(x^+)$ in which the linearization error $E(v, u) = \Phi(v) - \Phi(u) - \Phi'(u)(v-u)$ has to be contained for all $v \in B_{\varrho(x^+)}$

17 Regularization of Linear Ill-Posed Problems in Hilbert Spaces

Let X and Y be Hilbert spaces. Further, let $T \in \mathcal{L}(X, Y)$ and assume $\mathsf{R}(T)$ to be non-closed in Y. Here we recall the regularization of the ill-posed problem $Tx = y^\delta$ in a nutshell. For more details see standard textbooks, e.g., [58, 108, 121, 148].

The task is to find a meaningful approximation to $x^+ \in \mathsf{N}(T)^\perp$ knowing only $y^\delta \in Y$ with $\|Tx^+ - y^\delta\|_Y \leq \delta$ and the noise level $\delta \geq 0$.

The generalized inverse $T^+ \colon \mathsf{R}(T) \oplus \mathsf{R}(T)^\perp \subset Y \to X$ is unbounded (discontinuous) and should therefore not be applied to y^δ even if y^δ is in $\mathsf{R}(T) \oplus \mathsf{R}(T)^\perp$. A remedy for the unboundedness of T^+ are regularization schemes which we define below.

We call a family $\{\mathcal{R}_\ell\}_{\ell \in \mathbb{N}_0}$ with $\mathcal{R}_\ell \colon Y \to X$ continuous and $\mathcal{R}_\ell 0 = 0$ a family of *regularizing operators* for T^+ if

$$\lim_{\ell \to \infty} \mathcal{R}_\ell y = T^+ y \quad \text{for all } y \in \mathsf{R}(T) \oplus \mathsf{R}(T)^\perp.$$

In the following examples of regularizing operators all operators are linear.

Example 17.1 (Tikhonov-Phillips) Here, $\mathcal{R}_\ell = (T^*T + \alpha_\ell I)^{-1} T^*$ where $\lim_{\ell \to \infty} \alpha_\ell = 0$ (strictly decreasing).

Example 17.2 (Landweber Iteration) Here, $\mathcal{R}_\ell y := u_\ell$ where $u_0 = 0$ and

$$u_{k+1} = u_k + \vartheta T^*(y - Tu_k), \quad 0 < \vartheta < \frac{2}{\|T\|^2}.$$

W. Dörfler et al., *Wave Phenomena*, Oberwolfach Seminars 49,
https://doi.org/10.1007/978-3-031-05793-9_17

We have that

$$\mathcal{R}_\ell = q_\ell(T^*T)T^* \quad \text{with} \quad q_\ell(t) = \vartheta \sum_{j=0}^{\ell-1} (1 - \vartheta t)^j \in \Pi_{\ell-1}.$$

Example 17.3 (Iterated Tikhonov-Phillips/Implicit Iteration) Here, $\mathcal{R}_\ell y := u_\ell$ where $u_0 = 0$ and

$$u_\ell = (T^*T + \alpha_\ell I)^{-1}(T^*y + \alpha_\ell u_{\ell-1})$$

where $\{\alpha_\ell\} \subset [\alpha_{\min}, \alpha_{\max}]$ and $0 < \alpha_{\min} < \alpha_{\max} < \infty$.

There are also examples where all operators are non-linear.

Example 17.4 (Steepest Descent) Here, $\mathcal{R}_\ell y := u_\ell$ where $u_0 = 0$ and

$$u_{k+1} = u_k + \vartheta_k T^*(y - Tu_k), \quad \vartheta_k = \frac{\|T^*(y - Tu_k)\|_X^2}{\|TT^*(y - Tu_k)\|_Y^2},$$

Note that

$$\vartheta_k = \operatorname{argmin}\big\{\|y - T(u_k + \lambda T^*(y - Tu_k))\|_Y : \lambda \in \mathbb{R}\big\}.$$

Here,

$$\mathcal{R}_\ell y = q_\ell(T^*T, y)T^*y \quad \text{for a } q_\ell(\cdot, y) \in \Pi_{\ell-1}.$$

Example 17.5 (Conjugate Gradients (cg)) Here, $\mathcal{R}_\ell y := q_\ell(T^*T, y)T^*y$ where

$$q_\ell(\cdot, y) = \operatorname{argmin}\big\{\|y - Tp(T^*T)T^*y\|_Y : p \in \Pi_{\ell-1}\big\}.$$

Note that $u_\ell = \mathcal{R}_\ell y$ can be computed iteratively from $u_{\ell-1}$.

All regularizing operators from above examples satisfy

$$\lim_{\ell\to\infty} \mathcal{R}_\ell y = \begin{cases} T^+y & : \quad y \in \mathsf{R}(T) \oplus \mathsf{R}(T)^\perp, \\ \infty & : \quad \text{otherwise.} \end{cases}$$

Given noisy data y^δ we have to pick one index $\ell^\star = \ell^\star(\delta, y^\delta)$ such that $\mathcal{R}_{\ell^\star} y^\delta$ is a good approximation to T^+y.

Definition 17.6 Let $\{\mathcal{R}_\ell\}_{\ell\in\mathbb{N}_0}$ be a family of regularizing operators.

If there is a mapping $\gamma :]0,\infty[\times Y \to \mathbb{N}_0$ such that

$$\sup\left\{\left\|x^+ - \mathcal{R}_{\gamma(\delta,y^\delta)}y^\delta\right\|_X : y^\delta \in Y,\ \|Tx^+ - y^\delta\|_Y \le \delta\right\} \longrightarrow 0 \quad \text{as } \delta \to 0$$

for all $x^+ \in \mathsf{N}(T)^\perp$, then the pair $(\{\mathcal{R}_\ell\}_{\ell\in\mathbb{N}_0}, \gamma)$ is a **regularization scheme** for T^+. The mapping γ is called a **parameter choice rule**.

Morozov's discrepancy principle is a popular parameter choice rule: Choose $\gamma(\delta, y^\delta)$ such that

$$\left\|T\mathcal{R}_{\gamma(\delta,y^\delta)}y^\delta - y^\delta\right\|_Y \approx \delta.$$

For instance,

$$\gamma(\delta, y^\delta) := \min\left\{\ell \in \mathbb{N}_0 : \left\|T\mathcal{R}_\ell y^\delta - y^\delta\right\|_Y \le \tau\,\delta\right\}$$

where $\tau > 1$.

All families $\{\mathcal{R}_\ell\}_{\ell\in\mathbb{N}_0}$ from above examples furnished with the discrepancy principle yield regularization schemes.

Theorem 17.7 (Error Reducing Property [38]) *Let $\{u_k\}$, $u_0 = 0$, be generated by the Landweber method with $0 < \vartheta < 1/\|T\|^2$ or the steepest descent method w.r.t. $T \in \mathcal{L}(X, Y)$ and $0 \ne y \in Y$. For $x \in X$ we have that*

$$\|y - Tu_k\|_Y \ge 2\|y - Tx\|_Y \quad \Longrightarrow \quad \|u_{k+1} - x\|_X < \|u_k - x\|_X.$$

A similar result holds for the conjugate gradient method.

Proof Let $r_k := y - Tu_k$. From

$$\|u_{k+1} - x\|_X^2 - \|u_k - x\|_X^2 = \vartheta_k\big(2\langle y - Tx, r_k\rangle_Y - \|r_k\|^2 - \langle r_{k+1}, r_k\rangle_Y\big)$$

and $\langle r_{k+1}, r_k\rangle_Y > 0$ we find that

$$\|u_{k+1} - x\|_X^2 - \|u_k - x\|_X^2 < \vartheta_k\|r_k\|_Y \underbrace{\big(2\|y - Tx\|_Y - \|r_k\|_Y\big)}_{\le 0}.$$

The proof for the cg method is more involved and is given in [118]. □

The above theorem reveals the discrepancy principle to be especially suited to stop the Landweber, steepest descent, and cg methods.

Newton-Like Solvers for Non-linear Ill-Posed Problems

18

Let $F: \mathsf{D}(F) \subset X \to Y$ act continuously Fréchet-differentiable between the Hilbert spaces X and Y. We like to solve

$$F(x) = y^\delta$$

where $\|y - y^\delta\| \leq \delta$, $y = F(x^+)$, and $F(\cdot) = y$ is locally ill-posed at $x^+ \in \mathsf{D}(F)$.

Now we will derive an iterative procedure to approximate x^+. Assume x_n is already an approximation which we would like to improve by adding a correction s_n: $x_{n+1} = x_n + s_n$. Naturally, we want s_n to approximate the error $e_n := x^+ - x_n$ (exact Newton step) which satisfies the linear equation ($A_n := F'(x_n)$)

$$A_n e_n = y - F(x_n) - E(x^+, x_n) =: b_n$$

where

$$E(u, v) = F(u) - F(v) - F'(v)(u - v)$$

is the linearization error. The exact right hand side b_n is unknown. Therefore, we determine s_n from

$$A_n s = b_n^\delta \quad \text{with } b_n^\delta := y^\delta - F(x_n). \tag{18.1}$$

Observe that (18.1) is ill-posed under (16.12), see Theorem 16.18. We, hence, apply regularization operators to stably solve (18.1):

$$s_n = s_{n,\ell} = s_{\star,n} + \mathcal{R}_{n,\ell}\big(b_n^\delta - A_n s_{\star,n}\big)$$

W. Dörfler et al., *Wave Phenomena*, Oberwolfach Seminars 49,
https://doi.org/10.1007/978-3-031-05793-9_18

where $\{\mathcal{R}_{n,\ell}\}_\ell$ are regularizing operators for A_n^+ and $s_{\star,n} \in X$ is a starting guess. Recall that[1]

$$\lim_{\ell\to\infty} s_{n,\ell} = \begin{cases} A_n^+ b_n^\delta + P_{\mathsf{N}(A_n)} s_{\star,n} & : \ b_n^\delta \in \mathsf{R}(A_n) \oplus \mathsf{R}(A_n)^\perp, \\ \infty & : \quad \text{otherwise.} \end{cases}$$

Our generic Newton iteration accordingly reads

$$x_{n+1} = x_n + s_{\star,n} + \mathcal{R}_{n,m_n}\big(b_n^\delta - A_n s_{\star,n}\big)$$

To set up a specific scheme requires the choice of $\{\mathcal{R}_{n,\ell}\}$, the choice of $s_{\star,n}$, the choice of m_n, and the choice of a stopping criterion for the Newton iteration.

Here are some examples.

Example 18.1 (Nonlinear Landweber Method) This iteration is given by

$$x_{n+1} = x_n + A_n^* b_n^\delta, \qquad \|A_n\| \le 1.$$

It can be derived from the generic scheme when the choices are $\mathcal{R}_{n,\ell} = q_\ell(A_n^* A_n) A_n^*$ where $q_\ell(t) = \sum_{j=0}^{\ell-1}(1-t)^j$, $m_n = 1$, $s_{\star,n} = 0$.

Example 18.2 (Nonlinear Steepest Descent Method) The iteration

$$x_{n+1} = x_n + \lambda_n A_n^* b_n^\delta, \qquad \lambda_n = \frac{\|A_n^* b_n^\delta\|_X^2}{\|A_n A_n^* b_n^\delta\|_Y^2},$$

fits the generic scheme from above with $\mathcal{R}_{n,\ell} = q_\ell(A_n^* A_n, b_n^\delta) A_n^*$ where $q_\ell(\cdot, b_n^\delta) \in \Pi_{\ell-1}$, $m_n = 1$, $s_{\star,n} = 0$.

Example 18.3 (Iteratively Regularized Gauß-Newton Methods) . Here,

$$x_{n+1} = x_0 + \mathcal{R}_{n,m_n}\big(b_n^\delta - A_n(x_0 - x_n)\big)$$

where $\mathcal{R}_{n,\ell} = \big(A_n^* A_n + \alpha_\ell I\big)^{-1} A_n^*$ (in the original version), $\lim_{\ell\to\infty}\alpha_\ell = 0$ (strictly decreasing), $\{m_n\}$ is chosen a priori (strictly increasing), $s_{\star,n} = x_0 - x_n$.

Please see [104] for more details on the three examples from above as well as references to the original literature.

[1] By $P_W : X \to X$ we denote the orthogonal projector onto the closed subspace $W \subset X$.

Example 18.4 (Inexact Newton Methods) These iterations are of the form

$$x_{n+1} = x_n + \mathcal{R}_{n,m_n} b_n^\delta$$

where $\{\mathcal{R}_{n,\ell}\}$ may be chosen from the following methods: Landweber, steepest descent, conjugate gradients, implicit iteration. Moreover, $\{m_n\}$ is chosen a posteriori (using x_n) to allow adaptivity. Further, $s_{\star,n} = 0$.

In this chapter we study the inexact Newton method `REGINN` (REGularization by INexact Newton methods), see Algorithm 1.

Algorithm 1 `REGINN`

Input: x_N; (y^δ, δ); F; F'; $\{\mu_n\} \subset]0, 1[$; $\tau > 0$;
Output: x_N with $\|y^\delta - F(x_N)\| \le \tau\delta$;
  $n := 0$; $x_0 := x_N$;
  while $\|b_n^\delta\| > \tau\delta$ **do**
    $m := 0$; $s_{n,0} := 0$;
    repeat
      $m := m + 1$;
      $s_{n,m} := \mathcal{R}_{n,m} b_n^\delta$;
    until $\|b_n^\delta - A_n s_{n,m}\|_Y < \mu_n \|b_n^\delta\|_Y$
    $x_{n+1} := x_n + s_{n,m}$;
    $n := n + 1$;
  end while
  $x_N := x_n$;

Remark 18.5 The "while"-loop in Algorithm `REGINN` realizes the Newton or outer iteration whereas the "repeat"-loop computes the Newton step and is called inner iteration.

Observe that, if `REGINN` is well-defined and terminates then

$$m_n = \min\left\{m \in \mathbb{N} : \|A_n s_{n,m} - b_n^\delta\|_Y < \mu_n \|b_n^\delta\|_Y\right\} \tag{18.2}$$

and

$$\|y^\delta - F(x_N)\|_Y \le \tau\delta < \|y^\delta - F(x_n)\|_Y, \quad n = 0, \ldots, N-1. \tag{18.3}$$

18.1 Decreasing Error and Weak Convergence

For the analysis of REGINN we require two properties of the regularizing sequence $\{s_{n,m}\}$, $s_{n,m} = \mathcal{R}_{n,m} b_n^\delta$. If the iterate x_n is well defined then

$$\lim_{m\to\infty} A_n s_{n,m} = P_{\overline{\mathsf{R}(A_n)}} b_n^\delta, \tag{18.4}$$

and there exist a $\Theta \geq 1$ such that

$$\|A_n s_{n,m}\|_Y \leq \Theta \|b_n^\delta\|_Y \quad \text{for all } m. \tag{18.5}$$

The following methods satisfy (18.4) as well as (18.5): Landweber, Tikhonov, implicit iteration, conjugate gradients (all with $\Theta = 1$), and steepest descent ($\Theta \leq 2$ proven but $\Theta = 1$ conjectured).

Lemma 18.6 *Let x_n be well defined and assume* (18.4) *as well as* $\|P_{\mathsf{R}(A_n)^\perp} b_n^\delta\|_Y < \|b_n^\delta\|_Y$. *Then, for any tolerance*

$$\mu_n \in \left]\frac{\|P_{\mathsf{R}(A_n)^\perp} b_n^\delta\|_Y}{\|b_n^\delta\|_Y}, 1\right] \tag{18.6}$$

the stopping index m_n (18.2) *is well defined.*

Proof By (18.4),

$$\lim_{m\to\infty} \frac{\|A_n s_{n,m} - b_n^\delta\|_Y}{\|b_n^\delta\|_Y} = \frac{\|P_{\mathsf{R}(A_n)^\perp} b_n^\delta\|_Y}{\|b_n^\delta\|_Y} < \mu_n$$

which completes the proof. □

If the assumption $\|P_{\mathsf{R}(A_n)^\perp} b_n^\delta\|_Y < \|b_n^\delta\|_Y$ of the above lemma is violated then REGINN fails (as well as other Newton schemes): under $\|P_{\mathsf{R}(A_n)^\perp} b_n^\delta\|_Y = \|b_n^\delta\|_Y$, that is, $P_{\mathsf{R}(A_n)} b_n^\delta = 0$, we have $s_{n,m} = 0$ for all m.

Lemma 18.7 *Let x_n be well defined with $\|P_{\mathsf{R}(A_n)^\perp} b_n^\delta\|_Y < \|b_n^\delta\|_Y$. Assume that μ_n satisfies* (18.6). *Then, s_{n,m_n} is a descent direction for $\varphi\colon \mathsf{D}(F) \subset X \to \mathbb{R}$, $\varphi(\cdot) := \frac{1}{2}\|y^\delta - F(\cdot)\|_Y^2$, at x_n:*[2]

$$\langle \varphi'(x_n), s_{n,m_n}\rangle_X < 0.$$

[2] For the ease of presentation we identify $\varphi'(x_n) \in X'$ with its Riesz-representer in X.

Proof By $\varphi'(\cdot) = -F'(\cdot)^*\big(y^\delta - F(\cdot)\big)$ we find that

$$\begin{aligned}\langle \varphi'(x_n), s_{n,m_n}\rangle_X &= \langle b_n^\delta, -A_n s_{n,m_n}\rangle_Y = -\|b_n^\delta\|_Y^2 + \langle b_n^\delta, b_n^\delta - A_n s_{n,m_n}\rangle_Y \\ &\le \|b_n^\delta\|_Y\big(\|b_n^\delta - A_n s_{n,m_n}\| - \|b_n^\delta\|_Y\big) \\ &< \|b_n^\delta\|_Y^2(\mu_n - 1) < 0\end{aligned}$$

and the lemma is verified. □

Remark 18.8 If the assumptions of above lemma hold true then

$$\varphi(x_n + \lambda s_{n,m_n}) < \varphi(x_n) \quad \text{for all positive } \lambda \text{ being small enough.}$$

This property explains the term "descent direction". Indeed, by Theorem 16.17,

$$\varphi(x_n + \lambda s_{n,m_n}) - \varphi(x_n) = \lambda \int_0^1 \langle \varphi'(x_n + t\lambda s_{n,m_n}), s_{n,m_n}\rangle_X \,\mathrm{d}t.$$

As φ' is continuous in $x_n \in \operatorname{int}\big(\mathsf{D}(F)\big)$, the integrand is negative for small $\lambda > 0$.

We need a further property of the regularizing sequence $\{s_{n,m}\}$:

$$\left.\begin{array}{l}\text{If } x_n \text{ is well defined then for any } m \in \{1,\dots,m_n\} \text{ there is a } v_{n,m-1} \in Y \text{ such that} \\ \qquad\qquad s_{n,m} = s_{n,m-1} + A_n^* v_{n,m-1}. \\ \text{Further, let there be a continuous and monotonically increasing function} \\ \Psi\colon \mathbb{R} \to \mathbb{R} \text{ with } t \le \Psi(t) \text{ for } t \in [0,1] \text{ such that if} \\ \qquad\qquad \gamma_n := \dfrac{\|b_n^\delta - A_n e_n\|_Y}{\|b_n^\delta\|_Y} < 1 \\ \text{then} \\ \|s_{n,m} - e_n\|_X^2 - \|s_{n,m-1} - e_n\|_X^2 \\ \qquad < C_{\mathrm{M}}\|b_n^\delta\|_Y\|v_{n,m-1}\|_Y\left(\Psi(\gamma_n) - \dfrac{\|b_n^\delta - A_n s_{n,m-1}\|_Y}{\|b_n^\delta\|_Y}\right) \\ \text{for } m = 1,\dots,m_n \text{ where } C_{\mathrm{M}} > 0 \text{ is a constant.}\end{array}\right\} \tag{18.7}$$

A direct consequence of (18.7) is monotonicity, i.e.,

$$\frac{\|b_n^\delta - A_n s_{n,m-1}\|_Y}{\|b_n^\delta\|_Y} \ge \Psi(\gamma_n) \implies \|s_{n,m} - e_n\|_X < \|s_{n,m-1} - e_n\|_X. \tag{18.8}$$

Recall that s_{n,m_n} should approximate $e_n = x^+ - x_n$.

The following methods satisfy (18.7):

- Landweber/steepest descent: $\Psi(t) = 2t$ (see proof of Theorem 17.7),
- implicit iteration: $\Psi(t) = 2Ct$ with a constant $C > 1$,
- conjugate gradients: $\Psi(t) = \sqrt{2t}$.

For our analysis we require the TCC in the following version

$$\begin{gathered} \|E(v, w)\|_Y \leq L \, \|F'(w)(v - w)\|_Y \text{ for one } L < 1 \\ \text{and for all } v, w \in B_r(x^+) \subset \mathsf{D}(F). \end{gathered} \tag{18.9}$$

Lemma 18.9 *Assume* (18.9) *to hold with* $L < 1/2$. *Then, there is a* $0 < \varrho \leq L/(1-L) < 1$ *such that*

$$\left\| P_{\mathsf{R}(F'(u))^\perp}(y - F(u)) \right\|_Y \leq \varrho \|y - F(u)\|_Y \quad \textit{for all } u \in B_r(x^+).$$

Proof We have that

$$\begin{aligned} \left\| P_{\mathsf{R}(F'(u))^\perp}(y - F(u)) \right\|_Y &= \left\| P_{\mathsf{R}(F'(u))^\perp}(y - F(u) - F'(u)(x^+ - u)) \right\|_Y \\ &\leq L \, \|F'(u)(x^+ - u)\|_Y . \end{aligned}$$

Further,

$$\begin{aligned} \|F'(u)(x^+ - u)\|_Y &\leq \|E(x^+, u)\|_Y + \|y - F(u)\|_Y \\ &\leq L \, \|F'(u)(x^+ - u)\|_Y + \|F(x^+) - F(u)\|_Y \end{aligned}$$

yielding first

$$\|F'(u)(x^+ - u)\|_Y \leq \frac{1}{1-L} \|y - F(u)\|_Y$$

and then the assertion. □

Theorem 18.10 *Assume* (18.4), (18.5), (18.7). *Additionally, assume* (18.9) *with* L *satisfying*[3]

$$\Psi\Big(\frac{L}{1-L}\Big) + \Theta L < \Lambda \quad \textit{for one } \Lambda < 1.$$

Further, define

$$\mu_{\min} := \Psi\Big(\Big(\frac{1}{\tau} + L\Big)\frac{1}{1-L}\Big)$$

and choose τ *so large that*

$$\mu_{\min} + \Theta L < \Lambda.$$

Restrict all tolerances $\{\mu_n\}$ *to* $\mu_n \in [\mu_{\min}, \Lambda - \Theta L]$ *and start with* $x_0 \in B_r(x^+)$.

Then, there exists an $N(\delta)$ *such that all iterates* $\{x_1, \ldots, x_{N(\delta)}\}$ *of* REGINN *are well defined and stay in* $B_r(x^+)$. *We even have a strictly monotone error reduction:*

$$\|x^+ - x_n\|_X < \|x^+ - x_{n-1}\|_X, \quad n = 1, \ldots, N(\delta).$$

Moreover, only the final iterate satisfies the discrepancy principle (18.3) *and the nonlinear residuals decrease linearly at the estimated rate*

$$\frac{\|y^\delta - F(x_{n+1})\|_Y}{\|y^\delta - F(x_n)\|_Y} < \mu_n + \theta_n L \le \Lambda, \quad n = 0, \ldots, N(\delta) - 1, \tag{18.10}$$

where $\theta_n = \|A_n s_{n,m_n}\|_Y / \|b_n^\delta\|_Y \le \Theta$.

Proof We use an inductive argument: Assume the iterates $x_1, \ldots, x_n$ to be well defined in $B_\rho(x^+)$. If $\|b_n^\delta\|_Y < \tau\delta$ REGINN will be stopped with $N(\delta) = n$. Otherwise, $\|b_n^\delta\|_Y \ge \tau\delta$ and $\mu_n \in [\mu_{\min}, \Lambda - \Theta L]$ will provide a new Newton step. Indeed, by Lemma 18.9 we see that

$$\frac{\|P_{\mathsf{R}(A_n)^\perp} b_n^\delta\|_Y}{\|b_n^\delta\|_Y} < \frac{1+\varrho}{\tau} + \varrho \overset{(18.7)}{\le} \Psi\Big(\frac{1+\varrho}{\tau} + \varrho\Big) \le \mu_{\min}$$

[3] As $\frac{L}{1-L} + L \le \frac{L}{1-L} + \Theta L \le \Psi(\frac{L}{1-L}) + \Theta L < 1$ we have the necessary condition $L < (3 - \sqrt{5})/2 \approx 0.38$.

where the latter estimate holds true due to $\varrho \leq L/(1-L)$ (Lemma 18.9) and the monotonicity of Ψ. By Lemma 18.6 the Newton step s_{n,m_n} and hence $x_{n+1} = x_n + s_{n,m_n} \in X$ are well defined.

Next we validate the monotone error reduction relying on (18.8). By (18.9), we have

$$\|b_n - b_n^\delta\|_Y \leq \delta + L\|b_n\|_Y \leq \frac{1}{\tau}\|b_n^\delta\|_Y + L\big(\|b_n - b_n^\delta\|_Y + \|b_n^\delta\|_Y\big)$$

yielding first

$$\gamma_n = \frac{\|b_n - b_n^\delta\|_Y}{\|b_n^\delta\|_Y} \leq \Big(\frac{1}{\tau} + L\Big)\frac{1}{1-L} \tag{18.11}$$

and then

$$\Psi(\gamma_n) \leq \mu_{\min} \leq \mu_n.$$

Accordingly, $\|b_n^\delta - A_n s_{n,m-1}\|_Y \geq \mu_{\min}\|b_n^\delta\|_Y$, $m = 1, \ldots, m_n$, and we have by repeatedly applying (18.8)

$$\begin{aligned} \|x^+ - x_{n+1}\|_X &= \|e_n - s_{n,m_n}\|_X \\ &< \|e_n - s_{n,m_n-1}\|_X < \|e_n - s_{n,m_n-2}\|_X \\ &< \cdots < \|e_n - s_{n,0}\|_X = \|e_n\|_X = \|x^+ - x_n\|_X. \end{aligned} \tag{18.12}$$

See [118] for a proof of the rate (18.10). □

Corollary 18.11 *Adopt all assumptions and notations of Theorem* 18.10. *Additionally let F be weakly sequentially closed*[4] *and let $\{\delta_j\}_{j\in\mathbb{N}}$ be a positive zero sequence.*

Then, any subsequence of $\{x_{N(\delta_j)}\}_{j\in\mathbb{N}}$ contains a subsequence which converges weakly to a solution of $F(x) = y$. The whole family $\{x_{N(\delta_j)}\}_{j\in\mathbb{N}}$ converges weakly to x^+ if x^+ is the unique solution of $F(x) = y$ in $B_r(x^+)$.

Proof Any subsequence of the bounded family $\{x_{N(\delta_j)}\}_{j\in\mathbb{N}} \subset B_r(x^+)$ is bounded and, therefore, has a weakly convergent subsequence. Let ξ be its weak limit. By

$$\|y - F(x_{N(\delta_j)})\|_Y \leq \|y - y^{\delta_j}\|_Y + \|y^{\delta_j} - F(x_{N(\delta_j)})\|_Y \leq (1+\tau)\delta_j$$

[4] Weakly sequentially closed: If $x_n \to \xi$ weakly in X and $F(x_n) \to \eta$ weakly in Y then $F(\xi) = \eta$.

the images under F of this weakly convergent subsequence converge (weakly) to y. Due to the weak closedness of F we have that $y = F(\xi)$. □

Remark 18.12

(a) We have that

$$N(\delta) < \frac{\ln\left(\tau\delta/\|b_0^\delta\|_Y\right)}{\ln\Lambda} + 1$$

which follows from $\tau\delta < \|b_{N(\delta)-1}\|_Y \leq \Lambda^{N(\delta)-1}\|b_0^\delta\|_Y$.

(b) We can get rid of assumption (18.5) at the price of slightly more restrictive constraints on L and μ_n. Indeed, for (18.10) we used that

$$\|A_n s_{n,m_n}\|_Y = \theta_n\|b_n^\delta\|_Y \overset{(18.5)}{\leq} \Theta\|b_n^\delta\|_Y.$$

Without (18.5) we always have

$$\|A_n s_{n,m_n}\|_Y \leq \|b_n^\delta\|_Y + \|A_n s_{n,m_n} - b_n^\delta\|_Y < (1+\mu_n)\|b_n^\delta\|_Y.$$

We finish this section with a comment on how to choose the tolerances $\{\mu_n\}$ in REGINN.

18.1.1 A Heuristic for Choosing the Tolerances

The goal in choosing the μ_n's is to keep the overall number $\sum_{n=0}^{N(\delta)-1} m_n$ of passes through the 'repeat'-loop of REGINN rather small to reduce the numerical effort. To this end we try to minimize $N(\delta)$. The relation, see (18.10),

$$\frac{\|y^\delta - F(x_{n+1})\|_Y}{\|y^\delta - F(x_n)\|_Y} \approx \mu_n + \Theta L$$

suggest the choice of small tolerances. However, the tolerances should not be too small to avoid noise amplification while solving the linearization (18.1). In the starting phase of algorithm REGINN the nonlinear defect will be relatively large and the 'repeat'-loop will terminate in spite of a small tolerance.

We therefore start with a small tolerance and increase it during the Newton iteration. An increase of the tolerance will be indicated when the number of passes through the 'repeat'-loop of two successive Newton steps increases significantly. The tolerances shall be decreased by a constant factor whenever the consecutive numbers of passes through the 'repeat'-loop drop.

We propose the following strategy: Initialize $\mu_{\text{start}} \in]0, 1[$, $\gamma \in]0, 1]$, and let $\mu_0 = \mu_1 := \mu_{\text{start}}$. For $n = 2, \ldots, N(\delta) - 1$ define

$$\widetilde{\mu}_n := \begin{cases} 1 - \frac{m_{n-2}}{m_{n-1}} (1 - \mu_{n-1}) & : \quad m_{n-1} \geq m_{n-2}, \\ \gamma \, \mu_{n-1} & : \quad \text{otherwise}, \end{cases}$$

and set

$$\mu_n := \mu_{\max} \max \left\{ \tau\delta / \|F(x_{n-1}) - y^\delta\|_Y, \, \widetilde{\mu}_n \right\}, \quad n = 0, 1, \ldots, N(\delta) - 1, \qquad (18.13)$$

where $\mu_{\max} \in]\mu_{\text{start}}, 1[$ bounds the μ_n's away from 1. The parameter $\mu_{\max}$ should be very close to 1, for instance, $\mu_{\max} = 0.999$ is reasonable. We know that the 'repeat'-loop may not terminate if the tolerance is too small. A rapid decrease of the tolerances should be avoided therefore. Restricting γ to the interval $[0.9, 1]$ has proved quite satisfactory in numerical experiments.

Taking the maximum in (18.13) is a *safeguarding* technique to prevent over-regularization of in the final Newton step. The idea is obvious: if the nonlinear defect of $x_{N(\delta)-1}$ is already close to $\tau\delta$ it is superfluous to reduce it in the last step possibly far beyond the desired level by the factor $\widetilde{\mu}_{N(\delta)-1}$. An alternative safeguarding technique bounds m_n by $m_n^{\max}$ where, for instance,

$$m_n^{\max} := \begin{cases} 1 & : \quad n = 0, \\ 2 & : \quad n = 1, \\ m_{n-2} + m_{n-1} & : \quad n \geq 2. \end{cases}$$

18.2 Convergence Without Noise

From now on we differ clearly between the noisy ($\delta > 0$) and the noise-free situations.

Quantities with a superscript δ refer to the noisy situation: x_n^δ, b_n^δ, A_n^δ, etc. Thus, x_n, $b_n^0 := y - F(x_n)$, and A_n originate from exact data y. Note that $b_n = y - F(x_n) - E(x^+, x_n)$. The starting vector is independent of the noise: $x_0 = x_0^\delta$.

REGINN is well defined in the noise-free situation when we set $\delta = 0$, $\tau = \infty$, and $\tau\delta = 0$. Except for termination, all results of Theorem 18.10 hold true for even smaller $\mu_{\min} = \Psi(\frac{L}{1-L})$. Further, $N(\delta) = \infty$ if there is no iterate x_n with $F(x_n) = y$ (premature termination).

Theorem 18.13 *Let* $\delta = 0$ *and adopt all assumptions of Theorem* 18.10. *Restrict* $\{\mu_n\}_n$ *to* $[\underline{\mu}, \Lambda - \Theta L]$ *where* $\underline{\mu} > \Psi(\frac{L}{1-L})$. *Then,* REGINN *either stops after finitely many iterations with a solution of* $F(\cdot) = y$ *or the generated sequence* $\{x_n\}_n \subset B_r(x^+)$ *converges in norm* *to a solution of* $F(\cdot) = y$. *If* x^+ *is the unique solution in* $B_r(x^+)$ *then*

$$\lim_{n\to\infty} \|x_n - x^+\|_X = 0\,.$$

Proof If `REGINN` terminates prematurely with x_N then $\|F(x_N) - y\| = 0$, hence, x_N is a solution.

Otherwise, $\{x_n\}_n$ is a Cauchy sequence which we show now. Let $\ell, p \in \mathbb{N}$ with $\ell > p$. We have that

$$\|x_\ell - x_p\|^2 = \|e_\ell - e_p\|^2 = 2\langle e_\ell - e_p, e_\ell\rangle + \|e_p\|^2 - \|e_\ell\|^2.$$

From the monotonicity $0 \le \|e_{n+1}\| < \|e_n\|$ we infer that $\lim_{n\to\infty} \|e_n\| = \xi \ge 0$. Thus,

$$\|e_p\|^2 - \|e_\ell\|^2 \xrightarrow{p\to\infty} 0. \tag{18.14}$$

Further,

$$e_\ell - e_p = -(x_\ell - x_p) = -\sum_{i=p}^{\ell-1} s_{i,m_i} \overset{(18.7)}{=} \sum_{i=p}^{\ell-1} A_i^*\widetilde{v}_i \quad \text{where } \widetilde{v}_i = -\sum_{k=1}^{m_i} v_{i,k-1}.$$

Hence,

$$\langle e_\ell - e_p, e_\ell\rangle = \sum_{i=p}^{\ell-1}\langle \widetilde{v}_i, A_i e_\ell\rangle \le \sum_{i=p}^{\ell-1} \|\widetilde{v}_i\|\, \|A_i e_\ell\|$$

We proceed with

$$\begin{aligned}
\|A_i e_\ell\| &= \|F'(x_i)(x^+ - x_\ell)\| \le \|F'(x_i)(x^+ - x_i)\| + \|F'(x_i)(x_\ell - x_i)\| \\
&\overset{(18.9)}{\le} \frac{1}{1-L}\big(\|y - F(x_i)\| + \|F(x_i) - F(x_\ell)\|\big) \\
&\le \frac{1}{1-L}\big(2\|y - F(x_i)\| + \underbrace{\|y - F(x_\ell)\|}_{\le \|y - F(x_i)\| \text{ as } i \le \ell, \text{ see } (18.10)}\big) \\
&\le \frac{3}{1-L}\, \|y - F(x_i)\|
\end{aligned}$$

yielding

$$\langle e_\ell - e_p, e_\ell\rangle \le \frac{3}{1-L}\sum_{i=p}^{\ell-1} \|\widetilde{v}_i\|\, \|y - F(x_i)\|.$$

For bounding the summands we apply (18.7) once again. Setting $n = i$ in the inequality in (18.7) and summing both sides from $m = 1$ to m_i we end up in

$$\underbrace{\| e_i - s_{i,m_i} \|^2}_{=e_{i+1}} - \|e_i\|^2 < C_{\mathrm{M}} \|b_i^0\| \sum_{m=1}^{m_i} \|v_{i,m-1}\| \big(\Psi(\gamma_i) - \mu_i\big)$$

where we have used that $\|b_i^0 - A_i s_{i,m-1}\| / \|b_i^0\| \geq \mu_i$. The latter displayed inequality leads to

$$\|y - F(x_i)\| \sum_{m=1}^{m_i} \|v_{i,m-1}\| < \frac{\|e_i\|^2 - \|e_{i+1}\|^2}{C_{\mathrm{M}}\big(\mu_i - \Psi(\gamma_i)\big)}$$

$$\leq \frac{\|e_i\|^2 - \|e_{i+1}\|^2}{C_{\mathrm{M}}\big(\underline{\mu} - \Psi(\frac{L}{1-L})\big)} =: \widetilde{C}_{\mathrm{M}}\big(\|e_i\|^2 - \|e_{i+1}\|^2\big)$$

where the final estimate holds since $\mu_i \geq \underline{\mu} > \Psi(\frac{L}{1-L}) \overset{(18.11)}{\geq} \Psi(\gamma_i)$. We get

$$\|\widetilde{v}_i\| \, \|y - F(x_i)\| \leq \sum_{m=1}^{m_i} \|v_{i,m-1}\| \, \|y - F(x_i)\| \leq \widetilde{C}_{\mathrm{M}}\big(\|e_i\|^2 - \|e_{i+1}\|^2\big)$$

and then

$$\langle e_\ell - e_p, e_\ell \rangle \leq \frac{3\,\widetilde{C}_{\mathrm{M}}}{1 - L} \big(\|e_p\|^2 - \|e_\ell\|^2\big).$$

Finally,

$$\|x_\ell - x_p\|^2 \leq \Big(\frac{6\,\widetilde{C}_{\mathrm{M}}}{1 - L} + 1\Big)\big(\|e_p\|^2 - \|e_\ell\|^2\big) \xrightarrow[(18.14)]{p \to \infty} 0$$

which reveals $\{x_n\}$ to be a Cauchy sequence with limit $\widehat{x} \in B_r(x^+)$. The proof is finished by $0 = \lim_{n\to\infty} \|y - F(x_n)\| = \|y - F(\widehat{x})\|$. □

18.3 Regularization Property of REGINN

In this section we assume that x^+ is the unique solution $F(x) = y$ in $B_r(x^+)$.

To validate the regularization property of REGINN we follow a general scheme:

1. Show convergence in the noise-free setting: $\lim_{n\to\infty} \|x^+ - x_n\|_X = 0$. This has been established in the previous section.
2. Show a kind of stability for any $n \in \mathbb{N}_0$: $\lim_{\delta\to 0} x_n^\delta = \xi_n$ and $\lim_{n\to\infty} \|x^+ - \xi_n\|_X = 0$ (generically one would expect $\xi_n = x_n$).
3. Employ the triangle inequality and monotonicity of the REGINN iterates: For $m \leq N(\delta)$,

$$\|x^+ - x^\delta_{N(\delta)}\|_X \leq \|x^+ - x^\delta_m\|_X \leq \|x^+ - \xi_m\|_X + \|\xi_n - x^\delta_m\|_X.$$

For step 2 we introduce lists $\mathcal{X}_n$ with elements from $B_r(x^+)$, $n \in \mathbb{N}_0$. As we will see later $\mathcal{X}_n$ contains just all possible limits of x_n^δ for $\delta \to 0$.

Definition 18.14 Set $\mathcal{X}_0 := (x_0)$ and determine $\mathcal{X}_{n+1}$ from $\mathcal{X}_n$ as follows: for each $\xi \in \mathcal{X}_n$ compute the Newton step $s_{n,m_n(\xi)} = s_{n,m_n(\xi)}(\xi)$ by the repeat–loop of REGINN where, however, A_n is replaced by $F'(\xi)$ and b_n^δ by $y - F(\xi)$. If $y \neq F(\xi)$ then $\xi + s_{n,m_n(\xi)}(\xi)$ belongs to $\mathcal{X}_{n+1}$, otherwise include ξ into $\mathcal{X}_{n+1}$.

Further, if

$$\big\| F'(\xi) s_{n,m_n-i}(\xi) - \big(y - F(\xi)\big)\big\|_Y = \mu_n \|y - F(\xi)\|_Y \qquad \text{for } i = 1,\ldots,k_n < m_n = m_n(\xi) \tag{18.15}$$

then $\xi + s_{n,m_n-i}(\xi)$, $i = 1,\ldots,k_n$, belong to $\mathcal{X}_{n+1}$ as well.

We call ξ a **predecessor** of $\xi + s_{n,m_n-i}(\xi)$, $i = 1,\ldots,k_n$, and, in turn, call the latter **successors** of ξ.

Obviously, $x_n \in \mathcal{X}_n$ and generically, $\mathcal{X}_n$ will contain only this element. Assume that $\mathcal{X}_j = (x_j)$, $j = 0,\ldots,n$, and that $s_{n,m_n-1}(x_n)$ as well as $s_{n,m_n-2}(x_n)$ satisfy (18.15). Then, $\mathcal{X}_{n+1} = (x_{n+1}, x_n + s_{n,m_n-1}(x_n), x_n + s_{n,m_n-2}(x_n))$ and from now on all $\mathcal{X}_j$, $j \geq n+1$, will have three elements at least. By (18.12)

$$\|x^+ - \xi_{n+1}\|_X < \|x^+ - \xi_n\|_X \qquad \text{whenever } \xi_{n+1} \in \mathcal{X}_{n+1} \text{ is a successor of } \xi_n \in \mathcal{X}_n. \tag{18.16}$$

As $\mathcal{X}_{n+1}$ contains only successors of elements in $\mathcal{X}_n$ the whole sequence $\{\mathcal{X}_n\}_n$ comes with a tree structure. Figure 18.1 sketches such a possible tree. The bold dots indicate the sequence $\{x_n\}_n$ generated by REGINN with exact data ($\delta = 0$ and the standard stopping criterion for the repeat-loop).

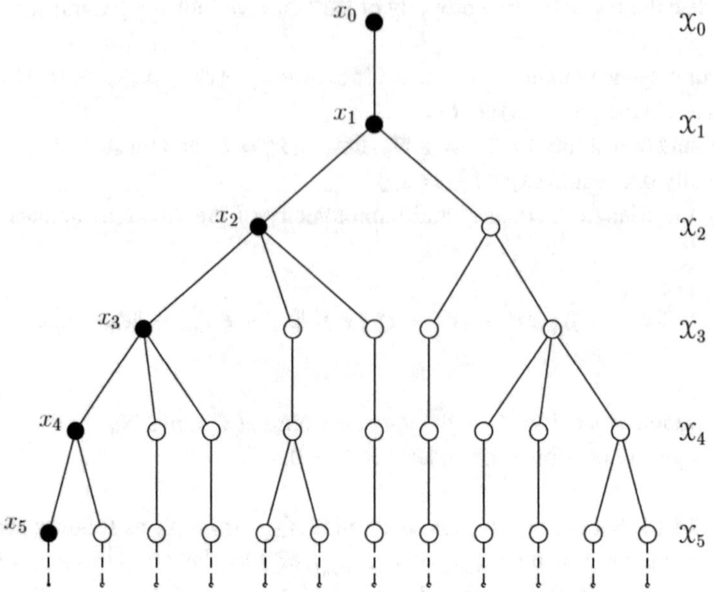

Fig. 18.1 A possible tree structure of the lists $\{\mathcal{X}_n\}_{n\in\mathbb{N}_0}$ from Definition 18.14

Next we require a stability property of the inner iteration of REGINN: Let $\{\delta_i\}_{i\in\mathbb{N}}$ be a positive zero sequence and let $\{x_n^{\delta_i}\}_{i\in\mathbb{N}}$ be well defined. Then,

$$\lim_{i\to\infty} x_n^{\delta_i} = \xi \in \mathcal{X}_n \quad \Longrightarrow \quad \lim_{i\to\infty} s_{n,k}^{\delta_i} = s_{n,k}(\xi), \quad k = 1, \ldots, m_n(\xi), \tag{18.17}$$

where $s_{n,k}^{\delta_i} = s_{n,k}(x_n^{\delta_i})$.

The following methods satisfy (18.17):

- For the Landweber and implicit iterations stability (18.17) can be shown by induction straightforwardly.
- For the nonlinear schemes steepest descent and conjugate gradients the situation is more delicate, see [151, Lemma 3.2] and [83, Lemma 3.4], respectively.

Lemma 18.15 *Let all assumptions of Theorem* 18.10 *hold true. Further, assume* (18.17). *If* $\lim_{i\to\infty}\delta_i = 0$ *and* $n \le N(\delta_i)$ *for* i *sufficiently large then* $\{x_n^{\delta_i}\}_{i\in\mathbb{N}}$ *splits into convergent subsequences, all of which converge to elements of* $\mathcal{X}_n$.

Proof We argue inductively. For $n = 0$ the assertion holds as $\mathcal{X}_0 = (x_0)$ and $x_0^\delta = x_0$.

Now assume that $\{x_n^{\delta_i}\}_{i\in\mathbb{N}}$, for an $n < N(\delta_i)$, i sufficiently large, splits into convergent subsequences with limits in $\mathcal{X}_n$. To simplify the notation, let $\{x_n^{\delta_i}\}_{i\in\mathbb{N}}$ itself converge to an element of $\mathcal{X}_n$, say, $\lim_{i\to\infty} x_n^{\delta_i} = \xi \in \mathcal{X}_n$. By (18.17) and continuity,

$$\lim_{i\to\infty} \|A_n^{\delta_i} s_{n,k}^{\delta_i} - b_n^{\delta_i}\|_Y = \|F'(\xi)s_{n,k}(\xi) - \widetilde{b}_n\|_Y, \quad k = 0, \dots, m_n(\xi),$$

where $\widetilde{b}_n = y - F(\xi)$. Let $y \neq F(\xi)$. As $\|F'(\xi)s_{n,m_n(\xi)}(\xi) - \widetilde{b}_n\|_Y < \mu_n\|\widetilde{b}_n\|_Y$ we get

$$\|A_n^{\delta_i} s_{n,m_n(\xi)}^{\delta_i} - b_n^{\delta_i}\|_Y < \mu_n\|b_n^{\delta_i}\|_Y \quad \text{for } i \text{ sufficiently large.}$$

This yields that $m_n(x_n^{\delta_i}) \le m_n(\xi)$ for i sufficiently large. In case of $\mu_n\|\widetilde{b}_n\|_Y < \|F'(\xi)s_{n,m_n(\xi)-1}(\xi) - \widetilde{b}_n\|_Y$ we also have that

$$\mu_n\|b_n^{\delta_i}\|_Y < \|A_n^{\delta_i} s_{n,m_n(\xi)-1}^{\delta_i} - b_n^{\delta_i}\|_Y \quad \text{for } i \text{ large enough.}$$

Thus, $m_n(\xi) - 1 < m_n(x_n^{\delta_i}) \le m_n(\xi)$, i.e., $m_n(x_n^{\delta_i}) = m_n(\xi)$ for i sufficiently large and

$$\lim_{i\to\infty} s_{n,m_n(x_n^{\delta_i})}(x_n^{\delta_i}) = s_{n,m_n(\xi)}(\xi).$$

Hence,

$$\lim_{i\to\infty} x_{n+1}^{\delta_i} = \lim_{i\to\infty} \left(x_n^{\delta_i} + s_{n,m_n(x_n^{\delta_i})}(x_n^{\delta_i})\right) = \xi + s_{n,m_n(\xi)}(\xi) \in \mathcal{X}_{n+1}.$$

In case (18.15) applies we have that $\mu_n\|\widetilde{b}_n\|_Y < \|F'(\xi)s_{n,m_n(\xi)-(k_n+1)}(\xi) - \widetilde{b}_n\|_Y$. Arguing as above we see that $m_n(\xi) - (k_n + 1) < m_n(x_n^{\delta_i})$ implying $m_n(\xi) - k_n \le m_n(x_n^{\delta_i}) \le m_n(\xi)$ for i large enough. Therefore, the set $\{m_n(x_n^{\delta_i})\}_{i\in\mathbb{N}}$ has limit points in $\{m_n(\xi) - k_n, \dots, m_n(\xi)\}$. In any case, all possible limit points of $x_{n+1}^{\delta_i} = x_n^{\delta_i} + s_{n,m_n(x_n^{\delta_i})}(x_n^{\delta_i})$ as $i \to \infty$ are in $\mathcal{X}_{n+1}$ by the very construction of this list.

It remains to consider the situation where $y = F(\xi)$. Since x^+ is the unique solution in $B_r(x^+)$ we obtain $x^+ = \xi$. Using the monotonicity $\|x^+ - x_{n+1}^{\delta_i}\|_X < \|x^+ - x_n^{\delta_i}\|_X$, see Theorem 18.10, the convergence $\lim_{i\to\infty} x_{n+1}^{\delta_i} = \xi \in \mathcal{X}_{n+1}$ follows immediately. □

The set

$$S := \left\{\xi \in X^{\mathbb{N}_0} \colon \ \xi_k \in \mathcal{X}_k \ \forall k \ \wedge \ \xi_{k+1} \text{ is a successor of } \xi_k\right\}$$

contains all sequences which can be generated following a continuous path in the tree of $\{\mathcal{X}_n\}_n$ from the root downwards, see Fig. 18.1. Each sequence in S can be generated by a run of REGINN with a slightly modified stopping criterion: The '<' sign is replaced by the '≤' sign. This modification does not affect the statement of Theorem 18.13, that is,

$$\lim_{n\to\infty} \xi_n = x^+ \quad \text{for any } \xi \in S. \tag{18.18}$$

However, the sequences in S converge even kind of uniformly to x^+, see Lemma 18.16 below.

We generalize the definitions of successor and predecessor: For $\xi \in S$ we call any ξ_k with $k > n$ a successor of ξ_n and any ξ_ℓ with $\ell < n$ a predecessor.

Lemma 18.16 *Let* $\{\xi^{(\ell)}\}_{\ell\in\mathbb{N}}$ *be a sequence in S. Under the assumptions of Theorem* 18.13, *for any* $\eta > 0$ *there is an* $M(\eta) \in \mathbb{N}_0$ *such that*

$$\|x^+ - \xi_n^{(\ell)}\|_X < \eta \quad \textit{for all } n \geq M(\eta) \textit{ and all } \ell \in \mathbb{N}. \tag{18.19}$$

Proof Assume (18.19) not to be true. Then, there are an $\eta > 0$ and sequences $\{n_j\}$ and $\{\ell_j\}$, where $\{n_j\}$ is strictly increasing, such that

$$\|x^+ - \xi_{n_j}^{(\ell_j)}\|_X \geq \eta \quad \text{for all } j \in \mathbb{N}.$$

We distinguish two cases:

1. Let $\{\ell_j\}$ be bounded. Then, $\{\ell_j\}$ has finitely many limit points. Let $\{\ell_{j_k}\}_k$ converge to ℓ^*, that is, $\ell^* = \ell_{j_k}$ for k large enough. Hence, $\|x^+ - \xi_{n_{j_k}}^{(\ell)}\|_X \geq \eta$ which contradicts (18.18).
2. Let $\{\ell_j\}$ be unbounded. Without loss of generality we may consider $\{\ell_j\}$ to be strictly increasing. We rearrange, that is, relabel the sequences in $\{\xi^{(i)}\}_{i\in\mathbb{N}}$ such that

$$\|x^+ - \xi_{n_\ell}^{(\ell)}\|_X \geq \eta \quad \text{for all } \ell \in \mathbb{N}. \tag{18.20}$$

As $\mathcal{X}_{n_1}$ is finite there are infinitely many $\ell \in \mathbb{N}$ such that $\xi_{n_1}^{(\ell)} = \rho_{n_1} \in \mathcal{X}_{n_1}$. We collect those ℓ in $\mathcal{L}_1 \subset \mathbb{N}$. As $\mathcal{X}_{n_2}$ is finite too there are infinitely many $\ell \in \mathcal{L}_1 \backslash \{1\}$ such that $\xi_{n_2}^{(\ell)} = \rho_{n_2} \in \mathcal{X}_{n_2}$. We collect those ℓ in $\mathcal{L}_2 \subset \mathbb{N}\backslash\{1\}$. Since $\mathcal{L}_2 \subset \mathcal{L}_1$, ρ_{n_2} is a successor of ρ_{n_1}.

Proceeding inductively we find $\{\rho_{n_k}\}_k$, $\rho_{n_k} \in \mathcal{X}_{n_k}$, and $\{\mathcal{L}_k\}_k$ with $\mathcal{L}_{k+1} \subset \mathcal{L}_k$, $\mathcal{L}_k \subset \mathbb{N}\backslash\{1, \ldots, k-1\}$ unbounded, such that

$$\xi_{n_k}^{(\ell)} = \rho_{n_k} \quad \text{for all } \ell \in \mathcal{L}_k. \tag{18.21}$$

Since $\mathcal{L}_{k+1} \subset \mathcal{L}_k$, $\rho_{n_{k+1}}$ is a successor of ρ_{n_k}. By (18.18), $\lim_{k\to\infty} \rho_{n_k} = x^+$, that is, there exists a $K = K(\eta)$:

$$\|x^+ - \rho_{n_k}\|_X < \eta \quad \text{for all } k \geq K. \tag{18.22}$$

For $\ell \in \mathcal{L}_K$ fixed, the errors $\|x^+ - \xi_{n_r}^{(\ell)}\|_X$ are decreasing in $r \in \mathbb{N}$, see (18.16). Since $\ell \geq K$ we have

$$\|x^+ - \xi_{n_\ell}^{(\ell)}\|_X \leq \|x^+ - \xi_{n_K}^{(\ell)}\|_X \overset{(18.21)}{=} \|x^+ - \rho_{n_K}\|_X \overset{(18.22)}{<} \eta$$

contradicting (18.20). □

Theorem 18.17 *Adopt all assumptions of Theorem* 18.10 *and let x^+ be the only solution of $F(x) = y$ in $B_r(x^+)$. Then,*

$$\lim_{\delta\to 0} \|x^+ - x_{N(\delta)}^\delta\|_X = 0.$$

Proof Let $\{\delta_j\}_{j\in\mathbb{N}}$ be a positive zero sequence. Set $n_j = N(\delta_j)$.

1. If $n_j = m$ as $j \to \infty$ then $x_{n_j}^{\delta_j} = x_m^{\delta_j}$ for j sufficiently large and $\|y^{\delta_j} - F(x_m^{\delta_j})\|_Y \leq \tau\delta_j$. By Lemma 18.15, $\{x_m^{\delta_j}\}$ splits into subsequences which converge to elements of $\mathcal{X}_m$. Let $\{x_m^{\delta_{jr}}\}_r$ converge to ξ_m. Thus, $F(\xi_m) = y$ and $\xi_m = x^+$ since $\mathcal{X}_m \subset B_r(\xi)$. This argumentation holds for any limit point of $\{x_m^{\delta_j}\}$ which yields $x_m^{\delta_j} \to x^+$ as $j \to \infty$.
2. If $n_j \leq m$ for $j \to \infty$. Then $\{n_j\}$ has limit points in $\{0, \ldots, m\}$ each of which can be dealt with as in case 1.
3. We begin with constructing a special sequence $\{\xi^{(\ell)}\}_{\ell\in\mathbb{N}} \subset S$. To this end we name the elements in $\mathcal{X}_n$ uniquely: for any $n \in \mathbb{N}_0$ let $\mathcal{X}_n = (\sigma_{n,1}, \ldots, \sigma_{n,k_n})$. With $\sigma_{n,i}$ we associate the sequence $\xi^{(\ell)}$, $\ell = \ell(n,i) = i + \sum_{m=0}^{n-1} k_m$, defined by

$$\xi_r^{(\ell)} = \begin{cases} \text{the predessor of } \sigma_{n,i} \text{ in } \mathcal{X}_r & : r < n, \\ \sigma_{n,i} & : r = n, \\ \text{a successor of } \sigma_{n,i} \text{ in } \mathcal{X}_r & : r > n. \end{cases}$$

Obviously, $\xi^{(\ell)} \in S$. Let $\eta > 0$. By Lemma 18.16 there is an $M = M(\eta)$ such that

$$\|x^+ - \xi_M^{(\ell)}\|_X < \eta/2 \quad \text{for all } \ell \in \mathbb{N}.$$

The sequence $\{x_M^{\delta_j}\}_j$ splits into subsequences which converge to elements of $\mathcal{X}_M$, see Lemma 18.15. Suppose that the subsequence $\{x_M^{\delta_{j_p}}\}_p$ converges to $\sigma_{M,i}$ for one $i \in \{1, \dots, k_M\}$. Then, there is a $P = P(\eta) \in \mathbb{N}$ such that

$$\|\sigma_{M,i} - x_M^{\delta_{j_p}}\|_X < \eta/2 \quad \text{for all } p \geq P.$$

Further, by the construction of $\{\xi^{(\ell)}\}_{\ell \in \mathbb{N}}$ there exists an $\ell^* = \ell(M, i)$ such that $\xi_M^{(\ell^*)} = \sigma_{M,i}$. Thus, for $p \geq P$ so large that $n_{j_p} \geq M$ we have

$$\|x^+ - x_{N(\delta_{j_p})}^{\delta_{j_p}}\|_X < \|x^+ - x_M^{\delta_{j_p}}\|_X \leq \|x^+ - \xi_M^{(\ell^*)}\|_X + \|\sigma_{M,i} - x_M^{\delta_{j_p}}\|_X < \eta$$

where the first estimate is due to the monotone error decrease of the REGINN-iterates. As this argumentation holds for any of the finitely many limit points of $\{x_M^{\delta_j}\}_j$ the stated convergence is validated. □

Remark 18.18

(a) From a practical point of view, as an additional safeguarding technique, one would like to limit the number of inner iterations of REGINN. To this end the stopping criterion for the repeat-loop might be replaced by $\|b_n^\delta - A_n s_{n,m}\| < \mu_n \|b_n^\delta\|$ **or** $m = m_{\max}$. In this case, Theorem 18.17 as well as the monotone error decrease still hold true if Landweber, implicit iteration, steepest descent or conjugate gradients are used as inner iterations.
(b) In [164, 165] REGINN is adapted to solve the inverse problem of electrical impedance tomography. For this application all assumptions of this section hold true, especially TCC (18.9) is satisfied [117] (in a semi-discrete setting). The implementation of [165] uses the conjugate gradient method as inner iteration and the tolerances are chosen according to (18.13). A variety of numerical experiments including some with measured data show the performance of the scheme.
(c) For variants of REGINN in a Banach space environment consult [124–126].

Inverse Problems Related to Abstract Evolution Equations 19

In this section we consider a special type of abstract evolution equations and related inverse problems which are motivated by the visco-elastic wave equation. This equation describes wave propagation in a realistic medium incorporating dispersion and attenuation. Thus, it serves as forward model in seismic tomography which entails the reconstruction of material parameters (wave speeds, density etc.) from measurements of reflected wave fields.

This section is mainly based on results presented originally in [111, 112, 167] where the underlying differential operators are independent of time. For a wave equation with time-dependent differential operators we refer to [73, 74].

19.1 Motivation: Full Waveform Inversion in Seismic Imaging

19.1.1 Elastic Wave Equation

The propagation of waves in an elastic medium $D \subset \mathbb{R}^3$ is governed by the elastic wave equation: Let $\boldsymbol{\sigma} : [0, \infty) \times D \to \mathbb{R}^{3\times 3}_{\text{sym}}$ be the stress tensor and $\mathbf{v} \colon [0, \infty) \times D \to \mathbb{R}^3$ be the velocity field. Then,

$$\partial_t \boldsymbol{\sigma}(t, x) = C\big(\mu(x), \pi(x)\big)\varepsilon(\mathbf{v}(t, x)) \qquad \text{in } [0, \infty) \times D, \tag{19.1}$$

$$\rho(x)\partial_t \mathbf{v}(t, x) = \operatorname{div} \boldsymbol{\sigma}(t, x) + \mathbf{f}(t, x) \qquad \text{in } [0, \infty) \times D, \tag{19.2}$$

where $\rho \colon D \to \mathbb{R}$ is the mass density, $\mathbf{f} \colon [0, \infty) \times D \to \mathbb{R}^3$ a volume force, and

$$C(m, p)\boldsymbol{\epsilon} = 2m\, \boldsymbol{\epsilon} + (p - 2m) \operatorname{trace}(\boldsymbol{\epsilon})\mathbf{I}, \quad \boldsymbol{\epsilon} \in \mathbb{R}^{3\times 3}_{\text{sym}}, \; m, p \in \mathbb{R},$$

W. Dörfler et al., *Wave Phenomena*, Oberwolfach Seminars 49,
https://doi.org/10.1007/978-3-031-05793-9_19

specifies Hooke's law. Further, $\mu, \pi : D \to \mathbb{R}$ are the relaxed P- and S-wave moduli. Finally,

$$\varepsilon(\mathbf{v}) := \frac{1}{2}\big[(\nabla_x \mathbf{v})^\top + \nabla_x \mathbf{v}\big]$$

is the linearized strain rate. Initial and boundary conditions for $\boldsymbol{\sigma}$ and $\mathbf{v}$ have to be specified.

Remark 19.1 The stress in a body is caused by the displacement field $\mathbf{u}$ and the body's material. One material law for a standard linear material is Hooke's law:

$$\boldsymbol{\sigma}(t, x) = C\big(\mu(x), \pi(x)\big)\varepsilon(\mathbf{u}(t, x)),$$

see, e.g., [76, Chap. 4]. As the velocity $\mathbf{v}$ is the time derivative of the displacement, $\mathbf{v} = \partial_t \mathbf{u}$, we get the first equation (19.1) of the elastic wave equation. The second one (19.2) is a balance law for the conservation of momentum, see Sect. 1.4.

Remark 19.2 The first order system (19.1) and (19.2) can equivalently be formulated as one second order equation:

$$\rho(x)\partial_t^2 \mathbf{v}(t, x) = \operatorname{div}\big[C\big(\mu(x), \pi(x)\big)\varepsilon\big(\mathbf{v}(t, x)\big)\big] + \mathbf{g}(t, x). \tag{19.3}$$

Set

$$\boldsymbol{\sigma}(t, x) := C\big(\mu(x), \pi(x)\big) \int_0^t \varepsilon\big(\mathbf{v}(s, x)\big)\, \mathrm{d}s.$$

Then, $(\boldsymbol{\sigma}, \mathbf{v})$ solves

$$\partial_t \boldsymbol{\sigma}(t, x) = C\big(\mu(x), \pi(x)\big)\varepsilon\big(\mathbf{v}(t, x)\big), \tag{19.4}$$

$$\rho(x)\, \partial_t \mathbf{v}(t, x) = \operatorname{div} \boldsymbol{\sigma}(t, x) + \int_0^t \mathbf{g}(s, x)\, \mathrm{d}s + \rho(x)\partial_t \mathbf{v}(0, x). \tag{19.5}$$

Conversely, if $(\boldsymbol{\sigma}, \mathbf{v})$ solves (19.4) and (19.5) then $\mathbf{v}$ is a solution of (19.3).

Initial and boundary conditions for both formulations have to be compatible.

19.1.2 Visco-Elastic Wave Equation

Waves propagating in the earth exhibit damping (loss of energy) which is not reflected by the elastic wave equation. Thus, the elastic wave equation has to be augmented by a mechanism which models dispersion and attenuation. Several of these mechanisms are known in the literature which are all closely related, see [70, Chap. 5], [167, Chap. 2] and Sect. 1.5 for an overview and references. Here, we focus upon the visco-elastic wave equation in the velocity stress formulation based on the generalized standard linear solid rheology: Using $L \in \mathbb{N}$ memory tensors $\boldsymbol{\eta}_l : [0, T] \times D \to \mathbb{R}^{3\times 3}_{\text{sym}}$, $l = 1, \ldots, L$, the new formulation reads

$$\rho\, \partial_t \mathbf{v} = \operatorname{div} \boldsymbol{\sigma} + \mathbf{f} \qquad \text{in }]0, T[\times D, \tag{19.6a}$$

$$\partial_t \boldsymbol{\sigma} = C\big((1 + L\tau_{\text{S}})\mu, (1 + L\tau_{\text{P}})\pi\big)\, \varepsilon(\mathbf{v}) + \sum_{l=1}^{L} \boldsymbol{\eta}_l \qquad \text{in }]0, T[\times D, \tag{19.6b}$$

$$-\tau_{\sigma,l}\partial_t \boldsymbol{\eta}_l = C\big(\tau_{\text{S}}\mu, \tau_{\text{P}}\pi\big)\, \varepsilon(\mathbf{v}) + \boldsymbol{\eta}_l, \quad l = 1, \ldots, L, \qquad \text{in }]0, T[\times D. \tag{19.6c}$$

The functions $\tau_{\text{P}}, \tau_{\text{S}} : D \to \mathbb{R}$ are scaling factors for the relaxed moduli π and μ, respectively. They have been introduced by Blanch et al. [12].

Wave propagation is frequency-dependent and the numbers $\tau_{\sigma,l} > 0$, $l = 1, \ldots, L$, are used to model this dependency over a frequency band with center frequency ω_0. Within this band the rate of the full energy over the dissipated energy remains nearly constant. This observation lets us determine the stress relaxation times $\tau_{\sigma,l}$ by a least-squares approach [13, 14].

Now the frequency-dependent phase velocities of P- and S-waves are given by

$$v_{\text{P}}^2 = \frac{\pi}{\rho}(1 + \tau_{\text{P}}\alpha) \text{ and } v_{\text{S}}^2 = \frac{\mu}{\rho}(1 + \tau_{\text{S}}\alpha) \text{ with } \alpha = \sum_{l=1}^{L} \frac{\omega_0^2 \tau_{\sigma,l}^2}{1 + \omega_0^2 \tau_{\sigma,l}^2}. \tag{19.7}$$

19.1.3 The Inverse Problem of Seismic Imaging in the Visco-Elastic Regime

Let $T > 0$ be the observation time and assume—for the time being—the parameter-to-solution map

$$\Phi : (\rho, v_{\text{S}}, \tau_{\text{S}}, v_{\text{P}}, \tau_{\text{P}}) \mapsto (\boldsymbol{\sigma}|_{[0,T]}, \mathbf{v}|_{[0,T]})$$

to be well defined. Further, we model the measurement process by the linear measurement operator R which records $(\boldsymbol{\sigma}|_{[0,T]}, \mathbf{v}|_{[0,T]})$ at finitely many times at the receiver positions, that is,

$$R \begin{pmatrix} \boldsymbol{\sigma}|_{[0,T]} \\ \mathbf{v}|_{[0,T]} \end{pmatrix} \in \mathbb{R}^N$$

where N is the overall number of data points (time points × number of receivers). Given a seismogram $w \in \mathbb{R}^N$ find five parameters $(\rho, v_S, \tau_S, v_P, \tau_P)$ such that

$$R\Phi(\rho, v_S, \tau_S, v_P, \tau_P) = w. \tag{19.8}$$

Solving above problem is called *full waveform inversion* (FWI) of seismic imaging in the visco-elastic regime.

In the remainder of this section we explore the properties of Φ which are essential for applying a Newton-like solver of Sect. 18 to FWI: We will show that Φ is well defined and Fréchet-differentiable. Moreover, we will characterize the adjoint operator of the Fréchet-derivative by an adjoint wave equation akin to (19.6). Finally, we will see that (19.8) is locally ill-posed indeed.

19.1.4 Visco-Elastic Wave Equation (Transformed)

By the transformation, see [167],

$$\begin{pmatrix} \mathbf{v} \\ \boldsymbol{\sigma}_0 \\ \boldsymbol{\sigma}_1 \\ \vdots \\ \boldsymbol{\sigma}_L \end{pmatrix} := \begin{pmatrix} \mathbf{v} \\ \boldsymbol{\sigma} + \sum_{l=1}^{L} \tau_{\sigma,l}\boldsymbol{\eta}_l \\ -\tau_{\sigma,1}\boldsymbol{\eta}_1 \\ \vdots \\ -\tau_{\sigma,L}\boldsymbol{\eta}_L \end{pmatrix}$$

we reformulate (19.6) equivalently into

$$\partial_t \mathbf{v} = \frac{1}{\rho}\operatorname{div}\Big(\sum_{l=0}^{L}\boldsymbol{\sigma}_l\Big) + \frac{1}{\rho}\mathbf{f} \qquad \text{in }]0,T[\times D, \tag{19.9a}$$

$$\partial_t \boldsymbol{\sigma}_0 = C\big(\mu,\pi\big)\,\varepsilon(\mathbf{v}) \qquad \text{in }]0,T[\times D, \tag{19.9b}$$

$$\partial_t \boldsymbol{\sigma}_l = C\big(\tau_S\mu,\tau_P\pi\big)\,\varepsilon(\mathbf{v}) - \frac{1}{\tau_{\sigma,l}}\boldsymbol{\sigma}_l, \qquad l = 1,\dots,L, \qquad \text{in }]0,T[\times D. \tag{19.9c}$$

We augment the above system by initial conditions

$$\mathbf{v}(0) = \mathbf{v}_0 \ \text{ and } \ \boldsymbol{\sigma}_l(0) = \boldsymbol{\sigma}_{l,0}, \ l = 0,\dots,L. \tag{19.9d}$$

For a suitable function space X and suitable[1] $w = (\mathbf{w}, \boldsymbol{\psi}_0, \ldots, \boldsymbol{\psi}_L) \in X$ we define operators A, B, and Q mapping into X by

$$Aw = -\begin{pmatrix} \operatorname{div}\left(\sum_{l=0}^{L} \boldsymbol{\psi}_l\right) \\ \varepsilon(\mathbf{w}) \\ \vdots \\ \varepsilon(\mathbf{w}) \end{pmatrix}, \quad B^{-1}w = \begin{pmatrix} \frac{1}{\rho}\mathbf{w} \\ C\big(\mu, \pi\big)\boldsymbol{\psi}_0 \\ C\big(\tau_\mathrm{S}\mu, \tau_\mathrm{P}\pi\big)\boldsymbol{\psi}_1 \\ \vdots \\ C\big(\tau_\mathrm{S}\mu, \tau_\mathrm{P}\pi\big)\boldsymbol{\psi}_L \end{pmatrix}, \quad Qw = \begin{pmatrix} \mathbf{0} \\ \mathbf{0} \\ \frac{1}{\tau_{\sigma,1}}\boldsymbol{\psi}_1 \\ \vdots \\ \frac{1}{\tau_{\sigma,L}}\boldsymbol{\psi}_L \end{pmatrix}. \tag{19.10}$$

Now (19.9) can be formulated as

$$Bu'(t) + Au(t) + BQu(t) = f(t)$$

where $u = (\mathbf{v}, \boldsymbol{\sigma}_0, \ldots, \boldsymbol{\sigma}_L)$ and $f = (\mathbf{f}, \mathbf{0}, \ldots, \mathbf{0})$.

We emphasize that the five parameters $(\rho, v_\mathrm{S}, \tau_\mathrm{S}, v_\mathrm{P}, \tau_\mathrm{P})$ to be reconstructed by FWI show up solely in the operator B since, by (19.7),

$$\pi = \frac{\rho\, v_\mathrm{P}^2}{1 + \tau_\mathrm{P}\alpha} \quad \text{and} \quad \mu = \frac{\rho\, v_\mathrm{S}^2}{1 + \tau_\mathrm{S}\alpha}. \tag{19.11}$$

19.2 Abstract Framework

Motivated by our previous considerations we investigate an abstract evolution equation in a Hilbert space X: Find $u\colon [0, T] \to X$ satisfying

$$Bu'(t) + Au(t) + BQu(t) = f(t), \quad t \in]0, T[, \quad u(0) = u_0. \tag{19.12}$$

We assume the following throughout (unless otherwise specified): $T > 0$, $u_0 \in X$,

$B \in \mathcal{L}^*(X) = \{J \in \mathcal{L}(X) : J^* = J\}$ satisfies $\langle Bx, x\rangle_X = \langle x, Bx\rangle_X \geq \beta\|x\|_X^2$ for some $\beta > 0$ and for all $x \in X$,
$A\colon \mathsf{D}(A) \subset X \to X$ is a maximal monotone operator: $\langle Ax, x\rangle_X \geq 0$ for all $x \in \mathsf{D}(A)$ and $\mathrm{Id} + A\colon \mathsf{D}(A) \to X$ is onto,
$Q \in \mathcal{L}(X)$, and $f \in L^1\big([0, T], X\big)$.

[1] A rigorous mathematical formulation will be given in Chap. 20 below.

Above, $L^1([0,T],X)$ denotes the space of X-valued functions over $[0,T]$ which are measurable and Bochner-integrable. In the sequel we also use other spaces, namely

$$\begin{aligned} C([0,T],\mathsf{D}(A)) &= \{v\colon [0,T]\to \mathsf{D}(A) : v \text{ continuous}\},^2\\ C^1([0,T],X) &= \{v\colon [0,T]\to X : v \text{ continuously Fréchet-differentiable}\},\\ W^{1,1}([0,T],X) &= \{v\in C([0,T],X) : v'\in L^1([0,T],X)\}. \end{aligned}$$

Recursively, we then define $C^k([0,T],X)$ and $W^{k,1}([0,T],X)$ for $k\in\mathbb{N}$, $k\geq 2$.

19.2.1 Existence, Uniqueness, and Regularity

Our first theorem is fundamental for our analysis of (19.12).

Theorem 19.3 (Hille-Yosida) *Let A be maximal monotone. If $u_0 \in \mathsf{D}(A)$ then the evolution problem*

$$u'(t)+Au(t)=0,\quad t>0,\quad u(0)=u_0, \tag{19.13}$$

has a unique classical solution $u\in C([0,\infty[,\mathsf{D}(A))\cap C^1([0,\infty[,X)$ *satisfying*

$$\|u(t)\|_X\leq\|u_0\|_X \quad \text{and} \quad \|u'(t)\|_X\leq\|Au_0\|_X \quad \text{for all } t>0.$$

Proof The proof consists of four steps: 1. Uniqueness follows from the monotonicity of A. Let u and w be two solutions of (19.13). Then, for $d=u-v$,

$$\frac{1}{2}\frac{\mathrm{d}}{\mathrm{d}t}\|d(t)\|_X^2=\langle d'(t),d(t)\rangle_X=-\langle Ad(t),d(t)\rangle_X\leq 0$$

and $d(0)=0$. Thus $d=0$.

2. For $\lambda>0$ replace A by the bounded operator (Yosida approximation)

$$A_\lambda=A(I+\lambda A)^{-1}.$$

3. The unique solution of $v'(t)+A_\lambda v(t)=0$, $t>0$, $v(0)=u_0$, is given by

$$u_\lambda(t)=\sum_{k=0}^{\infty}\frac{(-t)^k}{k!}A_\lambda^k u_0=\mathrm{e}^{-tA_\lambda}u_0.$$

[2] $\mathsf{D}(A)$ is equipped with the Graph norm: $\|v\|_{\mathsf{D}(A)}^2=\|v\|_X^2+\|Av\|_X^2$.

4. Pass to the limit $\lambda \searrow 0$, see, e.g., [17, Chap. 7.2] for all technical details. □

The bound $\|u(t)\|_X \le \|u_0\|_X$ implies that the linear operator

$$S(t) \colon u_0 \mapsto u(t)$$

has a bounded extension $S(t) \in \mathcal{L}(X)$ with $\|S(t)\| \le 1$. Thus the family $\{S(t)\}_{t\ge 0}$ constitutes a C^0-*semigroup of contractions* (see, e.g., [17, 99, 138]):

$$S(t+s) = S(t)S(s); \qquad \|S(t)\| \le 1; \qquad \lim_{t\to 0} S(t)x = x,\ x \in X. \tag{19.14}$$

Further,

$$\frac{\mathrm{d}}{\mathrm{d}t} S(t)x = AS(t)x = S(t)Ax, \qquad x \in \mathsf{D}(A).$$

For later reference we present the following result.

Lemma 19.4 *A maximal monotone operator A is closed, has a dense domain* $\mathsf{D}(A)$ *and the adjoint* A^* *is maximal monotone as well. Furthermore,*

$$\mathsf{D}(A) = \big\{w \in X : \lim_{h\to 0}[w - S(h)w]/h \in X\big\} \ \textit{and}\ Av = \lim_{h\to 0} \frac{1}{h}[v - S(h)v],\ v \in \mathsf{D}(A).$$

Theorem 19.5 *Let A be maximal monotone.*

(a) (Classical solution) If $u_0 \in \mathsf{D}(A)$ *and* $f \in W^{1,1}_{\mathrm{loc}}([0,\infty[, X)$ *then*

$$u'(t) + Au(t) = f(t), \quad t > 0, \quad u(0) = u_0, \tag{19.15}$$

has the unique solution $u \in C([0,\infty[, \mathsf{D}(A)) \cap C^1([0,\infty[, X)$ *given by the variation-of-constant formula*

$$u(t) = S(t)u_0 + \int_0^t S(t-s)\, f(s)\,\mathrm{d}s \tag{19.16}$$

and we have the stability estimate

$$\|u'(t)\|_X \le \|Au_0 - f(0)\|_X + \|f'\|_{L^1([0,t],X)} \quad \textit{for all } t > 0.$$

(b) *(Mild solution) If only* $u_0 \in X$ *and* $f \in L^1_{\mathrm{loc}}([0,\infty[,X)$ *then formula* (19.16) *defines the* mild solution *satisfying*

$$\|u(t)\|_X \lesssim \|u_0\|_X + \|f\|_{L^1([0,t],X)} \quad \textit{for all } t > 0. \tag{19.17}$$

Proof The estimate (19.17) for the mild solution is a direct consequence of (19.16), (16.9), and (19.14). This settles the proof of part (b).

To prove part (a) we first show that $v(t) := \int_0^t S(t-s)\, f(s)\, \mathrm{d}s = \int_0^t S(s)\, f(t-s)\, \mathrm{d}s$ is in $\mathsf{D}(A)$. By our assumptions,

$$v'(t) = S(t) f(0) + \int_0^t S(t-s)\, f'(s)\, \mathrm{d}s$$

and

$$\frac{1}{h}\big(v(t+h) - v(t)\big) = \frac{1}{h}\int_t^{t+h} S(t+h-s)\, f(s)\, \mathrm{d}s + \frac{1}{h}\big[S(h) - \mathrm{Id}\big]v(t).$$

Thus, by

$$\lim_{h\to 0} \frac{1}{h}\big[\mathrm{Id} - S(h)\big]v(t) = f(t) - v'(t)$$

we get $v(t) \in \mathsf{D}(A)$ (Lemma 19.4) and $v'(t) = f(t) - Av(t)$. Hence, if $u(t)$ denotes the right hand side of (19.16) then it solves the evolution problem:

$$u'(t) = -AS(t)u_0 + f(t) - Av(t) = f(t) - A\big(S(t)u_0 + v(t)\big) = f(t) - Au(t).$$

The stability estimate follows from

$$u'(t) = S(t)\big[-Au_0 + f(0)\big] + \int_0^t S(t-s)\, f'(s)\, \mathrm{d}s$$

and the proof is done. □

Note that the mild solution does not satisfy the differential equation of (19.15). However, it is the unique solution of the weak formulation: Find $u \in C([0,\infty[,X)$ such that $u(0) = u_0$ and

$$\frac{\mathrm{d}}{\mathrm{d}t}\langle u(t), v\rangle_X + \langle u(t), A^* v\rangle_X = \langle f(t), v\rangle_X \quad \text{for a.a. } t > 0, \quad \forall v \in \mathsf{D}(A^*).$$

This new notion of a solution is well defined since the test space $\mathsf{D}(A^*)$ is dense in X (Lemma 19.4).

Lemma 19.6 *Let $u_0 \in X$ and $f \in L^1_{\text{loc}}([0,\infty[, X)$. Then, the mild solution of* (19.15) *agrees with the weak solution.*

Proof See [99, Th. 2.18] for details. □

Next we consider evolution equations with an operator in front of the time derivative. To this end let

$$\mathcal{B} := \big\{ J \in \mathcal{L}(X) : J = J^*,\ \exists\beta > 0\colon\ \beta\|v\|_X^2 \le \langle Jv, v\rangle_X\ \forall v \in X \big\}.$$

Theorem 19.7 *Let A be maximal monotone and $B \in \mathcal{B}$.*

(a) If $u_0 \in \mathsf{D}(A)$ and $f \in W^{1,1}_{\text{loc}}([0,\infty[, X)$ then

$$Bu'(t) + Au(t) = f(t), \quad t > 0, \quad u(0) = u_0,$$

has a unique solution $u \in C([0,\infty[, \mathsf{D}(A)) \cap C^1([0,\infty[, X)$ satisfying

$$\|u'(t)\|_X \lesssim \|Au_0 - f(0)\|_X + \|f'\|_{L^1([0,t],X)} \quad \text{for all } t > 0.$$

The constant in the above estimate depends on $\|B\|$ and $\|B^{-1}\|$.

(b) If only $u_0 \in X$ and $f \in L^1_{\text{loc}}([0,\infty[, X)$ then the mild/weak solution $u \in C([0,\infty[, X)$ is given by $u(0) = u_0$,

$$\frac{\mathrm{d}}{\mathrm{d}t}\langle Bu(t), v\rangle_X + \langle u(t), A^*v\rangle_X = \langle f(t), v\rangle_X \quad \text{for a.a. } t > 0, \quad \forall v \in \mathsf{D}(A^*).$$

Proof We sketch the main steps and refer to [111, Lem. 3.1] for details. First we show that $A + B$ is onto: Indeed, A and A^* are maximal monotone (Lemma 19.4). Hence, $A^* + B$ is one-to-one yielding that $\mathsf{R}(A + B)$ is dense in X. Since $\beta\|v\|_X \le \|(A + B)v\|_X$ for $v \in \mathsf{D}(A)$ we see that $\mathsf{R}(A + B)$ is closed. Thus, $\mathsf{R}(A + B) = X$.

Now, $B^{-1}A$ is maximal monotone in $(X, \langle\cdot,\cdot\rangle_B)$ with the weighted inner product $\langle\cdot,\cdot\rangle_B = \langle B\cdot,\cdot\rangle_X$. Finally we apply the Hille-Yosida theorem to

$$u'(t) + B^{-1}Au(t) = B^{-1}f(t), \quad t > 0, \quad u(0) = u_0,$$

which proves part (a). □

Remark 19.8 The inner product $\langle\cdot,\cdot\rangle_B = \langle B\cdot,\cdot\rangle_X$ induces a norm on X which is equivalent to the original norm $\|\cdot\|_X$.

We are now ready to approach our original evolution equation (19.12).

Theorem 19.9 *Let A be maximal monotone, $B \in \mathcal{B}$, $Q \in \mathcal{L}(X)$, and $T > 0$.*

(a) If $u_0 \in \mathsf{D}(A)$ and $f \in W^{1,1}([0,T],X)$ then

$$Bu'(t) + Au(t) + BQu(t) = f(t), \quad u(0) = u_0,$$

has a unique solution $u \in C([0,T],\mathsf{D}(A)) \cap C^1([0,T],X)$ satisfying

$$\|u'(t)\|_X \lesssim \left\|(B^{-1}A + Q)u_0 - B^{-1}f(0)\right\|_X + \|f'\|_{L^1([0,t],X)} \tag{19.18}$$

for $t \in [0,T]$ where the constant depends on $\|B\|$, $\|B^{-1}\|$, $\|Q\|$, and T.

(b) If only $u_0 \in X$ and $f \in L^1([0,T],X)$ then the mild/weak solution $u \in C([0,T],X)$ is given by $u(0) = u_0$,

$$\frac{\mathrm{d}}{\mathrm{d}t}\langle Bu(t), v\rangle_X + \langle u(t), A^*v\rangle_X + \langle BQu(t), v\rangle_X = \langle f(t), v\rangle_X$$

for a.a. $t \in \,]0,T[$ and all $v \in \mathsf{D}(A^*)$. Further,

$$\|u(t)\|_X \lesssim \|u_0\|_X + \|f\|_{L^1([0,t],X)} \tag{19.19}$$

for $t \in [0,T]$ where the constant depends on $\|B\|$, $\|B^{-1}\|$, $\|Q\|$, and T.

Proof We look at the equivalent evolution equation

$$u'(t) + (B^{-1}A + Q)u(t) = B^{-1}f(t), \quad u(0) = u_0,$$

As above, $B^{-1}A$ with $\mathsf{D}(B^{-1}A) = \mathsf{D}(A)$ generates a contraction semigroup on $(X, \langle\cdot,\cdot\rangle_B)$. Further, $P = B^{-1}A + Q$, $\mathsf{D}(P) = \mathsf{D}(A)$, is the infinitesimal generator of a C_0-semigroup $\{R(t)\}_{t\geq 0}$ and

$$\|R(t)\|_B \leq \exp(\|Q\|_B t), \tag{19.20}$$

see, e.g., [138, Th. 3.1.1]. Thus, the above evolution equation has the unique classical/weak solution represented by

$$u(t) = R(t)u_0 + \int_0^t R(t-s)B^{-1}f(s)\,\mathrm{d}s$$

from which the stability estimate follows as before when using (19.20) and the equivalence of the norms $\|\cdot\|_B$ and $\|\cdot\|_X$. □

Higher regularity of u in time can be shown when the data u_0 and f are smoother and compatible with the domain of A.

Theorem 19.10 *Let A be maximal monotone, $B \in \mathcal{B}$, $Q \in \mathcal{L}(X)$, and $T > 0$. For $k \in \mathbb{N}$ let $f \in W^{k,1}([0,T], X)$ and*

$$u_{0,\ell} := (-P)^\ell u_0 + \sum_{j=0}^{\ell-1}(-P)^j B^{-1} f^{(\ell-1-j)}(0) \in \mathsf{D}(A), \quad \ell = 0, \ldots, k-1,$$

where $P = B^{-1}A + Q$.

Then, $u \in C^{k-1}([0,T], \mathsf{D}(A)) \cap C^k([0,T], X)$ satisfying

$$\|u^{(\ell)}(t)\|_X \lesssim \|u_{0,\ell}\|_X + \|f^{(\ell)}\|_{L^1([0,t],X)} \quad \text{for all } t \in [0,T], \quad \ell = 0, \ldots, k,$$

where $u_{0,k} := -Pu_{0,k-1} + B^{-1}f^{(k-1)}(0) \in X$. Note that $f^{(k-1)}$ is continuous.

Proof The $u_{0,\ell}$'s satisfy the recursion $u_{0,0} = u_0$ and

$$u_{0,\ell} = -Pu_{0,\ell-1} + B^{-1}f^{(\ell-1)}(0), \quad \ell = 1, \ldots, k.$$

By induction we are going to show

$$u^{(\ell)}(t) = R(t)u_{0,\ell} + \int_0^t R(t-s)\,B^{-1}f^{(\ell)}(s)\,\mathrm{d}s, \quad \ell = 0, \ldots, k. \tag{19.21}$$

The case $k = 0$ is just (19.19).

Assume above formula to be true for $\ell = 0, \ldots, k$. Let $f \in W^{k+1,1}$ and $u_{0,k} \in \mathsf{D}(A)$. Then (19.21) holds for $\ell = k$; that is,

$$u^{(k)}(t) = R(t)u_{0,k} + \int_0^t R(s)\,B^{-1}f^{(k)}(t-s)\,\mathrm{d}s.$$

The additional differentiability of f yields that $u^{(k)}$ is differentiable and thus,

$$\begin{aligned} u^{(k+1)}(t) &= -PR(t)u_{0,k} + R(t)B^{-1}f^{(k)}(0) + \int_0^t R(s)\,B^{-1}f^{(k+1)}(t-s)\,\mathrm{d}s \\ &= R(t)\left[-Pu_{0,k} + B^{-1}f^{(k)}(0)\right] + \int_0^t R(t-s)\,B^{-1}f^{(k+1)}(s)\,\mathrm{d}s \end{aligned}$$

which is formula (19.21) for $\ell = k+1$.

From (19.21) the assertion follows immediately in the usual way. □

19.2.2 Parameter-to-Solution Map

We consider the nonlinear map

$$F\colon \mathsf{D}(F) \subset \mathcal{L}^*(X) \to C([0,T],X), \quad B \mapsto u, \tag{19.22}$$

which maps B to the unique solution of (19.12) where A, Q, u_0, and f are fixed. Its domain is set to

$$\mathsf{D}(F) := \{B \in \mathcal{L}^*(X) : \beta_- \|x\|_X^2 \le \langle Bx, x\rangle_X \le \beta_+ \|x\|_X^2\}$$

for given $0 < \beta_- < \beta_+ < \infty$.

We first specify sufficient conditions for the corresponding inverse problem to be locally ill-posed, see Definition 16.1. But we do not need the results of Chap. 16 which rely on compactness and weak-⋆-to-weak continuity of F.

Theorem 19.11 *Let $u_0 \in \mathsf{D}(A)$ and $f \in W^{1,1}([0,T],X)$. The inverse problem*

$$F(\cdot) = u$$

is locally ill-posed at any $\widehat{B} \in \mathsf{D}(F)$ satisfying $F(\widehat{B}) = u$ if there are

- *an operator family $\{E_k\} \subset \mathcal{L}(X)$ with $\|E_k\| \sim 1$, $E_k \to 0$ point-wise, and*
- *an $\widehat{r} > 0$ and a $\gamma > 0$ such that*

$$\forall r \in \,]0,\widehat{r}]\colon \widehat{B} + rE_k \in \mathsf{D}(F), \quad \forall v \in X\colon \gamma\|v\|_X^2 \le \big\langle(\widehat{B} + rE_k)v, v\big\rangle_X.$$

Proof Set $B_k := \widehat{B} + rE_k$. Since $\|B_k - \widehat{B}\| \sim r$ it remains to show that

$$u_k := F(B_k) \xrightarrow{k\to\infty} u := F(\widehat{B}) \text{ in } C([0,T],X).$$

The difference $d_k := u - u_k$ solves

$$B_k d_k'(t) + (A + B_k Q)d_k(t) = rE_k\big(u'(t) + Qu(t)\big), \quad d_k(0) = 0.$$

As the above right hand side is in $L^1(]0,T[,X)$, d_k is the mild solution given by

$$d_k(t) = r\int_0^t R(t-s)B^{-1}E_k\big(u'(s) + Qu(s)\big)\mathrm{d}s.$$

Thus,

$$\sup_{t\in[0,T]} \|d_k(t)\|_X \le \frac{1}{\gamma}\sup_{t\in[0,T]} \|d_k(t)\|_{B_k} \lesssim r\int_0^T \big\|E_k\big(u'(s) + Qu(s)\big)\big\|_X\,\mathrm{d}s \xrightarrow{k\to\infty} 0.$$

The convergence follows by the dominated convergence theorem as the integrand converges point-wise to zero and is uniformly bounded. □

Remark 19.12 The local-illposedness result remains true if we furnish F with the larger image space $L^2([0,T],X)$ which is more appropriate in the context of ill-posed problems.

Our next goal is to show Fréchet differentiability of F. To this end we need continuity of F with respect to the stronger topology of $C^1([0,T],X)$.

Lemma 19.13 *Let $B \in \mathrm{int}(\mathsf{D}(F))$. Further, let $u_0 \in \mathsf{D}(A)$, $f \in W^{2,1}([0,T],X)$, and $u_{0,1} = (B^{-1}A + Q)u_0 - B^{-1}f(0) \in \mathsf{D}(A)$. Then, F is Lipschitz continuous at B, that is,*

$$\|F(B) - F(\widetilde{B})\|_{C^1([0,T],X)} \lesssim \big(1 + \|f\|_{W^{2,1}([0,T],X)}\big)\|B - \widetilde{B}\|_{\mathcal{L}(X)}$$

for all $\widetilde{B}$ in a neighborhood of B. The involved constant depends on u_0, $u_{0,1}$, $f(0)$, $f'(0)$, A, Q, T, β_-, and β_+.

Proof Set $\widetilde{B} = B + \delta B$ for $\delta B \in \mathcal{L}^*(X)$ sufficiently small and let $u = F(B)$ and $\widetilde{u} = F(\widetilde{B})$. According to our assumptions we have that $u \in C^2([0,T],X)$ (Theorem 19.10) and $\widetilde{u} \in C^1([0,T],X)$. Especially, both are classical solutions of their respective evolution equations. Hence, $d = \widetilde{u} - u$ solves

$$d'(t) + (\widetilde{B}^{-1}A + Q)d(t) = \widetilde{B}^{-1}\delta B(u'(t) + Qu(t)), \quad d(0) = 0.$$

Applying the stability estimate (19.17) for the mild solution yields

$$\|d(t)\|_X \lesssim \|\delta B\| \, \|u\|_{C^1([0,T],X)}, \quad t \in [0, T]. \tag{19.23}$$

By (19.18) for the strong solution we get

$$\|d'(t)\|_X \lesssim \|\delta B\| \big(\|u'(0) + Qu_0\|_X + \|u\|_{C^2([0,T],X)}\big), \quad t \in [0, T]. \tag{19.24}$$

Finally, Theorem 19.10 leads to

$$\|u^{(\ell)}(t)\|_X \lesssim \|u_{0,\ell}\|_X + \|f^{(\ell)}\|_{L^1([0,t],X)}, \quad t \in [0, T], \quad \ell = 0, 1, 2,$$

where $u_{0,0} = u_0$ and $u_{0,2} = (B^{-1}A + Q)u_{0,1} - f'(0)$. These three bounds imply the claimed Lipschitz continuity when plugged into (19.23) and (19.24). □

To prove continuity we need a further ingredient, namely the invariance of $\mathsf{D}(A)$ under Q:

$$Q\big(\mathsf{D}(A)\big) \subset \mathsf{D}(A). \tag{19.25}$$

Theorem 19.14 *Assume* (19.25). *Then, for $u_0 \in \mathsf{D}(A)$ and $f \in W^{1,1}([0, T], X)$, the map $F\colon \mathsf{D}(F) \subset \mathcal{L}^*(X) \to C^1([0, T], X)$ is continuous at interior points.*

Proof Let $\{B_k\}_k \subset \mathsf{D}(F)$ converge to $B \in \operatorname{int}(\mathsf{D}(F))$ and set $u_k = F(B_k)$, $u = F(B)$. Note that $u_k, u \in C^1([0, T], X)$. We define a sequence of linear operators $J_k\colon \mathsf{D}(A) \times W^{1,1}([0, T], X) \to C^1([0, T], X)$, $J_k(u_0, f) := u_k - u$. By the previous lemma we obtain $J_k(u_0, f) \to 0$ in $C^1([0, T], X)$ for $(u_0, f) \in \mathcal{D} := \big\{(u_0, f) \in \mathsf{D}(A) \times W^{2,1}([0, T], X) : B^{-1}Au_0 - B^{-1}f(0) \in \mathsf{D}(A)\big\}$. Further, in view of (19.18) and (19.17),

$$\begin{aligned} \|J_k(u_0, f)\|_{C^1([0,T],X)} &\leq \|u_k\|_{C^1([0,T],X)} + \|u\|_{C^1([0,T],X)} \\ &\lesssim \|u_0\|_{\mathsf{D}(A)} + \|f\|_{W^{1,1}([0,T],X)}, \end{aligned}$$

i.e. $\{\|J_k\|\}_k$ is uniformly bounded. The continuity of F at B follows now at once from the Banach-Steinhaus theorem since the space $\mathcal{D}$ is dense in $\mathsf{D}(A) \times W^{1,1}([0, T], X)$ as we show in the remainder of the proof.

Let $(u_0, f) \in \mathsf{D}(A) \times W^{1,1}([0, T], X)$. Choose sequences $\{u_0^j\}_j \subset \mathsf{D}((B^{-1}A)^2)$, $\{v_j\}_j \subset \mathsf{D}(A)$, and $\{\widetilde{f}_j\}_j \subset W^{2,1}([0, T], X)$ such that $u_0^j \to u_0$ in[3] $\mathsf{D}(A)$, $v_j \to$

[3] This is possible since $\mathsf{D}((B^{-1}A)^2)$ is dense in $\mathsf{D}(B^{-1}A) = \mathsf{D}(A)$, see, e.g., [17, Lemma 7.2].

$B^{-1}f(0)$ in X, and $\widetilde{f}_j \to f$ in $W^{1,1}([0,T],X)$. Additionally, choose $\varphi \in C^\infty(\mathbb{R})$ with $\varphi(0) = 1$ and $\varphi(t) = 0$ for $t \geq t_0$ and some fixed $t_0 \in]0,T[$. Define $\{f_j\}_j \subset W^{2,1}([0,T],X)$ by $f_j(t) = \widetilde{f}_j(t) + \varphi(t)\big(Bv_j - \widetilde{f}_j(0)\big)$ for $t \geq 0$. Then, $B^{-1}f_j(0) = v_j \in \mathsf{D}(A)$ and $f_j \to f$ in $W^{1,1}([0,T],X)$. Finally, $B^{-1}Au_0^j \in \mathsf{D}(B^{-1}A) = \mathsf{D}(A)$, that is, $(u_0^j, f_j) \in \mathcal{D}$ and $(u_0^j, f_j) \to (u_0, f)$ in $\mathsf{D}(A) \times W^{1,1}([0,T],X)$ which settles the proof. □

Theorem 19.15 *Let $u_0 \in \mathsf{D}(A)$ and $B \in \operatorname{int}(\mathsf{D}(F))$.*

(a) *Let $f \in W^{1,1}([0,T],X)$ and assume* (19.25). *Then, F as defined in* (19.22) *is Fréchet differentiable at B with $F'(B)H = \overline{u}$ where $\overline{u}$ is the mild/weak solution of*

$$B\overline{u}'(t) + (A + BQ)\overline{u}(t) = -H\big(u'(t) + Qu(t)\big), \quad \overline{u}(0) = 0, \tag{19.26}$$

with $u = F(B)$ being the classical solution of (19.12).

(b) *Let $u_0 = 0$ and $f \in W^{2,1}([0,T],X)$ with $f(0) = f'(0) = 0$. Then, F as defined in* (19.22) *is Fréchet differentiable at B with $F'(B)H = \overline{u}$ where $\overline{u}$ is now the classical solution of* (19.26). *Moreover, F' is Lipschitz continuous:*

$$\|F'(\widetilde{B}) - F'(B)\|_{\mathcal{L}(\mathcal{L}^*(X), C([0,T],X))} \lesssim \|\widetilde{B} - B\|_{\mathcal{L}(X)}$$

for all $\widetilde{B}$ in a neighborhood of B.

Proof

(a) As $H(u' + Qu)$ is in $C([0,T],X)$, the mild solution $\overline{u}$ of (19.26) exists.

Let $H \in \mathcal{L}^*(X)$ such that $B + H \in \mathsf{D}(F)$ and let $u_+ = F(B+H)$ be the classical solution of

$$(B+H)u_+'(t) + (A + (B+H)Q)u_+(t) = f(t), \quad u_+(0) = u_0,$$

i.e.

$$u_+'(t) + (B^{-1}A + Q)u_+(t) = B^{-1}\big(f(t) - H(u_+'(t) + Qu_+(t))\big).$$

Thus,

$$u_+(t) = R(t)u_0 + \int_0^t R(t-s)B^{-1}\big(f(s) - H(u_+'(s) + Qu_+(s))\big)\mathrm{d}s.$$

Since

$$u(t) = R(t)u_0 + \int_0^t R(t-s)B^{-1}f(t)\mathrm{d}s,$$

and

$$\overline{u}(t) = -\int_0^t R(t-s)B^{-1}H(u'(s) + Qu(s))\mathrm{d}s$$

we get, for $d := u_+ - u$,

$$u_+(t) - u(t) - \overline{u}(t) = \int_0^t R(t-s)B^{-1}H\big(d'(s) + Qd(s)\big)\mathrm{d}s.$$

Straightforward estimates yield

$$\frac{\|u_+ - u - \overline{u}\|_{C([0,T],X)}}{\|H\|} \lesssim \|u_+ - u\|_{C^1([0,T],X)} = \|F(B+H) - F(B)\|_{C^1([0,T],X)}.$$

The assertion follows from the previous theorem about continuity of F.

(b) Under our assumptions, u is $C^2([0,T],X)$ (Theorem 19.10) so that the source term of (19.26) is in $W^{1,1}([0,T],X)$. Hence, $\overline{u} \in C^1([0,T],X)$ by Theorem 19.9.

Now we check the Lipschitz continuity. To this end let $\overline{u} = F'(B)H$ and $\overline{v} = F'(B+\delta B)H$. By the regularity assumptions on f, $\overline{v}$ and $\overline{u}$ are the classical solutions of

$$(B+\delta B)\overline{v}'(t) + (A + (B+\delta B)Q)\overline{v}(t) = -H(v'(t) + Qv(t)),\ t \in]0,T[,\quad \overline{v}(0) = 0,$$

$$B\overline{u}'(t) + (A + BQ)\overline{u}(t) = -H(u'(t) + Qu(t)),\ t \in]0,T[,\quad \overline{u}(0) = 0,$$

where u solves (19.12) and v solves (19.12) with B replaced by $B + \delta B$. Hence, $\overline{d} = \overline{v} - \overline{u}$ mildly solves

$$B\overline{d}'(t) + (A+BQ)\overline{d}(t) = -H(d'(t) + Qd(t)) - \delta B(\overline{v}'(t) + Q\overline{v}(t)),\ t \in]0,T[,\quad \overline{d}(0) = 0,$$

where $d = v - u$. By the continuous dependency of $\overline{d}$ on the right hand side, see (19.19), we get

$$\|\overline{d}\|_{C([0,T],X)} \lesssim \|H\|_{\mathcal{L}(X)}\,\|d\|_{C^1([0,T],X)} + \|\delta B\|_{\mathcal{L}(X)}\,\|\overline{v}\|_{C^1([0,T],X)}. \tag{19.27}$$

Next we apply the regularity estimate (19.18) to d which solves

$$Bd'(t) + (A+BQ)d(t) = -\delta B(v'(t) + Qv(t))\quad t \in]0,T[,\quad d(0) = 0.$$

Thus,

$$\|d\|_{C^1([0,T],X)} \lesssim \|\delta B\|_{\mathcal{L}(X)} \|v\|_{C^2([0,T],X)} \lesssim \|\delta B\|_{\mathcal{L}(X)} \|f\|_{W^{2,1}([0,T],X)}$$

where the right bound comes from the regularity of v. In a similar way we get

$$\|\overline{v}\|_{C^1([0,T],X)} \lesssim \|H\|_{\mathcal{L}(X)} \|v\|_{C^2([0,T],X)} \lesssim \|H\|_{\mathcal{L}(X)} \|f\|_{W^{2,1}([0,T],X)}.$$

Plugging these bounds into (19.27) we end up with

$$\sup_{H \in \mathcal{L}^*(X)} \frac{\|\overline{v} - \overline{u}\|_{C([0,T],X)}}{\|H\|_{\mathcal{L}(X)}} \lesssim \|f\|_{W^{2,1}([0,T],X)} \|\delta B\|_{\mathcal{L}(X)}$$

which is the claimed Lipschitz continuity. □

Remark 19.16

(a) The conclusion of Remark 19.12 applies as well: F is Fréchet differentiable into $L^2([0, T], X)$.
(b) Fréchet differentiability of $F\colon \mathsf{D}(F) \subset \mathcal{L}^*(X) \to C^1([0, T], X)$ at B requires stronger regularity assumptions on f and u_0. We do not want to go into more details here but mention that $u_0 = 0$ and $f \in W^{3,1}([0, T], X)$ with $f(0) = f'(0) = f''(0) = 0$ are sufficient and yield $F(B) \in C^3([0, T], X)$.

In a general concrete setting B contains the parameters of a partial differential equation. So, the conditions of Theorem 19.10 for higher regularity encode smoothness assumptions on these parameters!

For the Newton-like solvers of Sect. 18 the adjoint of the Fréchet derivative is needed. We consider $F'(B)\colon \mathcal{L}^*(X) \to L^2([0, T], X)$. Since $\mathcal{L}^*(X)$ does not carry a canonical Hilbert space structure the adjoint (or dual) $F'(B)^*$ maps into the dual of $\mathcal{L}^*(X)$:

$$F'(B)^*\colon L^2([0, T], X) \to \mathcal{L}^*(X)'.$$

This adjoint is characterized as follows.

Theorem 19.17 *Under the notation and assumptions of Theorem* 19.15 (a) *we have that*

$$[F'(B)^* g]H = \int_0^T \big\langle H(u'(t) + Qu(t)), w(t)\big\rangle_X \, \mathrm{d}t$$

where w is the mild/weak solution of the adjoint equation

$$Bw'(t) - (A^* + Q^* B)w(t) = g(t), \quad t \in]0, T[, \quad w(T) = 0, \tag{19.28}$$

and $u = F(B)$, i.e., the classical solution of (19.12).

The adjoint equation (19.28) is furnished with an end condition which we turn into an initial condition by 'reversing' time: introducing $\widetilde{w}(t) = w(T-t)$ and $\widetilde{g}(t) = g(T-t)$ we get the initial value problem

$$B\widetilde{w}'(t) + A^*\widetilde{w}(t) + Q^* B\widetilde{w}(t) = -\widetilde{g}(t), \ t \in]0, T[, \quad \widetilde{w}(0) = 0.$$

Since A^* is maximal monotone as well (Lemma 19.4) the transformed equation inherits the structure of (19.12). For the visco-elastic wave equation we even have that $A^* = -A$ (see next section) so that the same subroutine can be used to solve both, the state and the adjoint state equation.

Proof *(of Theorem 19.17)* Recall that $F'(B)H = v$ mildly solves

$$Bv'(t) + (A + BQ)v(t) = -H(u'(t) + Qu(t)), \ t \in [0, T], \quad v(0) = 0.$$

First we note that $\langle F'(B)H, g\rangle_{L^2([0,T],X)} = \int_0^T \langle v(t), g(t)\rangle_X \mathrm{d}t$. To work with classical solutions we choose sequences $\{g_k\}_k, \{h_k\}_k \subset W^{1,1}([0, T], X)$ with $g_k \to g$ and $h_k \to H(u' + Qu)$ in $L^2([0, T], X)$. Further, let w_k and v_k be the classical solutions when replacing g by g_k and $H(u' + Qu)$ by h_k, respectively. Integrating by parts yields

$$\begin{aligned}\int_0^T \big\langle v_k(t), g_k(t)\big\rangle_X \mathrm{d}t &= \int_0^T \big\langle v_k(t), Bw_k'(t) - (A^* + Q^* B)w_k(t)\big\rangle_X \mathrm{d}t \\ &= -\int_0^T \big\langle Bv_k'(t) + (A + BQ)v_k(t), w_k(t)\big\rangle_X \mathrm{d}t = \int_0^T \big\langle h_k(t), w_k(t)\big\rangle_X \mathrm{d}t\end{aligned}$$

where the boundary terms vanish due to zero initial and end conditions.

Taking the limit in k proves the theorem. □

20 Applications

In this section we apply the abstract results to FWI (Sect. 19.1.3) and an inverse problem for Maxwell's equation (electromagnetic scattering in conducting media). For an application to the visco-acoustic wave equation including a multitude of numerical experiments see [15].

20.1 Full Waveform Inversion in the Visco-Elastic Regime

We consider visco-elastic wave equation in its transformed version (19.9). The underlying Hilbert space is

$$X = L^2(D, \mathbb{R}^3) \times L^2(D, \mathbb{R}^{3\times 3}_{\text{sym}})^{1+L}$$

furnished with the inner product

$$\big\langle (\mathbf{v}, \boldsymbol{\sigma}_0, \ldots, \boldsymbol{\sigma}_L), (\mathbf{w}, \boldsymbol{\psi}_0, \ldots, \boldsymbol{\psi}_L) \big\rangle_X = \int_D \Big(\mathbf{v} \cdot \mathbf{w} + \sum_{l=0}^{L} \boldsymbol{\sigma}_l : \boldsymbol{\psi}_l \Big)\, \mathrm{d}x$$

where the colon denotes the Frobenius inner product on $\mathbb{R}^{3\times 3}$ and D is a bounded Lipschitz domain. We split the boundary of D into disjoint parts, $\partial D = \partial D_D \,\dot\cup\, \partial D_N$, and define

$$\mathsf{D}(A) = \Big\{ (\mathbf{w}, \boldsymbol{\psi}_0, \ldots \boldsymbol{\psi}_L) \in H^1_D \times H(\operatorname{div}) : \sum_{l=0}^{L} \boldsymbol{\psi}_l \mathbf{n} = \mathbf{0} \text{ on } \partial D_N \Big\}$$

W. Dörfler et al., *Wave Phenomena*, Oberwolfach Seminars 49,
https://doi.org/10.1007/978-3-031-05793-9_20

with

$$H^1_D = \{\mathbf{v} \in H^1(D, \mathbb{R}^3) : \mathbf{v} = \mathbf{0} \text{ on } \partial D_D\}$$

and

$$H(\operatorname{div}) = \Big\{\boldsymbol{\sigma} \in L^2\big(D, \mathbb{R}^{3\times 3}_{\text{sym}}\big)^{1+L} : \operatorname{div}\Big(\sum_{l=0}^{L} \boldsymbol{\sigma}_l\Big) \in L^2(D, \mathbb{R}^3)\Big\}.$$

Lemma 20.1 *Under the settings of this subsection, the operator* $A \colon \mathsf{D}(A) \subset X \to X$ *from* (19.10) *is maximal monotone.*

Proof The operator A is skew-symmetric and, as such, is monotone:

$$\big\langle A(\mathbf{v}, \boldsymbol{\sigma}_0, \ldots, \boldsymbol{\sigma}_L), (\mathbf{v}, \boldsymbol{\sigma}_0, \ldots, \boldsymbol{\sigma}_L)\big\rangle_X = 0.$$

Indeed, using the identities ($\boldsymbol{\sigma} = \boldsymbol{\sigma}^\top$)

$$\operatorname{div}(\boldsymbol{\sigma}\mathbf{v}) = \operatorname{div}\boldsymbol{\sigma} \cdot \mathbf{v} + \boldsymbol{\sigma} : \nabla\mathbf{v} \text{ and } \varepsilon(\mathbf{v}) : \boldsymbol{\sigma} = \nabla\mathbf{v} : \boldsymbol{\sigma},$$

as well as the divergence theorem we find for $(\mathbf{v}, \boldsymbol{\sigma}_0, \ldots, \boldsymbol{\sigma}_L)$ and $(\mathbf{w}, \boldsymbol{\psi}_0, \ldots, \boldsymbol{\psi}_l) \in \mathsf{D}(A)$ that

$$\begin{aligned}
&\big\langle A(\mathbf{v}, \boldsymbol{\sigma}_0, \ldots, \boldsymbol{\sigma}_L), (\mathbf{w}, \boldsymbol{\psi}_0, \ldots, \boldsymbol{\psi}_L)\big\rangle_X \\
&\quad = -\int_D \Big[\operatorname{div}\underbrace{\Big(\sum_{l=0}^{L} \boldsymbol{\sigma}_l\Big)}_{=:\,\boldsymbol{\sigma}} \cdot \mathbf{w} + \varepsilon(\mathbf{v}) : \underbrace{\Big(\sum_{l=0}^{L} \boldsymbol{\psi}_l\Big)}_{=:\,\boldsymbol{\psi}}\Big] \mathrm{d}x \\
&\quad = -\int_D \big(\operatorname{div}(\boldsymbol{\sigma}\mathbf{w}) - \boldsymbol{\sigma} : \nabla\mathbf{w} + \varepsilon(\mathbf{v}) : \boldsymbol{\psi}\big) \mathrm{d}x \\
&\quad = -\int_D (\varepsilon(\mathbf{v}) : \boldsymbol{\psi} - \varepsilon(\mathbf{w}) : \boldsymbol{\sigma}) \mathrm{d}x - \underbrace{\int_{\partial D} (\boldsymbol{\sigma}\mathbf{w}) \cdot \mathbf{n} \, \mathrm{d}s}_{=0} \\
&\quad = \int_D (\varepsilon(\mathbf{w}) : \boldsymbol{\sigma} - \varepsilon(\mathbf{v}) : \boldsymbol{\psi}) \mathrm{d}x + \underbrace{\int_{\partial D} (\boldsymbol{\psi}\mathbf{v}) \cdot \mathbf{n} \, \mathrm{d}s}_{=0} \\
&\quad = -\int_D \big(\operatorname{div}(\boldsymbol{\sigma}) \cdot \mathbf{w} + \varepsilon(\mathbf{v}) : \boldsymbol{\psi})\big) \mathrm{d}x
\end{aligned}$$

$$= \int_D \big(\operatorname{div}(\boldsymbol{\psi}\mathbf{v}) - \boldsymbol{\psi} : \nabla\mathbf{v} + \varepsilon(\mathbf{w}) : \boldsymbol{\sigma}\big)\mathrm{d}x$$

$$= \int_D \big(\varepsilon(\mathbf{w}) : \boldsymbol{\sigma} + \operatorname{div}(\boldsymbol{\psi}) \cdot \mathbf{v}\big)\mathrm{d}x$$

$$= \big\langle(\mathbf{v}, \boldsymbol{\sigma}_0, \dots, \boldsymbol{\sigma}_L), -A(\mathbf{w}, \boldsymbol{\psi}_0, \dots, \boldsymbol{\psi}_L)\big\rangle_X.$$

Next we show that Id $+A$ is onto: For $(\mathbf{f}, \mathbf{g}_0, \dots, \mathbf{g}_L) \in X$ we need to find $(\mathbf{v}, \boldsymbol{\sigma}_0, \dots, \boldsymbol{\sigma}_L) \in \mathsf{D}(A)$ satisfying

$$\mathbf{v} - \operatorname{div}\Big(\sum_{l=0}^{L} \boldsymbol{\sigma}_l\Big) = \mathbf{f}, \qquad \boldsymbol{\sigma}_l - \varepsilon(\mathbf{v}) = \mathbf{g}_l, \quad l = 0, \dots, L.$$

Assume—for the time being—that the $\boldsymbol{\sigma}_l$'s are known. We multiply the equation on the left by a $\mathbf{w} \in H_D^1$, integrate over D and use the divergence theorem to get

$$\int_D \Big[\mathbf{v} \cdot \mathbf{w} + \Big(\sum_{l=0}^{L} \boldsymbol{\sigma}_l\Big) : \nabla\mathbf{w}\Big]\mathrm{d}x = \int_D \mathbf{f} \cdot \mathbf{w}\,\mathrm{d}x.$$

Now we sum up the $L + 1$ equations on the right and obtain

$$\int_D \big(\mathbf{v} \cdot \mathbf{w} + (L + 1)\varepsilon(\mathbf{v}) : \varepsilon(\mathbf{w})\big)\mathrm{d}x = \int_D \Big(\mathbf{f} \cdot \mathbf{w} - \sum_{l=0}^{L} \mathbf{g}_l : \nabla\mathbf{w}\Big)\mathrm{d}x \quad \forall \mathbf{w} \in H_D^1.$$

This is a standard variational problem (cf. displacement ansatz in elasticity) admitting a unique solution $\mathbf{v} \in H_D^1$ by Korn's inequality and the Lax-Milgram theorem, see, e.g., [16].

Finally, set $\boldsymbol{\sigma}_l := \mathbf{g}_l + \varepsilon(\mathbf{v})$. Thus, $\boldsymbol{\sigma}_l = \boldsymbol{\sigma}_l^\top$. Plugging $\mathbf{g}_l = \boldsymbol{\sigma}_l - \varepsilon(\mathbf{v})$ into the displacement ansatz yields

$$\int_D \mathbf{f} \cdot \mathbf{w}\,\mathrm{d}x = \int_D \Big[\mathbf{v} \cdot \mathbf{w} + \Big(\sum_{l=0}^{L} \boldsymbol{\sigma}_l\Big) : \nabla\mathbf{w}\Big]\mathrm{d}x \quad \forall \mathbf{w} \in H_D^1. \tag{20.1}$$

Since (20.1) holds especially for $\mathbf{w} \in C_0^\infty(D, \mathbb{R}^3)$ we have

$$\operatorname{div}\Big(\sum_{l=0}^{L} \boldsymbol{\sigma}_l\Big) = \mathbf{v} - \mathbf{f} \text{ in } L^2.$$

Hence, $(\boldsymbol{\sigma}_0, \ldots, \boldsymbol{\sigma}_L) \in H(\operatorname{div})$. Plugging this equality back into (20.1) leads to

$$0 = \int_D \operatorname{div}\Big(\sum_{l=0}^{L} \boldsymbol{\sigma}_l \mathbf{w}\Big) \mathrm{d}x \quad \forall \mathbf{w} \in H_D^1$$

which is the weak form of $\sum_{l=0}^{L} \boldsymbol{\sigma}_l \mathbf{n} = \mathbf{0}$ on ∂D_N. □

Next we show that $B \in \mathcal{L}(X)$ from (19.10) is well defined with the required properties. To this end we first study[1]

$$C : \mathsf{D}(C) \subset \mathbb{R}^2 \to \operatorname{Aut}(\mathbb{R}^{3\times 3}_{\mathrm{sym}}), \quad C(m, p)\mathbf{M} = 2m\mathbf{M} + (p - 2m)\operatorname{trace}(\mathbf{M})\operatorname{Id}, \tag{20.2}$$

with

$$\mathsf{D}(C) := \big\{(m, p) \in \mathbb{R}^2 : \underline{\mathrm{m}} \le m \le \overline{\mathrm{m}},\ \underline{\mathrm{p}} \le p \le \overline{\mathrm{p}}\big\}$$

where

$$0 < \underline{\mathrm{m}} < \overline{\mathrm{m}} \text{ and } 0 < \underline{\mathrm{p}} < \overline{\mathrm{p}} \text{ such that } 3\underline{\mathrm{p}} > 4\overline{\mathrm{m}}. \tag{20.3}$$

For $(m, p) \in \mathsf{D}(C)$,

$$\widetilde{C}(m, p) := C(m, p)^{-1} = C\left(\frac{1}{4m}, \frac{p - m}{m(3p - 4m)}\right). \tag{20.4}$$

Moreover,

$$C(m, p)\mathbf{M} : \mathbf{N} = \mathbf{M} : C(m, p)\mathbf{N}.$$

Lemma 20.2 *For $(p, m) \in \mathsf{D}(C)$ we have that*

$$\min\{2\underline{\mathrm{m}}, 3\underline{\mathrm{p}} - 4\overline{\mathrm{m}}\}\mathbf{M} : \mathbf{M} \le C(m, p)\mathbf{M} : \mathbf{M} \le \max\{2\overline{\mathrm{m}}, 3\overline{\mathrm{p}} - 4\underline{\mathrm{m}}\}\,\mathbf{M} : \mathbf{M}.$$

Proof The assertion follows readily from

$$\begin{aligned} C(m, p)\mathbf{M} : \mathbf{M} &= 2m\mathbf{M} : \mathbf{M} + (p - 2m)\operatorname{trace}(\mathbf{M})\operatorname{Id} : \mathbf{M} \\ &= 2m\Big(\mathbf{M} : \mathbf{M} - \frac{1}{3}\operatorname{trace}(\mathbf{M})^2\Big) + \frac{1}{3}(3p - 4m)\operatorname{trace}(\mathbf{M})^2 \end{aligned}$$

[1] $\operatorname{Aut}(\mathbb{R}^{3\times 3}_{\mathrm{sym}})$ is the space of linear maps from $\mathbb{R}^{3\times 3}_{\mathrm{sym}}$ into itself (space of automorphisms).

and

$$\mathbf{M} : \mathbf{M} \geq \frac{1}{3}\,\mathrm{trace}(\mathbf{M})^2. \tag{20.5}$$

The latter can be seen from

$$\mathrm{trace}(\mathbf{M})^2 = \Big(\sum_{i=1}^{3} 1 \cdot \mathbf{M}_{i,i}\Big)^2 \leq \Big(\sum_{i=1}^{3} 1^2\Big)\Big(\sum_{i=1}^{3} \mathbf{M}_{i,i}^2\Big) = 3\Big(\sum_{i=1}^{3} \mathbf{M}_{i,i}^2\Big)$$

yielding

$$\mathbf{M} : \mathbf{M} - \frac{1}{3}\,\mathrm{trace}(\mathbf{M})^2 \geq \mathbf{M} : \mathbf{M} - \Big(\sum_{i=1}^{3} \mathbf{M}_{i,i}^2\Big) = \sum_{\substack{i,j=1\\ i\neq j}}^{3} \mathbf{M}_{i,j}^2 \geq 0$$

which is (20.5). □

If

$$\rho(\cdot) > 0 \text{ and } \big(\mu(\cdot), \pi(\cdot)\big), \big(\tau_{\mathrm{S}}(\cdot)\mu(\cdot), \tau_{\mathrm{P}}(\cdot)\pi(\cdot)\big) \in \mathsf{D}(C) \text{ a.e. in } D \tag{20.6}$$

then B given by

$$B\begin{pmatrix} \mathbf{w} \\ \boldsymbol{\psi}_0 \\ \boldsymbol{\psi}_1 \\ \vdots \\ \boldsymbol{\psi}_L \end{pmatrix} = \begin{pmatrix} \rho\, \mathbf{w} \\ \widetilde{C}(\mu, \pi)\boldsymbol{\psi}_0 \\ \widetilde{C}\big(\tau_{\mathrm{S}}\mu, \tau_{\mathrm{P}}\pi\big)\boldsymbol{\psi}_1 \\ \vdots \\ \widetilde{C}\big(\tau_{\mathrm{S}}\mu, \tau_{\mathrm{P}}\pi\big)\boldsymbol{\psi}_L \end{pmatrix} \tag{20.7}$$

is in $\mathcal{L}^*(X)$. As Q from (19.10) is just a multiplication operator by real numbers we conclude that Q is in $\mathcal{L}(X)$ and satisfies (19.25). Hence, all hypotheses are fulfilled for the visco-acoustic wave equation and all abstract results of the previous subsection apply. Accordingly the following theorem holds true.

Theorem 20.3 *Under* (20.6) *we have the following.*

(a) *A unique mild/weak solution of* (19.9) *exists in* $C\big([0, T], X\big)$ *for*

$$\mathbf{f} \in L^1\big([0, T], L^2(D, \mathbb{R}^3)\big), \quad \mathbf{v}_0 \in L^2(D, \mathbb{R}^3), \quad \boldsymbol{\sigma}_{l,0} \in L^2(D, \mathbb{R}^{3\times 3}_{sym}),\ l = 0, \ldots, L.$$

(b) *A unique classical solution of* (19.9) *exists in* $C\big([0,T],\mathsf{D}(A)\big)\cap C^1\big([0,T],X\big)$ *for*

$$\mathbf{f}\in W^{1,1}\big([0,T],L^2(D,\mathbb{R}^3)\big),\quad \mathbf{v}_0\in H^1_D,\quad (\boldsymbol{\sigma}_{0,0},\ldots,\boldsymbol{\sigma}_{L,0})\in H(\mathrm{div}),$$

$$\sum_l \boldsymbol{\sigma}_{l,0}\mathbf{n}=\mathbf{0}\ \textit{on}\ \partial D_N.$$

(c) *A unique classical solution of* (19.9) *exists in* $C^1\big([0,T],\mathsf{D}(A)\big)\cap C^2\big([0,T],X\big)$ *for*

$$\mathbf{f}\in W^{2,1}\big([0,T],L^2(D,\mathbb{R}^3)\big),\quad (\mathbf{v}_0,\boldsymbol{\sigma}_{0,0},\ldots,\boldsymbol{\sigma}_{L,0})\in \mathsf{D}(A)(\textit{as in part(b)}),$$

$$\varrho^{-1}\big[\mathrm{div}\sum_l \boldsymbol{\sigma}_{l,0}+\mathbf{f}(0)\big]\in H^1_D,\quad \mathrm{div}\,\mathbf{C}\in L^2(D,\mathbb{R}^3),\quad \mathbf{Cn}=\mathbf{0}\ \textit{on}\ \partial D_N,$$

where $\mathbf{C}=C\big((1+L\tau_s)\mu,(1+L\tau_P)\pi\big)\varepsilon(\mathbf{v}_0)$.

Proof (a) and (b) are the concrete formulations of Theorem 19.9.
(c) We have to check the conditions of Theorem 19.10 for $k=2$. □

Remark 20.4 For the visco-elastic wave equation it even holds that $A+BQ\colon \mathsf{D}(A)\subset X\to X$ is maximal monotone (The proof of Lemma 20.1 is adapted straightforwardly to establish surjectivity of $A+BQ+\mathrm{Id}$, see also [167, Prop. 70]). As a consequence the solution of (19.9) exists for all times: meaning $T=\infty$ is allowed in the above theorem if $\mathbf{f}$ is locally in the respective space, see Theorem 19.7.

20.1.1 Full Waveform Forward Operator

FWI entails the reconstruction of the five parameters $(\rho, v_S, \tau_S, v_P, \tau_P)$ from wave field measurements. Therefore we will define a parameter-to-solution map Φ which takes these parameters as arguments with the physically meaningful domain of definition

$$\mathsf{D}(\Phi)=\big\{(\rho,v_S,\tau_S,v_P,\tau_P)\in L^\infty(D)^5:\rho_{\min}\le\rho(\cdot)\le\rho_{\max},\ v_{P,\min}\le v_P(\cdot)\le v_{P,\max},$$
$$v_{S,\min}\le v_S(\cdot)\le v_{S,\max},\ \tau_{P,\min}\le\tau_P(\cdot)\le\tau_{P,\max},\ \tau_{S,\min}\le\tau_S(\cdot)\le\tau_{S,\max}\ \text{a.e. in}\ D\big\}$$

where $0<\rho_{\min}<\rho_{\max}<\infty$, etc. are suitable positive bounds.
In view of (19.7) we set

$$\mu_{\min}:=\frac{\rho_{\min}\,v_{S,\min}^2}{1+\tau_{S,\max}\alpha}\quad\text{and}\quad \mu_{\max}:=\frac{\rho_{\max}\,v_{S,\max}^2}{1+\tau_{S,\min}\alpha}$$

which are lower and upper bounds for μ. We define the bounds $\pi_{\min}$ and $\pi_{\max}$ for π in the same way replacing $_{\mathrm{S}}$ by $_{\mathrm{P}}$. Next we define $\underline{\mathrm{p}}$, $\overline{\mathrm{p}}$, $\underline{\mathrm{m}}$, and $\overline{\mathrm{m}}$ such that (μ, π), $(\tau_{\mathrm{S}}\mu, \tau_{\mathrm{P}}\pi)$ as functions of $(\rho, v_{\mathrm{P}}, v_{\mathrm{S}}, \tau_{\mathrm{P}}, \tau_{\mathrm{S}}) \in \mathsf{D}(\Phi)$ are in $\mathsf{D}(C)$. Indeed,

$$\underline{\mathrm{p}} := \pi_{\min} \min\{1, \tau_{\mathrm{P},\min}\} \quad \text{and} \quad \overline{\mathrm{p}} := \pi_{\max} \max\{1, \tau_{\mathrm{P},\max}\}$$

with $\underline{\mathrm{m}}$ and $\overline{\mathrm{m}}$ set correspondingly will do the job. The condition $3\underline{\mathrm{p}} > 4\overline{\mathrm{m}}$, see (20.3), recasts into

$$\frac{4}{3} \frac{\rho_{\max}}{\rho_{\min}} \frac{1 + \tau_{\mathrm{P},\max}\alpha}{1 + \tau_{\mathrm{S},\min}\alpha} \frac{\max\{1, \tau_{\mathrm{S},\max}\}}{\min\{1, \tau_{\mathrm{P},\min}\}} < \frac{v_{\mathrm{P},\min}^2}{v_{\mathrm{S},\max}^2}$$

which is in agreement with the physical evidence that pressure waves propagate considerably faster than shear waves.

For $\mathbf{f} \in W^{1,1}([0, T], L^2(D, \mathbb{R}^3))$ and $u_0 = (\mathbf{v}_0, \boldsymbol{\sigma}_{0,0}, \ldots, \boldsymbol{\sigma}_{L,0}) \in \mathsf{D}(A)$ the *full waveform forward operator*

$$\Phi \colon \mathsf{D}(\Phi) \subset L^\infty(D)^5 \to L^2([0, T], X), \quad (\rho, v_{\mathrm{S}}, \tau_{\mathrm{S}}, v_{\mathrm{P}}, \tau_{\mathrm{P}}) \mapsto (\mathbf{v}, \boldsymbol{\sigma}_0, \ldots, \boldsymbol{\sigma}_L),$$

is well defined where $(\mathbf{v}, \boldsymbol{\sigma}_0, \ldots, \boldsymbol{\sigma}_L)$ is the unique classical solution of (19.9) with initial value u_0.

We factorize $\Phi = F \circ V$ with F from (19.22) and

$$V \colon \mathsf{D}(\Phi) \subset L^\infty(D)^5 \to \mathcal{L}^*(X), \quad (\rho, v_{\mathrm{S}}, \tau_{\mathrm{S}}, v_{\mathrm{P}}, \tau_{\mathrm{P}}) \mapsto B,$$

where B is defined in (20.7) by way of (19.11).

Remark 20.5 Note that V maps $\mathsf{D}(\Phi)$ into $\mathsf{D}(F)$ by an appropriate choice of β_- and β_+ in terms of $\rho_{\min}$, $\rho_{\max}$, $\underline{\mathrm{p}}$, $\overline{\mathrm{p}}$, $\underline{\mathrm{m}}$, and $\overline{\mathrm{m}}$.

FWI in the visco-elastic regime is locally ill-posed. We do not rely here on Theorem 19.11 but give a direct proof.

Theorem 20.6 *The inverse seismic problem* $\Phi(\cdot) = (\mathbf{v}, \boldsymbol{\sigma}_0, \ldots, \boldsymbol{\sigma}_L)$ *is locally ill-posed at any interior point* $\mathbf{p} = (\rho, v_S, \tau_S, v_P, \tau_P)$ *of* $\mathsf{D}(\Phi)$.

Proof For a point $\xi \in D$ define balls $K_n = \{y \in \mathbb{R}^3 : |y - \xi| \le \delta/n\}$, $n \in \mathbb{N}$, where $\delta > 0$ is so small that $K_1 \subset D$. Let χ_n be the characteristic function of K_n. Further, for any $r > 0$ such that $\mathbf{p}_n := \mathbf{p} + r(\chi_n, \chi_n, \chi_n, \chi_n, \chi_n) \in \mathsf{D}(\Phi)$ we have that $\|\mathbf{p}_n - \mathbf{p}\|_{L^\infty(D)^5} = r$ and $\mathbf{p}_n$ does not converge to $\mathbf{p}$. However, $\lim_{n\to\infty} \|\Phi(\mathbf{p}_n) - \Phi(\mathbf{p})\|_{L^2([0,T],X)} = 0$ as we demonstrate now.

Let $u_n = \Phi(\mathbf{p}_n)$ and $u = \Phi(\mathbf{p})$. Then, $d_n = u_n - u$ satisfies

$$V(\mathbf{p}_n)d_n' + Ad_n + V(\mathbf{p}_n)Qd_n = \big(V(\mathbf{p}) - V(\mathbf{p}_n)\big)(u' + Qu), \quad d_n(0) = 0.$$

By Theorem 19.9 (b), see (19.19), we obtain

$$\|d_n\|_{L^2([0,T],X)} \lesssim \big\|\big(V(\mathbf{p}) - V(\mathbf{p}_n)\big)(u' + Qu)\big\|_{L^1([0,T],X)}$$

where the constant is independent of n, see Remark 20.5. Next, $\lim_{n\to\infty} \|\big(V(\mathbf{p}) - V(\mathbf{p}_n)\big)v\|_X = 0$ for any $v \in X$ by the dominated convergence theorem since $\mathbf{p}_n \to \mathbf{p}$ point-wise a.e. in D. Because $\|V(\mathbf{p}_n)\|_X \lesssim 1$ uniformly in n a further application of the dominated convergence theorem with respect to the time domain yields

$$\int_0^T \big\|\big(V(\mathbf{p}) - V(\mathbf{p}_n)\big)\big(u'(t) + Qu(t)\big)\big\|_X \mathrm{d}t \xrightarrow{n\to\infty} 0$$

and finishes the proof. □

20.1.2 Differentiability and Adjoint

The chain rule determines the Fréchet derivative of Φ by the derivatives of F and V. The latter needs the derivative of $\widetilde{C}$ whose structure is closely related to the derivative of the matrix inverse from Example 16.14.

Lemma 20.7 *Let* $C\colon \mathsf{D}(C) \subset \mathbb{R}^2 \to \mathrm{Aut}(\mathbb{R}^{3\times 3}_{\mathrm{sym}})$ *be the mapping defined in* (20.2). *Its Fréchet derivative at* $(m, p) \in \mathrm{int}(\mathsf{D}(C))$ *is given by*

$$\widetilde{C}'(m, p)\begin{bmatrix} \widehat{m} \\ \widehat{p} \end{bmatrix} = -\widetilde{C}(m, p) \circ C(\widehat{m}, \widehat{p}) \circ \widetilde{C}(m, p) \tag{20.8}$$

for any $(\widehat{m}, \widehat{p}) \in \mathbb{R}^2$.

Proof Recall that C is a linear operator with $\widetilde{C}(m, p) \circ C(m, p) = C(m, p) \circ \widetilde{C}(m, p) = \mathrm{Id}$. For $\widehat{m}$ and $\widehat{p}$ sufficiently small we have

$$\begin{aligned}
&\widetilde{C}(m + \widehat{m}, p + \widehat{p}) - \widetilde{C}(m, p) + \widetilde{C}(m, p) \circ C(\widehat{m}, \widehat{p}) \circ \widetilde{C}(m, p) \\
&\quad = \widetilde{C}(m + \widehat{m}, p + \widehat{p}) \circ \big(C(m, p) - C(m + \widehat{m}, p + \widehat{p})\big) \circ \widetilde{C}(m, p) \\
&\qquad\qquad + \widetilde{C}(m, p) \circ C(\widehat{m}, \widehat{p}) \circ \widetilde{C}(m, p) \\
&\qquad = \big(\widetilde{C}(m, p) - \widetilde{C}(m + \widehat{m}, p + \widehat{p})\big) \circ C(\widehat{m}, \widehat{p}) \circ \widetilde{C}(m, p).
\end{aligned}$$

Thus,

$$\begin{aligned}\big\|\widetilde{C}(m+\widehat{m}, p+\widehat{p}) - \widetilde{C}(m,p) + \widetilde{C}(m,p)\circ C(\widehat{m},\widehat{p})\circ \widetilde{C}(m,p)\big\|_{\mathrm{Aut}} \\ \lesssim \big\|\widetilde{C}(m,p)\big\|_{\mathrm{Aut}}\big\|\widetilde{C}(m,p) - \widetilde{C}(m+\widehat{m}, p+\widehat{p})\big\|_{\mathrm{Aut}} \max\{|\widehat{m}|, |\widehat{p}|\} \\ = \mathrm{o}\big(\max\{|\widehat{m}|, |\widehat{p}|\}\big) \quad \text{as } \max\{|\widehat{m}|, |\widehat{p}|\} \to 0\end{aligned}$$

which proves the assertion. □

Let $\mathbf{p} = (\rho, v_{\mathrm{S}}, \tau_{\mathrm{S}}, v_{\mathrm{P}}, \tau_{\mathrm{P}}) \in \mathrm{int}(\mathsf{D}(\Phi))$ and $\widehat{\mathbf{p}} = (\widehat{\rho}, \widehat{v}_{\mathrm{S}}, \widehat{\tau}_{\mathrm{S}}, \widehat{v}_{\mathrm{P}}, \widehat{\tau}_{\mathrm{P}}) \in L^\infty(D)^5$. Then, $V'(\mathbf{p})\widehat{\mathbf{p}} \in \mathcal{L}^*(X)$ is given by

$$V'(\mathbf{p})\widehat{\mathbf{p}}\begin{pmatrix}\mathbf{w}\\ \boldsymbol{\psi}_0\\ \vdots\\ \boldsymbol{\psi}_L\end{pmatrix} = \begin{pmatrix}\widehat{\rho}\,\mathbf{w}\\ -\frac{\widehat{\rho}}{\rho}\widetilde{C}(\mu,\pi)\boldsymbol{\psi}_0 + \rho\,\widetilde{C}'(\mu,\pi)\begin{bmatrix}\widetilde{\mu}\\ \widetilde{\pi}\end{bmatrix}\boldsymbol{\psi}_0\\ -\frac{\widehat{\rho}}{\rho}\widetilde{C}(\tau_{\mathrm{S}}\mu,\tau_{\mathrm{P}}\pi)\boldsymbol{\psi}_1 + \rho\,\widetilde{C}'(\tau_{\mathrm{S}}\mu,\tau_{\mathrm{P}}\pi)\begin{bmatrix}\widehat{\mu}\\ \widehat{\pi}\end{bmatrix}\boldsymbol{\psi}_1\\ \vdots\\ -\frac{\widehat{\rho}}{\rho}\widetilde{C}(\tau_{\mathrm{S}}\mu,\tau_{\mathrm{P}}\pi)\boldsymbol{\psi}_L + \rho\,\widetilde{C}'(\tau_{\mathrm{S}}\mu,\tau_{\mathrm{P}}\pi)\begin{bmatrix}\widehat{\mu}\\ \widehat{\pi}\end{bmatrix}\boldsymbol{\psi}_L\end{pmatrix} \tag{20.9}$$

where $\widetilde{\mu}$, $\widetilde{\pi}$, $\widehat{\mu}$, and $\widehat{\pi}$ are abbreviations for the occurring inner derivatives:

$$\widetilde{\mu} = \frac{2v_{\mathrm{S}}}{1+\tau_{\mathrm{S}}\alpha}\widehat{v}_{\mathrm{S}} - \frac{\alpha\, v_{\mathrm{S}}^2}{(1+\tau_{\mathrm{S}}\alpha)^2}\widehat{\tau}_{\mathrm{S}}, \qquad \widetilde{\pi} = \frac{2v_{\mathrm{P}}}{1+\tau_{\mathrm{P}}\alpha}\widehat{v}_{\mathrm{P}} - \frac{\alpha\, v_{\mathrm{P}}^2}{(1+\tau_{\mathrm{P}}\alpha)^2}\widehat{\tau}_{\mathrm{P}}, \tag{20.10}$$

$$\widehat{\mu} = \frac{2\tau_{\mathrm{S}}\, v_{\mathrm{S}}}{1+\tau_{\mathrm{S}}\alpha}\widehat{v}_{\mathrm{S}} + \frac{v_{\mathrm{S}}^2}{(1+\tau_{\mathrm{S}}\alpha)^2}\widehat{\tau}_{\mathrm{S}}, \qquad \widehat{\pi} = \frac{2\tau_{\mathrm{P}}\, v_{\mathrm{P}}}{1+\tau_{\mathrm{P}}\alpha}\widehat{v}_{\mathrm{P}} + \frac{v_{\mathrm{P}}^2}{(1+\tau_{\mathrm{P}}\alpha)^2}\widehat{\tau}_{\mathrm{P}}. \tag{20.11}$$

With these preparations we are ready to characterize the Fréchet derivative of the full waveform forward operator.

Theorem 20.8 *Under the assumptions made in this section the full waveform forward operator Φ is Fréchet differentiable at any interior point $\mathbf{p} = (\rho, v_S, \tau_S, v_P, \tau_P)$ of $\mathsf{D}(\Phi)$: For $\widehat{\mathbf{p}} = (\widehat{\rho}, \widehat{v}_S, \widehat{\tau}_S, \widehat{v}_P, \widehat{\tau}_P) \in L^\infty(D)^5$ we have $\Phi'(\mathbf{p})\widehat{\mathbf{p}} = \overline{u}$ where $\overline{u} = (\overline{\mathbf{v}}, \overline{\boldsymbol{\sigma}}_0, \ldots, \overline{\boldsymbol{\sigma}}_L) \in C([0,T], X)$ with $\overline{u}(0) = 0$ is the mild solution of*

$$\rho\,\partial_t\overline{\mathbf{v}} = \mathrm{div}\Big(\sum_{l=0}^{L}\overline{\boldsymbol{\sigma}}_l\Big) - \widehat{\rho}\,\partial_t\mathbf{v}, \tag{20.12a}$$

$$\partial_t \overline{\boldsymbol{\sigma}}_0 = C(\mu, \pi)\varepsilon(\overline{\mathbf{v}}) + \Big(\frac{\widehat{\rho}}{\rho} C(\mu, \pi) + \rho\, C(\widetilde{\mu}, \widetilde{\pi})\Big)\varepsilon(\mathbf{v}), \tag{20.12b}$$

$$\begin{aligned}\partial_t \overline{\boldsymbol{\sigma}}_l = {} & C(\tau_S\mu, \tau_P\pi)\varepsilon(\overline{\mathbf{v}}) \\ & - \frac{1}{\tau_{\sigma,l}}\overline{\boldsymbol{\sigma}}_l + \Big(\frac{\widehat{\rho}}{\rho} C(\tau_S\mu, \tau_P\pi) + \rho\, C(\widehat{\mu}, \widehat{\pi})\Big)\varepsilon(\mathbf{v}), \quad l = 1, \ldots, L,\end{aligned} \tag{20.12c}$$

where $(\mathbf{v}, \boldsymbol{\sigma}_0, \ldots, \boldsymbol{\sigma}_L)$ *is the classical solution of* (19.9).

Proof The chain rule $\Phi'(\mathbf{p})\widehat{\mathbf{p}} = F'(V(\mathbf{p}))V'(\mathbf{p})\widehat{\mathbf{p}}$ together with part (a) of Theorem 19.15 yields

$$\begin{pmatrix} \rho\,\partial_t\overline{\mathbf{v}} \\ \widetilde{C}(\mu,\pi)\partial_t\overline{\boldsymbol{\sigma}}_0 \\ \widetilde{C}(\tau_S\mu, \tau_P\pi)\partial_t\overline{\boldsymbol{\sigma}}_1 \\ \vdots \\ \widetilde{C}(\tau_S\mu, \tau_P\pi)\partial_t\overline{\boldsymbol{\sigma}}_L \end{pmatrix} = \begin{pmatrix} \operatorname{div}\big(\sum_{l=0}^L \overline{\boldsymbol{\sigma}}_l\big) \\ \varepsilon(\overline{\mathbf{v}}) \\ \vdots \\ \varepsilon(\overline{\mathbf{v}}) \end{pmatrix} - \begin{pmatrix} \mathbf{0} \\ \mathbf{0} \\ \frac{1}{\tau_{\sigma,1}}\widetilde{C}\big(\tau_S\mu, \tau_P\pi\big)\overline{\boldsymbol{\sigma}}_1 \\ \vdots \\ \frac{1}{\tau_{\sigma,L}}\widetilde{C}\big(\tau_S\mu, \tau_P\pi\big)\overline{\boldsymbol{\sigma}}_L \end{pmatrix}$$
$$- V'(\mathbf{p})\widehat{\mathbf{p}}\left[\begin{pmatrix} \partial_t\mathbf{v} \\ \partial_t\boldsymbol{\sigma}_0 \\ \partial_t\boldsymbol{\sigma}_1 \\ \vdots \\ \partial_t\boldsymbol{\sigma}_L \end{pmatrix} + \begin{pmatrix} \mathbf{0} \\ \mathbf{0} \\ \frac{1}{\tau_{\sigma,1}}\boldsymbol{\sigma}_1 \\ \vdots \\ \frac{1}{\tau_{\sigma,L}}\boldsymbol{\sigma}_L \end{pmatrix}\right].$$

This system can be rewritten as (20.12) using (19.9b), (19.9c), (20.9), and (20.8). □

Now we are able to apply Theorem 19.17 to obtain the following representation of the dual of the Fréchet derivative $\Phi'(\mathbf{p})^*$.

Theorem 20.9 *The assumptions are as in Theorem* 20.8. *Then, the adjoint* $\Phi'(\mathbf{p})^* \in \mathcal{L}\big(L^2([0,T], X), (L^\infty(D)^5)'\big)$ *at* $\mathbf{p} = (\rho, v_S, \tau_S, v_P, \tau_P) \in \operatorname{int}\big(\mathsf{D}(\Phi)\big)$ *is given by*

$$\Phi'(\mathbf{p})^*\mathbf{g} = \begin{pmatrix} \int_0^T \big(\partial_t\mathbf{v}\cdot\mathbf{w} - \frac{1}{\rho}\,\varepsilon(\mathbf{v}) : (\boldsymbol{\varphi}_0 + \boldsymbol{\Sigma})\big)\mathrm{d}t \\ \frac{2}{v_S}\int_0^T \big(-\varepsilon(\mathbf{v}) : (\boldsymbol{\varphi}_0 + \boldsymbol{\Sigma}) + \pi\,\operatorname{trace}(\boldsymbol{\Sigma}^v)\operatorname{div}\mathbf{v}\big)\mathrm{d}t \\ \frac{1}{1+\alpha\tau_S}\int_0^T \big(\varepsilon(\mathbf{v}) : \boldsymbol{\Sigma}^\tau_{s,2} + \pi\,\operatorname{trace}(\boldsymbol{\Sigma}^\tau_{s,1})\operatorname{div}\mathbf{v}\big)\mathrm{d}t \\ -\frac{2\pi}{v_P}\int_0^T \operatorname{trace}(\boldsymbol{\Sigma}^v)\operatorname{div}\mathbf{v}\,\mathrm{d}t \\ \frac{\pi}{1+\alpha\tau_P}\int_0^T \operatorname{trace}(\boldsymbol{\Sigma}^\tau_p)\operatorname{div}\mathbf{v}\,\mathrm{d}t \end{pmatrix} \in L^1(D)^5 \tag{20.13}$$

for $\mathbf{g} = (\mathbf{g}_{-1}, \mathbf{g}_0, \dots, \mathbf{g}_L) \in L^2\big([0,T], L^2(D,\mathbb{R}^3) \times L^2(D,\mathbb{R}^{3\times 3}_{\text{sym}})^{1+L}\big)$ *where* $\mathbf{v}$ *is the first component of the solution of* (19.9), $\boldsymbol{\Sigma} = \sum_{l=1}^{L} \boldsymbol{\varphi}_l$, *and*

$$\boldsymbol{\Sigma}^{v} = \frac{1}{3\pi - 4\mu}\,\boldsymbol{\varphi}_0 + \frac{\tau_P}{3\tau_P\pi - 4\tau_S\mu}\,\boldsymbol{\Sigma},$$

$$\boldsymbol{\Sigma}^{\tau}_{S,1} = -\frac{\alpha}{3\pi - 4\mu}\,\boldsymbol{\varphi}_0 + \frac{\tau_P}{\tau_S(3\tau_P\pi - 4\tau_S\mu)}\,\boldsymbol{\Sigma}, \quad \boldsymbol{\Sigma}^{\tau}_{S,2} = \alpha\,\boldsymbol{\varphi}_0 - \frac{1}{\tau_S}\,\boldsymbol{\Sigma},$$

$$\boldsymbol{\Sigma}^{\tau}_{P} = \frac{\alpha}{3\pi - 4\mu}\,\boldsymbol{\varphi}_0 - \frac{1}{3\tau_P\pi - 4\tau_S\mu}\,\boldsymbol{\Sigma}.$$

Further, $w = (\mathbf{w}, \boldsymbol{\varphi}_0, \dots, \boldsymbol{\varphi}_L) \in C([0,T], X)$ *the mild/weak solution of*

$$\partial_t \mathbf{w} = \frac{1}{\rho}\operatorname{div}\Big(\sum_{l=0}^{L} \boldsymbol{\varphi}_l\Big) + \frac{1}{\rho}\mathbf{g}_{-1}, \tag{20.14a}$$

$$\partial_t \boldsymbol{\varphi}_0 = C(\mu,\pi)\big(\varepsilon(\mathbf{w}) + \mathbf{g}_0\big), \tag{20.14b}$$

$$\partial_t \boldsymbol{\varphi}_l = C(\tau_S\mu, \tau_P\pi)\big(\varepsilon(\mathbf{w}) + \mathbf{g}_l\big) + \frac{1}{\tau_{\sigma,l}}\boldsymbol{\varphi}_l, \qquad l = 1,\dots,L, \tag{20.14c}$$

with $w(T) = 0$.

Remark 20.10 Please note that each component of the right hand side of (20.13) is still a function depending on the spatial variable and is in $L^1(D)$ as product of two L^2-functions. Hence, $\Phi'(\mathbf{p})^*$ indeed maps into $L^1(D)^5$ which is a subspace of $(L^\infty(D)^5)'$.

Proof *(of Theorem 20.9)* By $A^* = -A$ (skew-symmetry), $Q^* = Q$, and $QB = BQ$ we see that (20.14) is the visco-elastic formulation of (19.28). Further, by Theorem 19.17,

$$\begin{aligned}\big\langle \Phi'(\mathbf{p})^*\mathbf{g}, \widehat{\mathbf{p}}\big\rangle_{(L^\infty(D)^5)' \times L^\infty(D)^5} &= \big\langle F'(V(\mathbf{p}))^*\mathbf{g}, V'(\mathbf{p})\widehat{\mathbf{p}}\big\rangle_{\mathcal{L}(X)' \times \mathcal{L}(X)} \\ &= \int_0^T \big\langle V'(\mathbf{p})\widehat{\mathbf{p}}\big(u'(t) + Qu(t)\big), w(t)\big\rangle_X \,\mathrm{d}t \end{aligned} \tag{20.15}$$

where $u = (\mathbf{v}, \boldsymbol{\sigma}_0, \dots, \boldsymbol{\sigma}_L)$ is the classical solution of (19.9). We will now evaluate the above inner product suppressing its dependence on time. Using (20.9) and (20.8) we find for $\widehat{\mathbf{p}} = (\widehat{\rho}, \widehat{v}_S, \widehat{\tau}_S, \widehat{v}_P, \widehat{\tau}_P)$ that

$$\big\langle V'(\mathbf{p})\widehat{\mathbf{p}}\big(u' + Qu\big), w\big\rangle_X = \int_D \big(\widehat{\rho}\,\partial_t\mathbf{v}\cdot\mathbf{w} + S_0 + S_1 + \dots + S_L\big)\mathrm{d}x \tag{20.16}$$

with

$$S_0 = \Big[-\frac{\widehat{\rho}}{\rho}\widetilde{C}(\mu,\pi)\partial_t\boldsymbol{\sigma}_0 - \rho\,\widetilde{C}(\mu,\pi)C(\widetilde{\mu},\widetilde{\pi})\widetilde{C}(\mu,\pi)\partial_t\boldsymbol{\sigma}_0\Big] : \boldsymbol{\varphi}_0$$

and, for $l = 1, \ldots, L$,

$$S_l = \Big[-\frac{\widehat{\rho}}{\rho}\widetilde{C}(\tau_{\mathrm{S}}\mu, \tau_{\mathrm{P}}\pi)\Big(\partial_t\boldsymbol{\sigma}_l + \frac{\boldsymbol{\sigma}_l}{\tau_{\sigma,l}}\Big) - \rho\widetilde{C}(\tau_{\mathrm{S}}\mu, \tau_{\mathrm{P}}\pi)C(\widehat{\mu},\widehat{\pi})\widetilde{C}(\tau_{\mathrm{S}}\mu, \tau_{\mathrm{P}}\pi)\Big(\partial_t\boldsymbol{\sigma}_l + \frac{\boldsymbol{\sigma}_l}{\tau_{\sigma,l}}\Big)\Big] : \boldsymbol{\varphi}_l.$$

In view of (19.9b) we may write

$$\begin{aligned} S_0 &= \Big[-\frac{\widehat{\rho}}{\rho}\varepsilon(\mathbf{v}) - \rho\,\widetilde{C}(\mu,\pi)C(\widetilde{\mu},\widetilde{\pi})\varepsilon(\mathbf{v})\Big] : \boldsymbol{\varphi}_0 \\ &= -\frac{\widehat{\rho}}{\rho}\varepsilon(\mathbf{v}) : \boldsymbol{\varphi}_0 - \rho\,C(\widetilde{\mu},\widetilde{\pi})\varepsilon(\mathbf{v}) : \widetilde{C}(\mu,\pi)\boldsymbol{\varphi}_0 \end{aligned}$$

and, similarly by (19.9c),

$$S_l = -\frac{\widehat{\rho}}{\rho}\varepsilon(\mathbf{v}) : \boldsymbol{\varphi}_l - \rho\,C(\widehat{\mu},\widehat{\pi})\varepsilon(\mathbf{v}) : \widetilde{C}(\tau_{\mathrm{S}}\mu, \tau_{\mathrm{P}}\pi)\boldsymbol{\varphi}_l, \quad l = 1, \ldots, L.$$

Next, via (20.4), we obtain

$$\begin{aligned} &C(\widetilde{\mu},\widetilde{\pi})\varepsilon(\mathbf{v}) : \widetilde{C}(\mu,\pi)\boldsymbol{\varphi}_0 \\ &\quad = \big(2\widetilde{\mu}\,\varepsilon(\mathbf{v}) + (\widetilde{\pi} - 2\widetilde{\mu})\operatorname{div}\mathbf{v}\,\mathbf{I}\big) : \Big(\frac{1}{2\mu}\boldsymbol{\varphi}_0 + \frac{2\mu - \pi}{2\mu(3\pi - 4\mu)}\operatorname{trace}(\boldsymbol{\varphi}_0)\mathbf{I}\Big) \\ &\quad = \widetilde{\mu}\Big(\frac{1}{\mu}\varepsilon(\mathbf{v}) : \boldsymbol{\varphi}_0 - \frac{\pi}{\mu(3\pi - 4\mu)}\operatorname{div}\mathbf{v}\operatorname{trace}(\boldsymbol{\varphi}_0)\Big) \\ &\qquad + \frac{\widetilde{\pi}}{3\pi - 4\mu}\operatorname{div}\mathbf{v}\operatorname{trace}(\boldsymbol{\varphi}_0) \end{aligned}$$

yielding

$$\begin{aligned} S_0 = -\frac{\widehat{\rho}}{\rho}\varepsilon(\mathbf{v}) : \boldsymbol{\varphi}_0 + \rho\widetilde{\mu}\Big(-\frac{1}{\mu}\varepsilon(\mathbf{v}) : \boldsymbol{\varphi}_0 + \frac{\pi}{\mu(3\pi - 4\mu)}\operatorname{div}\mathbf{v}\operatorname{trace}(\boldsymbol{\varphi}_0)\Big) \\ - \frac{\rho\widetilde{\pi}}{3\pi - 4\mu}\operatorname{div}\mathbf{v}\operatorname{trace}(\boldsymbol{\varphi}_0). \end{aligned}$$

Analogously,

$$S_l = -\frac{\widehat{\rho}}{\rho}\varepsilon(\mathbf{v}) : \boldsymbol{\varphi}_l + \rho\widehat{\mu}\Big(- \frac{1}{\tau_S\mu}\varepsilon(\mathbf{v}) : \boldsymbol{\varphi}_l + \frac{\tau_P\pi}{\tau_S\mu(3\tau_P\pi - 4\tau_S\mu)} \operatorname{div}\mathbf{v} \operatorname{trace}(\boldsymbol{\varphi}_l)\Big)$$
$$- \frac{\rho\widehat{\pi}}{3\tau_P\pi - 4\tau_S\mu} \operatorname{div}\mathbf{v} \operatorname{trace}(\boldsymbol{\varphi}_l).$$

Now we group the terms in the sum of (20.16) with respect to the components of $\widehat{\mathbf{p}}$. To this end we replace $\rho\widetilde{\mu}$, $\rho\widetilde{\pi}$, $\rho\widehat{\mu}$, and $\rho\widehat{\pi}$ by their expressions from (20.10) and (20.11) which we slightly rewrite introducing μ and π. Thus,

$$\rho\widetilde{\mu} = \frac{2\mu}{v_S}\widehat{v}_S - \frac{\alpha\,\mu}{1+\tau_S\alpha}\widehat{\tau}_S, \qquad \rho\widetilde{\pi} = \frac{2\pi}{v_P}\widehat{v}_P - \frac{\alpha\,\pi}{1+\tau_P\alpha}\widehat{\tau}_P,$$

$$\rho\widehat{\mu} = \frac{2\tau_S\,\mu}{v_S}\widehat{v}_S + \frac{\mu}{1+\tau_S\alpha}\widehat{\tau}_S, \qquad \rho\widehat{\pi} = \frac{2\tau_P\,\pi}{v_P}\widehat{v}_P + \frac{\pi}{1+\tau_P\alpha}\widehat{\tau}_P.$$

A few algebraic rearrangements lead to

$$\begin{aligned}\big\langle V'(\mathbf{p})\widehat{\mathbf{p}}\big(u' + Qu\big), \overline{u}\rangle_X = \int_D \Big[\widehat{\rho}\Big(\partial_t\mathbf{v}\cdot\mathbf{w} - \frac{1}{\rho}\,\varepsilon(\mathbf{v}) : (\boldsymbol{\varphi}_0 + \boldsymbol{\Sigma})\Big)\\ + \widehat{v}_S\,\frac{2}{v_S}\Big(- \varepsilon(\mathbf{v}) : (\boldsymbol{\varphi}_0 + \boldsymbol{\Sigma}) + \pi \operatorname{trace}(\boldsymbol{\Sigma}^v) \operatorname{div}\mathbf{v}\Big)\\ + \frac{\widehat{\tau}_S}{1+\alpha\tau_S}\Big(\varepsilon(\mathbf{v}) : \boldsymbol{\Sigma}^\tau_{S,2} + \pi \operatorname{trace}(\boldsymbol{\Sigma}^\tau_{S,1}) \operatorname{div}\mathbf{v}\Big)\\ - \widehat{v}_P\,\frac{2\pi}{v_P} \operatorname{trace}(\boldsymbol{\Sigma}^v) \operatorname{div}\mathbf{v} + \widehat{\tau}_P\,\frac{\pi}{1+\alpha\tau_P} \operatorname{trace}(\boldsymbol{\Sigma}^\tau_P) \operatorname{div}\mathbf{v}\Big]\mathrm{d}x\end{aligned}$$

from which (20.13) follows. □

The expression for the Fréchet derivative and its adjoints cannot directly be applied to the visco-elastic wave equation in two spatial dimensions. There are two differences to the 3D case:

$$\operatorname{trace}(\mathbf{I}) = 2 \quad \text{and} \quad \widetilde{C}(m,p)\mathbf{M} = C^{-1}(m,p)\mathbf{M} = \frac{1}{2m}\mathbf{M} + \frac{2m-p}{4m(p-m)} \operatorname{trace}(\mathbf{M})\mathbf{I}.$$

Taking into account these adjustments we get the following 2D version of Theorem 20.9.

Theorem 20.11 *Re-defining the three quantities*

$$\boldsymbol{\Sigma}^v = \frac{1}{2(\pi - \mu)}\,\varphi_0 + \frac{\tau_p}{2(\tau_p\pi - \tau_s\mu)}\,\boldsymbol{\Sigma},$$

$$\boldsymbol{\Sigma}^\tau_{s,1} = -\frac{\alpha}{2(\pi - \mu)}\,\varphi_0 + \frac{\tau_p}{2\,\tau_s(\tau_p\pi - \tau_s\mu)}\,\boldsymbol{\Sigma},$$

$$\boldsymbol{\Sigma}^\tau_p = \frac{\alpha}{2(\pi - \mu)}\,\varphi_0 - \frac{1}{2(\tau_p\pi - \tau_s\mu)}\,\boldsymbol{\Sigma},$$

the statement of Theorem 20.9 *can be copied without any further changes.*

Proof In 2D we have that

$$\begin{aligned}
&C(\widetilde{\mu},\widetilde{\pi})\varepsilon(\mathbf{v}) : \widetilde{C}(\mu,\pi)\boldsymbol{\varphi}_0 \\
&\quad = \big(2\widetilde{\mu}\,\varepsilon(\mathbf{v}) + (\widetilde{\pi} - 2\widetilde{\mu})\,\mathrm{div}\,\mathbf{v}\,\mathbf{I}\big) : \Big(\frac{1}{2\mu}\,\varphi_0 + \frac{2\mu - \pi}{4\mu(\pi - \mu)}\,\mathrm{trace}(\boldsymbol{\varphi}_0)\mathbf{I}\Big) \\
&\quad = \widetilde{\mu}\Big(\frac{1}{\mu}\varepsilon(\mathbf{v}) : \boldsymbol{\varphi}_0 - \frac{\pi}{2\mu(\pi - \mu)}\,\mathrm{div}\,\mathbf{v}\ \mathrm{trace}(\boldsymbol{\varphi}_0)\Big) + \frac{\widetilde{\pi}}{2(\pi - \mu)}\,\mathrm{div}\,\mathbf{v}\ \mathrm{trace}(\boldsymbol{\varphi}_0)
\end{aligned}$$

and this is the only difference compared to the 3D proof. □

20.2 Maxwell's Equation: Inverse Electromagnetic Scattering

Let $\mathbf{E}, \mathbf{H}\colon [0,\infty) \times D \to \mathbb{R}^3$ be the electric and magnetic fields in a bounded Lipschitz domain $D \subset \mathbb{R}^3$. We consider the following Maxwell system

$$\varepsilon\,\partial_t\mathbf{E} - \mathrm{curl}\,\mathbf{H} = -\mathbf{J}_{\mathrm{e}} - \sigma\,\mathbf{E} \qquad \text{in } [0,\infty) \times D, \tag{20.17a}$$

$$\mu\,\partial_t\mathbf{H} + \mathrm{curl}\,\mathbf{E} = \mathbf{J}_{\mathrm{m}} \qquad \text{in } [0,\infty) \times D, \tag{20.17b}$$

with boundary condition

$$\mathbf{n} \times \mathbf{E} = \mathbf{0} \quad \text{in } [0,\infty) \times \partial D \tag{20.17c}$$

and initial conditions

$$\mathbf{E}(0,\cdot) = \mathbf{E}_0(\cdot) \quad \text{and} \quad \mathbf{H}(0,\cdot) = \mathbf{H}_0(\cdot) \quad \text{in } D. \tag{20.17d}$$

The scalar functions $\varepsilon, \mu, \sigma : D \to \mathbb{R}$ are the (electric) permittivity, the (magnetic) permeability, and the conductivity which characterize an isotropic electromagnetic medium. Further, $\mathbf{J}_{\mathrm{e}}, \mathbf{J}_{\mathrm{m}} : [0, \infty) \times D \to \mathbb{R}^3$ are the current and magnetic densities.

Remark 20.12 In case of a dielectric medium, that is, $\sigma = 0$, the conservation equations

$$\partial_t \operatorname{div}(\varepsilon\, \mathbf{E}) + \operatorname{div} \mathbf{J}_{\mathrm{e}} = 0 \quad \text{and} \quad \partial_t \operatorname{div}(\mu\, \mathbf{H}) - \operatorname{div} \mathbf{J}_{\mathrm{m}} = 0$$

follow directly from (20.17a) and (20.17b), respectively. If $\operatorname{div} \mathbf{J}_{\mathrm{e}} = 0$ then $\operatorname{div}(\varepsilon\, \mathbf{E}) = 0$ provided the initial field satisfies $\operatorname{div}(\varepsilon \mathbf{E}_0) = 0$. The analog result holds for the magnetic field as well. The additional boundary condition $\mathbf{n} \cdot \mathbf{H} = 0$ on ∂D (in the physically relevant case $\mathbf{J}_{\mathrm{m}} = \mathbf{0}$) originates from

$$\partial_t (\mu\, \mathbf{n} \cdot \mathbf{H}) = -\mathbf{n} \cdot \operatorname{curl} \mathbf{E} = -\operatorname{Div}(\mathbf{n} \times \mathbf{E}) \quad \text{on } \partial D$$

and the boundary condition (20.17c). Here, Div denotes the surface divergence, see, e.g., [109, Sec. A.5].

If we define

$$A = \begin{pmatrix} \sigma\, \mathrm{Id} & -\operatorname{curl} \\ \operatorname{curl} & 0 \end{pmatrix}, \quad B = \begin{pmatrix} \varepsilon\, \mathrm{Id} & 0 \\ 0 & \mu\, \mathrm{Id} \end{pmatrix}, \quad Q = 0, \; f = \begin{pmatrix} -\mathbf{J}_{\mathrm{e}} \\ \mathbf{J}_{\mathrm{m}} \end{pmatrix}, \quad \text{and } u = \begin{pmatrix} \mathbf{E} \\ \mathbf{H} \end{pmatrix}$$

then the system (20.17a), (20.17b) is of the abstract form (19.12) with $u_0 = (\mathbf{E}_0, \mathbf{H}_0)^\top$.

As Hilbert space and domain of A we choose

$$X = L^2(D, \mathbb{R}^3) \times L^2(D, \mathbb{R}^3) \quad \text{and} \quad \mathsf{D}(A) = H_0(\operatorname{curl}, D) \times H(\operatorname{curl}, D), \qquad (20.18)$$

respectively. Here, $H(\operatorname{curl}, D)$ is the space of all vector fields which do have a weak curl in L^2, see, e.g., [109, Sec. 4.1.2]. By $H_0(\operatorname{curl}, D)$ we denote the closure of the compactly supported C^∞ vector fields in $H(\operatorname{curl}, D)$. Note that fields $\mathbf{w} \in H_0(\operatorname{curl}, D)$ do not necessarily vanish on ∂D, however, their tangential components do: $\mathbf{w} \times \mathbf{n} = \mathbf{0}$ on ∂D.

We assume that

$$0 < \mathrm{c} \le \varepsilon, \mu \in L^\infty(D) \;\text{ and }\; 0 \le \sigma \in L^\infty(D) \;\text{ a.e.} \qquad (20.19)$$

Lemma 20.13 *Under* (20.19) *we have* $B \in \mathcal{L}^*(X)$ *and* $A : \mathsf{D}(A) \subset X \to X$ *is maximal monotone.*

Proof The first assertion is clear. We prove now the second.

For $(\mathbf{E},\mathbf{H})^\top \in \mathsf{D}(A)$ we have

$$\left\langle A\begin{pmatrix}\mathbf{E}\\ \mathbf{H}\end{pmatrix}, \begin{pmatrix}\mathbf{E}\\ \mathbf{H}\end{pmatrix}\right\rangle_X = \int_D \big((\sigma\mathbf{E} - \operatorname{curl}\mathbf{H})\cdot\mathbf{E} + \operatorname{curl}\mathbf{E}\cdot\mathbf{H}\big)\mathrm{d}x = \int_D \sigma\,|\mathbf{E}|^2\mathrm{d}x \geq 0$$

by Green's theorem, see, e.g., [128, Remark 3.28] (no boundary term appears due to $\mathbf{E} \in H_0(\mathrm{curl}, D)$).

It remains to show surjectivity of $A + \mathrm{Id}$. For any $\mathbf{J}_\mathrm{e}, \mathbf{J}_\mathrm{m} \in L^2(D, \mathbb{R}^3)$ we have to find $\mathbf{E} \in H_0(\mathrm{curl}, D)$ and $\mathbf{H} \in H(\mathrm{curl}, D)$ with

$$\sigma\,\mathbf{E} - \operatorname{curl}\mathbf{H} + \mathbf{E} = \mathbf{J}_\mathrm{e} \quad\text{and}\quad \operatorname{curl}\mathbf{E} + \mathbf{H} = \mathbf{J}_\mathrm{m}. \tag{20.20}$$

We multiply the first equation by $\boldsymbol{\psi} \in H_0(\mathrm{curl}, D)$ and the second by $\operatorname{curl}\boldsymbol{\psi}$, add the equations and integrate over D. Taking into account that $\int_D(\boldsymbol{\psi}\cdot\operatorname{curl}\mathbf{H} - \mathbf{H}\cdot\operatorname{curl}\boldsymbol{\psi})\mathrm{d}x = 0$ we arrive at

$$\int_D (\operatorname{curl}\mathbf{E}\cdot\operatorname{curl}\boldsymbol{\psi} + (\sigma+1)\mathbf{E}\cdot\boldsymbol{\psi})\,\mathrm{d}x = \int_D \big(\mathbf{J}_\mathrm{m}\cdot\operatorname{curl}\boldsymbol{\psi} + \mathbf{J}_\mathrm{e}\cdot\boldsymbol{\psi}\big)\mathrm{d}x.$$

The theorem of Lax-Milgram in $H_0(\mathrm{curl}, D)$ implies existence of a unique solution $\mathbf{E} \in H_0(\mathrm{curl}, D)$. Finally, we define $\mathbf{H} = \mathbf{J}_\mathrm{m} - \operatorname{curl}\mathbf{E}$. Then the equation on the right of (20.20) is satisfied and the variational equation becomes

$$\int_D \big((\sigma+1)\mathbf{E}\cdot\boldsymbol{\psi} - \mathbf{H}\cdot\operatorname{curl}\boldsymbol{\psi}\big)\mathrm{d}x = \int_D \mathbf{J}_\mathrm{e}\cdot\boldsymbol{\psi}\,\mathrm{d}x$$

which is the weak form of the equation on the left of (20.20). □

Here we are in a situation considered in Theorem 19.7.

Theorem 20.14 *Under* (20.19) *we have the following.*

(a) *A unique mild/weak solution* $\mathbf{E},\mathbf{H} \in C\big([0,\infty[, L^2(D,\mathbb{R}^3)\big)$ *of* (20.17) *exists for*

$$\mathbf{J}_\mathrm{e}, \mathbf{J}_\mathrm{m} \in L^1_\mathrm{loc}\big([0,\infty[, L^2(D,\mathbb{R}^3)\big), \quad \mathbf{E}_0, \mathbf{H}_0 \in L^2(D,\mathbb{R}^3).$$

(b) *A unique classical solution* $\mathbf{E} \in C^1\big([0,\infty[, L^2(D,\mathbb{R}^3)\big) \cap C\big([0,\infty[, H_0(\mathrm{curl}, D)\big)$, $\mathbf{H} \in C^1\big([0,\infty[, L^2(D,\mathbb{R}^3)\big) \cap C\big([0,\infty[, H(\mathrm{curl}, D)\big)$ *of* (20.17) *exists for*

$$\mathbf{J}_\mathrm{e}, \mathbf{J}_\mathrm{m} \in W^{1,1}_\mathrm{loc}\big([0,\infty[, L^2(D,\mathbb{R}^3)\big), \quad \mathbf{E}_0 \in H_0(\mathrm{curl}, D), \quad \mathbf{H}_0 \in H(\mathrm{curl}, D).$$

(c) *A unique classical solution* $\mathbf{E} \in C^2\big([0,\infty[, L^2(D,\mathbb{R}^3)\big) \cap C^1\big([0,\infty[, H_0(\mathrm{curl}, D)\big)$, $\mathbf{H} \in C^2\big([0,\infty[, L^2(D,\mathbb{R}^3)\big) \cap C^1\big([0,\infty[, H(\mathrm{curl}, D)\big)$ *of* (20.17) *exists for*

$$\mathbf{J}_e, \mathbf{J}_m \in W^{2,1}_{\mathrm{loc}}\big([0,\infty[, L^2(D,\mathbb{R}^3)\big), \quad \mathbf{E}_0 \in H_0(\mathrm{curl}, D), \quad \mathbf{H}_0 \in H(\mathrm{curl}, D),$$

$$\varepsilon^{-1}\big[\mathrm{curl}\,\mathbf{H}_0 - \sigma\mathbf{E}_0 - \mathbf{J}_e(0)\big] \in H_0(\mathrm{curl}, D), \quad \mu^{-1}\big[\mathbf{J}_m(0) - \mathrm{curl}\,\mathbf{E}_0\big] \in H(\mathrm{curl}, D).$$

Proof Parts (a) and (b) directly follow from Theorem 19.7 whereas (c) needs the application of Theorem 19.10 which holds true for (20.17) even for $T = \infty$. □

20.2.1 Inverse Electromagnetic Scattering

We consider the inverse problem of determining the permittitivity and permeability of the medium D from measurements of the electric and magnetic fields at parts of D, the boundary for instance. The involved linear measurement operator is of no interest in what follows and will be neglected therefore, cf. Sect. 19.1.3.

20.2.2 The Electromagnetic Forward Map

We define the corresponding *electromagnetic forward map* by

$$\Phi\colon \mathsf{D}(\Phi) \subset L^\infty(D)^2 \to L^2\big([0,T], L^2(D,\mathbb{R}^3)^2\big), \quad (\varepsilon,\mu) \mapsto (\mathbf{E},\mathbf{H}),$$

where $T > 0$ is the observation period and $(\mathbf{E},\mathbf{H})$ is the (classical) solution of the Maxwell system (20.17) under the assumptions that $\mathbf{J}_e, \mathbf{J}_m \in W^{1,1}\big([0,T], L^2(D,\mathbb{R}^3)\big)$, $\mathbf{E}_0 \in H_0(\mathrm{curl}, D)$, $\mathbf{H}_0 \in H(\mathrm{curl}, D)$, and that σ satisfies (20.19). Further,

$$\mathsf{D}(\Phi) = \big\{(\varepsilon,\mu) \in L^\infty(D)^2 : \mathrm{c}_- \le \varepsilon, \mu \le \mathrm{c}_+ \text{ a.e.}\big\}$$

with suitable constants $0 < \mathrm{c}_- < \mathrm{c}_+ < \infty$.

Again it will be convenient to factorize $\Phi = F \circ V$ where F is the mapping from (19.22) adapted to the Maxwell system and

$$V\colon \mathsf{D}(\Phi) \subset L^\infty(D)^2 \to \mathcal{L}^*\big(L^2(D,\mathbb{R}^3)^2\big), \quad (\varepsilon,\mu) \mapsto \begin{pmatrix} \varepsilon\,\mathrm{Id} & 0 \\ 0 & \mu\,\mathrm{Id} \end{pmatrix}.$$

Here, $V\big(\mathsf{D}(\Phi)\big) \subset \mathsf{D}(F)$ when setting $\beta_\pm = \mathrm{c}_\pm$ (under the canonical Hilbert norm on $L^2(D,\mathbb{R}^3)^2$).

Theorem 20.15 *The inverse electromagnetic scattering problem* $\Phi(\cdot,\cdot) = (\mathbf{E},\mathbf{H})$ *is locally ill-posed at any interior point* (ε,μ) *of* $\mathsf{D}(\Phi)$.

Proof For a point $\xi \in D$ define balls $K_k = \{y \in \mathbb{R}^3 : |y-\xi| \le \delta/k\}$, $k \in \mathbb{N}$, where $\delta > 0$ is so small that $K_1 \subset D$. Let χ_k be the characteristic function of K_k. Define

$$E_k := \begin{pmatrix} \chi_k \,\mathrm{Id} & 0 \\ 0 & \chi_k \,\mathrm{Id} \end{pmatrix} = V(\chi_k, \chi_k).$$

Then, E_k is monotone (non-negative) with $\|E_k\|_{\mathcal{L}(L^2(D,\mathbb{R}^3)^2)} = 1$, $E_k \to 0$ point-wise as $k \to \infty$ by the dominated convergence theorem. Further, for any $r > 0$ such that $(\varepsilon_k, \mu_k) := (\varepsilon + r\chi_k, \mu + r\chi_k) \in \mathsf{D}(\Phi)$, we have that

$$\Phi(\varepsilon_k, \mu_k) = F\big(V(\varepsilon,\mu) + rE_k\big) \xrightarrow{k\to\infty} F\big(V(\varepsilon,\mu)\big) = \Phi(\varepsilon,\mu)$$

where the convergence can been shown exactly as in the proof of Theorem 19.11. But $\|(\varepsilon_k,\mu_k) - (\varepsilon,\mu)\|_{L^\infty(D)^2} = r$ for all k. □

20.2.3 Differentiability and Adjoint

Since V is a linear operator we immediately get

$$V'(\varepsilon,\mu)\begin{bmatrix} \widehat{\varepsilon} \\ \widehat{\mu} \end{bmatrix} = V(\widehat{\varepsilon}, \widehat{\mu}) \quad \text{for all } (\varepsilon,\mu) \in \mathrm{int}\big(\mathsf{D}(\Phi)\big). \tag{20.21}$$

Theorem 20.16 *Under the assumptions made in this section the electromagnetic forward operator* Φ *is Fréchet differentiable at any interior point* (ε,μ) *of* $\mathsf{D}(\Phi)$*: For* $(\widehat{\varepsilon},\widehat{\mu}) \in L^\infty(D)^2$ *we have* $\Phi'(\varepsilon,\mu)\begin{bmatrix} \widehat{\varepsilon} \\ \widehat{\mu} \end{bmatrix} = (\overline{\mathbf{E}}, \overline{\mathbf{H}})$ *where* $\overline{\mathbf{E}}, \overline{\mathbf{H}} \in C\big([0,T], L^2(D,\mathbb{R}^3)\big)$ *is the mild solution of*

$$2\varepsilon\, \partial_t \overline{\mathbf{E}} + \sigma\, \overline{\mathbf{E}} - \operatorname{curl} \overline{\mathbf{H}} = -\widehat{\varepsilon}\, \partial_t \mathbf{E}, \qquad \overline{\mathbf{E}}(0) = \mathbf{0}, \tag{20.22a}$$

$$\mu\, \partial_t \overline{\mathbf{H}} + \operatorname{curl} \overline{\mathbf{E}} = -\widehat{\mu}\, \partial_t \mathbf{H}, \qquad \overline{\mathbf{H}}(0) = \mathbf{0}, \tag{20.22b}$$

with $(\mathbf{E},\mathbf{H}) = \Phi(\varepsilon,\mu)$ *being the classical solution of* (20.17) *in* $[0,T] \times D$.

Proof The chain rule and (20.21) yield $\Phi'(\varepsilon, \mu)\begin{bmatrix} \widehat{\varepsilon} \\ \widehat{\mu} \end{bmatrix} = F'(V(\varepsilon, \mu))V(\widehat{\varepsilon}, \widehat{\mu})$. Now, we apply part (a) of Theorem 19.15: the abstract system (19.26) formulated in the present case reads (recall that $Q = 0$)

$$\begin{pmatrix} \varepsilon\,\mathrm{Id} & 0 \\ 0 & \mu\,\mathrm{Id} \end{pmatrix}\begin{pmatrix} \partial_t\overline{\mathbf{E}} \\ \partial_t\overline{\mathbf{H}} \end{pmatrix} + \begin{pmatrix} \sigma\,\mathrm{Id} & -\mathrm{curl} \\ \mathrm{curl} & 0 \end{pmatrix}\begin{pmatrix} \overline{\mathbf{E}} \\ \overline{\mathbf{H}} \end{pmatrix} = -\begin{pmatrix} \widehat{\varepsilon}\,\mathrm{Id} & 0 \\ 0 & \widehat{\mu}\,\mathrm{Id} \end{pmatrix}\begin{pmatrix} \partial_t\mathbf{E} \\ \partial_t\mathbf{H} \end{pmatrix}, \quad \begin{pmatrix} \overline{\mathbf{E}}(0) \\ \overline{\mathbf{H}}(0) \end{pmatrix} = \begin{pmatrix} \mathbf{0} \\ \mathbf{0} \end{pmatrix},$$

which is the system we stated. □

Remark 20.17 Under the additional regularity assumption of part (c) of Theorem 20.14 the mild solution of (20.22) is indeed a classical solution.

Theorem 20.18 *The assumptions are as in Theorem* 20.16. *Then, the adjoint* $\Phi'(\varepsilon, \mu)^* \in \mathcal{L}\big(L^2([0, T], L^2(D, \mathbb{R}^3)^2), (L^\infty(D)^2)'\big)$ *at* $(\varepsilon, \mu) \in \mathrm{int}\big(\mathsf{D}(\Phi)\big)$ *is given by*

$$\Phi'(\varepsilon, \mu)^* \begin{pmatrix} \mathbf{g}_E \\ \mathbf{g}_H \end{pmatrix} = \begin{pmatrix} \int_0^T \partial_t\mathbf{E}(t, \cdot) \cdot \overline{\mathbf{E}}(t, \cdot)\,\mathrm{d}t \\ \int_0^T \partial_t\mathbf{H}(t, \cdot) \cdot \overline{\mathbf{H}}(t, \cdot)\,\mathrm{d}t \end{pmatrix} \in L^1(D)^2 \tag{20.23}$$

where $(\overline{\mathbf{E}}, \overline{\mathbf{H}}) \in C\big([0, T], L^2(D, \mathbb{R}^3)^2\big)$ *is the mild/weak solution of*

$$\varepsilon\,\partial_t\overline{\mathbf{E}} - \sigma\,\overline{\mathbf{E}} - \mathrm{curl}\,\overline{\mathbf{H}} = \mathbf{g}_E, \quad \overline{\mathbf{E}}(T) = \mathbf{0}, \tag{20.24a}$$

$$\mu\,\partial_t\overline{\mathbf{H}} + \mathrm{curl}\,\overline{\mathbf{E}} = \mathbf{g}_H, \quad \overline{\mathbf{H}}(T) = \mathbf{0}, \tag{20.24b}$$

and $(\mathbf{E}, \mathbf{H}) = \Phi(\varepsilon, \mu)$ *is the classical solution of* (20.17) *in* $[0, T] \times D$.

Proof By $A^* = \left(\begin{smallmatrix} \sigma\,\mathrm{Id} & \mathrm{curl} \\ -\mathrm{curl} & 0 \end{smallmatrix}\right)$ and $Q = 0$ we see that (20.24) is the formulation of (19.28) in the Maxwell setting. Further, by Theorem 19.17,

$$\begin{aligned} &\left\langle \Phi'(\varepsilon, \mu)^* \begin{pmatrix} \mathbf{g}_E \\ \mathbf{g}_H \end{pmatrix}, (\widehat{\varepsilon}, \widehat{\mu}) \right\rangle_{(L^\infty(D)^2)' \times L^\infty(D)^2} \\ &\qquad = \big\langle F'(V(\varepsilon, \mu))^*\mathbf{g}, V(\widehat{\varepsilon}, \widehat{\mu}) \big\rangle_{\mathcal{L}(L^2(D,\mathbb{R}^3)^2)' \times \mathcal{L}(L^2(D,\mathbb{R}^3)^2)} \\ &\qquad = \int_0^T \left\langle V(\widehat{\varepsilon}, \widehat{\mu})) \begin{pmatrix} \partial_t\mathbf{E} \\ \partial_t\mathbf{H} \end{pmatrix}, \begin{pmatrix} \overline{\mathbf{E}} \\ \overline{\mathbf{H}} \end{pmatrix} \right\rangle_{L^2(D,\mathbb{R}^3)^2} \mathrm{d}t \\ &\qquad = \int_D \left(\widehat{\varepsilon} \int_0^T \partial_t\mathbf{E} \cdot \overline{\mathbf{E}}\,\mathrm{d}t + \widehat{\mu} \int_0^T \partial_t\mathbf{H} \cdot \overline{\mathbf{H}}\,\mathrm{d}t \right) \mathrm{d}x \end{aligned}$$

This equation proves the stated representation (20.23) of $\Phi'(\varepsilon, \mu)^*$. □

Remark 20.19 Remark 20.10 applies to (20.23) accordingly.

Remark 20.20 Differentiability of the electric and magnetic fields with respect to the conductivity σ cannot directly be achieved by our abstract theory. A slightly different evolution equation has to be considered, namely,

$$u'(t) + B^{-1}(A+Q)u(t) = B^{-1}f(t), \quad t \in]0,T[, \quad u(0) = u_0, \tag{20.25}$$

where B, A, and Q are as in the abstract theory, see beginning of Sect. 19.2. Setting

$$A = \begin{pmatrix} 0 & -\operatorname{curl} \\ \operatorname{curl} & 0 \end{pmatrix}, \quad B = \begin{pmatrix} \varepsilon \operatorname{Id} & 0 \\ 0 & \mu \operatorname{Id} \end{pmatrix}, \quad Q = \begin{pmatrix} \sigma \operatorname{Id} & 0 \\ 0 & 0 \end{pmatrix}, \quad f = \begin{pmatrix} -\mathbf{J}_{\mathrm{e}} \\ \mathbf{J}_{\mathrm{m}} \end{pmatrix}, \quad \text{and } u = \begin{pmatrix} \mathbf{E} \\ \mathbf{H} \end{pmatrix}$$

we can frame the Maxwell system in the abstract evolution equation (20.25). The spaces X and $\mathsf{D}(A)$ are as in (20.18). Then $B^{-1}A$ is maximal monotone in $(X, \langle\cdot,\cdot\rangle_B)$ and, by the Hille-Yosida theorem and Theorem 3.1.1 of [138], $B^{-1}A + B^{-1}Q$ generates a C^0-semigroup even for arbitrary $\sigma \in L^\infty(D)$ (no sign condition is required). Now, the mapping $F\colon \mathcal{L}(X) \ni Q \mapsto u \in C([0,T],X)$ can be approached with the techniques of Sect. 19.2. It is Fréchet differentiable under conditions which are weaker than those of Theorem 19.15. The following holds true: Let $u_0 \in X$ and $f \in L^1([0,T],X)$. Then, $F'(Q)H = \overline{u}$ with $\overline{u}$ being the mild solution of

$$B\overline{u}'(t) + (A+Q)\overline{u}(t) = -Hu(t), \quad t \in]0,T[, \quad \overline{u}(0) = 0,$$

where u mildly solves (20.25). We apply this result to the mapping $\Phi\colon L^\infty(D) \ni \sigma \mapsto (\mathbf{E},\mathbf{H}) \in C([0,T],X)$ where $(\mathbf{E},\mathbf{H})$ solves (20.17) in $[0,T] \times D$ under the conditions of Theorem 20.14 (a) restricted to $[0,T]$. Thus, $\Phi'(\sigma)\widehat{\sigma} = (\overline{\mathbf{E}},\overline{\mathbf{H}})$ where $(\overline{\mathbf{E}},\overline{\mathbf{H}})$ is the mild solution of

$$\begin{aligned} \varepsilon\, \partial_t \overline{\mathbf{E}} + \sigma\, \overline{\mathbf{E}} - \operatorname{curl} \overline{\mathbf{H}} &= -\widehat{\sigma}\, \mathbf{E}, \quad & \overline{\mathbf{E}}(0) &= \mathbf{0}, \\ \mu\, \partial_t \overline{\mathbf{H}} + \operatorname{curl} \overline{\mathbf{E}} &= \mathbf{0}, & \overline{\mathbf{H}}(0) &= \mathbf{0}. \end{aligned}$$

References

1. Abramowitz, M., Stegun, I.A.: Handbook of Mathematical Functions: With Formulas, Graphs, and Mathematical Tables. Applied Mathematics Series. Dover Publications (1964). https://doi.org/10.1119/1.15378
2. Adams, R.A., Fournier, J.J.F.: Sobolev Spaces, 2nd edn. Elsevier/Academic Press, Amsterdam (2003)
3. Aregba-Driollet, D., Hanouzet, B.: Kerr-Debye relaxation shock profiles for Kerr equations. Commun. Math. Sci. **9**(1), 1–31 (2011). http://projecteuclid.org/euclid.cms/1294170323
4. Assous, F., Ciarlet, P., Labrunie, S.: Mathematical Foundations of Computational Electromagnetism. Applied Mathematical Sciences, vol. 198. Springer, Cham (2018). https://doi.org/10.1007/978-3-319-70842-3
5. Astala, K., Päivärinta, L.: Calderón's inverse conductivity problem in the plane. Ann. Math. **163**(1), 265–299 (2006). https://doi.org/10.4007/annals.2006.163.265
6. Banjai, L., Georgoulis, E.H., Lijoka, O.: A Trefftz polynomial space-time discontinuous Galerkin method for the second order wave equation. SIAM J. Numer. Anal. **55**(1), 63–86 (2017). https://doi.org/10.1137/16M1065744
7. Bahouri, H., Chemin, J.Y., Danchin, R.: Fourier Analysis and Nonlinear Partial Differential Equations. Springer, Heidelberg (2011). https://doi.org/10.1007/978-3-642-16830-7
8. Bansal, P., Moiola, A., Perugia, I., Schwab, C.: Space–time discontinuous Galerkin approximation of acoustic waves with point singularities. IMA J. Numer. Anal. **41**(3), 2056–2109 (2021). https://doi.org/10.1093/imanum/draa088
9. Beauchard, K., Zuazua, E.: Large time asymptotics for partially dissipative hyperbolic systems. Arch. Ration. Mech. Anal. **199**(1), 177–227 (2011). https://doi.org/10.1007/s00205-010-0321-y
10. Becker, R., Rannacher, R.: An optimal control approach to a posteriori error estimation in finite element methods. Acta Numer. **10**(1), 1–102 (2001). https://doi.org/10.1017/S0962492901000010
11. Benzoni-Gavage, S., Serre, D.: Multidimensional Hyperbolic Partial Differential Equations. The Clarendon Press/Oxford University Press, Oxford (2007)
12. Blanch, J.O., Robertsson, J.O.A., Symes, W.W.: Modeling of a constant Q: Methodology and algorithm for an efficient and optimally inexpensive viscoelastic technique. Geophysics **60**(1), 176–184 (1995). https://doi.org/10.1190/1.1443744
13. Bohlen, T.: Viskoelastische FD-Modellierung seismischer Wellen zur Interpretation gemessener Seismogramme. Ph.D. thesis, Christian-Albrechts-Universität zu Kiel (1998). https://bit.ly/2LM0SWr

W. Dörfler et al., *Wave Phenomena*, Oberwolfach Seminars 49,
https://doi.org/10.1007/978-3-031-05793-9

14. Bohlen, T.: Parallel 3-D viscoelastic finite difference seismic modelling. Comput. Geosci. **28**(8), 887–899 (2002). https://doi.org/10.1016/S0098-3004(02)00006-7
15. Bohlen, T., Fernandez, M.R., Ernesti, J., Rheinbay, C., Rieder, A., Wieners, C.: Visco-acoustic full waveform inversion: from a DG forward solver to a Newton-CG inverse solver. Comput. Math. Appl. **100**, 126–140 (2021). https://doi.org/10.1016/j.camwa.2021.09.001
16. Braess, D.: Finite Elements. Cambridge University Press, Cambridge (2001)
17. Brezis, H.: Functional analysis, Sobolev spaces and partial differential equations. Universitext. Springer, New York (2011). https://link.springer.com/book/10.1007/978-0-387-70914-7
18. Busch, K., von Freymann, G., Linden, S., Mingaleev, S., Tkeshelashvili, L., Wegener, M.: Periodic nanostructures for photonics. Phys. Rep. **444**, 101–202 (2007). https://doi.org/10.1016/j.physrep.2007.02.011
19. Butcher, P.N., Cotter, D.: The Elements of Nonlinear Optics. Cambridge University Press, Cambridge (1990)
20. Burazin, K., Erceg, M.: Non-stationary abstract Friedrichs systems. Mediterr. J. Math. **13**(6), 3777–3796 (2016). https://doi.org/10.1007/s00009-016-0714-8
21. Cai, Z., Lazarov, R., Manteuffel, T.A., McCormick, S.F.: First-order system least squares for second-order partial differential equations: Part I. SIAM J. Numer. Anal. **31**(6), 1785–1799 (1994). https://doi.org/10.1137/0731091
22. Cai, Z., Manteuffel, T.A., McCormick, S.F., Ruge, J.: First-order system LL^* (FOSLL*): Scalar elliptic partial differential equations. SIAM J. Numer. Anal. **39**(4), 1418–1445 (2001). https://doi.org/10.1137/S0036142900388049
23. Carcione, J.M.: Wave Fields in Real Media: Wave Propagation in Anisotropic, Anelastic, Porous and Electromagnetic Media. Elsevier (2014). https://doi.org/10.1016/C2013-0-18893-9
24. Cessenat, M.: Mathematical Methods in Electromagnetism. World Scientific Publishing Co., River Edge, NJ (1996). https://doi.org/10.1142/2938
25. Chabassier, J., Imperiale, S.: Construction and convergence analysis of conservative second order local time discretisation for linear wave equations. ESAIM Math. Model. Numer. Anal. **55**(4), 1507–1543 (2021). https://doi.org/10.1051/m2an/2021030
26. Chazarain, J., Piriou, A.: Introduction to the Theory of Linear Partial Differential Equations. North-Holland Publishing Co., Amsterdam (1982)
27. Ciarlet, P.G.: Mathematical Elasticity. Vol. I: Three-dimensional Elasticity. Studies in Mathematics and its Applications, vol. 20. North-Holland Publishing Co., Amsterdam (1988). https://doi.org/10.1002/crat.2170250509
28. Colombini, F., De Giorgi, E., Spagnolo, S.: Sur les équations hyperboliques avec des coefficients qui ne dépendent que du temps. Ann. Scuola Norm. Sup. Pisa Cl. Sci. (4) **6**(3), 511–559 (1979)
29. Collino, F., Fouquet, T., Joly, P.: A conservative space-time mesh refinement method for the 1-D wave equation. I. Construction. Numer. Math. **95**(2), 197–221 (2003). https://doi.org/10.1007/s00211-002-0446-5
30. Collino, F., Fouquet, T., Joly, P.: A conservative space-time mesh refinement method for the 1-D wave equation. II. Analysis. Numer. Math. **95**(2), 223–251 (2003). https://doi.org/10.1007/s00211-002-0447-4
31. Collino, F., Fouquet, T., Joly, P.: Conservative space-time mesh refinement methods for the FDTD solution of Maxwell's equations. J. Comput. Phys. **211**(1), 9–35 (2006). https://doi.org/10.1016/j.jcp.2005.03.035
32. Costabel, M., Dauge, M., Nicaise, S.: Corner singularities and analytic regularity for linear elliptic systems. Part I (2010). See http://hal.archives-ouvertes.fr/hal-00453934/en/

33. Courant, R., Friedrichs, K., Lewy, H.: über die partiellen Differenzengleichungen der mathematischen Physik. Math. Ann. **100**(1), 32–74 (1928). https://doi.org/10.1007/BF01448839
34. D'Ancona, P., Nicaise, S., Schnaubelt, R.: Blow-up for nonlinear Maxwell equations. Electron. J. Differential Equations, pp. 9, Paper No. 73 (2018)
35. Dautray, R., Lions, J.L.: Mathematical analysis and numerical methods for science and technology, vol. 1. Springer-Verlag, Berlin (1990). Physical origins and classical methods, With the collaboration of Philippe Bénilan, Michel Cessenat, André Gervat, Alain Kavenoky and Hélène Lanchon
36. Dautray, R., Lions, J.L.: Mathematical analysis and numerical methods for science and technology, vol. 3. Springer-Verlag, Berlin (1990). Spectral theory and applications, With the collaboration of Michel Artola and Michel Cessenat
37. Davis, J.L.: Wave propagation in solids and fluids. Springer Science & Business Media (2012). https://doi.org/10.1007/978-1-4612-3886-7
38. Defrise, M., De Mol, C.: A note on stopping rules for iterative regularization methods and filtered SVD. In: Inverse Problems: An Interdisciplinary Study (Montpellier, 1986), Adv. Electron. Electron Phys., Suppl. 19, pp. 261–268. Academic Press, London (1987)
39. Descombes, S., Lanteri, S., Moya, L.: Locally implicit time integration strategies in a discontinuous Galerkin method for Maxwell's equations. J. Sci. Comput. **56**(1), 190–218 (2013). https://doi.org/10.1007/s10915-012-9669-5
40. Descombes, S., Lanteri, S., Moya, L.: Locally implicit discontinuous Galerkin time domain method for electromagnetic wave propagation in dispersive media applied to numerical dosimetry in biological tissues. SIAM J. Sci. Comput. **38**(5), A2611–A2633 (2016). https://doi.org/10.1137/15M1010282
41. Descombes, S., Lanteri, S., Moya, L.: Temporal convergence analysis of a locally implicit discontinuous Galerkin time domain method for electromagnetic wave propagation in dispersive media. J. Comput. Appl. Math. **316**, 122–132 (2017). https://doi.org/10.1016/j.cam.2016.09.038
42. Di Pietro, D.A., Ern, A.: Mathematical Aspects of Discontinuous Galerkin Methods, vol. 69. Springer Science & Business Media (2011). https://doi.org/10.1007/978-3-642-22980-0
43. Di Pietro, D.A., Ern, A.: Mathematical Aspects of Discontinuous Galerkin Methods. Mathématiques & Applications (Berlin) [Mathematics & Applications], vol. 69. Springer, Heidelberg (2012). https://doi.org/10.1007/978-3-642-22980-0
44. Diaz, J., Grote, M.J.: Energy conserving explicit local time stepping for second-order wave equations. SIAM J. Sci. Comput. **31**(3), 1985–2014 (2009). https://doi.org/10.1137/070709414
45. Diehl, R., Busch, K., Niegemann, J.: Comparison of low-storage Runge-Kutta schemes for discontinuous Galerkin time-domain simulations of Maxwell's equations. J. Comput. Theor. Nanosci. **7**(8), 1572–1580 (2010). http://www.ingentaconnect.com/content/asp/jctn/2010/00000007/00000008/art00035
46. Diestel, J.: Sequences and Series in Banach Spaces. Graduate Texts in Mathematics, vol. 92. Springer-Verlag, New York (1984). https://doi.org/10.1007/978-1-4612-5200-9
47. Dörfler, W., Lechleiter, A., Plum, M., Schneider, G., Wieners, C.: Photonic Crystals: Mathematical Analysis and Numerical Approximation, vol. 42. Springer Science & Business Media (2011). https://doi.org/10.1007/978-3-0348-0113-3
48. Dörfler, W., Findeisen, S., Wieners, C.: Space-time discontinuous Galerkin discretizations for linear first-order hyperbolic evolution systems. Comput. Methods Appl. Math. **16**(3), 409–428 (2016). https://doi.org/10.1515/cmam-2016-0015

49. Dörfler, W., Findeisen, S., Wieners, C., Ziegler, D.: Parallel adaptive discontinuous Galerkin discretizations in space and time for linear elastic and acoustic waves. In: Langer, U., Steinbach, O. (eds.) Space-Time Methods. Applications to Partial Differential Equations. Radon Series on Computational and Applied Mathematics, vol. 25, pp. 61–88. Walter de Gruyter (2019). https://doi.org/10.1515/9783110548488-002
50. Dörfler, W., Wieners, C., Ziegler, D.: Space-time discontinuous Galerkin methods for linear hyperbolic systems and the application to the forward problem in seismic imaging. In: Klöfkorn, R., Keilegavlen, E., Radu, F., Fuhrmann, J. (eds.) Finite Volumes for Complex Applications IX – Methods, Theoretical Aspects, Examples. Springer Proceedings in Mathematics & Statistics, vol. 323, pp. 477–485. Springer (2020). https://doi.org/10.1007/978-3-030-43651-3_44
51. Eilinghoff, J., Schnaubelt, R.: Error analysis of an ADI splitting scheme for the inhomogeneous Maxwell equations. Discrete Contin. Dyn. Syst. **38**(11), 5685–5709 (2018). https://doi.org/10.3934/dcds.2018248
52. Eller, M.: Continuous observability for the anisotropic Maxwell system. Appl. Math. Optim. **55**(2), 185–201 (2007). https://doi.org/10.1007/s00245-006-0886-x
53. Eller, M.: On symmetric hyperbolic boundary problems with nonhomogeneous conservative boundary conditions. SIAM J. Math. Anal. **44**(3), 1925–1949 (2012). https://doi.org/10.1137/110834652
54. Eller, M.: Stability of the anisotropic Maxwell equations with a conductivity term. Evol. Equ. Control Theory **8**(2), 343–357 (2019). https://doi.org/10.3934/eect.2019018
55. Eller, M., Rieder, A.: Tangential cone condition and Lipschitz stability for the full waveform forward operator in the acoustic regime. Inverse Problems **37**, 085011 (2021). https://doi.org/10.1088/1361-6420/ac11c5
56. Eller, M., Lagnese, J.E., Nicaise, S.: Decay rates for solutions of a Maxwell system with nonlinear boundary damping. Comput. Appl. Math. **21**(1), 135–165 (2002)
57. Engel, K.J., Nagel, R.: One-Parameter Semigroups for Linear Evolution Equations. Graduate Texts in Mathematics, vol. 194. Springer-Verlag, New York (2000). https://doi.org/10.1007/b97696
58. Engl, H.W., Hanke, M., Neubauer, A.: Regularization of Inverse Problems. Mathematics and Its Applications, vol. 375. Kluwer Academic Publishers Group, Dordrecht (1996)
59. Ern, A., Guermond, J.L.: Discontinuous galerkin methods for Friedrichs' systems. I. General theory. SIAM J. Numer. Anal. **44**(2), 753–778 (2006). https://doi.org/10.1137/050624133
60. Ern, A., Guermond, J.L.: Discontinuous Galerkin methods for Friedrichs' systems. II. Second-order elliptic PDEs. SIAM J. Numer. Anal. **44**(6), 2363–2388 (2006). https://doi.org/10.1137/05063831X
61. Ern, A., Guermond, J.L.: Discontinuous Galerkin methods for Friedrichs' systems. III. Multifield theories with partial coercivity. SIAM J. Numer. Anal. **46**(2), 776–804 (2008). https://doi.org/10.1137/060664045
62. Ern, A., Guermond, J.L.: A converse to Fortin's lemma in Banach spaces. C. R. Math. **354**(11), 1092–1095 (2016). https://doi.org/10.1016/j.crma.2016.09.013
63. Ern, A., Vohralík, M.: Polynomial-degree-robust a posteriori estimates in a unified setting for conforming, nonconforming, discontinuous Galerkin, and mixed discretizations. SIAM J. Numer. Anal. **53**(2), 1058–1081 (2015). https://doi.org/10.1137/130950100
64. Ern, A., Guermond, J.L., Caplain, G.: An intrinsic criterion for the bijectivity of Hilbert operators related to Friedrichs' systems. Commun. Partial Differential Equations **32**(1–3), 317–341 (2007). https://doi.org/10.1080/03605300600718545

65. Ernesti, J., Wieners, C.: Space-time discontinuous Petrov-Galerkin methods for linear wave equations in heterogeneous media. Comput. Methods Appl. Math. **19**(3), 465–481 (2019). https://doi.org/10.1515/cmam-2018-0190
66. Ernesti, J., Wieners, C.: A space-time DPG method for acoustic waves. In: Langer, U., Steinbach, O. (eds.) Space-Time Methods. Applications to Partial Differential Equations. Radon Series on Computational and Applied Mathematics, vol. 25, pp. 89–116. Walter de Gruyter (2019). https://doi.org/10.1515/9783110548488-003
67. Evans, L.C.: Partial differential equations. Graduate Studies in Mathematics, vol. 19, 2nd edn. American Mathematical Society, Providence, RI (2010)
68. Fabrizio, M., Morro, A.: Electromagnetism of Continuous Media. Oxford University Press, Oxford (2003). https://doi.org/10.1093/acprof:oso/9780198527008.001.0001
69. Faragó, I., Horváth, R., Schilders, W.H.: Investigation of numerical time-integrations of Maxwell's equations using the staggered grid spatial discretization. Int. J. Numer. Model. Electron. Netw. Dev. Fields **18**(2), 149–169 (2005). https://doi.org/10.1002/jnm.570
70. Fichtner, A.: Full Seismic Waveform Modelling and Inversion. Advances in Geophysical and Environmental Mechanics and Mathematics. Springer-Verlag, Berlin Heidelberg (2011). https://doi.org/10.1007/978-3-642-15807-0
71. Findeisen, S.: A parallel and adaptive space-time method for Maxwell's equations. Ph.D. thesis, Dept. of Mathematics, Karlsruhe Institute of Technology (2016). https://doi.org/10.5445/IR/1000056876
72. Friedrichs, K.O.: Symmetric positive linear differential equations. Commun. Pure Appl. Math. **11**, 333–418 (1958). https://doi.org/10.1002/cpa.3160110306
73. Gerken, T.: Dynamic inverse wave problems—part II: operator identification and applications. Inverse Problems **36**(2), 024005 (2020). https://doi.org/10.1088/1361-6420/ab47f4
74. Gerken, T., Grützner, S.: Dynamic inverse wave problems—part I: regularity for the direct problem. Inverse Problems **36**(2), 024004 (2020). https://doi.org/10.1088/1361-6420/ab47ec
75. Gopalakrishnan, J., Schöberl, J., Wintersteiger, C.: Mapped tent pitching schemes for hyperbolic systems. SIAM J. Sci. Comput. **39**(6), B1043–B1063 (2017). https://doi.org/10.1137/16M1101374
76. Gould, P.L., Feng, Y.: Introduction to linear elasticity, 4th edn. Springer, Cham (2018). https://doi.org/10.1007/978-3-319-73885-7
77. Grafakos, L.: Modern Fourier Analysis, 2nd edn. Springer, New York (2009). https://doi.org/10.1007/978-0-387-09434-2
78. Grote, M.J., Michel, S., Sauter, S.A.: Stabilized leapfrog based local time-stepping method for the wave equation. Math. Comput. **90**(332), 2603–2643 (2021). https://doi.org/10.1090/mcom/3650
79. Guès, O.: Problème mixte hyperbolique quasi-linéaire caractéristique. Comm. Partial Differential Equations **15**(5), 595–645 (1990). https://doi.org/10.1080/03605309908820701
80. Hairer, E., Wanner, G.: Solving ordinary differential equations II: Stiff and differential-algebraic problems. Springer Series in Computational Mathematics, vol. 14, 2nd edn. Springer-Verlag, Berlin (1996). https://doi.org/10.1007/978-3-642-05221-7
81. Hairer, E., Lubich, C., Wanner, G.: Geometric numerical integration illustrated by the Störmer-Verlet method. Acta Numer. **12**, 399–450 (2003). https://doi.org/10.1017/S0962492902000144
82. Hairer, E., Lubich, C., Wanner, G.: Geometric Numerical Integration: Structure-Preserving Algorithms for Ordinary Differential Equations. Springer Series in Computational Mathematics, vol. 31, 2nd edn. Springer-Verlag, Berlin (2006). https://doi.org/10.1007/3-540-30666-8

83. Hanke, M.: Regularizing properties of a truncated Newton-CG algorithm for nonlinear inverse problems. Numer. Funct. Anal. Optim. **18**(9-10), 971–993 (1997). https://doi.org/10.1080/01630569708816804
84. Hansen, E., Henningsson, E.: A convergence analysis of the Peaceman-Rachford scheme for semilinear evolution equations. SIAM J. Numer. Anal. **51**(4), 1900–1910 (2013). https://doi.org/10.1137/120890570
85. Hesthaven, J.S.: Numerical methods for conservation laws: From analysis to algorithms. SIAM (2017). https://doi.org/10.1137/1.9781611975109
86. Hesthaven, J.S., Warburton, T.: Nodal Discontinuous Galerkin Methods. Algorithms, Analysis, and Applications. Texts in Applied Mathematics, vol. 54 (Springer, New York, 2008). https://doi.org/10.1007/978-0-387-72067-8
87. Hochbruck, M., Köhler, J.: On the efficiency of the Peaceman–Rachford ADI-dG method for wave-type problems. In: Radu, F., Kumar, K., Berre, I., Nordbotten, J., Pop, I. (eds.) Numerical Mathematics and Advanced Applications ENUMATH 2017. Lecture Notes in Computational Science and Engineering, vol. 126, pp. 135–144. Springer International Publishing (2019). https://doi.org/10.1007/978-3-319-96415-7
88. Hochbruck, M., Köhler, J.: Error Analysis of Discontinuous Galerkin Discretizations of a Class of Linear Wave-type Problems. In: Dörfler, W., et al. (eds.) Mathematics of Wave Phenomena, Trends in Mathematics, pp. 197–218. Birkhäuser, Cham (2020). https://doi.org/10.1007/978-3-030-47174-3_12
89. Hochbruck, M., Köhler, J.: Error analysis of a fully discrete discontinuous Galerkin alternating direction implicit discretization of a class of linear wave-type problems. Numer. Math. **150**, 893–927 (2022). https://doi.org/10.1007/s00211-021-01262-z
90. Hochbruck, M., Pažur, T.: Implicit Runge-Kutta methods and discontinuous Galerkin discretizations for linear Maxwell's equations. SIAM J. Numer. Anal. **53**(1), 485–507 (2015). https://doi.org/10.1137/130944114
91. Hochbruck, M., Sturm, A.: Error analysis of a second-order locally implicit method for linear Maxwell's equations. SIAM J. Numer. Anal. **54**(5), 3167–3191 (2016). https://doi.org/10.1137/15M1038037
92. Hochbruck, M., Pažur, T., Schulz, A., Thawinan, E., Wieners, C.: Efficient time integration for discontinuous Galerkin approximations of linear wave equations [Plenary lecture presented at the 83rd Annual GAMM Conference, Darmstadt, 26th–30th March, 2012]. ZAMM Z. Angew. Math. Mech. **95**(3), 237–259 (2015). https://doi.org/10.1002/zamm.201300306
93. Hochbruck, M., Jahnke, T., Schnaubelt, R.: Convergence of an ADI splitting for Maxwell's equations. Numer. Math. **129**(3), 535–561 (2015). https://doi.org/10.1007/s00211-014-0642-0
94. Hofmann, B., Plato, R.: On ill-posedness concepts, stable solvability and saturation. J. Inverse Ill-Posed Probl. **26**(2), 287–297 (2018). https://doi.org/10.1515/jiip-2017-0090
95. Hofmann, B., Scherzer, O.: Local ill-posedness and source conditions of operator equations in Hilbert spaces. Inverse Problems **14**(5), 1189–1206 (1998). https://doi.org/10.1088/0266-5611/14/5/007
96. Hytönen, T., van Neerven, J., Veraar, M., Weis, L.: Analysis in Banach spaces. Vol. I. Martingales and Littlewood-Paley Theory. Springer, Cham (2016)
97. Ifrim, M., Tataru, D.: Local well-posedness for quasilinear problems: a primer (2020). Preprint, arXiv:2008.05684
98. Imbert-Gérard, L.M., Moiola, A., Stocker, P.: A space-time quasi-Trefftz DG method for the wave equation with piecewise-smooth coefficients. Math. Comp. **92**, 1211–1249 (2023). https://doi.org/10.1090/mcom/3786

99. Ito, K., Kappel, F.: Evolution equations and approximations. Series on Advances in Mathematics for Applied Sciences, vol. 61. World Scientific Publishing Co., Inc., River Edge, NJ (2002). https://doi.org/10.1142/9789812777294
100. Jackson, J.D.: Classical Electrodynamics, 3rd edn. Wiley, New York (1999). https://doi.org/10.1002/3527600434.eap109
101. Jacob, B., Zwart, H.J.: Linear port-Hamiltonian systems on infinite-dimensional spaces. Operator Theory: Advances and Applications, vol. 223. Birkhäuser/Springer Basel AG, Basel (2012). https://doi.org/10.1007/978-3-0348-0399-1
102. Jensen, M.: Discontinuous Galerkin methods for Friedrichs systems with irregular solutions. Ph.D. thesis, University of Oxford (2004). http://sro.sussex.ac.uk/45497/
103. Joly, P., Rodríguez, J.: An error analysis of conservative space-time mesh refinement methods for the one-dimensional wave equation. SIAM J. Numer. Anal. **43**(2), 825–859 (2005). https://doi.org/10.1137/040603437
104. Kaltenbacher, B., Neubauer, A., Scherzer, O.: Iterative regularization methods for nonlinear ill-posed problems. Radon Series on Computational and Applied Mathematics, vol. 6. Walter de Gruyter GmbH & Co. KG, Berlin (2008). https://doi.org/10.1515/9783110208276
105. Kato, T.: The Cauchy problem for quasi-linear symmetric hyperbolic systems. Arch. Rational Mech. Anal. **58**(3), 181–205 (1975). https://doi.org/10.1007/BF00280740
106. Kato, T.: Quasi-linear equations of evolution, with applications to partial differential equations. In: William N. Everitt (eds.) Spectral Theory and Differential Equations (Proc. Sympos., Dundee, 1974). Lecture Notes in Math., vol. 448, pp. 25–70 (1975)
107. Kato, T.: Abstract Differential Equations and Nonlinear Mixed Problems. Scuola Normale Superiore/Accademia Nazionale dei Lincei, Pisa/Rome (1985)
108. Kirsch, A.: An introduction to the mathematical theory of inverse problems. Applied Mathematical Sciences, vol. 120, 2nd edn. Springer, New York (2011). https://doi.org/10.1007/978-1-4419-8474-6
109. Kirsch, A., Hettlich, F.: The mathematical theory of time-harmonic Maxwell's equations. Applied Mathematical Sciences, vol. 190. Springer, Cham (2015). https://doi.org/10.1007/978-3-319-11086-8
110. Kirsch, A., Rieder, A.: On the linearization of operators related to the full waveform inversion in seismology. Math. Methods Appl. Sci. **37**, 2995–3007 (2014). https://doi.org/10.1002/mma.3037
111. Kirsch, A., Rieder, A.: Inverse problems for abstract evolution equations with applications in electrodynamics and elasticity. Inverse Problems **32**(8), 085001, 24 (2016). https://doi.org/10.1088/0266-5611/32/8/085001
112. Kirsch, A., Rieder, A.: Inverse problems for abstract evolution equations II: higher order differentiability for viscoelasticity. SIAM J. Appl. Math. **79**(6), 2639–2662 (2019). https://doi.org/10.1137/19M1269403. Corrigendum under https://doi.org/10.48550/arXiv.2203.01309
113. Köhler, J.: The Peaceman–Rachford ADI-dG method for linear wave-type problems. Ph.D. thesis, Karlsruhe Institute of Technology (2018). https://doi.org/10.5445/IR/1000089271
114. Komornik, V.: Boundary stabilization, observation and control of Maxwell's equations. PanAmer. Math. J. **4**(4), 47–61 (1994)
115. Langer, U., Steinbach, O.: Space-Time Methods: Applications to Partial Differential Equations. Radon Series on Computational and Applied Mathematics, vol. 25. Walter de Gruyter GmbH & Co KG (2019). https://doi.org/10.1515/9783110548488
116. Lasiecka, I., Pokojovy, M., Schnaubelt, R.: Exponential decay of quasilinear Maxwell equations with interior conductivity. NoDEA Nonlinear Differential Equations Appl. **26**(6), Paper No. 51, 34 (2019). https://doi.org/10.1007/s00030-019-0595-1

117. Lechleiter, A., Rieder, A.: Newton regularizations for impedance tomography: convergence by local injectivity. Inverse Problems **24**(6), 065009, 18 (2008). https://doi.org/10.1088/0266-5611/24/6/065009
118. Lechleiter, A., Rieder, A.: Towards a general convergence theory for inexact Newton regularizations. Numer. Math. **114**(3), 521–548 (2010). https://doi.org/10.1007/s00211-009-0256-0
119. Leis, R.: Initial Boundary Value Problems in Mathematical Physics. Courier Corporation (2013). https://doi.org/10.1007/978-3-663-10649-4
120. Liu, Y., Shu, C.W., Zhang, M.: Sub-optimal convergence of discontinuous galerkin methods with central fluxes for linear hyperbolic equations with even degree polynomial approximations. J. Comput. Math. **39**(4), 518–537 (2021). https://doi.org/10.4208/jcm.2002-m2019-0305
121. Louis, A.K.: Inverse und schlecht gestellte Probleme. B. G. Teubner, Stuttgart (1989). https://doi.org/10.1007/978-3-322-84808-6
122. Lucente, S., Ziliotti, G.: Global existence for a quasilinear Maxwell system. Rend. Istit. Mat. Univ. Trieste **31**(suppl. 2), 169–187 (2000)
123. Majda, A.: Compressible Fluid Flow and Systems of Conservation Laws in Several Space Variables. Springer-Verlag, New York (1984). https://doi.org/10.1007/978-1-4612-1116-7
124. Margotti, F.: Inexact Newton regularization combined with gradient methods in Banach spaces. Inverse Problems **34**(7), 075007, 26 (2018). https://doi.org/10.1088/1361-6420/aac21f
125. Margotti, F., Rieder, A.: An inexact Newton regularization in Banach spaces based on the nonstationary iterated Tikhonov method. J. Inverse Ill-Posed Probl. **23**(4), 373–392 (2015). https://doi.org/10.1515/jiip-2014-0035
126. Margotti, F., Rieder, A., Leitão, A.: A Kaczmarz version of the `reginn`-Landweber iteration for ill-posed problems in Banach spaces. SIAM J. Numer. Anal. **52**(3), 1439–1465 (2014). https://doi.org/10.1137/130923956
127. Moiola, A., Perugia, I.: A space-time Trefftz discontinuous Galerkin method for the acoustic wave equation in first-order formulation. Numer. Math. **138**(2), 389–435 (2018). https://doi.org/10.1007/s00211-017-0910-x
128. Monk, P.: Finite Element Methods for Maxwell's Equations. Numerical Mathematics and Scientific Computation. Oxford University Press, New York (2003). https://doi.org/10.1093/acprof:oso/9780198508885.001.0001
129. Montseny, E., Pernet, S., Ferriéres, X., Cohen, G.: Dissipative terms and local time-stepping improvements in a spatial high order discontinuous Galerkin scheme for the time-domain Maxwell's equations. J. Comput. Phys. **227**(14), 6795–6820 (2008). https://doi.org/10.1016/j.jcp.2008.03.032
130. Moya, L.: Temporal convergence of a locally implicit discontinuous Galerkin method for Maxwell's equations. ESAIM Math. Model. Numer. Anal. **46**(5), 1225–1246 (2012). https://doi.org/10.1051/m2an/2012002
131. Muñoz Rivera, J.E., Racke, R.: Mildly dissipative nonlinear Timoshenko systems—global existence and exponential stability. J. Math. Anal. Appl. **276**(1), 248–278 (2002). https://doi.org/10.1016/S0022-247X(02)00436-5
132. Namiki, T.: A new FDTD algorithm based on alternating-direction implicit method. IEEE Trans. Microw. Theory Tech. **47**(10), 2003–2007 (1999). https://doi.org/10.1109/22.795075
133. Nicaise, S., Pignotti, C.: Boundary stabilization of Maxwell's equations with space-time variable coefficients. ESAIM Control Optim. Calc. Var. **9**, 563–578 (2003). https://doi.org/10.1051/cocv:2003027
134. Nicaise, S., Pignotti, C.: Internal stabilization of Maxwell's equations in heterogeneous media. Abstr. Appl. Anal. **7**, 791–811 (2005). https://doi.org/10.1155/AAA.2005.791

135. Nirenberg, L.: On elliptic partial differential equations. Ann. Scuola Norm. Sup. Pisa Cl. Sci. (3) **13**, 115–162 (1959)
136. Ostermann, A., Schratz, K.: Error analysis of splitting methods for inhomogeneous evolution equations. Appl. Numer. Math. **62**, 1436–1446 (2012). https://doi.org/10.1016/j.apnum.2012.06.002
137. Pažur, T.: Error analysis of implicit and exponential time integration of linear Maxwell's equations. Ph.D. thesis, Karlsruhe Institute of Technology (2013). https://doi.org/10.5445/IR/1000038617
138. Pazy, A.: Semigroups of Linear Operators and Applications to Partial Differential Equations. Applied Mathematical Sciences, vol. 44. Springer, New York (1983). https://doi.org/10.1007/978-1-4612-5561-1
139. Perugia, I., Schöberl, J., Stocker, P., Wintersteiger, C.: Tent pitching and Trefftz-DG method for the acoustic wave equation. Comput. Math. Appl. (2020). https://doi.org/10.1016/j.camwa.2020.01.006
140. Peaceman, D.W., Rachford Jr., H.H.: The numerical solution of parabolic and elliptic differential equations. J. Soc. Indust. Appl. Math. **3**, 28–41 (1955). https://doi.org/10.1137/0103003
141. Phillips, R.S.: Dissipative operators and hyperbolic systems of partial differential equations. Trans. Amer. Math. Soc. **90**, 193–254 (1959). https://doi.org/10.2307/1993202
142. Phung, K.D.: Contrôle et stabilisation d'ondes électromagnétiques. ESAIM Control Optim. Calc. Var. **5**, 87–137 (2000). https://doi.org/10.1051/cocv:2000103
143. Picard, R.H., Zajączkowski, W.M.: Local existence of solutions of impedance initial-boundary value problem for non-linear Maxwell equations. Math. Methods Appl. Sci. **18**(3), 169–199 (1995). https://doi.org/10.1002/mma.1670180302
144. Piperno, S.: Symplectic local time-stepping in non-dissipative DGTD methods applied to wave propagation problems. M2AN Math. Model. Numer. Anal. **40**(5), 815–841 (2006). https://doi.org/10.1051/m2an:2006035
145. Pokojovy, M., Schnaubelt, R.: Boundary stabilization of quasilinear Maxwell equations. J. Differential Equations **268**(2), 784–812 (2020). https://doi.org/10.1016/j.jde.2019.08.032
146. Racke, R.: Lectures on Nonlinear Evolution Equations. Friedr. Vieweg, Braunschweig (1992). https://doi.org/10.1007/978-3-663-10629-6
147. Rauch, J.: $\mathcal{L}_2$ is a continuable initial condition for Kreiss' mixed problems. Comm. Pure Appl. Math. **25**, 265–285 (1972). https://doi.org/10.1002/cpa.3160250305
148. Rieder, A.: Keine Probleme mit inversen Problemen. Friedr. Vieweg & Sohn, Braunschweig (2003). https://doi.org/10.1007/978-3-322-80234-7
149. Rudin, W.: Real and Complex Analysis, 3rd edn. McGraw-Hill Book Co., New York (1987)
150. Scherzer, O.: The use of Morozov's discrepancy principle for Tikhonov regularization for solving nonlinear ill-posed problems. Computing **51**(1), 45–60 (1993). https://doi.org/10.1007/BF02243828
151. Scherzer, O.: A convergence analysis of a method of steepest descent and a two-step algorithm for nonlinear ill-posed problems. Numer. Funct. Anal. Optim. **17**(1–2), 197–214 (1996). https://doi.org/10.1080/01630569608816691
152. Schnaubelt, R., Spitz, M.: Local wellposedness of quasilinear Maxwell equations with absorbing boundary conditions. Evol. Equ. Control Theory **10**(1), 155–198 (2021). https://doi.org/10.3934/eect.2020061
153. Schnaubelt, R., Spitz, M.: Local wellposedness of quasilinear Maxwell equations with conservative interface conditions. Commun. Math. Sci. **20**(8), 2265–2313 (2022). https://doi.org/10.48550/arXiv.1811.08714

154. Schulz, E., Hiptmair, R.: First-kind boundary integral equations for the Dirac operator in 3-dimensional Lipschitz domains. SIAM J. Math. Anal. **54**(1), 616–648 (2022). https://doi.org/10.1137/20M1389224
155. Secchi, P.: Well-posedness of characteristic symmetric hyperbolic systems. Arch. Rational Mech. Anal. **134**(2), 155–197 (1996). https://doi.org/10.1007/BF00379552
156. Speck, J.: The nonlinear stability of the trivial solution to the Maxwell-Born-Infeld system. J. Math. Phys. **53**(8), 083703, 83 (2012). https://doi.org/10.1063/1.4740047
157. Spitz, M.: Local wellposedness of nonlinear Maxwell equations. Ph.D. thesis, Karlsruhe Institute of Technology (2017). https://doi.org/10.5445/IR/1000078030
158. Spitz, M.: Local wellposedness of nonlinear Maxwell equations with perfectly conducting boundary conditions. J. Differential Equations **266**(8), 5012–5063 (2019). https://doi.org/10.1016/j.jde.2018.10.019
159. Spitz, M.: Regularity theory for nonautonomous Maxwell equations with perfectly conducting boundary conditions. J. Math. Anal. Appl. **506**(1), Paper No. 125646, 43 (2022). https://doi.org/10.1016/j.jmaa.2021.125646
160. Steinbach, O., Zank, M.: Coercive space-time finite element methods for initial boundary value problems. Electron. Trans. Numer. Anal. **52**, 154–194 (2020). https://doi.org/10.1553/etna_vol52s154
161. Sturm, A.: Locally implicit time integration for linear Maxwell's equations. Ph.D. thesis, Karlsruhe Institute of Technology (2017). https://doi.org/10.5445/IR/1000069341
162. Taylor, M.E.: Pseudodifferential Operators and Nonlinear PDE. Birkhäuser, Boston, MA (1991). https://doi.org/10.1007/978-1-4612-0431-2
163. Verwer, J.G.: Component splitting for semi-discrete Maxwell equations. BIT **51**(2), 427–445 (2011). https://doi.org/10.1007/s10543-010-0296-y
164. Winkler, R.: Tailored interior and boundary parameter transformations for iterative inversion in electrical impedance tomography. Inverse Problems **35**(11), 114007 (2019). https://doi.org/10.1088/1361-6420/ab2783
165. Winkler, R., Rieder, A.: Model-aware Newton-type inversion scheme for electrical impedance tomography. Inverse Problems **31**(4), 045009, 27 (2015). https://doi.org/10.1088/0266-5611/31/4/045009
166. Zeidler, E.: Nonlinear Functional Analysis and Its Applications. I. Springer-Verlag, New York (1986)
167. Zeltmann, U.: The viscoelastic seismic model: Existence, uniqueness and differentiability with respect to parameters. Ph.D. thesis, Karlsruhe Institute of Technology (2018). https://doi.org/10.5445/IR/1000093989
168. Zhen, F., Chen, Z., Zhang, J.: Toward the development of a three-dimensional unconditionally stable finite-difference time-domain method. IEEE Trans. Microw. Theory Technol. **48**(9), 1550–1558 (2000). https://doi.org/10.1109/22.869007
169. Ziegler, D.: A parallel and adaptive space-time discontinuous Galerkin method for visco-elastic and visco-acoustic waves. Ph.D. thesis, Karlsruhe Institute of Technology (KIT) (2019). https://doi.org/10.5445/IR/1000110469

Printed in the United States
by Baker & Taylor Publisher Services